ADVANCES IN CHEMICAL PHYSICS

VOLUME 112

Advances in
CHEMICAL PHYSICS

Edited by

I. PRIGOGINE

Center for Studies in Statistical Mechanics
and Complex Systems
The University of Texas
Austin, Texas
and
International Solvay Institutes
Université Libre de Bruxelles
Brussels, Belgium

and

STUART A. RICE

Department of Chemistry
and
The James Franck Institute
The University of Chicago
Chicago, Illinois

VOLUME 112

AN INTERSCIENCE® PUBLICATION
JOHN WILEY & SONS, INC.
NEW YORK • CHICHESTER • WEINHEIM • BRISBANE • SINGAPORE • TORONTO

This book is printed on acid-free paper. ⊗

An Interscience® Publication

Copyright © 2000 by John Wiley & Sons, Inc. All rights reserved.

Published simultaneously in Canada.

For ordering and customer service, call 1-800-CALL-WILEY.

Library of Congress Catalog Number: 58-9935

ISBN 0-471-38002-4

Printed in the United States of America

10 9 8 7 6 5 4 3 2 1

CONTRIBUTORS TO VOLUME 112

ANDRZEJ R. ALTENBERGER, Department of Chemistry, University of Minnesota, Institute of Technology, Minneapolis, MN 55455-0431

DAVID L. ANDREWS, School of Chemical Sciences, University of East Anglia, Norwich NR4 7TJ, England

JOHN S. DAHLER, University of Minnesota, Institute of Technology, Minneapolis, MN 55455-0431

J. L. GARCÍA-PALACIOS, Department of Materials Science — Division of Solid State Physics, Uppsala University, Uppsala, Sweden

GEDIMINAS JUZELIŪNAS, Institute of Theoretical Physics and Astronomy, A. Goštauto 12, 2600 Vilnius, Lithuania

E. E. C. KENNEDY, School of Mathematics and Physics, The Queen's University of Belfast, Belfast BT7, 1NN, Co. Antrim, Northern Ireland

INTRODUCTION

Few of us can any longer keep up with the flood of scientific literature, even in specialized subfields. Any attempt to do more and be broadly educated with respect to a large domain of science has the appearance of tilting at windmills. Yet the synthesis of ideas drawn from different subjects into new, powerful, general concepts is as valuable as ever, and the desire to remain educated persists in all scientists. This series, Advances in Chemical Physics, is devoted to helping the reader obtain general information about a wide variety of topics in chemical physics, a field that we interpret very broadly. Our intent is to have experts present comprenhensive analyses of subjects of interest and to encourage the expression of individual points of view. We hope that this approach to the presentation of an overview of a subject will both stimulate new research and serve as a personalized learning text for beginners in a field.

I. Prigogine
Stuart A. Rice

CONTENTS

ON THE STATICS AND DYNAMICS OF MAGNETOANISOTROPIC NANOPARTICLES

J. L. GARCÍA-PALACIOS*

Department of Materials Science — Division of Solid State Physics, Uppsala University, Uppsala, Sweden

CONTENTS

*On leave from Instituto de Ciencia de Materiales de Aragón, Consejo Superior de Investigaciones Científicas–Universidad de Zaragoza, 50015 Zaragoza, Spain. Electronic addresses: jose.garcia@angstrom.uu.se and jlgarcia@posta.unizar.es.

Advances in Chemical Physics, Volume 112, Edited by I. Prigogine and Stuart A. Rice
ISBN 0-471-38002-4.   © 2000 John Wiley & Sons, Inc.

I. INTRODUCTION

Small, magnetically ordered particles are ubiquitous in both naturally occurring and manufactured forms. On the one hand, the wide spectrum of applications of these systems — which range from magnetic recording media, catalysts, magnetic fluids, filtering and phase separation in the mineral processing industry, magnetic imaging and magnetic refrigeration, to numerous geophysical, biological, and medical uses — is remarkable. On the other hand, the *nanometric* magnetic particles can be considered as model systems for the study of various basic physical phenomena. Among others, we can mention rotational Brownian motion and thermally activated processes in multistable systems, mesoscopic quantum phenomena, dipole–dipole interaction effects, and the dependence of the properties of solids on their size.

Magnetically ordered particles of nanometric size generally consist of a single domain, whose constituent spins, at temperatures well below the Curie temperature, rotate in unison. The magnetic energy of a nanometric particle is then determined by its magnetic moment orientation, and has a number of stable directions separated by potential barriers (created by the magnetic anisotropy). As a result of the coupling of the magnetic moment of the particle $\vec{m}$ with the microscopic degrees of freedom of its "environment" (phonons, conducting electrons, nuclear spins, etc.), the magnetic moment is subjected to thermal fluctuations and may undergo a Brownian-type rotation surmounting the potential barriers. This solid-state relaxation process was proposed and studied by Néel (1949), and later on reexamined by Brown (1963), by dint of the theory of stochastic processes.

In the high potential-barrier range $\Delta U / k_B T \gg 1$, the characteristic time for the overbarrier rotation process, $\tau_\parallel$, can approximately be written in the Arrhenius form $\tau_\parallel \simeq \tau_0 \exp(\Delta U / k_B T)$, where τ_0 ($\sim 10^{-10}$–10^{-12} s) is related with the intra-potential-well dynamics. For $\tau_\parallel \ll t_m$ (t_m is the measurement or observation time), $\vec{m}$ maintains the equilibrium distribution of orientations as in a classical paramagnet; because $m = |\vec{m}|$ is much larger than a typical microscopic magnetic moment ($m \sim 10^3$–$10^5 \mu_B$), this phenomenon is termed *superparamagnetism*. In contrast, when $\tau_\parallel \gg t_m$, the magnetic moment rotates about a potential minimum whereas the overbarrier relaxation mechanism is *blocked*. This corresponds to the state of stable magnetization in a bulk magnet. Finally, under intermediate conditions ($\tau_\parallel \sim t_m$), *nonequilibrium phenomena*, accompanied by magnetic "relaxation," are observed. It is to be noted that, in the Arrhenius range mentioned, the system may pass through all these regimes in a relatively narrow temperature interval.

We shall describe a nanoparticle as a classical magnetic moment with magnetic-anisotropy energy. This brings certain generality to the results and the connection with other physical systems that can approximately be described as ensembles of "rotators" in certain orientational potentials. Examples include molecular magnetic clusters with high spin in their ground state (in the ranges where a classical description of their spins is reasonable), nematic liquid crystals, relaxor ferroelectrics, certain high-spin dilutely doped glasses described by the random axial anisotropy model, and superparamagnetic-like spin glasses.

The analogies between the macroscopic behavior of certain electric and magnetic "glassy" systems and that of ensembles of small magnetic particles have received recurrent attention during the past 20 years. The magnetic nanoparticle systems exhibit glassy-like phenomena associated with the distribution of particle parameters (anisotropy constants, volumes, magnetic moments, etc.), which lead to more or less wide distributions of relaxation times. On the other hand, ensembles of interacting nanoparticles apparently exhibit genuine glassy properties, due mainly to the extremely anisotropic character of the dipole–dipole interaction. Therefore, it is important to determine which phenomena are intrinsically due to the presence of interactions in the nanoparticle ensemble and which others are not. In this connection, owing to an insufficient knowledge about some properties of *independent* magnetic particles, it is not always known from which "laws" the corresponding quantities would depart as a consequence of the interparticle interactions. Similar considerations also apply to the study of the effects associated with quantum phenomena in small magnetic particles; as complete a knowledge as possible of the *classical* regime is a prerequisite for the study of, for example, quantum tunneling and coherence in these systems.

We finally mention that the study of the dynamics of noninteracting classical magnetic moments is an interesting strand of research per se, which seems to be far from exhausted. Indeed, relevant developments of the pioneering works of the 1960s and 1970s have been performed during the past 15 years.

The purpose of this chapter is to gain a deeper insight into the statical (thermal equilibrium) and dynamical (nonequilibrium) properties of *noninteracting* magnetically anisotropic nanoparticles in the framework of *classical* physics. In Sections II and III some thermal equilibrium properties of classical magnetic moments are studied. Section II is devoted to the obtainment of general results for the basic thermodynamical functions (partition function and thermodynamical potentials), some of which are subsequently used in Section III to calculate various important thermal equilibrium quantities. Some known results are reobtained (presenting in some cases alternative expressions and/or derivations), whereas the superparamagnetic theory is extended by calculating a number of other quantities. The central issue along these two sections is the study of the effects of the magnetic anisotropy on the thermal equilibrium properties of superparamagnetic systems. These effects are sometimes ignored because superparamagnetism is *restrictively* associated with the temperature range where the anisotropy energy is smaller than the thermal energy.

In Sections IV–VI we concentrate on the dynamical properties of classical magnetic moments. The heuristic approach to the dynamics of these systems is considered in Section IV, where analyses of the corresponding models in order to extract certain parameters of ensembles of magnetic nanoparticles are revised and developed. In Section V the dynamical properties of classical spins are studied by the methods of the *theory of stochastic processes*. The Brown–Kubo–Hashitsume stochastic model is presented in a unified way, and Langevin dynamics simulations are performed to study the non-zero-temperature dynamical properties. Both the study of individual stochastic trajectories and the response of ensembles of magnetic moments are undertaken. Section VI is finally devoted to the foundation of the dynamical equations that are the basis of Section V. The techniques of the formalism of the *independent-oscillator environment* are employed to derive dynamical equations for the magnetic moment that account for the effects of its interaction with the surrounding medium.

II. EQUILIBRIUM PROPERTIES: GENERALITIES AND METHODOLOGY

A. Introduction

Throughout this chapter we concentrate on the study of magnetic moments whose physical support (the crystal lattice in magnetic nanoparticles), to which they are linked by the magnetic anisotropy, is fastened in space. In small-particle magnetism, this corresponds to particles dispersed in a solid matrix. Although this apparently excludes the so-called magnetic fluids (where the physical rotation of the particles plays a fundamental role), these belong to the class of solid dispersions when the liquid carrier is frozen (which is besides the case of experimental interest when studying low-temperature properties). On the other hand, we also restrict our study to systems with axially symmetric magnetic anisotropy. This choice makes the problem mathematically tractable while provides valuable insight into more complex situations.

As was mentioned in Section I, the thermal equilibrium (superparamagnetic) behavior is observed when the measurement or observation time t_m is much longer than the characteristic relaxation times of the system (this is, of course, a general statement). The measurement times of various experimental techniques are displayed in Table I.

The thermal equilibrium range can extend down to temperatures where the heights of the energy barriers (created by the magnetic anisotropy) are much larger than the thermal energy. To illustrate, for a system with an axially symmetric Hamiltonian and in the high-barrier range, the mean time for the

TABLE I

Characteristic Measurement Times of Various
Experimental Techniques

Experimental Technique	Measurement Time, s
Magnetization	$1-100$
AC susceptibility	$10^{-6}-100$
Mössbauer spectroscopy	$10^{-9}-10^{-7}$
Ferromagnetic resonance	10^{-9}
Neutron scattering	$10^{-12}-10^{-8}$

overbarrier rotation process $\tau_{\parallel}$ can be written in the Arrhenius form

$$\tau_{\parallel} = \tau_0 \exp\left(\frac{\Delta U}{k_B T}\right) \qquad (2.1)$$

Besides, the "high-barrier" range where this expression for the relaxation time holds extends down to $\Delta U/k_B T \gtrsim 2$; moreover, for $\Delta U/k_B T \lesssim 2$, the relaxation time $\tau_{\parallel}$ is of the order of τ_0 ($\sim 10^{-10}-10^{-12}$ s for magnetic nanoparticles). Therefore, the exponential decrease of $\tau_{\parallel}$ with increasing T yields the range

$$\ln\left(\frac{t_m}{\tau_0}\right) > \frac{\Delta U}{k_B T} \geq 0$$

as the thermal equilibrium range ($\tau_{\parallel} \ll t_m$) for a given measurement time t_m. For instance, for magnetic measurements with $t_m \sim 1-100$ s, this range is extremely wide ($25 > \Delta U/k_B T \geq 0$). This indicates that the frequently encountered statement "superparamagnetism occurs when the thermal energy is comparable to or larger than the energy barriers" is unnecessarily restrictive.

Let us further illustrate this important point, which rests essentially on the magnitude of τ_0 and the exponential dependence of $\tau_{\parallel}$ on T in Eq. (2.1). For an experiment with measurement time t_m, the *blocking temperature* T_b, defined as the temperature where $t_m = \tau_{\parallel}$, is given by $t_m = \tau_0 \exp(\Delta U/k_B T_b)$. Accordingly, one has $\ln(t_m/\tau_0) = \Delta U/k_B T_b$ so that, if $t_m = \tau_0 10^{12}$ (a typical value for standard magnetic measurements), it follows that $\Delta U/k_B T_b = \ln(10^{12}) \simeq 27.6$. However, for $\Delta U/(k_B 1.1 T_b) \simeq 25$, one already finds $\tau_{\parallel} = 0.08 t_m$, while for $\Delta U/(k_B 1.2 T_b) \simeq 23$, one has $\tau_{\parallel} = 0.01 t_m$, thus, *the system clearly shows its thermal equilibrium behavior, whereas ΔU is still much larger than $k_B T$.*

Therefore, there exists an extremely wide range where superparamagnetism occurs ($\tau_{\parallel} \ll t_m$) and, simultaneously, the "naive condition of superparamagnetism" $\Delta U/k_B T \lesssim 1$ is not necessarily obeyed. Consequently, in that range, the effects of the anisotropy energy on the equilibrium quantities can be sizable. Indeed, for any thermal equilibrium quantity, prior to the observation of the corresponding "blocking" (departure from thermal equilibrium behavior) when the temperature is sufficiently lowered, one can clearly observe a crossover from the isotropic-type behavior at high temperatures (where the anisotropy potential plays a minor role) to either a discrete-orientation- or plane-rotator-type behavior at low temperatures (where the magnetic moment stays most of the time in the potential-minima regions), *without leaving the thermal equilibrium range.*

The organization of the remainder of this section is as follows. In Section II.B we introduce and discuss the Hamiltonian for a small magnetic particle. In Section II.C the partition function and free energy are introduced. In Section II.D we carry out the expansion of the partition function in powers of either the external field or the anisotropy constant, along with an asymptotic expansion for strong anisotropy. Finally, in Section II.E, we derive the corresponding expansions of the free energy.

B. Hamiltonian

First, we discuss the concept of *effective* Hamiltonian for a small, magnetically ordered particle. Then we introduce the basic form of the Hamiltonian that will be studied along this work, to conclude with the study of the energy barriers in the illustrative longitudinal-field case.

1. Effective Hamiltonian of a Nanoparticle

A basic assumption in small-particle magnetism is that a single-domain particle, with a given physical orientation, is in *internal* thermodynamical equilibrium at temperature T. Not too close to the Curie temperature, its constituent spins rotate in unison (coherent rotation), so the only relevant degree of freedom left is the orientation of the net magnetic moment. With respect to this variable, the thermal equilibration can take place in a timescale that can be considerably longer than that of the internal equilibration. Under such conditions, the internal free energy (for a given instantaneous orientation) can be considered as an effective energy (Hamiltonian) for the orientational degrees of freedom.

The consideration of an internal free energy as an effective Hamiltonian for the remaining degrees of freedom is indeed general, and it is based on the very statistical-mechanical definition of the free energy. Let (p, q) be the canonical variables "of interest" and $(\mathbf{P}, \mathbf{Q})$ the set of "internal" variables. The partition function $\mathcal{Z}$ and the free energy $\mathcal{F}$ are defined in terms of the total Hamiltonian of the system $\mathcal{H}_{\mathrm{T}}$ as

$$\mathcal{Z} = \int \mathrm{d}p \, \mathrm{d}q \, \mathrm{d}\mathbf{P} \, \mathrm{d}\mathbf{Q} \exp[-\beta \mathcal{H}_{\mathrm{T}}(p, q; \mathbf{P}, \mathbf{Q})], \qquad \mathcal{F} = -\frac{1}{\beta} \ln \mathcal{Z}$$

where $\beta = 1/k_\mathrm{B}T$.* One can define *internal* quantities (marked by a tilde) for given values of the variables p and q as follows:

$$\tilde{\mathcal{Z}}(p,q) = \int \mathrm{d}\mathbf{P}\,\mathrm{d}\mathbf{Q}\exp[-\beta\mathcal{H}_\mathrm{T}(p,q;\mathbf{P},\mathbf{Q})], \qquad \tilde{\mathcal{F}}(p,q) = -\frac{1}{\beta}\ln\tilde{\mathcal{Z}}(p,q)$$

Note that, by definition, the internal free energy obeys the relation

$$\exp[-\beta\tilde{\mathcal{F}}(p,q)] = \int \mathrm{d}\mathbf{P}\,\mathrm{d}\mathbf{Q}\exp[-\beta\mathcal{H}_\mathrm{T}(p,q;\mathbf{P},\mathbf{Q})]$$

Therefore, the total partition function $\mathcal{Z}$, from which all the equilibrium quantities of the system can be derived, can be written as

$$\mathcal{Z} = \int \mathrm{d}p\,\mathrm{d}q\exp[-\beta\tilde{\mathcal{F}}(p,q)]$$

This equation demonstrates the above statement: the so-defined internal free energy $\tilde{\mathcal{F}}(p,q)$ plays the role of an effective Hamiltonian for the variables (p,q) when studying the *equilibrium* properties of the system. Note that this effective Hamiltonian may have, by definition, terms dependent on T.

Naturally, this approach is in principle applicable to any chosen pair of variables (p,q). However, for this procedure to be useful, a timescale separation between some internal "fast" variables and certain "slow" ones must occur. In our case, the orientation of the total magnetic moment plays the role of the latter and, in what follows, we shall refer to the so-introduced internal free energy as the *magnetic energy (Hamiltonian) of the nanoparticle*, and denote it simply by $\mathcal{H}(\vec{m})$.

Similar considerations can, in principle, be applied to a magnetic domain in a bulk magnet; however, for such a macroscopic system, the timescale separation mentioned is so huge that the probability of thermally activated magnetization reversal is almost zero over astronomical timescales; the system is then effectively confined in a restricted region of the phase space. Note finally that the separation procedure between "internal" and "relevant" variables would lead to *exact* results if one in fact uses the preceding definitions to calculate the internal free energy by "integrating out" the internal variables.

---

* In these preliminary considerations, we omit in $\mathcal{Z}$ a factor $(2\pi\hbar)^{-s}$, where s is the number of degrees of freedom (see Landau and Lifshitz, 1980, Section 31). This factor, which renders $\mathcal{Z}$ dimensionless, when multiplied by the volume element in the phase space $\mathrm{d}p_1\ldots\mathrm{d}q_s$ gives the semiclassical "number of states" in this volume element, providing in this way the proper link with the quantum-mechanical expression for the partition function.

However, this is not possible in general, but one determines $\tilde{\mathcal{F}}(p, q)$ on the basis of series truncations, symmetry arguments, etc. (Brown, 1979).

2. Hamiltonian Studied

The magnetic energy of a nanoparticle has a number of different contributions, including magnetostatic self-energy ("demagnetization" or "shape" energy), magnetocrystalline energy, surface terms, and magnetoelastic energy. All these contributions give rise to a dependence of the energy of the nanoparticle on the orientation of its magnetic moment; in other words, in the absence of an external magnetic field, the magnetic properties of the system are anisotropic. We shall consider mainly systems where the *magnetic-anisotropy energy* has the simplest axial symmetry. Then, if an external field $\vec{B}$ is applied (assumed to be uniform over the volume of the system), the total magnetic energy reads

$$\mathcal{H}(\vec{m}) = -\frac{Kv}{m^2}(\vec{m} \cdot \hat{n})^2 - \vec{m} \cdot \vec{B} \qquad (2.2)$$

where K is the magnetic-anisotropy energy constant, v is the volume of the nanoparticle, and $\hat{n}$ is a unit vector along the symmetry axis of the magnetic-anisotropy term (hereafter referred to as the *anisotropy axis*).

On introducing the unit vectors $\vec{e}$ in the direction of the magnetic moment ($\vec{e} = \vec{m}/m$) and $\hat{b}$ in the direction of the external magnetic field ($\hat{b} = \vec{B}/B$), as well as the dimensionless anisotropy and field parameters

$$\sigma = \frac{Kv}{k_{\mathrm{B}}T}, \qquad \xi = \frac{mB}{k_{\mathrm{B}}T} \qquad (2.3)$$

we can express the Hamiltonian (2.2) as

$$-\beta\mathcal{H} = \sigma(\vec{e} \cdot \hat{n})^2 + \xi(\vec{e} \cdot \hat{b}) \qquad (2.4)$$

For $K > 0$ the anisotropy is of "easy axis" type, since the two existing minima of the anisotropy term point along $\pm\hat{n}$ (the "poles"). For $K < 0$ the anisotropy is of "easy plane" type, as the minima of the anisotropy term are then continuously distributed over the plane perpendicular to $\hat{n}$ (the "equatorial" region).

The adopted expression for the magnetic anisotropy is the leading term in the expansion of a general uniaxial magnetocrystalline anisotropy energy with respect to the direction cosines of the magnetization.[*] Besides, such a form

---

[*] For instance, directions of easy magnetization in the equatorial plane would be determined by higher-order terms in the expansion for $K < 0$ (see Landau and Lifshitz, 1984, Section 40).

is also the appropriate one for the shape anisotropy (internal magnetostatic energy) of an ellipsoid of revolution

$$\mathcal{H}_{\text{dem}} = \tfrac{1}{2}v\mu_0 M_s^2 (D_a \cos^2 \vartheta + D_b \sin^2 \vartheta)$$

where ϑ is the angle between the magnetic moment and the polar axis of the ellipsoid, $M_s = m/v$ is the spontaneous magnetization, D_a the demagnetization factor along the polar axis, and D_b the demagnetization factor along an equatorial axis. Indeed, except for a constant term, we can write the preceding expression as $\mathcal{H}_{\text{dem}} = -\tfrac{1}{2}v\mu_0 M_s^2 (D_b - D_a)\cos^2 \vartheta$, so that the corresponding anisotropy constant reads

$$K_{\text{dem}} = \tfrac{1}{2}\mu_0 M_s^2 (D_b - D_a) \tag{2.5}$$

In this case easy-axis and easy-plane anisotropy correspond, respectively, to prolate and oblate ellipsoids of revolution.

For many materials, slight deviations from spherical shape make the shape anisotropy to dominate the remainder contributions to the magnetic anisotropy. On the other hand, as was shown by Brown and Morrish (1957), a single-domain particle with an *arbitrary* shape is equivalent to a suitably chosen general ellipsoid of the same volume, as far as the total energy is concerned. Therefore, after these results, the seemingly specialized study of ellipsoids of revolution (i.e., of uniaxial anisotropy) can be of great importance to account for the effects of a general shape anisotropy.

In what follows we shall phrase our discussion in the language of classical magnetic moments. Nevertheless, the results obtained will be applicable to systems consisting of classical dipole moments that could approximately be described by Hamiltonians akin to (2.2), namely, Hamiltonians comprising a coupling term to an (electric or magnetic) external field plus an axially symmetric orientational potential.

3. Energy Barriers in the Longitudinal-Field Case

We shall now study the behavior of the Hamiltonian in the illustrative $\vec{B}\|\hat{n}$ case, determining its extrema and how they change as a function of the several parameters in the Hamiltonian.

Before proceeding, let us introduce two useful quantities: the maximum *anisotropy field* B_K and h, the external field measured in units of B_K:

$$B_K = \frac{2Kv}{m}, \qquad h = \frac{B}{B_K} = \frac{\xi}{2\sigma} \tag{2.6}$$

Let us now write the energy in terms of σ, the reduced field h, and the angle ϑ between $\vec{m}$ and the anisotropy axis [cf. Eq. (2.4)]

$$\beta\mathcal{H} = -\sigma(\cos^2\vartheta + 2h\cos\vartheta) \qquad (2.7)$$

To fix ideas, we shall assume $\sigma > 0$, that is, anisotropy of easy-axis type. The results for $\sigma < 0$, will be analogous but what is a maximum for $\sigma > 0$, becomes a minimum for $\sigma < 0$, and vice versa. The extrema of $\mathcal{H}$ are obtained by equating to zero the ϑ derivative $\partial(\beta\mathcal{H})/\partial\vartheta = 2\sigma\sin\vartheta(\cos\vartheta + h)$, obtaining

$$\frac{\partial(\beta\mathcal{H})}{\partial\vartheta} = 0 \Rightarrow \begin{cases} \sin\vartheta = 0 & \Leftrightarrow \vartheta = 0, \pi \\ \cos\vartheta = -h & \text{if } |h| \leq 1 \end{cases}$$

The type of extrema is obtained by evaluating the second derivative at the extrema:

$$\frac{\partial^2(\beta\mathcal{H})}{\partial\vartheta^2} = \begin{cases} 2\sigma(1+h) & \text{for } \vartheta = 0 \\ 2\sigma(1-h) & \text{for } \vartheta = \pi \\ -2\sigma(1-h^2) & \text{for } \cos\vartheta = -h \quad (\text{if } |h| \leq 1) \end{cases}$$

so that one gets the following results:

	Minima	Maxima
$|h| < 1$	$\vartheta = 0, \pi$	$\vartheta = \arccos(-h)$
$h > 1$	$\vartheta = 0$	$\vartheta = \pi$
$h < -1$	$\vartheta = \pi$	$\vartheta = 0$

Thus, for $|h| < 1$ (i.e., for $|B| < B_K$), the energy has minima at $\vartheta = 0$ and $\vartheta = \pi$, with a maximum between them (see the upper panel of Fig. 1). On the other hand, for $|h| > 1$ (i.e., for fields higher than the maximum anisotropy field B_K), the upper (shallower) energy minimum ($\vartheta = \pi$ for $h > 0$) turns into a maximum as it merges with the intermediate maximum, which disappears (lower panel of Fig. 1).

Finally, from the values of the energy at $\vartheta = 0, \pi$, and, when it exists, at the intermediate maximum $\vartheta_M = \arccos(-h)$, one gets the energy-barrier heights ($|h| < 1$)

$$\beta[\mathcal{H}(\vartheta_M) - \mathcal{H}(0)] = \sigma_+, \qquad \beta[\mathcal{H}(\vartheta_M) - \mathcal{H}(\pi)] = \sigma_-$$

where

$$\sigma_\pm = \sigma(1 \pm h)^2 \qquad (2.8)$$

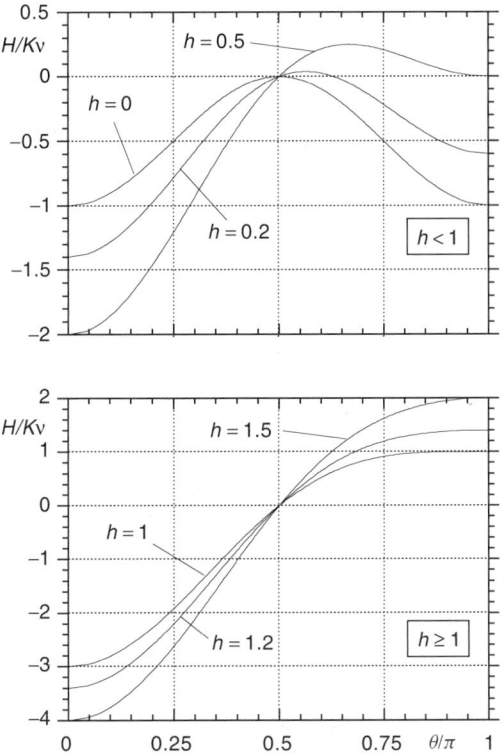

FIGURE 1. Magnetic energy in the longitudinal-field case for a number of values of the reduced field parameter $h = B/B_K$. Upper panel: $0 \leq h < 1$, so that the potential has two minima with an energy barrier between them. Lower panel: $h \geq 1$, so that no potential barrier exists.

C. Partition Function and Free Energy

1. General Definitions

The statistical independence of noninteracting magnetic moments allows one to express the thermodynamical quantities as sums over one-dipole contributions. Consequently, we shall study these contributions and obtain the results for the whole system by summing (or integrating) them over the ensemble of dipoles, taking their different anisotropy constants, orientations about the external field, magnitude of their dipole moments, etc., into account.

The *partition function* associated with a Hamiltonian $\mathcal{H}(\vartheta, \varphi)$, where ϑ, φ are the angular coordinates of $\vec{m}$ in a spherical coordinate system, can be defined as

$$\mathcal{Z} = \frac{1}{2\pi} \int_0^\pi d\vartheta \sin \vartheta \int_0^{2\pi} d\varphi \exp[-\beta\mathcal{H}(\vartheta, \varphi)] \qquad (2.9)$$

while the associated free energy is then given by

$$\mathcal{F} = -k_{\mathrm{B}}T \ln \mathcal{Z}$$

The definition (2.9) deserves some discussion. First, as was mentioned above, the definition of $\mathcal{Z}$ for a system with one degree of freedom is $\mathcal{Z} = \int (\mathrm{d}p\,\mathrm{d}q/2\pi\hbar)\exp(-\beta\mathcal{H})$ (see Landau and Lifshitz, 1980, Section 31). On the other hand, for a classical magnetic moment a convenient pair of conjugate canonical variables is $p = m_z/\gamma$ and $q = \varphi$ [see Eq. (6.11) in Section VI], where $m_z = m\cos\vartheta$ and γ is the magnetomechanical ratio. Therefore

$$\int \frac{\mathrm{d}p\,\mathrm{d}q}{2\pi\hbar}(\cdot) = \frac{m}{\gamma\hbar}\frac{1}{2\pi}\int_{-1}^{1}\mathrm{d}(\cos\vartheta)\int_{0}^{2\pi}\mathrm{d}\varphi(\cdot)$$

$$= S \times \frac{1}{2\pi}\int_{0}^{\pi}\mathrm{d}\vartheta\sin\vartheta\int_{0}^{2\pi}\mathrm{d}\varphi(\cdot)$$

where $S = (m/\gamma)/\hbar$ is the quantum number associated with the angular momentum m/γ. This expression yields $\mathcal{Z} = 2S$ for $\mathcal{H} \equiv 0$, which is the correct semiclassical case ($S \gg 1$) of the corresponding quantum expression $\mathcal{Z} = \Sigma_{S_z=-S}^{S}1 = 2S + 1$. Therefore the definition (2.9) corresponds to the proper statistical-mechanical definition, except for the factor S, which when required can be introduced by hand.

The *equilibrium probability distribution* of magnetic moment orientations is given by the Boltzmann distribution

$$P_{\mathrm{e}}(\cos\vartheta, \varphi) = \mathcal{Z}^{-1}\exp[-\beta\mathcal{H}(\vartheta, \varphi)]$$

so that the *statistical-mechanical average* of any observable $A = A(\vec{m}) = A(\vartheta, \varphi)$ reads

$$\langle A \rangle_{\mathrm{e}} = \int \mathrm{d}\Omega\, A(\vartheta, \varphi)P_{\mathrm{e}}(\vartheta, \varphi) = \frac{\int \mathrm{d}\Omega\, A(\vartheta, \varphi)\exp[-\beta\mathcal{H}(\vartheta, \varphi)]}{\int \mathrm{d}\Omega\,\exp[-\beta\mathcal{H}(\vartheta, \varphi)]} \qquad (2.10)$$

where $\int \mathrm{d}\Omega(\cdot) \equiv (1/2\pi)\int_{-1}^{1}\mathrm{d}(\cos\vartheta)\int_{0}^{2\pi}\mathrm{d}\varphi(\cdot)$. The relevant thermodynamical quantities can be written as the statistical-mechanical average of a certain function $A = A(\vartheta, \varphi)$ as in Eq. (2.10). Besides, they can be obtained as combinations of $\mathcal{Z}$ (or $\mathcal{F}$) and its derivatives. Table II summarizes some of those celebrated relations, which illustrate the pivotal role that the calculation of

TABLE II

Definition of Various Thermodynamical Quantities and Their Expressions in Terms of the
Partition Function $\mathcal{Z}$ and the Free Energy $\mathcal{F}$

	$\mathcal{A}$	Definition	$\mathcal{A}(\mathcal{Z})$	$\mathcal{A}(\mathcal{F})$
Energy	$\mathcal{U}$	$\langle \mathcal{H} \rangle_e$	$-\dfrac{\partial}{\partial \beta}(\ln \mathcal{Z})$	$\mathcal{F} + \beta \dfrac{\partial}{\partial \beta}\mathcal{F}$
Entropy	$\mathcal{S}$	$-\langle \ln P_e \rangle_e$	$\ln \mathcal{Z} - \beta \dfrac{\partial}{\partial \beta}(\ln \mathcal{Z})$	$\beta^2 \dfrac{\partial}{\partial \beta}\mathcal{F}$
Magnetization	M_B	$\langle \vec{m} \cdot \hat{b} \rangle_e$	$m\dfrac{\partial}{\partial \xi}(\ln \mathcal{Z})$	$-m\beta \dfrac{\partial}{\partial \xi}\mathcal{F}$

the partition function (or the free energy) plays in equilibrium statistical
mechanics.

2. Partition Function for the Simplest Axially Symmetric Anisotropy Potential

We shall usually choose the anisotropy axis $\hat{n}$ as the polar axis of a spherical
coordinate system. Then, if (ϑ, φ) and $(\alpha, 0)$ denote the angular coordinates
of $\vec{m}$ and $\vec{B}$, respectively (see Fig. 2), the Hamiltonian (2.4) can be written as

$$-\beta \mathcal{H} = \sigma \cos^2 \vartheta + \xi_\| \cos \vartheta + \xi_\perp \sin \vartheta \cos \varphi \qquad (2.11)$$

where we have introduced the longitudinal and transverse components (with
respect to the anisotropy-axis direction) of the dimensionless field $\vec{\xi} = m\vec{B}/k_B T$, namely

$$\xi_\| = \xi \cos \alpha, \qquad \xi_\perp = \xi \sin \alpha \qquad (2.12)$$

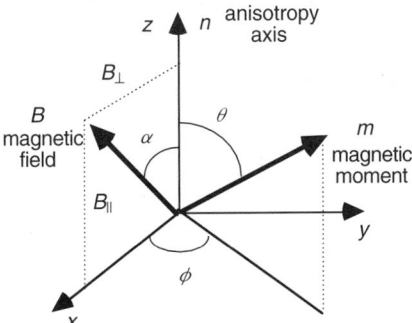

FIGURE 2. Coordinate system used in the calculation of the thermal equilibrium quantities.
The plane determined by $\hat{n}$ and $\vec{B}$ is chosen as the xz plane.

In order to analyze the partition function we, following Shcherbakova (1978), first do the integral over φ in the expression for $\mathcal{Z}$ associated with the Hamiltonian (2.11), getting [cf. Eq. (11) by Cregg and Bessais, 1999b]

$$\mathcal{Z} = \int_0^\pi d\vartheta \, \sin \vartheta \, \exp(\sigma \cos^2 \vartheta + \xi_\| \cos \vartheta) I_0(\xi_\perp \sin \vartheta) \qquad (2.13)$$

where

$$I_n(y) = \frac{1}{\pi} \int_0^\pi dt \, e^{y \cos t} \cos(nt) = \sum_{k=0}^\infty \frac{1}{k!(k+n)!} \left(\frac{y}{2}\right)^{2k+n}, \qquad n \geq 0$$
$$(2.14)$$

is the modified Bessel function of the first kind of order n (see, for example, Arfken, 1985, Section 11.5).

Equation (2.13) gives the partition function in terms of an integral over ϑ only. Therefore, the integrand (divided by $\mathcal{Z}$) can be interpreted as an effective probability distribution of the polar angle. Indeed, on introducing the substitution $z = \cos \vartheta$ one can first write Eq. (2.13) as

$$\mathcal{Z} = \int_{-1}^1 dz \, \exp(\sigma z^2 + \xi_\| z) I_0(\xi_\perp \sqrt{1 - z^2}) \qquad (2.15)$$

Then, the thermal equilibrium average of functions of $\cos \vartheta$ *only* can be obtained through $\langle A \rangle_e = \int_{-1}^1 dz \, A(z) P_e^{\text{eff}}(z)$, where

$$P_e^{\text{eff}}(z) = \frac{1}{\mathcal{Z}} \exp(\sigma z^2 + \xi_\| z) I_0(\xi_\perp \sqrt{1 - z^2}) \qquad (2.16)$$

is the effective or averaged (over the azimuthal angle), probability distribution.*

3. Particular Cases and Limiting Regimes

In various special cases, one can write the partition function and the free energy in a closed analytic form. Accordingly, along with being relevant to get insight into the thermal equilibrium properties of the system, those expressions will be used as reference for the general or approximate formulas derived along this section.

---

*Naturally $P_e^{\text{eff}}(z)$ coincides with the actual probability distribution when the total $\mathcal{H}(\vec{m})$ is axially symmetric.

a. Isotropic Case. We first consider the case $\sigma = 0$. This *isotropic* or *Langevin* regime will be attained if the anisotropy constant is identically zero or at high temperatures where $|\sigma| \ll 1$. Then, the partition function does not depend on α ($\cos \alpha = \hat{n} \cdot \hat{b}$), so we can choose α at will in Eq. (2.15). On setting $\alpha = 0$ (so that $\xi_{\perp} = 0$ and $\xi_{\parallel} = \xi$) and using $I_0(0) = 1$, Eq. (2.15) reduces to $\mathcal{Z}_{\mathrm{Lan}} = \int_{-1}^{1} dz \exp(\xi z)$. Therefore, the partition function and free energy in the isotropic case can be written as

$$\mathcal{Z}_{\mathrm{Lan}} = \frac{2}{\xi} \sinh \xi, \qquad \mathcal{F}_{\mathrm{Lan}} = k_{\mathrm{B}}T[\ln(\xi) - \ln(2 \sinh \xi)] \qquad (2.17)$$

Similarly, the probability distribution (2.16) reduces in this case to

$$P_{e,\mathrm{Lan}}(z) = \frac{\exp(\xi z)}{(2/\xi) \sinh \xi} \qquad (2.18)$$

which is displayed in Fig. 3.

b. Zero-Field Case. In the absence of an external field (unbiased case), one can use again $I_0(0) = 1$ in Eq. (2.15), to get $\mathcal{Z}_{\mathrm{unb}} = 2 \int_0^1 dz \exp(\sigma z^2)$. It will be then convenient to introduce the function (Raĭkher and Shliomis, 1975)

$$R(\sigma) \equiv \int_0^1 dz \exp(\sigma z^2) \qquad (2.19)$$

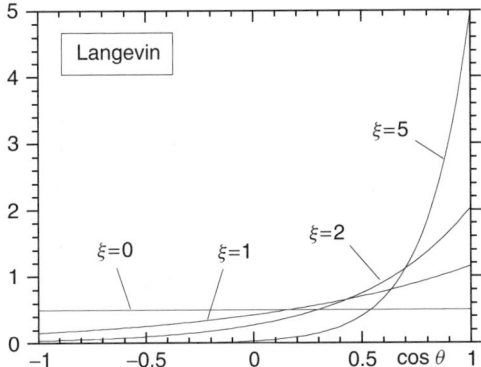

FIGURE 3. Probability distribution of the z component of the magnetic moment for $\sigma = Kv/k_{\mathrm{B}}T = 0$ in a magnetic field [Eq. (2.18)], for various values of the dimensionless field parameter $\xi = mB/k_{\mathrm{B}}T$. The value 0.5 corresponds to the uniform probability distribution ($\sigma = \xi = 0$).

in terms of which one can simply write the partition function and the free energy in the unbiased case as

$$\mathcal{Z}_{\mathrm{unb}} = 2R(\sigma), \qquad \mathcal{F}_{\mathrm{unb}} = -k_B T \ln[2R(\sigma)] \qquad (2.20)$$

while the probability distribution (2.16) then takes the form

$$P_{\mathrm{e,unb}}(z) = \frac{\exp(\sigma z^2)}{2R(\sigma)} \qquad (2.21)$$

In the easy-axis anisotropy case ($\sigma > 0$), this probability distribution evolves from uniform for $\sigma \ll 1$, to be quite concentrated around the poles for $\sigma \gg 1$ (see Fig. 4). Here the system approaches and effective Ising spin, since most of the time the magnetic moment stays close to the potential minima ($\vec{m} = \pm m\hat{n}$). For $\sigma < 0$ (easy-plane anisotropy), the probability distribution evolves from

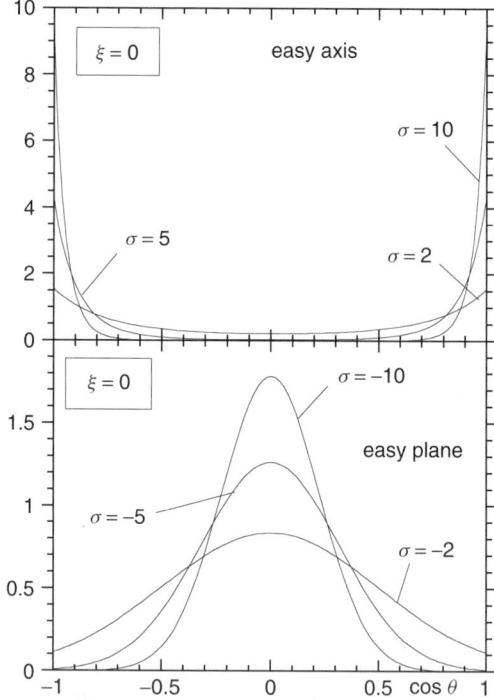

FIGURE 4. Probability distribution of the z component of the magnetic moment in zero field [Eq. (2.21)], for different values of the dimensionless anisotropy parameter $\sigma = Kv/k_B T$. The value 0.5 corresponds to the uniform probability distribution.

uniform for $|\sigma| \ll 1$, to be concentrated close to the equatorial circle for $\sigma \ll -1$ ("plane rotator" regime). Note that, in contrast to the easy-axis anisotropy case, where for $\sigma \sim 5$–10 the distribution of magnetic moment orientations is rather concentrated around the poles, for easy-plane anisotropy the corresponding concentration of the probability distribution around the equatorial region is less steep as a function of $|\sigma|$.

c. Ising Regime. We shall now consider in more detail the $\sigma \gg 1$ range. Here, the function $\exp(\sigma z^2)$ in the integrand of Eq. (2.15) is sharply peaked at the poles (see Fig. 4), so it can be approximated as a sum of two (nonnormalized) delta functions centered around $z = \pm 1$. Consequently, one has

$$\mathcal{Z} \simeq \left[e^{\xi_\parallel z} I_0(\xi_\perp \sqrt{1 - z^2}) \right]_{z=1} \int_0^1 dz\, e^{\sigma z^2} + \left[e^{\xi_\parallel z} I_0(\xi_\perp \sqrt{1 - z^2}) \right]_{z=-1} \int_{-1}^0 dz\, e^{\sigma z^2}$$

$$\overset{I_0(0)=1}{=} R(\sigma)(e^{\xi_\parallel} + e^{-\xi_\parallel}), \qquad \sigma \gg 1$$

Then, on using the leading asymptotic result $R(\sigma) \simeq e^\sigma/2\sigma$ (see Appendix A), the partition function and free energy in the "Ising" regime, can be written as

$$\mathcal{Z}_{\text{Ising}} = \frac{e^\sigma}{\sigma} \cosh \xi_\parallel, \qquad \mathcal{F}_{\text{Ising}} = -Kv + k_B T \left[\ln(\sigma) - \ln(\cosh \xi_\parallel) \right] \quad (2.22)$$

Note, however, that for an Ising spin, the factor $e^\sigma/2\sigma$ is absent in the corresponding $\mathcal{Z}$, which is equal to $e^{\xi_\parallel} + e^{-\xi_\parallel} = 2\cosh \xi_\parallel$. This factor does not alter quantities as the magnetization or the linear and nonlinear susceptibilities, because they are obtained as ξ derivatives of $\ln \mathcal{Z}$ (see Section III). Nevertheless, the occurrence of the factor $e^\sigma/2\sigma$ moves the "thermal" quantities (thermodynamical energy, entropy, and specific heat) from those of the archetypal Ising case. Note finally that the employed replacement of the factor $\exp(\sigma z^2)$ by a sum of Dirac deltas will work if the remainder terms in the integrand vary slowly enough with z. Naturally, this condition will not be obeyed for sufficiently high external fields (specifically, for $\xi \gtrsim \sigma$).

d. Plane-Rotator Regime. For $\sigma \ll -1$ the term $\exp(\sigma z^2)$ in the integrand of Eq. (2.15) is peaked at the equator (see Fig. 4). It can therefore be approximated by a Dirac delta located at $z = 0$, to get

$$\mathcal{Z} \simeq \left[e^{\xi_\parallel z} I_0(\xi_\perp \sqrt{1 - z^2}) \right]_{z=0} \int_{-1}^1 dz\, e^{\sigma z^2} = 2R(\sigma)I_0(\xi_\perp), \qquad \sigma \ll -1$$

Now, on employing the asymptotic ($\sigma \ll -1$) result $R(\sigma) \simeq (-\pi/4\sigma)^{1/2}$ (Appendix A), we obtain the following expressions for the partition function and free energy in the plane-rotator regime

$$\mathcal{Z}_{rot} = \left(-\frac{\pi}{\sigma}\right)^{1/2} I_0(\xi_\perp),$$

$$\mathcal{F}_{rot} = -k_B T \left\{ \frac{1}{2} \ln\left(-\frac{\pi}{\sigma}\right) + \ln[I_0(\xi_\perp)] \right\} \qquad (2.23)$$

The factor $(-\pi/\sigma)^{1/2}$ is absent in the partition function of the archetypal plane rotator, which is merely given by $(1/2\pi) \int_0^{2\pi} d\varphi\, e^{\xi_\perp \cos\varphi} = I_0(\xi_\perp)$. Again, this factor is irrelevant for the quantities obtained as ξ derivatives of $\ln \mathcal{Z}$, whereas is important for the calculation of the thermal quantities. Similarly, the replacement of the factor $\exp(\sigma z^2)$ by a Dirac delta will work only for external fields that are not very high.

e. Longitudinal-Field Case. We finally consider the situation in which the external field points along the anisotropy axis. In this case, without making assumptions concerning the magnitudes of the anisotropy energy or the field, one can write a closed analytic formula for the partition function (and accordingly for all the thermodynamical quantities).

When the external field is applied along the anisotropy axis, one has $\xi_\parallel = \xi$ and $\xi_\perp = 0$, so that the general partition function (2.15) reduces to

$$\mathcal{Z}_\parallel = \int_{-1}^{1} dz \exp\left(\sigma z^2 + \xi z\right) \qquad (2.24)$$

Then, on completing the square in the argument of the exponential and taking the definition (2.6) of h into account, one gets $\mathcal{Z}_\parallel = \exp(-\sigma h^2) \int_{-1}^{1} dz\, \exp[\sigma(z+h)^2]$. If we now introduce the substitution $t = z + h$, the partition function reads

$$\mathcal{Z}_\parallel = e^{-\sigma h^2} \int_{h-1}^{h+1} dt\, e^{\sigma t^2} = e^{-\sigma h^2} \left[\int_0^{h+1} dt\, e^{\sigma t^2} - \int_0^{h-1} dt\, e^{\sigma t^2} \right]$$

so that, on using the substitutions $u = t/(h+1)$ in the first integral after the last equal sign, and $u = t/(h-1)$ in the second one, we find

$$\mathcal{Z}_\parallel = e^{-\sigma h^2} \left\{ (1+h) \int_0^1 du\, e^{\sigma(1+h)^2 u^2} + (1-h) \int_0^1 du\, e^{\sigma(1-h)^2 u^2} \right\}$$

However, these integrals are merely the R function (2.19) evaluated at $\sigma_\pm = \sigma(1 \pm h)^2$ [the energy-barrier heights for $h < 1$, Eq. (2.8)], so that we can finally write the desired closed analytic formula for $\mathcal{Z}_{\parallel}$ as

$$\mathcal{Z}_{\parallel} = e^{-\sigma h^2} \left[(1+h)R(\sigma_+) + (1-h)R(\sigma_-) \right] \tag{2.25}$$

Besides, the probability distribution of $z = \cos \vartheta$ is in this case given by

$$P_{e,\parallel}(z) = \frac{\exp\left(\sigma z^2 + \xi z\right)}{\mathcal{Z}_{\parallel}(\sigma, \xi)} \tag{2.26}$$

which is displayed in Fig. 5 for various values of the longitudinal field.*

Let us finally consider some particular cases and approximations. On taking the $h \to 0$ limit in the expression (2.25), one again gets the unbiased partition function $\mathcal{Z}_{\mathrm{unb}} = 2R(\sigma)$ [Eq. (2.20)]. The $\sigma \to 0$ limit can also be taken, but this should be done with some care. One must first realize that, since $h = \xi/2\sigma$, the arguments of the R functions in Eq. (2.25) are large in this case. Accordingly, on assuming, for example, $\sigma > 0$ and using the leading term in the asymptotic expansion of R [see Appendix A, Eq. (A.16)], one has $R(\sigma_+) \simeq e^{\sigma_\pm}/2\sigma_\pm$, whence [cf. Eq. (3.12) by Garanin, 1996]

$$\mathcal{Z}_{\parallel} \simeq e^{-\sigma h^2} \left[(1+h)\frac{e^{\sigma_+}}{2\sigma_+} + (1-h)\frac{e^{\sigma_-}}{2\sigma_-} \right] = e^\sigma \left[\frac{e^\xi}{2\sigma + \xi} + \frac{e^{-\xi}}{2\sigma - \xi} \right]$$

where we have used $\sigma_\pm = \sigma(1 \pm h)^2$ and $\exp(\sigma_\pm) = \exp[\sigma(1+h^2)]\exp(\pm\xi)$. On further manipulating this expression, one eventually gets the approximate result

$$\mathcal{Z}_{\parallel} \simeq \frac{2e^\sigma}{4\sigma^2 - \xi^2} (2\sigma \cosh \xi - \xi \sinh \xi), \qquad (K > 0) \tag{2.28}$$

Note that we have obtained more than we were initially looking for. Taking the limit $\sigma \to 0$ in this expression, we indeed get the isotropic partition function $\mathcal{Z}_{\mathrm{Lan}} = (2/\xi)\sinh\xi$ [Eq. (2.17)]. However, on considering the $\sigma \gg 1$ range of Eq. (2.28), we get as a bonus the Ising partition function $\mathcal{Z}_{\mathrm{Ising}} = (e^\sigma/\sigma)\cosh\xi$

---

* An alternative expression for $\mathcal{Z}_{\parallel}$ can be obtained by using the relation (A.10) between $R(\sigma)$ and the Dawson integral $D(\cdot)$ [Eq. (A.9)], getting

$$\mathcal{Z}_{\parallel} = \frac{e^\sigma}{\sqrt{\sigma}} \left[e^\xi D(\sqrt{\sigma_+}) + e^{-\xi} D(\sqrt{\sigma_-}) \right] \tag{2.27}$$

Note, however, that, since the relation employed holds only for $\sigma > 0$, this formula for $\mathcal{Z}_{\parallel}$ is also subjected to the same restriction.

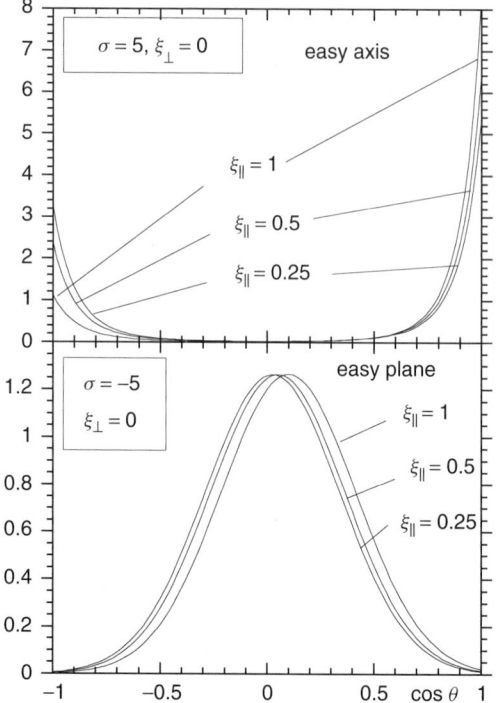

FIGURE 5. Probability distribution of the z component of the magnetic moment [Eq. (2.26)] for $|\sigma| = |Kv/k_B T| = 5$ and various values of the longitudinal-field parameter $\xi_\| = \xi = mB/k_B T$.

[Eq. (2.22)]. We have also obtained this result since, for $\sigma \gg 1$, the arguments of the functions $R(\sigma_\pm)$ in $\mathcal{Z}_\|$ are also large and positive. Note finally that Eq. (2.28) can also be written in terms of $h = \xi/2\sigma$ as

$$\mathcal{Z}_\| \simeq \frac{e^\sigma}{2\sigma} \frac{1}{(1-h^2)} \left[(1-h)e^{2\sigma h} + (1+h)e^{-2\sigma h} \right], \qquad (K > 0) \qquad (2.29)$$

which will be used in Section III.

D. Series Expansions of the Partition Function

We shall now carry out the expansion of the partition function in powers of either the external field or the anisotropy parameter, as well as an asymptotic expansion for strong anisotropy. These expansions will enable us to derive the first few terms in the corresponding expansions of the free energy in Section II.E. From these expressions one can obtain formulas for the linear and first nonlinear susceptibilities, as well as deviations of the magnetization from the Langevin or Ising-type curves.*

1. Field Expansion of the Partition Function

Let us first consider the expansion of $\mathcal{Z}$ in powers of the external field (García-Palacios and Lázaro, 1997). First, we insert the power expansions of the functions $\exp(\xi_\parallel z)$ and $I_0(\xi_\perp \sqrt{1 - z^2})$ [see Eq. (2.14)], into the partition function (2.15), to get

$$\mathcal{Z} = \sum_{i,k=0}^{\infty} \frac{\xi_\parallel^i}{i!} \left(\frac{\xi_\perp}{2}\right)^{2k} \frac{1}{(k!)^2} \int_{-1}^{1} dz\, z^i (\sqrt{1 - z^2})^{2k} \exp(\sigma z^2)$$

$$= 2 \sum_{i,k=0}^{\infty} \frac{\xi_\parallel^{2i} \xi_\perp^{2k}}{(2i)! 2^{2k} (k!)^2} \int_{0}^{1} dz\, z^{2i} (1 - z^2)^k \exp(\sigma z^2)$$

Note that the terms with odd powers of z have vanished on integration, while the integration of the terms with even powers of z has been reduced to the interval [0,1], by taking the symmetry of the corresponding integrands into account. Next, on recalling the definitions (2.12) of $\xi_\parallel$ and $\xi_\perp$ and introducing the angular coefficients

$$b_{i,k}(\alpha) = \frac{1}{(2i)! 2^{2k} (k!)^2} \cos^{2i} \alpha \sin^{2k} \alpha \qquad (2.30)$$

we can express the partition function as

$$\mathcal{Z} = 2 \sum_{i,k=0}^{\infty} b_{i,k}(\alpha) \xi^{2(i+k)} \int_{0}^{1} dz\, z^{2i} (1 - z^2)^k \exp(\sigma z^2) \qquad (2.31)$$

Now, expanding $(1 - z^2)^k$ by means of the binomial formula, we obtain

$$\mathcal{Z} = 2 \sum_{i,k=0}^{\infty} b_{i,k}(\alpha) \xi^{2(i+k)} \sum_{m=0}^{k} (-1)^m \binom{k}{m} R^{(i+m)}(\sigma) \qquad (2.32)$$

where the $\binom{k}{m} = k!/[m!(k - m)!]$ are *binomial coefficients* and we have introduced the derivatives $R^{(l)}(\sigma) = d^l R/d\sigma^l$ of the function $R(\sigma)$ [Eq. (2.19)],

---

*Cregg and Bessais (1999a) have obtained alternative series expansions for the partition function.

namely

$$R^{(l)}(\sigma) = \int_0^1 dz\, z^{2l} \exp(\sigma z^2), \qquad l = 0, 1, 2, \ldots, \qquad R^{(0)} \equiv R \qquad (2.33)$$

Finally, on collecting the terms with the same power of ξ by means of the identity

$$\sum_{i,k=0}^{\infty} A_{i,k} y^{(i+k)} = \sum_{j=0}^{\infty} \left(\sum_{l=0}^{j} A_{j-l,l} \right) y^j \qquad (2.34)$$

we can rewrite the expansion (2.32) as follows

$$\mathcal{Z} = 2R(\sigma) \sum_{i=0}^{\infty} \frac{C_i(\sigma, \alpha)}{i!} \xi^{2i} \qquad (2.35)$$

where the coefficients C_i are given by

$$C_i(\sigma, \alpha) = i! \sum_{k=0}^{i} b_{i-k,k}(\alpha) \sum_{m=0}^{k} (-1)^m \binom{k}{m} \frac{R^{(i-k+m)}(\sigma)}{R(\sigma)} \qquad (2.36)$$

For the sake of convenience later, we have extracted the factor $R(\sigma)$ in Eq. (2.35) [recall that $2R(\sigma)$ is the partition function at zero external field] and introduced the factor $i!$ in the definition of the coefficients C_i.

The functions $R^{(l)}$ are directly related with known special functions — such as confluent hypergeometric (Kummer) functions, error functions, and the Dawson integral — and their properties are summarized in Appendix A. All the quotients $R^{(l)}/R$ occurring in the coefficients above are nonnegative and increase monotonically in the whole σ range. Thus, $R^{(l)}/R$ tends to 0 as $\sigma \to -\infty$, takes the value $1/(2l+1)$ at $\sigma = 0$, and tends to 1 as $\sigma \to \infty$ [Eqs. (A.21), (A.4), and (A.17), respectively]. The first two quotients $R^{(l)}/R$ (R'/R and R''/R) are shown in Fig. 6. Note that, as we can write $R'/R = \langle \cos^2 \vartheta \rangle_e$, the quantity R'/R is a measure of the "degree of polarization" of $\vec{m}$ along the anisotropy axis in the absence of an external field.

a. Alternative Expressions for the Coefficients C_i. The coefficients C_i can also be written in terms of the Kummer functions $M(a, c; x)$. First, on using the integral representation (A.5) for $M(a, c; x)$, we can write the integral occurring

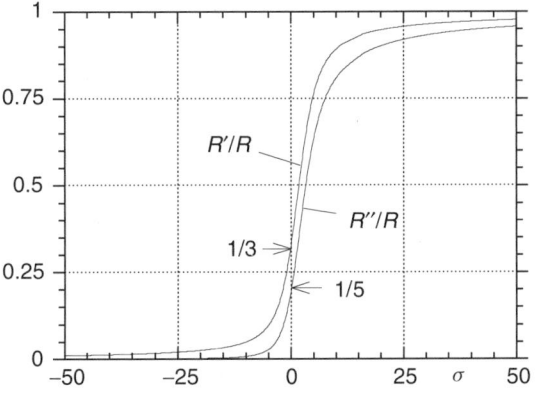

FIGURE 6.   The functions R'/R and R''/R.

in the expression (2.31) as

$$\int_0^1 dz\, z^{2i}(1-z^2)^k e^{\sigma z^2} = \frac{\Gamma\left(i+\frac{1}{2}\right)\Gamma(k+1)}{2\Gamma\left(i+k+\frac{3}{2}\right)} M\left(i+\frac{1}{2}, i+k+\frac{3}{2}; \sigma\right) \quad (2.37)$$

where $\Gamma(z)$ is the gamma (factorial) function [Eq. (A.2)]. If we introduce this expression into the expansion (2.31), we find the numerical coefficient

$$\frac{1}{(2i)!2^{2k}(k!)^2}\frac{\Gamma\left(i+\frac{1}{2}\right)\Gamma(k+1)}{2\Gamma\left(i+k+\frac{3}{2}\right)} = \frac{1}{[2(i+k)+1]!}\binom{i+k}{k}$$

where the basic property of the gamma function, $\Gamma(z+1) = z\Gamma(z)$, has been used. Then, on gathering the terms with the same power of ξ in the resulting $\mathcal{Z}$ by dint of Eq. (2.34), we obtain

$$\mathcal{Z} = 2\sum_{i=0}^{\infty} \frac{\xi^{2i}}{(2i+1)!}\left\{\sum_{k=0}^{i} d_{i-k,k}(\alpha)\, M\left(i-k+\frac{1}{2}, i+\frac{3}{2}; \sigma\right)\right\}$$

where the new angular coefficients $d_{i,k}(\alpha)$ are given by

$$d_{i,k}(\alpha) = \binom{i+k}{k}\cos^{2i}\alpha\,\sin^{2k}\alpha \quad (2.38)$$

Consequently, on comparing with Eq. (2.35), we can finally express the coefficients C_i in terms of Kummer functions as

$$C_i(\sigma, \alpha) = \frac{i!}{(2i+1)!} \sum_{k=0}^{i} d_{i-k,k}(\alpha) \frac{M(i-k+\frac{1}{2}, i+\frac{3}{2}; \sigma)}{M(\frac{1}{2}, \frac{3}{2}; \sigma)} \qquad (2.39)$$

where we have used (see Appendix A)

$$R(\sigma) = M(\tfrac{1}{2}, \tfrac{3}{2}; \sigma) \qquad (2.40)$$

Let us finally write in full the first few coefficients for future reference. If we introduce the first few angular coefficients $d_{i,k}(\alpha)$

$$d_{0,0} = 1 \qquad d_{1,0} = \cos^2 \alpha \qquad d_{0,1} = \sin^2 \alpha$$

$$d_{2,0} = \cos^4 \alpha \qquad d_{1,1} = 2\cos^2 \alpha \sin^2 \alpha \qquad d_{0,2} = \sin^4 \alpha$$

into Eq. (2.39), we get

$$C_1 = \frac{1}{3!} \left[\frac{M(\frac{3}{2}, \frac{5}{2}; \sigma)}{M(\frac{1}{2}, \frac{3}{2}; \sigma)} \cos^2 \alpha + \frac{M(\frac{1}{2}, \frac{5}{2}; \sigma)}{M(\frac{1}{2}, \frac{3}{2}; \sigma)} \sin^2 \alpha \right] \qquad (2.41)$$

as well as

$$C_2 = \frac{1}{60} \left[\frac{M(\frac{5}{2}, \frac{7}{2}; \sigma)}{M(\frac{1}{2}, \frac{3}{2}; \sigma)} \cos^4 \alpha + \frac{M(\frac{3}{2}, \frac{7}{2}; \sigma)}{M(\frac{1}{2}, \frac{3}{2}; \sigma)} 2\cos^2 \alpha \sin^2 \alpha \right.$$

$$\left. + \frac{M(\frac{1}{2}, \frac{7}{2}; \sigma)}{M(\frac{1}{2}, \frac{3}{2}; \sigma)} \sin^4 \alpha \right]$$

The coefficients C_i can also be expressed in terms of the averages of $\vec{m}$ in zero field. To this end, let us begin from the definition of the partition function

$$\mathcal{Z} = \int d\Omega \exp \left[\sigma(\vec{e} \cdot \hat{n})^2 + \xi(\vec{e} \cdot \hat{b}) \right]$$

where $\int d\Omega(\cdot) = (1/2\pi) \int_{-1}^{1} d(\cos \vartheta) \int_{0}^{2\pi} d\varphi(\cdot)$ and the expression (2.4) for $-\beta \mathcal{H}$ have been used. Next, on expanding $\exp[\xi(\vec{e} \cdot \hat{b})]$ in powers of ξ, we obtain

$$\mathcal{Z} = \sum_{i=0}^{\infty} \frac{\xi^{2i}}{(2i)!} \int d\Omega(\vec{e} \cdot \hat{b})^{2i} \exp \left[\sigma(\vec{e} \cdot \hat{n})^2 \right]$$

where, to eliminate the odd powers of ξ, we have merely considered that $(\vec{e} \cdot \hat{b})^{2i+1}$ reverses its sign when the transformation $\vec{e} \to -\vec{e}$ is applied, whereas the term $\exp[\sigma(\vec{e} \cdot \hat{n})^2]$ is invariant against such transformation (whence $\int d\Omega (\vec{e} \cdot \hat{b})^{2i+1} \exp[\sigma(\vec{e} \cdot \hat{n})^2] \equiv 0$). Finally, on comparing this expansion of $\mathcal{Z}$ with $\mathcal{Z} = 2R \sum_{i=0}^{\infty} (C_i/i!) \xi^{2i}$, noting that $2R(\sigma)$ can be written as $2R(\sigma) = \int d\Omega \exp[\sigma(\vec{e} \cdot \hat{n})^2]$, and introducing the thermal equilibrium averages in zero field [cf. Eq. (2.10)]

$$\langle (\vec{e} \cdot \hat{b})^n \rangle_e \big|_{B=0} = \frac{\displaystyle\int d\Omega (\vec{e} \cdot \hat{b})^n \exp[\sigma(\vec{e} \cdot \hat{n})^2]}{\displaystyle\int d\Omega \exp[\sigma(\vec{e} \cdot \hat{n})^2]}$$

we arrive at the desired relation

$$\frac{C_i(\sigma, \alpha)}{i!} = \frac{1}{(2i)!} \left\langle (\vec{e} \cdot \hat{b})^{2i} \right\rangle_e \big|_{B=0} \tag{2.42}$$

b. Particular Cases of the Coefficients C_i. Let us briefly consider the form that the coefficients appearing in the field expansion of the partition function take in the particular cases considered in Section II.C. To this end, the alternative expression for those coefficients in terms of Kummer functions [Eq. (2.39)] proves to be more convenient.

1. On noting that $M(a, c; x = 0) = 1$ [see the definition (A.1)], one immediately gets for C_i in the *isotropic* case

$$\frac{1}{i!} C_i \big|_{\sigma=0} = \frac{1}{(2i+1)!} \sum_{k=0}^{i} \binom{i}{k} \cos^{2(i-k)} \alpha \sin^{2k} \alpha = \frac{1}{(2i+1)!}$$

since by the binomial formula the sum is equal to $(\cos^2 \alpha + \sin^2 \alpha)^i = 1$.

2. In the $\sigma \to \infty$ limit, on employing the asymptotic expansion (A.15) of $M(a, c; x)$ for large positive argument, one finds

$$\frac{M(i - k + \frac{1}{2}, i + \frac{3}{2}; \sigma)}{M(\frac{1}{2}, \frac{3}{2}; \sigma)} \Bigg|_{\sigma \gg 1} = \frac{\Gamma(i + \frac{3}{2})}{\Gamma(i - k + \frac{1}{2})} \frac{2}{\sigma^k} \xrightarrow{\sigma \to \infty} (2i + 1)\delta_{k,0}$$

where we have used $\Gamma(i + \frac{3}{2}) = (i + \frac{1}{2}) \Gamma(i + \frac{1}{2})$. Therefore, the general expression (2.39) reduces in the *Ising* case to

$$\frac{1}{i!} C_i \big|_{\sigma \to \infty} = \frac{\cos^{2i} \alpha}{(2i)!}$$

3. To get the $\sigma \to -\infty$ limit of C_i, we can now use the asymptotic expansion (A.18) of $M(a, c; x)$ for large negative argument. On so doing, one first finds

$$
\left. \frac{M\left(i - k + \frac{1}{2}, i + \frac{3}{2}; \sigma\right)}{M\left(\frac{1}{2}, \frac{3}{2}; \sigma\right)} \right|_{\sigma \ll -1}
$$

$$
= \frac{\Gamma\left(i + \frac{3}{2}\right)}{\frac{1}{2}\pi^{1/2}k!} \frac{1}{(-\sigma)^{i-k}} \xrightarrow{\sigma \to -\infty} \frac{2\Gamma\left(i + \frac{3}{2}\right)}{\pi^{1/2}i!} \delta_{i,k}
$$

Therefore, by using Eq. (A.20) for the gamma function of half-odd-integer argument, the *plane-rotator* C_i reads

$$
\frac{1}{i!} C_i \big|_{\sigma \to -\infty} = \left(\frac{\sin \alpha}{2} \right)^{2i} \frac{1}{(i!)^2}
$$

4. The *longitudinal-field* case corresponds to set $\alpha = 0$ in the expression (2.39) for $C_i(\sigma, \alpha)$. On doing this and using $d_{i-k,k}|_{\alpha=0} = \delta_{k,0}$ [see Eq. (2.38)], one gets

$$
\frac{1}{i!} C_i \big|_{\alpha=0} = \frac{1}{(2i + 1)!} \frac{M\left(i + \frac{1}{2}, i + \frac{3}{2}; \sigma\right)}{M\left(\frac{1}{2}, \frac{3}{2}; \sigma\right)} = \frac{1}{(2i)!} \frac{R^{(i)}(\sigma)}{R(\sigma)}
$$

where the relations (A.3) between the functions $R^{(l)}$ and Kummer functions have been taken into account.

All these particular cases of the coefficients C_i are summarized in Table III, while the first few ones are displayed in Table IV.

TABLE III
Expressions for the Coefficients $C_i/i!$ of the Field
Expansion of the Partition Function in the Isotropic,
Ising, Plane-Rotator, and Longitudinal-Field Cases

	$\sigma = 0$	$\sigma \to \infty$	$\sigma \to -\infty$	$\vec{B} \| \hat{n}$
$\dfrac{C_i}{i!}$	$\dfrac{1}{(2i+1)!}$	$\dfrac{\cos^{2i} \alpha}{(2i)!}$	$\dfrac{\sin^{2i} \alpha}{2^{2i}(i!)^2}$	$\dfrac{1}{(2i)!} \dfrac{R^{(i)}}{R}$

TABLE IV

Coefficients C_1, C_2, and C_3 in the Isotropic, Ising, Plane-Rotator, and Longitudinal-Field Cases

	$\sigma = 0$	$\sigma \to \infty$	$\sigma \to -\infty$	$\vec{B} \| \hat{n}$
C_1	$\dfrac{1}{6}$	$\dfrac{1}{2}\cos^2 \alpha$	$\dfrac{1}{4}\sin^2 \alpha$	$\dfrac{1}{2}\dfrac{R'}{R}$
C_2	$\dfrac{1}{60}$	$\dfrac{1}{12}\cos^4 \alpha$	$\dfrac{1}{32}\sin^4 \alpha$	$\dfrac{1}{12}\dfrac{R''}{R}$
C_3	$\dfrac{1}{840}$	$\dfrac{1}{120}\cos^6 \alpha$	$\dfrac{1}{384}\sin^6 \alpha$	$\dfrac{1}{120}\dfrac{R'''}{R}$

2. Expansion of the Partition Function in Powers of the Anisotropy Parameter

We now derive the first few terms in the expansion of $\mathcal{Z}$ in powers of $\sigma = Kv/k_B T$. This expansion will provide a suitable description of the thermodynamical properties when the anisotropy energy is sufficiently small in comparison to the thermal energy.

In order to perform this expansion, it is more convenient to rotate the spherical coordinate system to set the polar axis pointing along the external field $\vec{B}$ (see Fig. 2; the anisotropy axis $\hat{n}$ is now in the xz plane and α is its polar angle). With this choice of coordinates, the partition function reads

$$\mathcal{Z} = \frac{1}{2\pi} \int_0^\pi d\vartheta \, \sin \vartheta \exp(\xi \cos \vartheta)$$
$$\times \int_0^{2\pi} d\varphi \exp\left[\sigma(\cos \alpha \cos \vartheta + \sin \alpha \sin \vartheta \cos \varphi)^2\right]$$

If we now expand the second exponential, we get an expression of the form

$$\mathcal{Z} = \sum_{i=0}^{\infty} \frac{\sigma^i}{i!} \mathcal{Z}_i \tag{2.43}$$

where

$$\mathcal{Z}_i = \frac{1}{2\pi} \int_0^\pi d\vartheta \, \sin \vartheta \exp(\xi \cos \vartheta) \int_0^{2\pi} d\varphi (\cos \alpha \cos \vartheta + \sin \alpha \sin \vartheta \cos \varphi)^{2i} \tag{2.44}$$

Note that the zeroth-order coefficient is naturally the partition function in the isotropic case $\mathcal{Z}_0 = (2/\xi) \sinh \xi$ [Eq. (2.17)].

On using the binomial expansion in the second integrand of Eq. (2.44), and employing the result (Arfken, 1985, p. 318)

$$\frac{1}{2\pi} \int_0^{2\pi} d\varphi \cos^n \varphi = \begin{cases} 0 & \text{for odd } n \\ \dfrac{(2k)!}{2^{2k}(k!)^2} & \text{for } n = 2k \end{cases} \tag{2.45}$$

to do the integrals over the azimuthal angle, we see that only even powers of $\cos\alpha$ and $\sin\alpha$ appear in $\mathcal{Z}_i$. Besides, $\sin^{2k}\vartheta$ can always be expressed as a sum of powers of the form $\cos^{2l}\vartheta$, with $l \le k$:

$$\sin^{2k}\vartheta = \left(1 - \cos^2\vartheta\right)^k = \sum_{l=0}^{k} \binom{k}{l}(-1)^l \cos^{2l}\vartheta$$

Accordingly, on introducing once more the substitution $z = \cos\vartheta$ and noting that

$$\int_{-1}^{1} dz\, z^n \exp(\xi z) = \frac{d^n}{d\xi^n} \int_{-1}^{1} dz \exp(\xi z) = \frac{d^n}{d\xi^n} \mathcal{Z}_0 \tag{2.46}$$

one realizes that all the functions $\mathcal{Z}_i$ can be expressed in terms of the isotropic partition function $\mathcal{Z}_0$ and its ξ derivatives. For instance, $\mathcal{Z}_1$ reads

$$\mathcal{Z}_1 = \cos^2\alpha \int_{-1}^{1} dz\, z^2 \exp(\xi z) + \tfrac{1}{2}\sin^2\alpha \int_{-1}^{1} dz \left(1 - z^2\right) \exp(\xi z)$$

$$= \mathcal{Z}_0'' \cos^2\alpha + \tfrac{1}{2}(\mathcal{Z}_0 - \mathcal{Z}_0'')\sin^2\alpha \tag{2.47}$$

where the prime denotes differentiation with respect to ξ. On the other hand, since $\mathcal{Z}_0 = (2/\xi)\sinh\xi$, the derivative $\mathcal{Z}_0'$ is given by

$$\mathcal{Z}_0' = L(\xi)\mathcal{Z}_0 \tag{2.48}$$

where

$$L(\xi) = \coth\xi - \frac{1}{\xi} \tag{2.49}$$

is the celebrated Langevin function. On taking a further ξ derivative and using the relation between L' and L, namely

$$L' = 1 - \frac{2}{\xi}L - L^2 \tag{2.50}$$

we get for the combinations of $\mathcal{Z}_0$ and $\mathcal{Z}_0''$ occurring in Eq. (2.47):

$$\mathcal{Z}_0'' = \mathcal{Z}_0\left(1 - \frac{2}{\xi}L\right), \qquad \frac{1}{2}(\mathcal{Z}_0 - \mathcal{Z}_0'') = \mathcal{Z}_0\frac{1}{\xi}L \tag{2.51}$$

Therefore, on introducing these results in Eq. (2.47), we finally get

$$\frac{\mathcal{Z}_1}{\mathcal{Z}_0} = \left(1 - \frac{2}{\xi}L\right)\cos^2\alpha + \frac{1}{\xi}L\sin^2\alpha \qquad (2.52)$$

The calculation of $\mathcal{Z}_2$ proceeds similarly. On taking the definition (2.44) into account and using $z = \cos\vartheta$, one obtains

$$\mathcal{Z}_2 = \cos^4\alpha \int_{-1}^{1} dz\, z^4 e^{\xi z} + \tfrac{6}{2}\cos^2\alpha\sin^2\alpha \int_{-1}^{1} dz\, z^2 \left(1 - z^2\right) e^{\xi z}$$

$$+ \tfrac{3}{8}\sin^4\alpha \int_{-1}^{1} dz \left(1 - z^2\right)^2 e^{\xi z}$$

where Eq. (2.45) has been used for doing the integrals over φ. Consequently, in terms of $\mathcal{Z}_0$ and its derivatives, $\mathcal{Z}_2$ is given by

$$\mathcal{Z}_2 = \mathcal{Z}_0''''\cos^4\alpha + 3(\mathcal{Z}_0'' - \mathcal{Z}_0'''')\cos^2\alpha\sin^2\alpha$$

$$+ \tfrac{3}{8}(\mathcal{Z}_0 - 2\mathcal{Z}_0'' + \mathcal{Z}_0'''')\sin^4\alpha \qquad (2.53)$$

In order to take the fourth-order derivative $\mathcal{Z}_0''''$, one can repeatedly use Eqs. (2.48) and (2.50). However, it significantly simplifies the calculations to obtain first the derivative $(L/\xi)'$, which can be written as

$$\left(\frac{1}{\xi}L\right)' = -\frac{1}{\xi}\left[L^2 - \left(1 - \frac{3}{\xi}L\right)\right] \qquad (2.54)$$

Then, after some manipulation, one gets the expression

$$\mathcal{Z}_0'''' = \mathcal{Z}_0 \left[1 - \frac{4}{\xi}L + \frac{8}{\xi^2}\left(1 - \frac{3}{\xi}L\right)\right]$$

which, along with Eqs. (2.51), gives

$$\mathcal{Z}_0'''' - \mathcal{Z}_0'' = 2\mathcal{Z}_0\left[\frac{4}{\xi^2}\left(1 - \frac{3}{\xi}L\right) - \frac{1}{\xi}L\right],$$

$$\mathcal{Z}_0 - 2\mathcal{Z}_0'' + \mathcal{Z}_0'''' = \mathcal{Z}_0\frac{8}{\xi^2}\left(1 - \frac{3}{\xi}L\right)$$

On introducing all these results into Eq. (2.53), we finally find for $\mathcal{Z}_2$:

$$
\begin{aligned}
\frac{\mathcal{Z}_2}{\mathcal{Z}_0} &= \left[1 - \frac{4}{\xi}L + \frac{8}{\xi^2}\left(1 - \frac{3}{\xi}L\right)\right]\cos^4\alpha \\
&\quad + 6\left[\frac{1}{\xi}L - \frac{4}{\xi^2}\left(1 - \frac{3}{\xi}L\right)\right]\cos^2\alpha\sin^2\alpha \\
&\quad + \left[\frac{3}{\xi^2}\left(1 - \frac{3}{\xi}L\right)\right]\sin^4\alpha
\end{aligned}
\tag{2.55}
$$

This formula completes the explicit expansion of the partition function in powers of the anisotropy parameter up to second order.

3. Asymptotic Expansion of the Partition Function for Strong Anisotropy

In order to complement the weak-anisotropy expansion derived above, we now carry out an asymptotic expansion of the partition function for strong anisotropy (easy-axis case only). As will be seen below, the approximate thermal equilibrium quantities obtained from the combined use of those expansions well approximate the exact results in almost the whole temperature range. Therefore we shall be able to get simple analytic expressions for the thermodynamical quantities that reasonably avoid the necessity of their computation by numerical methods.

In order to perform an expansion of the partition function for large $\sigma = Kv/k_BT$, we start from the field expansion (2.35) of $\mathcal{Z}$ and use the asymptotic results for its coefficients. Then, we shall obtain a number of infinite series of powers of $\xi = mB/k_BT$, which will be identified as certain elementary functions, obtaining in this way a closed asymptotic expression for the partition function.

We start by recalling that the whole coefficient of ξ^{2i} in the general ξ expansion of $\mathcal{Z}$ reads [see Eqs. (2.35) and (2.39)]

$$
\frac{2R(\sigma)C_i}{i!} = \frac{2}{(2i+1)!}\sum_{k=0}^{i} d_{i-k,k}(\alpha)\,M\left(i-k+\tfrac{1}{2}, i+\tfrac{3}{2}; \sigma\right)
\tag{2.56}
$$

where $R(\sigma) = M\left(\tfrac{1}{2}, \tfrac{3}{2}; \sigma\right)$ has been used [Eq. (2.40)], and $d_{i-k,k}(\alpha)$ is explicitly given by [see Eq. (2.38)]

$$
d_{i-k,k}(\alpha) = \binom{i}{k}\cos^{2(i-k)}\alpha\sin^{2k}\alpha
\tag{2.57}
$$

On the other hand, the asymptotic expansion (A.15) of the confluent hypergeometric functions yields the following equation for $\sigma \gg 1$:

$$M\left(i - k + \tfrac{1}{2}, i + \tfrac{3}{2}; \sigma\right) = \frac{e^{\sigma}}{2\sigma} \frac{2\Gamma\left(i + \tfrac{3}{2}\right)}{\Gamma\left(i - k + \tfrac{1}{2}\right)} \frac{1}{\sigma^k}$$

$$\times \left[1 + \frac{(2k - 2i + 1)(k + 1)}{2\sigma} + \frac{(2k - 2i + 3)(2k - 2i + 1)}{\times (k + 1)(k + 2)}{8\sigma^2} + \cdots \right]$$

Now, considering that the sum in Eq. (2.56) begins at $k = 0$, and that we shall carry out the expansion of $\mathcal{Z}$ through order $1/\sigma^2$, we write

$$\sum_{k=0}^{i} d_{i-k,k} M\left(i - k + \tfrac{1}{2}, i + \tfrac{3}{2}; \sigma\right) \simeq \cos^{2i} \alpha\, M\left(i + \tfrac{1}{2}, i + \tfrac{3}{2}; \sigma\right)$$

$$+ \tfrac{1}{2}(2i) \cos^{2(i-1)} \alpha \sin^2 \alpha\, M\left(i - \tfrac{1}{2}, i + \tfrac{3}{2}; \sigma\right)$$

$$+ \tfrac{1}{8}(2i)(2i - 2) \cos^{2(i-2)} \alpha \sin^4 \alpha\, M\left(i - \tfrac{3}{2}, i + \tfrac{3}{2}; \sigma\right)$$

where we have taken Eq. (2.57) into account. Then, on using $\Gamma(z + 1) = z\Gamma(z)$, we get for the quotients of gamma functions occurring in this equation via the Kummer functions

$$\frac{2\Gamma\left(i + \tfrac{3}{2}\right)}{\Gamma\left(i - k + \tfrac{1}{2}\right)} = \begin{cases} (2i + 1) & \text{for } k = 0 \\ \tfrac{1}{2}(2i + 1)(2i - 1) & \text{for } k = 1 \\ \tfrac{1}{4}(2i + 1)(2i - 1)(2i - 3) & \text{for } k = 2 \end{cases}$$

On collecting all these intermediate results, we can approximately write the ith term in the ξ expansion of $\mathcal{Z}$ in the form

$$\frac{\sigma}{e^{\sigma}} \frac{2R(\sigma)C_i}{i!} \xi^{2i} \simeq \frac{\xi_{\parallel}^{2i}}{(2i)!}\left[1 - \frac{(2i - 1)}{2\sigma} + \frac{(2i - 1)(2i - 3)}{4\sigma^2}\right]$$

$$+ \frac{\xi_{\parallel}^{2(i-1)}\xi_{\perp}^2}{[2(i - 1)]!}\left[\frac{1}{4\sigma} - \frac{(2i - 3)}{4\sigma^2}\right] + \frac{\xi_{\parallel}^{2(i-2)}\xi_{\perp}^4}{[2(i - 2)]!}\frac{1}{32\sigma^2} \qquad (2.58)$$

where we have multiplied across by σ/e^{σ} to avoid writing e^{σ}/σ in all the right-hand sides of the subsequent equations. In addition, in this equation we have introduced the longitudinal and transverse components of the dimensionless field: $\xi_\parallel = \xi \cos \alpha$ and $\xi_\perp = \xi \sin \alpha$. Note, however, that Eq. (2.58) holds only for the terms with $i \geq 2$. For $i = 1$, the sum in k in the expression (2.39) only runs over $k = 0$ and $k = 1$; therefore, the last term on the right-hand side of Eq. (2.58) is absent. Similarly, for $i = 0$, only the first term remains. Taking these considerations into account by properly adjusting the summation limits in the following expression, we can already write the partition function $\mathcal{Z} = 2R \sum_{i=0}^{\infty} (C_i/i!) \xi^{2i}$ as

$$\frac{\sigma}{e^\sigma} \mathcal{Z} \simeq \sum_{i=0}^{\infty} \frac{\xi_\parallel^{2i}}{(2i)!} \left[1 - \frac{(2i-1)}{2\sigma} + \frac{(2i-1)(2i-3)}{4\sigma^2} \right]$$

$$+ \frac{1}{4\sigma} \sum_{i=1}^{\infty} \frac{\xi_\parallel^{2(i-1)} \xi_\perp^2}{[2(i-1)]!} \left[1 - \frac{(2i-3)}{\sigma} \right] + \frac{1}{32\sigma^2} \sum_{i=2}^{\infty} \frac{\xi_\parallel^{2(i-2)} \xi_\perp^4}{[2(i-2)]!}$$

If we now redefine the summation indices in order to force all these series to start at the value zero of the corresponding new index and gather the terms multiplying the same type of series, we get

$$\frac{\sigma}{e^\sigma} \mathcal{Z} \simeq \left(1 + \frac{1}{4\sigma} \xi_\perp^2 + \frac{1}{32\sigma^2} \xi_\perp^4 \right) \sum_{i=0}^{\infty} \frac{\xi_\parallel^{2i}}{(2i)!}$$

$$- \left(\frac{1}{2\sigma} + \frac{1}{4\sigma^2} \xi_\perp^2 \right) \sum_{i=0}^{\infty} \frac{\xi_\parallel^{2i}}{(2i)!} (2i-1)$$

$$+ \frac{1}{4\sigma^2} \sum_{i=0}^{\infty} \frac{\xi_\parallel^{2i}}{(2i)!} (2i-1)(2i-3) \tag{2.59}$$

Our final goal is to identify all the power series occurring in Eq. (2.59). The series in the first term on the right-hand side is precisely that of the hyperbolic cosine, $\cosh x = \sum_{i=0}^{\infty} x^{2i}/(2i)!$. The other two series can also be identified after some redefinition of the summation indices ($k = i - 1$):

$$\sum_{i=0}^{\infty} \frac{x^{2i}}{(2i)!} (2i-1) = \sum_{k=0}^{\infty} \frac{x^{2k+2}}{(2k+1)!} - \cosh x = x \sinh x - \cosh x$$

while

$$\sum_{i=0}^{\infty} \frac{x^{2i}}{(2i)!} (2i-1)(2i-3) = \sum_{k=0}^{\infty} \frac{x^{2(k+1)}}{(2k)!} - 3(x \sinh x - \cosh x)$$

$$= (x^2 + 3) \cosh x - 3x \sinh x$$

Finally, we insert these results into Eq. (2.59), gather the terms with the same power of $1/\sigma$, and extract a factor $\cosh \xi_\parallel$, obtaining

$$\mathcal{Z} \simeq \frac{e^\sigma}{\sigma} \cosh \xi_\parallel \left\{ 1 + \frac{1}{4\sigma} [(2 + \xi_\perp^2) - 2\xi_\parallel \tanh \xi_\parallel] \right.$$
$$\left. + \frac{1}{4\sigma^2} \left[\left(3 + \xi_\parallel^2 + \xi_\perp^2 + \frac{1}{8}\xi_\perp^4 \right) - (3 + \xi_\perp^2)\xi_\parallel \tanh \xi_\parallel \right] \right\} \quad (2.60)$$

This equation is the desired asymptotic expansion of the partition function. Note that, as could be expected, the leading term in this equation is precisely the Ising partition function (2.22).

E. Series Expansions of the Free Energy

Once one has obtained an expansion of the partition function in a series of powers of a given quantity, one needs to construct the corresponding expansion of $\ln \mathcal{Z}$ in order to obtain the relevant thermal equilibrium quantities (see Table II). Here, we shall derive the expansions of the free energy $\mathcal{F} = -k_B T \ln \mathcal{Z}$ corresponding to those developed above for the partition function.

1. Expansion of the Logarithm of a Function

The problem of constructing the series expansion of the logarithm of a function with a given series representation appears in a number of physical and mathematical problems (e.g., in the construction of the *cumulants* of a probability distribution in terms of the known *moments* of such distribution; see Risken, 1989). Thus, if one has derived an expansion of the partition function of the type

$$\mathcal{Z}(y) = \mathcal{Z}(0) \sum_{i=0}^{\infty} \frac{A_i}{i!} y^i \quad (2.61)$$

(note that $A_0 = 1$), the first few terms in the corresponding expansion of $\ln \mathcal{Z}$ are given by

$$\ln \mathcal{Z}(y) = \ln \mathcal{Z}(0) + A_1 y + \frac{1}{2} \left(A_2 - A_1^2 \right) y^2 + \frac{1}{6} \left(A_3 - 3A_2 A_1 + 2A_1^3 \right) y^3$$
$$+ \frac{1}{24} \left(A_4 - 4A_3 A_1 - 3A_2^2 + 12A_2 A_1^2 - 6A_1^4 \right) y^4 + \cdots \quad (2.62)$$

This formula, when multiplied by $-k_B T$, gives the first few terms of the y expansion of the free energy.

2. Averages for Anisotropy Axes Distributed at Random

In what follows we shall frequently consider the values of the relevant quantities for an ensemble of spins whose anisotropy axes are distributed at random.

Note that averaging, in the sense of keeping some parameters fixed and then *summing* over the remaining ones (e.g., anisotropy-axis orientations), does not make sense for the partition function since, *for independent entities*, $\mathcal{Z}$ is a multiplicative quantity. However, averaging makes sense for the customary thermodynamical functions (free energy, entropy, energy, etc.) as they are additive quantities.

When averaging the thermodynamical quantities over assemblies of equivalent magnetic moments (i.e., with the same characteristic parameters) whose anisotropy axes are distributed at random, we shall need to calculate integrals of the general form

$$\langle f(\varphi_{\hat{n}}, \alpha) \rangle_{\text{ran}} = \frac{1}{4\pi} \int_0^{2\pi} d\varphi_{\hat{n}} \int_0^{\pi} d\alpha \sin \alpha \, f(\varphi_{\hat{n}}, \alpha)$$

where $\varphi_{\hat{n}}$ and α are, respectively, the azimuthal and polar angles of the unit vector along the anisotropy axis $\hat{n}$. We shall be mainly interested in the cases where $f(\varphi_{\hat{n}}, \alpha) = \cos^{2i} \alpha \sin^{2k} \alpha$, which does not depend on the azimuthal angle. For these functions, one finds

$$\left\langle \cos^{2i} \alpha \sin^{2k} \alpha \right\rangle_{\text{ran}} = \frac{1}{2} \int_0^{\pi} d\alpha \sin \alpha \cos^{2i} \alpha \sin^{2k} \alpha \overset{x = \cos \alpha}{=} \int_0^1 dx \, x^{2i} (1 - x^2)^k$$

Now, by comparing with the relation (2.37) between integrals of $z^{2i}(1 - z^2)^k$ weighted by $\exp(\sigma z^2)$, and Kummer functions, we get the expression

$$\left\langle \cos^{2i} \alpha \sin^{2k} \alpha \right\rangle_{\text{ran}} = \frac{\Gamma\left(i + \frac{1}{2}\right) k!}{2\Gamma\left(i + k + \frac{3}{2}\right)} \tag{2.63}$$

where we have employed $M(a, c; x = 0) = 1$ [see Eq. (A.1)] and $\Gamma(k + 1) = k!$. Alternatively, on using $\Gamma(z + 1) = z\Gamma(z)$ to expand the quotient of gamma functions we obtain

$$\left\langle \cos^{2i} \alpha \sin^{2k} \alpha \right\rangle_{\text{ran}} = \frac{2^k k!}{\underbrace{(2i + 1)[(2i + 1) + 2] \cdots [(2i + 1) + 2k]}_{k+1 \text{ terms}}}$$

To conclude, we explicitly write the particular cases of these results that, in what follows, will be used more frequently:

$$\left\langle \cos^2 \alpha \right\rangle_{\text{ran}} = \frac{1}{3} \qquad \left\langle \sin^2 \alpha \right\rangle_{\text{ran}} = \frac{2}{3}$$

$$\left\langle \cos^4 \alpha \right\rangle_{\text{ran}} = \frac{1}{5} \qquad \left\langle \cos^2 \alpha \sin^2 \alpha \right\rangle_{\text{ran}} = \frac{2}{15} \qquad \left\langle \sin^4 \alpha \right\rangle_{\text{ran}} = \frac{8}{15} \tag{2.64}$$

3. Field Expansion of the Free Energy

On considering the expansion (2.35) of the partition function in powers of $\xi = mB/k_BT$, one realizes that the function $2R(\sigma)$, ξ^2, and C_i play the roles, respectively, of $\mathcal{Z}(0)$, y, and A_i in the generic y expansion (2.61). Consequently, the corresponding general series (2.62) for $\ln \mathcal{Z}$ yields in this case

$$\ln \mathcal{Z} = \ln[2R(\sigma)] + C_1(\sigma, \alpha)\xi^2 + \tfrac{1}{2}\left[C_2(\sigma, \alpha) - C_1(\sigma, \alpha)^2\right]\xi^4 + \cdots \quad (2.65)$$

This result shows the convenience of the introduction of the factor $i!$ in the definition (2.36) of the coefficients C_i: the general expansion (2.62) can then be directly used by merely replacing the coefficients A_i by the C_i ones.
    Now, on introducing the first few angular terms $b_{i,k}(\alpha)$ [Eq. (2.30)]

$$b_{0,0} = 1 \qquad b_{1,0} = \tfrac{1}{2}\cos^2\alpha \qquad b_{0,1} = \tfrac{1}{4}\sin^2\alpha$$

$$b_{2,0} = \tfrac{1}{24}\cos^4\alpha \quad b_{1,1} = \tfrac{1}{8}\cos^2\alpha\sin^2\alpha \quad b_{0,2} = \tfrac{1}{64}\sin^4\alpha$$

into the definition (2.36), one gets for the first coefficients C_i: $C_0 = 1$

$$C_1 = \frac{1}{2}\left(\frac{R'}{R}\cos^2\alpha + \frac{R - R'}{2R}\sin^2\alpha\right) \quad (2.66)$$

and

$$C_2 = \frac{1}{4}\left(\frac{1}{3}\frac{R''}{R}\cos^4\alpha + \frac{R' - R''}{R}\cos^2\alpha\sin^2\alpha + \frac{R - 2R' + R''}{8R}\sin^4\alpha\right)$$

In these expressions, instead of superscripts, we have used primes to indicate derivatives of $R(\sigma)$ with respect to its argument. On using these formulas we get for the coefficient of ξ^4 in the expansion (2.65)

$$\frac{1}{2}(C_2 - C_1^2) = \frac{1}{8}\left\{\left[\frac{1}{3}\frac{R''}{R} - \left(\frac{R'}{R}\right)^2\right]\cos^4\alpha\right.$$

$$+ \left[\left(\frac{R'}{R}\right)^2 - \frac{R''}{R}\right]\cos^2\alpha\sin^2\alpha$$

$$\left. + \frac{1}{8}\left[-1 + 2\frac{R'}{R} - 2\left(\frac{R'}{R}\right)^2 + \frac{R''}{R}\right]\sin^4\alpha\right\} \quad (2.67)$$

Equations (2.66) and (2.67), along with (2.65), yield the desired ξ expansion of the free energy up to the fourth order.

In Section III we shall introduce the *reduced* linear and nonlinear susceptibilities. These quantities, which incorporate the *anisotropy-induced* temperature dependence of the susceptibilities, are directly related to C_1 and $(C_2 - C_1^2)$, respectively.

Averaging for anisotropy axes distributed at random is as follows. On introducing the values of the averaged trigonometric coefficients (2.64) into Eq. (2.66), we get $\langle C_1 \rangle_{\mathrm{ran}} = \frac{1}{6}$. Proceeding similarly with the expression (2.67) for $(C_2 - C_1^2)/2$, one obtains

$$\frac{1}{2}\langle C_2 - C_1^2 \rangle_{\mathrm{ran}} = \frac{1}{120}\left[2\frac{R'}{R} - 3\left(\frac{R'}{R}\right)^2 - 1\right] \tag{2.68}$$

If we introduce these results into the ξ expansion of $\ln \mathcal{Z}$ [Eq. (2.65)], we finally obtain the following expression for the free energy of an ensemble of equivalent dipoles with anisotropy axes distributed at random:

$$\langle \mathcal{F} \rangle_{\mathrm{ran}} = -k_{\mathrm{B}}T \left\{ \ln[2R(\sigma)] + \frac{1}{6}\xi^2 \right.$$
$$\left. + \frac{1}{120}\left[2\frac{R'}{R} - 3\left(\frac{R'}{R}\right)^2 - 1\right]\xi^4 + \cdots \right\} \tag{2.69}$$

It is to be noted that the first correction, $-k_{\mathrm{B}}T\xi^2/6$, to the unbiased free energy $-k_{\mathrm{B}}T\ln[2R(\sigma)]$, does not depend on the magnetic anisotropy. This will take its reflection in, for example, the independence of the linear susceptibility of the anisotropy energy for systems with axes distributed at random (see Section III.D).

4. Expansion of the Free Energy in Powers of the Anisotropy Parameter

The expansion of the free energy in powers of $\sigma = Kv/k_{\mathrm{B}}T$ can be obtained similarly. Let us first rewrite the expansion (2.43) of the partition function in powers of σ as

$$\mathcal{Z} = \mathcal{Z}_0 \left(1 + \frac{\mathcal{Z}_1}{\mathcal{Z}_0}\sigma + \frac{1}{2}\frac{\mathcal{Z}_2}{\mathcal{Z}_0}\sigma^2 + \cdots\right)$$

where $\mathcal{Z}_0$ is a shorthand for $\mathcal{Z}_{\mathrm{Lan}} = (2/\xi)\sinh\xi$. If one compares this expansion with the general one (2.61), one sees that $\mathcal{Z}_0$, σ, and $\mathcal{Z}_i/\mathcal{Z}_0$ play the roles, respectively, of $\mathcal{Z}(0)$, y, and A_i there. Accordingly, we can immediately write an equation similar to that obtained for the ξ expansion of $\ln \mathcal{Z}$:

$$\ln \mathcal{Z} \simeq \ln \mathcal{Z}_{\mathrm{Lan}} + \frac{\mathcal{Z}_1}{\mathcal{Z}_0}\sigma + \frac{1}{2}\left[\frac{\mathcal{Z}_2}{\mathcal{Z}_0} - \left(\frac{\mathcal{Z}_1}{\mathcal{Z}_0}\right)^2\right]\sigma^2 \tag{2.70}$$

Concerning the coefficients in this expansion, $\mathcal{Z}_1/\mathcal{Z}_0$ was already written in Eq. (2.52), namely

$$\frac{\mathcal{Z}_1}{\mathcal{Z}_0} = \left(1 - \frac{2}{\xi}L\right)\cos^2\alpha + \frac{1}{\xi}L\sin^2\alpha \qquad (2.71)$$

while, taking Eq. (2.55) into account, one obtains after some algebra

$$\frac{1}{2}\left[\frac{\mathcal{Z}_2}{\mathcal{Z}_0} - \left(\frac{\mathcal{Z}_1}{\mathcal{Z}_0}\right)^2\right] = \frac{2}{\xi^2}\left\{\left[2\left(1 - \frac{3}{\xi}L\right) - L^2\right]\cos^4\alpha\right.$$
$$-\left[6\left(1 - \frac{3}{\xi}L\right) - L^2 - \xi L\right]\cos^2\alpha\sin^2\alpha$$
$$\left.+\frac{1}{4}\left[3\left(1 - \frac{3}{\xi}L\right) - L^2\right]\sin^4\alpha\right\} \qquad (2.72)$$

Equations (2.71) and (2.72), together with Eq. (2.70), yield the desired expansion of the free energy in powers of the anisotropy parameter up to second order.

Let us average the free energy for anisotropy axes distributed at random. On introducing the averages (2.64) of the trigonometric coefficients into the expression for $\mathcal{Z}_1/\mathcal{Z}_0$, one gets $\langle \mathcal{Z}_1/\mathcal{Z}_0\rangle_{\mathrm{ran}} = \frac{1}{3}$. Analogously, by averaging Eq. (2.72), one arrives at

$$\frac{1}{2}\left\langle\frac{\mathcal{Z}_2}{\mathcal{Z}_0} - \left(\frac{\mathcal{Z}_1}{\mathcal{Z}_0}\right)^2\right\rangle_{\mathrm{ran}} = \frac{2}{15}\left(2 - \frac{3}{\xi}L\right)\frac{1}{\xi}L$$

On introducing these results into the expansion (2.70) of $\ln\mathcal{Z}$, one gets for $\langle\mathcal{F}\rangle_{\mathrm{ran}}$ the approximate result

$$\langle\mathcal{F}\rangle_{\mathrm{ran}} = -k_B T\left\{\ln\left(\frac{2}{\xi}\sinh\xi\right) + \frac{1}{3}\sigma + \frac{2}{15}\left[\left(2 - \frac{3}{\xi}L\right)\frac{1}{\xi}L\right]\sigma^2 + \cdots\right\} \qquad (2.73)$$

As $k_B T\sigma = Kv$ [see Eqs. (2.3)] is a constant (neglecting the possible temperature dependence of K), we get the important result that, for anisotropy axes distributed at random, the corrections due to the magnetic anisotropy to the isotropic free energy, begin at order σ^2. This will lead to, for example, a dramatic decrease of the anisotropy effects on the magnetization curves for weakly anisotropic systems ($\sigma \lesssim 2$) with a random distribution of anisotropy axes (see Section III.C).

5. Asymptotic Expansion of the Free Energy for Strong Anisotropy

Finally, the $1/\sigma$ expansion of the free energy can be obtained similarly. If we compare the asymptotic expansion (2.60) of the partition function with the general one (2.61), we see that $(e^\sigma/\sigma)\cosh\xi_\parallel$ and $1/\sigma$ play the roles, respectively, of $\mathcal{Z}(0)$ and y in that general formula. Therefore, we can directly write for $\ln\mathcal{Z}$

$$\ln\mathcal{Z} \simeq \ln\left(\frac{e^\sigma}{\sigma}\cosh\xi_\parallel\right) + \frac{1}{\sigma}\times\frac{1}{4}\left[(2+\xi_\perp^2) - 2\xi_\parallel\tanh\xi_\parallel\right]$$
$$+ \frac{1}{2\sigma^2}\left\{\frac{1}{2}\left[\left(3+\xi_\parallel^2+\xi_\perp^2+\frac{1}{8}\xi_\perp^4\right) - (3+\xi_\perp^2)\xi_\parallel\tanh\xi_\parallel\right]\right.$$
$$\left. -\frac{1}{16}\left[(2+\xi_\perp^2) - 2\xi_\parallel\tanh\xi_\parallel\right]^2\right\}$$

where, to get the coefficient of $1/\sigma^2$ [i.e., $(A_2 - A_1^2)/2$ in the general expansion], we have subtracted from the corresponding coefficient in the expansion of $\mathcal{Z}$ the square of the coefficient of $1/\sigma$ (i.e., A_1). Then, on explicitly squaring this term, we finally get

$$\ln\mathcal{Z} \simeq \ln\left(\frac{e^\sigma}{\sigma}\cosh\xi_\parallel\right) + \frac{1}{4\sigma}\left[(2+\xi_\perp^2) - 2\xi_\parallel\tanh\xi_\parallel\right]$$
$$+ \frac{1}{8\sigma^2}\left[5 + (2\xi_\parallel^2+\xi_\perp^2) - (4+\xi_\perp^2)\xi_\parallel\tanh\xi_\parallel - \xi_\parallel^2\tanh^2\xi_\parallel\right] \quad (2.74)$$

Note that this expansion has as leading term the Ising-type free energy (2.22) (this corresponds to a potential with two deep minima), while the next terms are corrections associated with the finite curvature of the potential at the minima.

Note finally that, because of the presence of $\cos\alpha$ (via $\xi_\parallel$) in the arguments of the hyperbolic trigonometric functions, we cannot write an explicit analytic formula for the average of the expansion (2.74) for anisotropy axes distributed at random.

III. EQUILIBRIUM PROPERTIES: SOME IMPORTANT QUANTITIES

A. Introduction

In this section we shall use some of the general results of the previous one, in order to calculate a number of thermodynamical quantities for independent classical magnetic moments with axially symmetric magnetic anisotropy.

The results obtained would also apply to systems approximately described as assemblies of classical dipole moments with Hamiltonians such as (2.2), that is, Hamiltonians comprising a coupling term to an external field plus an axially symmetric orientational potential.

The organization of this section is as follows. In Section III.B we study the thermal or caloric quantities — energy, entropy, and specific heat — in a number of particular situations. Sections III.C, III.D and III.E are devoted, respectively, to the study of the magnetization, the linear susceptibility, and the nonlinear susceptibilities. We shall be interested mainly in the effects of the magnetic anisotropy on these quantities.

B.   Thermal (Caloric) Quantities

We begin with a brief study of the thermal properties of noninteracting classical spins. We merely consider the particular cases of zero anisotropy and finite anisotropy in a zero field or in a constant longitudinal field.

1.   General Definitions

The thermodynamical energy $\mathcal{U}$ is defined as the statistical-mechanical average of the Hamiltonian $\mathcal{H}$ [cf. Eq. (2.10)]

$$\mathcal{U} = \langle \mathcal{H} \rangle_e = \frac{\int d\Omega\, \mathcal{H}(\vartheta, \varphi) \exp[-\beta \mathcal{H}(\vartheta, \varphi)]}{\int d\Omega \exp[-\beta \mathcal{H}(\vartheta, \varphi)]} \tag{3.1}$$

where $\int d\Omega(\cdot) = (1/2\pi) \int_{-1}^{1} d(\cos \vartheta) \int_{0}^{2\pi} d\varphi(\cdot)$. From this definition one immediately gets the relation written in Table II

$$\mathcal{U} = -\frac{\partial}{\partial \beta}(\ln \mathcal{Z}) \tag{3.2}$$

between $\mathcal{U}$ and the logarithm of the partition function $\mathcal{Z} = \int d\Omega \exp(-\beta \mathcal{H})$ (or the free energy $\mathcal{F} = -\beta^{-1} \ln \mathcal{Z}$).

The entropy $\mathcal{S}$ can formally be defined as minus the average of the logarithm of the equilibrium probability distribution $P_e = \exp(-\beta \mathcal{H})/\mathcal{Z}$:

$$\frac{\mathcal{S}}{k_B} = -\langle \ln P_e \rangle_e = \frac{-\int d\Omega \ln P_e(\vartheta, \varphi) \exp[-\beta \mathcal{H}(\vartheta, \varphi)]}{\int d\Omega \exp[-\beta \mathcal{H}(\vartheta, \varphi)]} \tag{3.3}$$

Note, however, that this quantity, in contrast to other thermodynamical quantities, is not defined as the average of a physical quantity of the system — it is an intrinsic *thermal* quantity. On the other hand, by using $-\langle \ln P_e \rangle_e = \beta \mathcal{U} + \ln \mathcal{Z}$, which is essentially the celebrated thermodynamical relation $\mathcal{F} = \mathcal{U} - T\mathcal{S}$, one gets from Eqs. (3.2) and (3.3) the entropy expressed in terms of the partition function as

$$\frac{\mathcal{S}}{k_B} = \ln \mathcal{Z} - \beta \frac{\partial}{\partial \beta}(\ln \mathcal{Z}) \tag{3.4}$$

The last thermal quantity that we shall consider is the specific heat at constant field, namely

$$c_B = \left. \frac{\partial \mathcal{U}}{\partial T} \right|_B \tag{3.5}$$

Taking into account the relation (3.2) between $\mathcal{U}$ and $\mathcal{Z}$, one obtains from this definition the well-known results

$$\frac{c_B}{k_B} = -\beta^2 \frac{\partial \mathcal{U}}{\partial \beta} = \beta^2 \frac{\partial^2}{\partial \beta^2}(\ln \mathcal{Z}) \tag{3.6}$$

Let us finally consider a quantity $A = A(\sigma, \xi)$ that is a function of $\sigma = Kv/k_B T$ and $\xi = mB/k_B T$ [the dimensionless anisotropy and field parameters (2.3)]. Then, on using $\beta(\partial \sigma/\partial \beta) = \sigma$ and $\beta(\partial \xi/\partial \beta) = \xi$, one gets for the β derivatives of A

$$\beta \frac{\partial A}{\partial \beta} = \frac{\partial A}{\partial \sigma}\sigma + \frac{\partial A}{\partial \xi}\xi, \qquad \beta^2 \frac{\partial^2 A}{\partial \beta^2} = \frac{\partial^2 A}{\partial \sigma^2}\sigma^2 + 2\frac{\partial^2 A}{\partial \sigma \, \partial \xi}\sigma\xi + \frac{\partial^2 A}{\partial \xi^2}\xi^2$$

Note that, when taking the β derivatives, we have implicitly assumed that the only dependence of σ and ξ on T enters via β; that is, we neglect the possible dependence on the temperature of both K and m, which otherwise might be relevant in systems of magnetic nanoparticles at sufficiently high temperatures. Next, if we put $A = \ln \mathcal{Z}$ and take the relations (3.2), (3.4), and (3.6) into account, we can express the thermal quantities for a system described by σ and ξ, as

$$\mathcal{U} = -\left(\frac{\partial \mathcal{Z}/\partial \sigma}{\mathcal{Z}}Kv + \frac{\partial \mathcal{Z}/\partial \xi}{\mathcal{Z}}mB \right) \tag{3.7}$$

$$\frac{\mathcal{S}}{k_B} = \ln \mathcal{Z} - \left(\frac{\partial \mathcal{Z}/\partial \sigma}{\mathcal{Z}}\sigma + \frac{\partial \mathcal{Z}/\partial \xi}{\mathcal{Z}}\xi \right) \tag{3.8}$$

$$\frac{c_B}{k_B} = \left[\frac{\partial^2 \mathcal{Z}/\partial\sigma^2}{\mathcal{Z}} - \left(\frac{\partial\mathcal{Z}/\partial\sigma}{\mathcal{Z}} \right)^2 \right] \sigma^2$$

$$+ 2 \left[\frac{\partial^2 \mathcal{Z}/\partial\sigma\partial\xi}{\mathcal{Z}} - \frac{(\partial\mathcal{Z}/\partial\sigma)(\partial\mathcal{Z}/\partial\xi)}{\mathcal{Z}^2} \right] \sigma\xi$$

$$+ \left[\frac{\partial^2 \mathcal{Z}/\partial\xi^2}{\mathcal{Z}} - \left(\frac{\partial\mathcal{Z}/\partial\xi}{\mathcal{Z}} \right)^2 \right] \xi^2 \tag{3.9}$$

These formulas allow one to identify the contribution of the anisotropy and Zeeman energies to the thermal quantities. However, one does not need to use them in their general forms since, when both types of energies are present, one can write $\xi = 2\sigma h$ and differentiate with respect to σ keeping $h = B/B_K$, which is assumed to be independent of the temperature, constant.

2. Thermal Quantities: Particular Cases

a. Isotropic Case. When the anisotropy energy is absent, the partition function reads $\mathcal{Z}_{\text{Lan}} = (2/\xi) \sinh \xi$ [Eq. (2.17)]. The σ derivatives of this partition function are identically zero, while the required ξ derivatives are given by Eqs. (2.48) and (2.51). Therefore, on taking Eq. (3.7) into account, one obtains for the mean energy

$$\mathcal{U}_{\text{Lan}} = -m \left(\coth \xi - \frac{1}{\xi} \right) B = -mL(\xi)B \tag{3.10}$$

where $L(\xi)$ is the Langevin function. This is the natural result considering that in this case $\mathcal{H} = -m_z B$ and that the Langevin result for the magnetization is $\langle m_z \rangle_e = mL(\xi)$. Similarly, Eq. (3.8) yields the following expression for the entropy:

$$\frac{\mathcal{S}_{\text{Lan}}}{k_B} = \ln \left(\frac{2}{\xi} \sinh \xi \right) - \xi L(\xi) \tag{3.11}$$

Finally, on introducing Eqs. (2.48) and (2.51) into Eq. (3.9), we can express the Langevin specific heat as

$$\frac{c_{B,\text{Lan}}}{k_B} = 1 - \frac{\xi^2}{\sinh^2 \xi} = \xi^2 L'(\xi) \tag{3.12}$$

At high temperatures, that is, when $\xi \ll 1$, we can approximate the square of the hyperbolic sine in Eq. (3.12) by $\sinh^2 \xi \simeq \xi^2 + \xi^4/3$, while at low temperatures ($\xi \gg 1$) we have $\xi^2/\sinh^2 \xi \simeq 0$. Consequently, in these limiting ranges c_B approximately reads

$$c_{B,\text{Lan}} \simeq \begin{cases} \frac{1}{3}k_B\xi^2 & \text{for} \quad \xi \ll 1 \\ k_B & \text{for} \quad \xi \gg 1 \end{cases} \tag{3.13}$$

Thus, the specific heat obeys a customary T^{-2} law in the high-temperature range, whereas it tends to k_B at low temperatures. This last limit does not obey Nernst's theorem, which states that $c_B \to 0$ as $T \to 0$, and this is due to the classical character of the magnetic moment (the energy levels of $\vec{m}$ are not discrete, which is a proviso for the result mentioned, but they are continuously distributed).

Figure 7 shows the specific heat in the isotropic case. This increases monotonically from 0 at high temperatures to k_B at low temperatures, where the curve exhibits a plateau. This region corresponds to the high-field ($\xi \gg 1$) range where the average magnetic moment is close to saturation $[1 - L(\xi)] \propto \xi^{-1}$; the thermodynamical energy, which is proportional to $L(\xi)$, then increases linearly with T, yielding a constant c_B.

b. Zero-Field Case. In the absence of an external field (unbiased case), the partition function is given by $\mathcal{Z}_{\text{unb}} = 2R(\sigma)$ [Eq. (2.20)]. Because the ξ derivatives of $\mathcal{Z}_{\text{unb}}$ are identically zero, the mean energy in the absence of an external field, obtained from Eq. (3.7), reads

$$\mathcal{U}_{\text{unb}} = -Kv\frac{R'}{R} \tag{3.14}$$

This expression provides another simple physical interpretation for the familiar quotient R'/R—it is essentially minus the thermodynamical energy in the absence of an external field.

The zero-field entropy and specific heat, as derived from Eqs. (3.8) and (3.9), read

$$\frac{\mathcal{S}_{\text{unb}}}{k_B} = \ln(2R) - \sigma\frac{R'}{R} \tag{3.15}$$

and

$$\frac{c_{B,\text{unb}}}{k_B} = \left[\frac{R''}{R} - \left(\frac{R'}{R}\right)^2\right]\sigma^2 \tag{3.16}$$

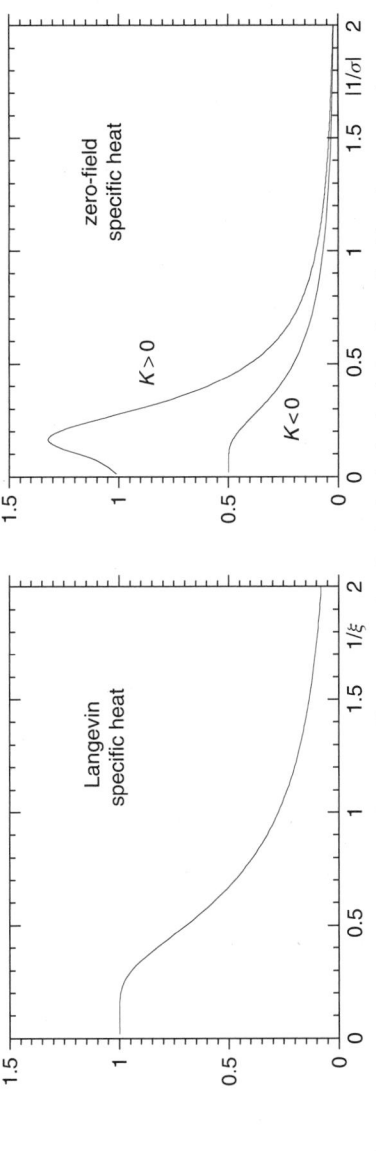

FIGURE 7. Temperature dependence of the specific heat c_B of a classical spin in the isotropic and unbiased cases; c_B is measured in units of k_B and the dimensionless temperatures are $1/\xi = k_B T/mB$ and $|1/\sigma| = k_B T/|K|v$, respectively.

45

In the high- ($|\sigma| \ll 1$) and low-temperature ($|\sigma| \gg 1$) ranges, we can use the approximate Eq. (A.31) for $R''/R - (R'/R)^2$, to get the limit behaviors of the zero-field specific heat:

$$c_{B,\text{unb}} \simeq \begin{cases} \frac{1}{2}k_B & \text{for} \quad \sigma \ll -1 \\ \frac{4}{45}k_B\sigma^2 & \text{for} \quad |\sigma| \ll 1 \\ k_B & \text{for} \quad \sigma \gg 1 \end{cases} \qquad (3.17)$$

As it should, the specific heat obeys a T^{-2} law at high temperatures. At low temperatures, owing to the classical nature of the spin (cf. Jacobs and Bean, 1963), c_B tends to k_B and $k_B/2$, for easy-axis and easy-plane anisotropy, respectively. The factor $\frac{1}{2}$ originates from the different geometry of the region of the minima; for easy-axis anisotropy the minima lay at the poles of the unit sphere, whereas for easy-plane anisotropy, the minima are continuously distributed along the equatorial circle.

Figure 7 also shows the specific heat in the unbiased case. The specific heat in the easy-plane case does not exhibit any peak, but it also has a plateau at low temperatures. In contrast, in the easy-axis case, the specific heat exhibits a maximum. This peak (located at $\sigma \sim 5$) can be interpreted in terms of the crossover from the isotropic behavior at high temperatures to the two-state (Ising-type) behavior at low temperatures. This is supported by Fig. 4, where it was shown that, whereas at $\sigma \simeq 2$, $P_{e,\text{unb}}(m_z)$ is not far from uniform, for $\sigma \simeq 5$, the probability distribution is quite concentrated close to the poles. These features of the specific heat resemble the Schottky effect, and, in this context, they could be attributed to the "depopulation" of the high-energy "equatorial levels."

c. Longitudinal-Field Case. We shall finally consider the caloric quantities when an external field is applied along the anisotropy axis. The corresponding partition function is given by Eq. (2.25), where $\sigma_\pm = \sigma(1 \pm h)^2$ and $h = \xi/2\sigma$. As mentioned previously, in order to calculate the thermal quantities, we do not need to make use of Eqs. (3.7)–(3.9) in their general forms; in this case we only need to take σ derivatives of $\mathcal{Z}_\parallel$ (denoted by primes) keeping $h = B/B_K$ constant.

On calculating $\mathcal{Z}'_\parallel/\mathcal{Z}_\parallel$, we get

$$\frac{\mathcal{Z}'_\parallel}{\mathcal{Z}_\parallel} = -h^2 + \frac{(1+h)^3 R'(\sigma_+) + (1-h)^3 R'(\sigma_-)}{(1+h)R(\sigma_+) + (1-h)R(\sigma_-)} \qquad (3.18)$$

where we have used $\partial\sigma_\pm/\partial\sigma = (1 \pm h)^2$. Equation (3.18) yields, essentially, minus the mean energy. However, before writing an equation for $\mathcal{U}$, we shall manipulate slightly the above expression in order to eliminate the derivatives $R'(\sigma_\pm)$. To this end, we can use $R' = (e^\sigma - R)/2\sigma$ [Eq. (A.12)], getting

$$(1 \pm h)^3 R'(\sigma_\pm) = \frac{1 \pm h}{2\sigma} \left\{ \exp[\sigma(1 + h^2) \pm 2\sigma h] - R(\sigma_\pm) \right\}$$

Then, on introducing the function

$$J(\sigma, h) = 2[\cosh(2\sigma h) + h \sinh(2\sigma h)]$$

one can write the thermodynamical energy in a longitudinal field as

$$\mathcal{U}_\| = Kv \left[h^2 + \frac{1}{2\sigma} \left(1 - \frac{e^\sigma J}{\mathcal{Z}_\|} \right) \right]. \tag{3.19}$$

The entropy can then be derived by merely using $\mathcal{F} = \mathcal{U} - TS$, to get

$$\frac{\mathcal{S}_\|}{k_B} = \sigma h^2 + \ln(\mathcal{Z}_\|) + \frac{1}{2} \left(1 - \frac{e^\sigma J}{\mathcal{Z}_\|} \right) \tag{3.20}$$

Note that, since $(e^\sigma J/\mathcal{Z}_\|)|_{h=0} = e^\sigma/R$ and $1 - e^\sigma/R = -2\sigma R'/R$, Eqs. (3.19) and (3.20) duly reduce for $h = 0$ to Eq. (3.14) and (3.15), respectively.

Let us finally derive the specific heat in the longitudinal-field case. On taking the σ derivative of Eq. (3.18) by using again $\partial\sigma_\pm/\partial\sigma = (1 \pm h)^2$, we find

$$\frac{c_{B,\|}}{k_B} = \left\{ \frac{(1 + h)^5 R''(\sigma_+) + (1 - h)^5 R''(\sigma_-)}{(1 + h)R(\sigma_+) + (1 - h)R(\sigma_-)} \right.$$
$$\left. - \left[\frac{(1 + h)^3 R'(\sigma_+) + (1 - h)^3 R'(\sigma_-)}{(1 + h)R(\sigma_+) + (1 - h)R(\sigma_-)} \right]^2 \right\} \sigma^2 \tag{3.21}$$

which generalizes the zero-field expression (3.16). An alternative formula, more suitable for computation, can be obtained by differentiating $\mathcal{U}$ in Eq. (3.19), getting

$$\frac{c_{B,\|}}{k_B} = \frac{1}{2} \left\{ 1 + \frac{e^\sigma J}{\mathcal{Z}_\|} \left[\sigma(1 + h^2) - \frac{1}{2} \left(1 + \frac{e^\sigma J}{\mathcal{Z}_\|} \right) + \sigma \frac{J'}{J} \right] \right\} \tag{3.22}$$

where the prime in J' stands for σ derivative (keeping h constant):

$$J'(\sigma, h) = 4h[\sinh(2\sigma h) + h \cosh(2\sigma h)]$$

In order to get the high-temperature behavior of c_B, we can expand Eq. (3.21) in powers of σ [to first order we evaluate $R^{(l)}(\sigma_\pm)$ at zero with help from Eq. (A.4)], obtaining

$$\left.\frac{c_{B,\parallel}}{k_B}\right|_{|\sigma|\ll 1} \simeq \left\{ \frac{1}{5}\frac{(1+h)^5 + (1-h)^5}{(1+h) + (1-h)} - \left[\frac{1}{3}\frac{(1+h)^3 + (1-h)^3}{(1+h) + (1-h)}\right]^2 \right\}\sigma^2$$

The low temperature behavior (case $K < 0$) can also be obtained by introducing the asymptotic Eq. (A.19) into Eq. (3.21), whereas for $K > 0$ it is more easily found by differentiating twice the approximate partition function (2.29) with respect to σ (keeping h constant). Thus, one arrives at the following limit behaviors of the specific heat

$$c_{B,\parallel} \simeq \begin{cases} \frac{1}{2}k_B & \text{for } \sigma \ll -1 \\ \frac{4}{45}k_B(1 + 15h^2)\sigma^2 & \text{for } |\sigma| \ll 1 \\ k_B & \text{for } \sigma \gg 1 \end{cases} \qquad (3.23)$$

Again, the specific heat obeys a T^{-2} law at high temperatures while, as a result of the classical character of the spin, c_B tends to nonzero values at low temperatures.

Figure 8 displays the specific heat in the longitudinal-field case. The c_B curves exhibit a maximum, the height and location of which depend on the magnitude of the applied field. For $h \leq 1$, these maxima can again be interpreted in terms of the crossover from the isotropic regime at high temperatures

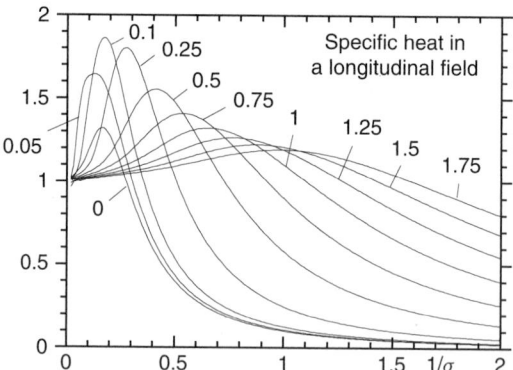

FIGURE 8. Temperature dependence of the specific heat, c_B, in various longitudinal fields $h = B/B_K$ for easy-axis anisotropy. c_B is measured in units of k_B and the dimensionless temperature is $1/\sigma = k_B T/Kv$.

to the low-temperature regime in which the magnetic moments are concentrated close to the potential minima. The height of the maximum steeply increases for $h \leq 0.15$ and then decreases monotonically with increasing h. At high fields, the maximum is actually rather smeared and its height is small, approaching a plateau. This occurs because the Zeeman energy dominates the magnetic-anisotropy energy for such high fields, so the specific heat approaches to the zero-anisotropy $c_{B,\text{Lan}}$, which, after exhibiting a plateau, decreases monotonically with increasing T (Fig. 7)

C. Magnetization

We shall now study the magnetization of classical spins with axially symmetric magnetic anisotropy. The magnetization along the external field direction, $M_B \equiv \langle \vec{m} \cdot \hat{b} \rangle_e$, where $\hat{b} = \vec{B}/B$, can in general be derived from the partition function as follows. Consider that $-\beta \mathcal{H}$ contains among others a Zeeman term $\xi(\vec{e} \cdot \hat{b})$, where $\xi = mB/k_B T$ and $\vec{e} = \vec{m}/m$. Then, because $\mathcal{Z} = \int d\Omega \exp(-\beta \mathcal{H})$, one has

$$\langle \vec{m} \cdot \hat{b} \rangle_e \equiv \mathcal{Z}^{-1} \int d\Omega \, m(\vec{e} \cdot \hat{b}) e^{-\beta \mathcal{H}} = m \mathcal{Z}^{-1} \frac{\partial}{\partial \xi} \int d\Omega \, e^{-\beta \mathcal{H}}$$

whence one gets the known statistical-mechanical relation

$$M_B = m \frac{\partial}{\partial \xi} (\ln \mathcal{Z}) \qquad (3.24)$$

The magnetization of an ensemble of noninteracting superparamagnetic particles without magnetic anisotropy can be obtained by means of a simple translation of the classical Langevin theory of paramagnetism, and it is given by $M_{B,\text{Lan}} = mL(\xi)$, where $L(\xi)$ is the Langevin function [Eq. (2.49)]. (Note that the magnetization then depends on the field and temperature via B/T.) A related salient result is that in a liquid suspension of magnetic particles (usually called *magnetic fluid* or *ferrofluid*) with a *general* single-particle magnetic anisotropy, the magnetization is also given by the Langevin result (Krueger, 1979). This holds essentially because the physical rotation of the particles in the liquid decouples the anisotropy from the magnetization process. In fact, the same result holds for a molecular beam of single-domain magnetic clusters, such as those deflected in Stern–Gerlach experiments (Maiti and Falicov, 1993). However, the rotational degrees of freedom are fastened in solid dispersions, giving rise to effects of the magnetic anisotropy on the equilibrium quantities.

West (1961) studied the magnetization of an ensemble of noninteracting classical spins with uniaxial anisotropy in a *longitudinal* constant field. He derived an equation for the magnetization (see Eq. (3.30) below) and studied the anisotropy-induced non-B/T superposition of the magnetization curves. Unfortunately, his analytic calculation cannot be easily extended to situations where the field and the anisotropy axis are not collinear, where only more or less complicated expressions have been derived.

Lin (1961), Chantrell (see, for example, Williams et al., 1993), and Cregg and Bessais (1999b) expressed the magnetization for an arbitrary orientation of the magnetic field as quotients of two infinite series. On the other hand, Mørup (1983) derived an approximate expression for the magnetization valid when K_BT is much smaller than $\mathcal{H}$, which holds *irrespective* of the symmetry the Hamiltonian. However, inasmuch as is assumed that the magnetic moment is effectively confined to *one* of the potential wells, Mørup's formula does not hold for the full equilibrium (superparamagnetic) range.*

In what follows, we first consider the form of the magnetization in various simple cases. Then, we briefly analyze a general expression derived from the field expansion (2.35) of the partition function (this is our contribution to the above-mentioned class of "more or less complicated expressions"). Finally, we study the expressions for the magnetization derived from the weak- and strong-anisotropy expansions of the free energy obtained in Section II.E

1. Magnetization: Particular Cases

We now study the expressions that emerge from Eq. (3.24) when one introduces into it the particular cases of the partition function considered in Section II.C.

a. Isotropic Case. For $\sigma = 0$ the partition function is given by $\mathcal{Z}_{\mathrm{Lan}} = (2/\xi)\sinh\xi$ [Eq. (2.17)], so that the magnetization reads

$$M_{B,\mathrm{Lan}} = m\left(\coth\xi - \frac{1}{\xi}\right) = mL(\xi) \qquad (3.25)$$

where $L(\xi)$ is the Langevin function (2.49).

b. Ising Regime. For $\sigma \to \infty$, the partition function is $\mathcal{Z}_{\mathrm{Ising}} \simeq (e^\sigma/\sigma)\cosh\xi_\parallel$ [Eq. (2.22)]. As $\xi_\parallel = \xi\cos\alpha$, the magnetization derived from Eq. (3.24) reads

$$M_{B,\mathrm{Ising}} = m\cos\alpha\tanh(\xi_\parallel) \qquad (3.26)$$

which naturally vanishes when $\vec{B}$ is perpendicular to the "Ising axis" $\hat{n}$.

---

* The approximation mentioned is different from what we are calling the Ising regime, where the magnetic moment stays most of the time around the potential minima, but it is still in complete equilibrium, and performs a sufficiently large number of inter-potential-well rotations during a typical observation time.

c. *Plane-Rotator Regime.* The $\sigma \to -\infty$ partition function is $\mathcal{Z}_{\text{rot}} \simeq (-\pi/\sigma)^{1/2} I_0(\xi_\perp)$ [Eq. (2.23)], so that the plane-rotator magnetization is given by

$$M_{B,\text{rot}} = m \sin\alpha \frac{I_1(\xi_\perp)}{I_0(\xi_\perp)} \tag{3.27}$$

where we have used $I_0'(y) = I_1(y)$ [see the integral representation (2.14) for $I_n(y)$]. In this case, M_B is zero when $\vec{B}$ is perpendicular to the rotator plane.

Note that, when the magnitude of the magnetic moment is independent of the temperature, M_B depends on B and T via ξ ($\propto B/T$) in all three considered cases. This is called the B/T *superposition of* M_B; the magnetization vs field curves corresponding to different temperatures, when plotted against B/T, collapse onto a single master curve. However, outside those limit ranges, T does not enter in $M_B(B, T)$ via B/T only, but M_B depends on ξ as well as on σ. This is illustrated now with the magnetization in a longitudinal field.

d. *Longitudinal-Field Case.* When $\vec{B} \parallel \hat{n}$, the partition function is given by Eq. (2.25). In order to derive the associated magnetization, we need to take the derivatives ($h = \xi/2\sigma$)

$$\frac{\partial}{\partial\xi} \left[(1 \pm h) R(\sigma_\pm) \right] = \pm \left[\frac{1}{2\sigma} R(\sigma_\pm) + (1 \pm h)^2 R'(\sigma_\pm) \right] = \pm \frac{e^{\sigma_\pm}}{2\sigma}$$

where we have used $\partial\sigma_\pm/\partial\xi = \pm(1 \pm h)$ and the derivatives $R'(\sigma_\pm)$ have been eliminated by dint of Eq. (A.12). Then, with help from $\exp(\sigma_\pm) = \exp[\sigma(1 + h^2)] \exp(\pm\xi)$, we get from Eq. (3.24) the magnetization in a longitudinal field as

$$\frac{M_{B,\parallel}}{m} = \frac{e^{\sigma(1+h^2)}}{\sigma} \frac{\sinh\xi}{(1+h)R(\sigma_+) + (1-h)R(\sigma_-)} - h \tag{3.28}$$

which, by using Eq. (2.25), can more compactly be written as

$$\frac{M_{B,\parallel}}{m} = \frac{e^\sigma}{\sigma} \frac{\sinh\xi}{\mathcal{Z}_\parallel} - \frac{\xi}{2\sigma} \tag{3.29}$$

Figure 9 displays the magnetization against the longitudinal field, showing that $M_{B,\parallel}$ does not depend on B and T via ξ only. As T decreases, one finds the crossover, induced by the uniaxial magnetic anisotropy, from the high-temperature ($|\sigma| \ll 1$) isotropic regime, to the low-temperature ($\sigma \gg 1$) Ising regime. Note that, even for $\sigma \sim 20$, the typical measurement times for the magnetization ($\sim 1-100$ s) would be much longer than the relaxation times of

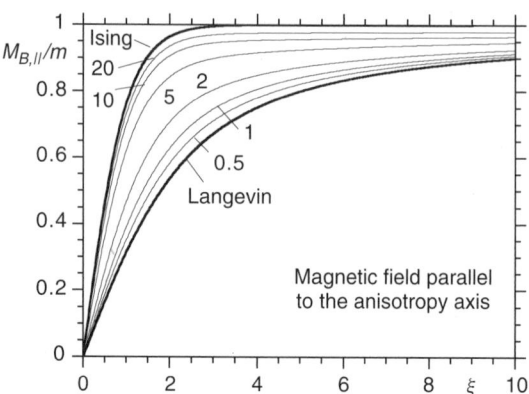

FIGURE 9. Magnetization versus longitudinal field $\xi = mB/k_\mathrm{B}T$ [Eq. (3.29)] for various values of the dimensionless anisotropy parameter $\sigma = Kv/k_\mathrm{B}T$, showing the anisotropy-induced non-B/T superposition of the magnetization curves.

the magnetic moment. Therefore, all the displayed curves could be observed experimentally without leaving the equilibrium (superparamagnetic) range.

Finally, we compare the preceding results with other expressions derived for the magnetization. For $\sigma > 0$, Eq. (3.29) reduces to the expression obtained by West (1961). Indeed, if we use the alternative expression (2.27) for $\mathcal{Z}_\|$ in terms of the Dawson integral [Eq. (A.9)], we get

$$\frac{M_{B,\|}}{m} = \frac{1}{\sqrt{\sigma}} \frac{\sinh \xi}{e^\xi D(\sqrt{\sigma_+}) + e^{-\xi} D(\sqrt{\sigma_-})} - \frac{\xi}{2\sigma} \tag{3.30}$$

which is the result of West almost in its original form. Another formula was derived by Coffey, Cregg and Kalmykov (1993) when calculating relaxation times for magnetic nanoparticles by the effective eigenvalue method:

$$\frac{M_{B,\|}}{m} = \frac{\xi}{\sqrt{\sigma}\left[(\xi L(\xi)+1+\xi)D\left(\sqrt{\sigma}+\frac{\xi}{2\sqrt{\sigma}}\right) + (\xi L(\xi)+1-\xi)D\left(\sqrt{\sigma}-\frac{\xi}{2\sqrt{\sigma}}\right)\right]}$$
$$- \frac{\xi}{2\sigma}$$

where $L(\xi)$ is the Langevin function. However, on merely noting that $\sqrt{\sigma} \pm \xi/2\sqrt{\sigma} = \sqrt{\sigma_\pm}$, and using

$$\xi L(\xi) + 1 \pm \xi = \frac{\xi}{\sinh \xi}(\cosh \xi \pm \sinh \xi) = \frac{\xi}{\sinh \xi} e^{\pm \xi}$$

their formula can be cast into the form (3.30) of West. As occurs with the latter, the preceding alternative expression for $M_{B,\parallel}$ is written by assuming easy-axis anisotropy implicitly [recall the discussion regarding the validity of the expression (2.27) for $\mathcal{Z}_{\parallel}$].

2. General Formula for the Magnetization

On inserting the field expansion of the partition function (2.35) into the statistical-mechanical relation (3.24), the magnetization emerges in the form

$$M_B = m \frac{\displaystyle\sum_{i=1}^{\infty} \frac{2C_i}{(i-1)!}\xi^{2i-1}}{\displaystyle\sum_{i=0}^{\infty} \frac{C_i}{i!}\xi^{2i}} \tag{3.31}$$

This formula gives a general expression for M_B as a quotient of two series of powers of ξ whose coefficients are expressible in terms of Kummer functions [Eq. (2.39)]. Such a mathematical object is clearly not easy to deal with. Nevertheless, one can check by an explicit identification of the corresponding series that, when the limit cases of the coefficients C_i (see Table III) are introduced into Eq. (3.31), one gets the isotropic, Ising, and plane-rotator results for the magnetization. Indeed, for the series in the numerator and the denominator (the magnetic-field-dependent factor in the partition function) we obtain

	$\sigma = 0$	$\sigma \to \infty$	$\sigma \to -\infty$
$\displaystyle\sum_{i=1}^{\infty} \frac{2C_i}{(i-1)!}\xi^{2i-1}$	$\dfrac{1}{\xi}\left(\cosh\xi - \dfrac{1}{\xi}\sinh\xi\right)$	$\cos\alpha\,\sinh(\xi_{\parallel})$	$\sin\alpha\,I_1(\xi_{\perp})$
$\displaystyle\sum_{i=0}^{\infty} \frac{C_i}{i!}\xi^{2i}$	$\dfrac{1}{\xi}\sinh\xi$	$\cosh(\xi_{\parallel})$	$I_0(\xi_{\perp})$

Therefore, Eq. (3.31) contains, as particular cases, the limit formulas for the magnetization discussed above.

3. Series Expansions of the Magnetization

a. Expansion of the Magnetization in Powers of the Anisotropy Parameter. Here we derive the magnetization from the weak-anisotropy expansion of the free energy obtained in Section II.E. In this way, we shall arrive at an approximate analytic expression for M_B that contains the first corrections to the Langevin magnetization due to nonzero magnetic anisotropy.

To this end, we must differentiate the expansion of $\mathcal{F}$ in powers of $\sigma = Kv/k_BT$ [Eq. (2.70)] with respect to the field. Prior to taking the ξ

derivatives of the first two coefficients of that expansion, we can rewrite them in alternative forms. Equation (2.71) for $\mathcal{Z}_1/\mathcal{Z}_0$ can be written as

$$\frac{\mathcal{Z}_1}{\mathcal{Z}_0} = \cos^2\alpha - (3\cos^2\alpha - 1)\frac{1}{\xi}L$$

while Eq. (2.72) for the coefficient in σ^2 can be cast into the form

$$\frac{1}{2}\left[\frac{\mathcal{Z}_2}{\mathcal{Z}_0} - \left(\frac{\mathcal{Z}_1}{\mathcal{Z}_0}\right)^2\right] = (35\cos^4\alpha - 30\cos^2\alpha + 3)\frac{1}{2\xi^2}\left(1 - \frac{3}{\xi}L\right)$$

$$- (9\cos^4\alpha - 6\cos^2\alpha + 1)\frac{1}{2\xi^2}L^2 - (\cos^4\alpha - \cos^2\alpha)\frac{2}{\xi}L$$

On taking now the derivatives of these coefficients with help form Eq. (2.54) for $d(L/\xi)/d\xi$, we get

$$\frac{d}{d\xi}\left(\frac{\mathcal{Z}_1}{\mathcal{Z}_0}\right) = (3\cos^2\alpha - 1)\frac{1}{\xi}\left[L^2 - \left(1 - \frac{3}{\xi}L\right)\right] \tag{3.32}$$

$$\frac{1}{2}\frac{d}{d\xi}\left[\frac{\mathcal{Z}_2}{\mathcal{Z}_0} - \left(\frac{\mathcal{Z}_1}{\mathcal{Z}_0}\right)^2\right] = (35\cos^4\alpha - 30\cos^2\alpha + 3)\frac{1}{2\xi^3}\left[3L^2 - 5\left(1 - \frac{3}{\xi}L\right)\right]$$

$$+ (9\cos^4\alpha - 6\cos^2\alpha + 1)\frac{1}{\xi^2}L\left[L^2 - \left(1 - \frac{3}{\xi}L\right)\right]$$

$$+ (\cos^4\alpha - \cos^2\alpha)\frac{2}{\xi}\left[L^2 - \left(1 - \frac{3}{\xi}L\right)\right] \tag{3.33}$$

These expressions, when introduced into

$$\frac{M_B}{m} \simeq L(\xi) + \frac{d}{d\xi}\left(\frac{\mathcal{Z}_1}{\mathcal{Z}_0}\right)\sigma + \frac{1}{2}\frac{d}{d\xi}\left[\frac{\mathcal{Z}_2}{\mathcal{Z}_0} - \left(\frac{\mathcal{Z}_1}{\mathcal{Z}_0}\right)^2\right]\sigma^2 \tag{3.34}$$

yield the first terms of the desired weak-anisotropy expansion of the magnetization.

Some relevant particular cases are those where the field points along the anisotropy axis, perpendicular to it, and when the anisotropy axes are distributed at random. In the first two cases we find

$$\frac{M_{B,\parallel}}{m} \simeq L(\xi) + \frac{2}{\xi}\left[L^2 - \left(1 - \frac{3}{\xi}L\right)\right]\sigma$$

$$+ \frac{4}{\xi^3}\left\{\left[3L^2 - 5\left(1 - \frac{3}{\xi}L\right)\right] + \xi L\left[L^2 - \left(1 - \frac{3}{\xi}L\right)\right]\right\}\sigma^2 \tag{3.35}$$

$$\frac{M_{B,\perp}}{m} \simeq L(\xi) - \frac{1}{\xi}\left[L^2 - \left(1 - \frac{3}{\xi}L\right)\right]\sigma$$

$$+ \frac{1}{\xi^3}\left\{\frac{3}{2}\left[3L^2 - 5\left(1 - \frac{3}{\xi}L\right)\right] + \xi L\left[L^2 - \left(1 - \frac{3}{\xi}L\right)\right]\right\}\sigma^2 \quad (3.36)$$

while $\langle M_B\rangle_{\text{ran}}$ is obtained by introducing the averages (2.64) into Eqs. (3.32) and (3.33) and reads*

$$\frac{\langle M_B\rangle_{\text{ran}}}{m} \simeq L(\xi) - \frac{4}{15}\left(1 - \frac{3}{\xi}L\right)\frac{1}{\xi}\left[L^2 - \left(1 - \frac{3}{\xi}L\right)\right]\sigma^2 \quad (3.37)$$

Naturally, one can also obtain this result by taking the ξ derivative of the σ expansion of $\langle\mathcal{F}\rangle_{\text{ran}}$ [Eq. (2.73)]. As was anticipated there, *for anisotropy axes distributed at random, the corrections to the Langevin magnetization due to the magnetic anisotropy begin at second order.*

In order to estimate the range of validity of the weak-anisotropy expansion of the magnetization obtained, this has been compared with the exact analytic formula (3.29) for the longitudinal magnetization. It is shown in Fig. 10 that the approximate (3.35) works reasonably well up to $\sigma \sim 2$. Considering that the expansion has been performed by assuming σ as the small parameter, the range of validity obtained is quite wide.

The effect of the orientation of the field with respect to the anisotropy axis is shown in Fig. 11. In contrast to the longitudinal-field case, where the anisotropy energy favors the alignment of the magnetic moment in the field direction, in the transverse-field case the anisotropy hinders the magnetization process, and the magnetization curve goes below the Langevin curve. In addition, for an ensemble of spins with anisotropy axes distributed at random, this phenomenon slightly dominates the favored alignment of the longitudinal-field case, so that the corresponding magnetization is slightly lower than the Langevin magnetization.

The anisotropy-induced contribution to the magnetization, $M_B(\xi) - mL(\xi)$, has been isolated in the lower panel of Fig. 11. This representation neatly shows that the random orientation of the anisotropy axes significantly reduces the anisotropy-induced contribution to the magnetization process. In the range of low fields, moreover, that significant reduction becomes an exact cancellation. This is due to the fact that the *linear* susceptibility is independent of the anisotropy energy for randomly distributed anisotropy axes. This result,

---

*Note that $\langle 3\cos^2\alpha - 1\rangle_{\text{ran}} = \langle 35\cos^4\alpha - 30\cos^2\alpha + 3\rangle_{\text{ran}} = 0$, where the terms into the brackets are proportional, respectively, to the second and fourth Legendre polynomials [see Eq. (3.68) below].

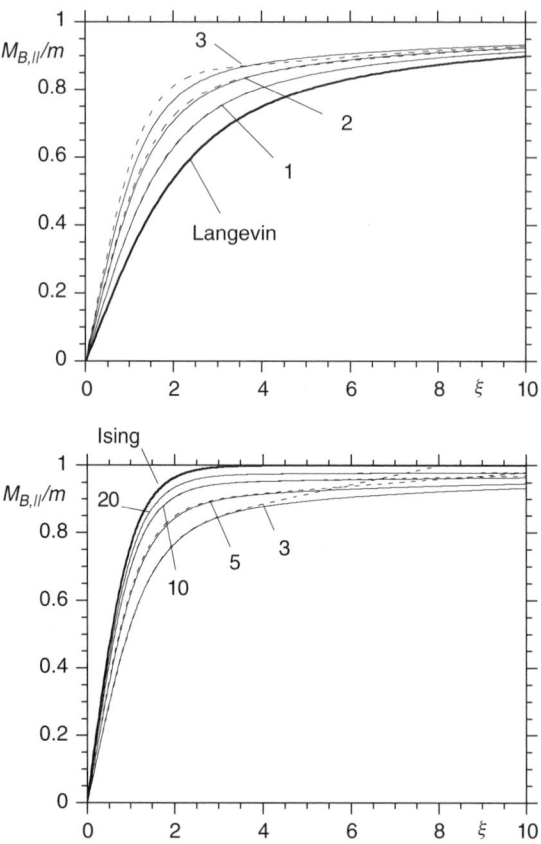

FIGURE 10. Magnetization versus longitudinal field $\xi = mB/k_BT$ [Eq. (3.29); solid lines], along with the weak-anisotropy formula (3.35) (upper panel, dashes) and the asymptotic formula (3.39) (lower panel, dashes), for various values of the dimensionless anisotropy parameter $\sigma = Kv/k_BT$.

which was advanced when considering such an average of the field expansion of the free energy [Eq. (2.69)], is not restricted to the weak-anisotropy range (see Section III.D).

We finally remark that for easy-plane anisotropy ($\sigma < 0$), the results described are only slightly modified. Here, the longitudinal- and transverse-field cases interchange their roles in some sense. For $\vec{B} \| \hat{n}$ and $\sigma < 0$, the magnetic anisotropy hinders the magnetization process, whereas this is naturally favored in the transverse-field case. However, for anisotropy axes distributed at random, the net magnetization curve again goes slightly below the Langevin curve. The mathematical reason for this behavior is that the term

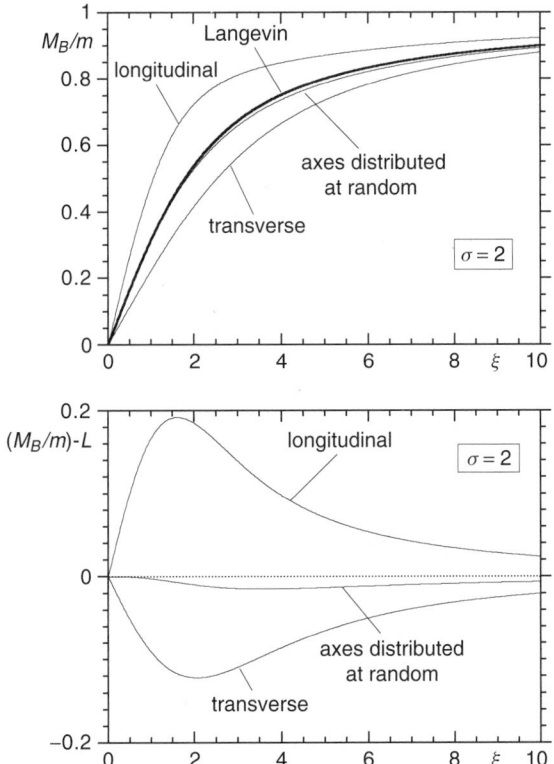

FIGURE 11.   Upper panel: magnetization versus longitudinal field [Eq. (3.35)] and transverse field [Eq. (3.36)], and for anisotropy axes distributed at random [Eq. (3.37)]. Lower panel: anisotropy-induced contribution to the magnetization.

linear in σ in the expansion of $\langle M_B \rangle_{\rm ran}$ vanishes, while the quadratic term is *always* negative.

b. Asymptotic Expansion of the Magnetization for Strong Anisotropy. We now derive the magnetization from the asymptotic expansion of the free energy for large $\sigma = Kv/k_{\rm B}T$. In this way, we shall obtain an analytic formula that contains the first corrections to the Ising-type magnetization due to finite magnetic anisotropy.

We proceed by differentiating the $1/\sigma$ expansion of $\ln \mathcal{Z}$ [Eq. (2.74)] with respect to the field. The ξ derivative of the coefficient of $1/\sigma$ reads

$$\frac{\mathrm{d}}{\mathrm{d}\xi}\left[(2+\xi_\perp^2) - 2\xi_\parallel \tanh\xi_\parallel\right] = -2\left[-\sin\alpha\xi_\perp + \cos\alpha\tanh\xi_\parallel + \frac{\cos\alpha\xi_\parallel}{\cosh^2\xi_\parallel}\right]$$

where $d\xi_\parallel / d\xi = \cos\alpha$ and $d\xi_\perp / d\xi = \sin\alpha$ have been used. The ξ derivative of the coefficient of $1/\sigma^2$ is taken similarly, yielding

$$
\frac{d}{d\xi}\left[5 + (2\xi_\parallel^2 + \xi_\perp^2) - (4 + \xi_\perp^2)\xi_\parallel \tanh\xi_\parallel - \xi_\parallel^2 \tanh^2 \xi_\parallel\right]
$$

$$
= -\cos\alpha \tanh\xi_\parallel \left\{4 + 3\xi_\perp^2 - 2\xi_\parallel \left[\tanh\xi_\parallel - \frac{\xi_\parallel}{\cosh^2 \xi_\parallel}\right]\right\}
$$

$$
+ 2\sin\alpha\,\xi_\perp - \cos\alpha\,\xi_\perp^2 \frac{\xi_\parallel}{\cosh^2 \xi_\parallel}
$$

On collecting these results and using $M_B = m(\partial \ln \mathcal{Z}/\partial\xi)$, the approximate magnetization can finally be written as

$$
\frac{M_B}{m} \simeq \cos\alpha \tanh\xi_\parallel \left\{1 - \frac{1}{2\sigma}\left[1 + \frac{2\xi_\parallel}{\sinh(2\xi_\parallel)}\right] - \frac{1}{8\sigma^2}\left[4 - \xi_\parallel \frac{\sinh(2\xi_\parallel) - 2\xi_\parallel}{\cosh^2 \xi_\parallel}\right]\right\}
$$

$$
+ \sin\alpha\,\xi_\perp \left(\frac{1}{2\sigma} + \frac{1}{4\sigma^2}\right) - \cos\alpha\,\xi_\perp^2 \frac{3\sinh(2\xi_\parallel) + 2\xi_\parallel}{\cosh^2 \xi_\parallel} \frac{1}{16\sigma^2} \tag{3.38}
$$

This formula extends the asymptotic result of Garanin (1996, Eq. (3.13)) for the longitudinal-field case ($\xi_\parallel = \xi$, $\xi_\perp = 0$) to an arbitrary orientation of the field.

Let us explicitly write Eq. (3.38) when the field points along and perpendicular to the anisotropy axis:

$$
\frac{M_{B,\parallel}}{m} \simeq \tanh\xi \left\{1 - \frac{1}{2\sigma}\left[1 + \frac{2\xi}{\sinh(2\xi)}\right] - \frac{1}{8\sigma^2}\left[4 - \xi\frac{\sinh(2\xi) - 2\xi}{\cosh^2 \xi}\right]\right\}
$$

$$
\tag{3.39}
$$

$$
\frac{M_{B,\perp}}{m} \simeq \xi\left(\frac{1}{2\sigma} + \frac{1}{4\sigma^2}\right) \tag{3.40}
$$

Note that in the transverse-field case the leading (Ising) result is identically zero and one gets a linear increase of the magnetization with ξ. On the other hand, the occurrence of α in the arguments of the hyperbolic functions in Eq. (3.38) precludes the obtainment of a simple formula for $M_B(\sigma \gg 1)$ when the anisotropy axes are distributed at random.

As we did when studying the magnetization for weak anisotropy, we may estimate the range of validity of the asymptotic expansion of M_B, by comparing it with the exact analytic formula for $M_{B,\parallel}$. Figure 10 also displays such a comparison showing that, for the field range considered, the approximation derived works reasonably well down to quite small values of σ. There is however an important difference with the weak-anisotropy formula for M_B, the accuracy of which was not significantly sensitive to the magnitude of the field. Here, all the approximate curves depart from the exact results above certain value of the field, which decreases as the anisotropy does. The breaking down of the asymptotic expansion at high fields is apparent in the transverse-field case (3.40), which yields a linear dependence of M_B on ξ, whereas at high fields the magnetization must saturate ($M_B \simeq m$).

These limitations occur because of the $\sigma \gg 1$ expansions have as leading terms Ising-type results (i.e., they correspond to a potential with two deep minima), and the next-order terms are corrections associated with the finite curvature of the potential at the bottom of the wells. However, at sufficiently high fields the two-minima structure of the potential disappears (for example, for $B = B_K$ in a longitudinal field), and the expansion must break down. In fact, already for $B \sim B_K/2$, which corresponds to $\xi \sim \sigma$ [see Eq. (2.6)], the upper potential well is quite shallow (see Fig. 1) and the inverse of the potential curvature at the minimum is therefore large, so the expansion must already fail. This is consistent with the asymptotic results shown in Fig. 10: the approximate M_B departs from the exact one at $\xi \sim 3$ for $\sigma = 3$, at $\xi \sim 4$ for $\sigma = 5$, at $\xi \sim 8$ for $\sigma = 10$, and so on.

We finally mention that, as Fig. 10 suggests, the use of the weak-anisotropy formula, replaced at some point between $\sigma = 2$ and $\sigma = 4$ by the asymptotic expression, yields a reasonable approximation of the exact magnetization, except for the discrepancies discussed for the asymptotic $\xi \gtrsim \sigma$ results. In this connection, as the $\sigma = 3$ curve suggests, one could replace the asymptotic expansion for $\xi \gtrsim \sigma$ by the weak-anisotropy formula in order to improve the overall approximation.

c. Field Expansion of the Magnetization. Let us finally discuss the low-field expansion of the magnetization ($H = B/\mu_0$)

$$M_B = \chi_1 H + \chi_3 H^3 + \chi_5 H^5 + \cdots \qquad (3.41)$$

which defines the linear, χ_1 (or simply χ), and nonlinear, χ_{2n+1}, $n = 1, 2, 3, \ldots$, susceptibilities. In order to derive general expressions for these quantities, we can take the ξ derivative of the low-ξ expansion of $\ln \mathcal{Z}$ [Eq. (2.65)], getting

$$M_B = m \left[2C_1\xi + 2(C_2 - C_1^2)\xi^3 + (C_3 - 3C_2C_1 + 2C_1^3)\xi^5 \right.$$
$$\left. + \tfrac{1}{3}(C_4 - 4C_3C_1 - 3C_2^2 + 12C_2C_1^2 - 6C_1^4)\xi^7 + \cdots \right] \quad (3.42)$$

where the coefficients C_i are given by Eq. (2.36) or (2.39).*

---

* One also arrives at Eq. (3.42) by expanding in powers of ξ the inverse of the denominator of the general formula (3.31), and multiplying this expansion by the first terms of the series in the numerator.

TABLE V
Combinations of the Coefficients C_i Occurring in the First Terms of the Expansion (3.42) of
the Magnetization in Powers of the Magnetic Field, in the Isotropic, Ising, Plane-Rotator, and
Longitudinal-Field Cases

	$\sigma = 0$	$\sigma \to \infty$	$\sigma \to -\infty$	$\vec{B} \| \hat{n}$
$2C_1$	$\dfrac{1}{3}$	$\cos^2 \alpha$	$\dfrac{1}{2} \sin^2 \alpha$	$\dfrac{R'}{R}$
$2(C_2 - C_1^2)$	$-\dfrac{1}{45}$	$-\dfrac{1}{3} \cos^4 \alpha$	$-\dfrac{1}{16} \sin^4 \alpha$	$\dfrac{1}{2}\left[\dfrac{1}{3}\dfrac{R''}{R} - \left(\dfrac{R'}{R}\right)^2\right]$
$C_3 - 3C_2 C_1 + 2C_1^3$	$\dfrac{2}{945}$	$\dfrac{2}{15} \cos^6 \alpha$	$\dfrac{1}{96} \sin^6 \alpha$	$\dfrac{1}{4}\left[\dfrac{1}{30}\dfrac{R'''}{R} - \dfrac{1}{2}\dfrac{R''}{R}\dfrac{R'}{R} + \left(\dfrac{R'}{R}\right)^3\right]$

The expansion (3.42) embodies χ, χ_3, χ_5, and χ_7; in general, χ_{2n+1} can be
obtained by inserting the appropriate C_i into the expression for the nth-order
cumulant. The coefficients of the first three terms at $\sigma \to 0$, $\pm\infty$, and for $\vec{B} \| \hat{n}$,
are given in Table V (they can be obtained from the expressions of Table IV).
On inserting those coefficients into the above expansion of M_B, one gets the
approximate formulas

$$M_{B,\mathrm{Lan}} = m\left[\frac{1}{3}\xi - \frac{1}{45}\xi^3 + \frac{2}{945}\xi^5 + \cdots\right] \tag{3.43}$$

$$M_{B,\mathrm{Ising}} = m\cos\alpha\left[\xi_\| - \frac{1}{3}\xi_\|^3 + \frac{2}{15}\xi_\|^5 + \cdots\right] \tag{3.44}$$

$$M_{B,\mathrm{rot}} = m\sin\alpha\left[\frac{1}{2}\xi_\perp - \frac{1}{16}\xi_\perp^3 + \frac{1}{96}\xi_\perp^5 + \cdots\right] \tag{3.45}$$

$$M_{B,\|} = m\left\{\frac{R'}{R}\xi + \frac{1}{2}\left[\frac{1}{3}\frac{R''}{R} - \left(\frac{R'}{R}\right)^2\right]\xi^3 \right.$$
$$\left. + \frac{1}{4}\left[\frac{1}{30}\frac{R'''}{R} - \frac{1}{2}\frac{R''}{R}\frac{R'}{R} + \left(\frac{R'}{R}\right)^3\right]\xi^5 + \cdots\right\} \tag{3.46}$$

Note that in the first three cases χ_{2n+1} depends on T with a $T^{-(2n+1)}$ law. This
is the translation to linear and nonlinear susceptibilities of the B/T superposition of the corresponding magnetization curves. Outside these limit ranges,
however, the temperature dependence of $C_i(\sigma, \alpha)$ through σ, provokes that
$\chi_{2n+1}(T)$ no longer satisfies such a simple $T^{-(2n+1)}$ law. This is illustrated by

the expansion (3.46) of $M_{B,\parallel}$, in which can be recognized the extra dependence of the susceptibilities on T, provided by the functions $R^{(l)}(\sigma)/R(\sigma)$ via σ.

These points are further investigated in the following two subsections devoted to the linear and nonlinear susceptibilities, respectively.

D. Linear Susceptibility

We shall now study the linear susceptibility of classical spins with axially symmetric magnetic anisotropy. The linear susceptibility χ can be defined as the coefficient of the linear term in the expansion of the magnetization in powers of the external field. On comparing the H expansion of M_B (3.41) with the ξ expansion (3.42), and using $\xi = \mu_0 mH/k_B T$, one gets the following expression for χ

$$\chi = \frac{\mu_0 m^2}{k_B T} 2C_1(\sigma, \alpha) \qquad (3.47)$$

which involves the first coefficient in the expansion of the partition function in powers of ξ. (Recall that α is the angle between the anisotropy axis $\hat{n}$ and the field, while $\sigma = Kv/k_B T$.)

1. Linear Susceptibility: Particular Cases

Let us first consider the expressions that emerge from Eq. (3.47) when one inserts the particular cases of $2C_1$ into it (Table V).

a. Isotropic Case. For $\sigma \to 0$, $2C_1 = \frac{1}{3}$, from which one gets the Curie law for the susceptibility:

$$\chi_{\text{Lan}} = \frac{\mu_0 m^2}{3k_B T} \qquad (3.48)$$

For classical spins, this result naturally follows from the absence of anisotropy.

b. Ising Regime. For $\sigma \to \infty$, $2C_1 = \cos^2 \alpha$, so that

$$\chi_{\text{Ising}} = \frac{\mu_0 m^2}{k_B T} \cos^2 \alpha \qquad (3.49)$$

which is analogous to the susceptibility of an Ising spin. Thus, when the field points along a direction perpendicular to the "Ising axis" ($\cos \alpha = 0$), χ vanishes.

c. *Plane-Rotator Regime.* For $\sigma \to -\infty$, $2C_1 = \sin^2 \alpha/2$, so that the plane-rotator linear susceptibility is given by

$$\chi_{\text{rot}} = \frac{\mu_0 m^2}{2k_\text{B}T} \sin^2 \alpha \qquad (3.50)$$

In this case the linear response is identically zero when the field points perpendicular to the rotator plane.

d. *Longitudinal-Field Case.* On introducing $2C_1|_{\alpha=0} = R'/R$ in Eq. (3.47) one gets the longitudinal susceptibility

$$\chi_\| = \frac{\mu_0 m^2}{k_\text{B}T} \frac{R'}{R} \qquad (3.51)$$

where the factor R'/R induces an extra dependence on T via σ, "interpolating" between the isotropic $(R'/R|_{\sigma=0} = \frac{1}{3})$ and Ising $(R'/R|_{\sigma\to\infty} = 1)$ results.

2. Formulas for the Linear Susceptibility

When the general expression (2.66) for C_1 is introduced into Eq. (3.47), the linear susceptibility emerges in the form*

$$\chi = \frac{\mu_0 m^2}{k_\text{B}T} \left(\frac{R'}{R} \cos^2 \alpha + \frac{R - R'}{2R} \sin^2 \alpha \right) \qquad (3.52)$$

It is convenient to introduce the longitudinal and transverse components of χ (which are related with the diagonal elements of the susceptibility tensor; see text below)

$$\chi_\| = \frac{\mu_0 m^2}{k_\text{B}T} \frac{R'}{R}, \qquad \chi_\perp = \frac{\mu_0 m^2}{k_\text{B}T} \frac{R - R'}{2R} \qquad (3.53)$$

so that χ can be written as

$$\chi = \chi_\| \cos^2 \alpha + \chi_\perp \sin^2 \alpha \qquad (3.54)$$

The quantities $\chi_\|$ and $\chi_\perp$ characterize, respectively, the equilibrium response to a longitudinal (parallel to $\hat{n}$) and transverse (perpendicular to $\hat{n}$) probing

---

* Other derivations of the equilibrium linear susceptibility of a dipole moment in the simplest axially symmetric anisotropy potential were carried out by Lin (1961), Raĭkher and Shliomis (1975) (see also Shliomis and Stepanov, 1993), Shcherbakova (1978), and Chantrell et al. (1985).

field. Because of the linearity of the response, when the probing field points along an arbitrary direction, the projection of the response along the probing-field direction in given by the weighted sum (3.54) of the longitudinal and transverse responses.

a. Average Linear Susceptibility for Anisotropy Axes Distributed at Random. For an ensemble of equivalent magnetic moments (i.e., with the same characteristic parameters) whose anisotropy axes are distributed at random, one finds

$$\langle \chi \rangle_{\mathrm{ran}} = \frac{\mu_0 m^2}{k_B T} \left(\frac{R'}{R} \frac{1}{3} + \frac{R - R'}{2R} \frac{2}{3} \right) = \frac{\mu_0 m^2}{3 k_B T} \tag{3.55}$$

which is merely the Curie law for the linear susceptibility. This equation entails that, irrespective of the magnitude of the magnetic anisotropy as compared with the thermal energy, the linear susceptibility of the randomly oriented ensemble is equal to the susceptibility of isotropic spins. This also holds in the extreme anisotropy cases: for an ensemble of Ising spins, with Ising axes distributed at random, $\langle \chi_{\mathrm{Ising}} \rangle_{\mathrm{ran}} = \mu_0 m^2 / 3 k_B T$; likewise, for an ensemble of plane rotators, whose axes of rotation are distributed at random, $\langle \chi_{\mathrm{rot}} \rangle_{\mathrm{ran}}$ is given by the Curie law.

We shall see later that Eq. (3.55) is in fact rather general; it holds whenever the Hamiltonian of the spin (in the absence of the probing field) has inversion symmetry ($\vec{m} \leftrightarrow -\vec{m}$).

b. Reduced Linear Susceptibility. An informative quantity is the *reduced* linear susceptibility defined as $\chi^{\mathrm{red}} = \chi(k_B T / \mu_0 m^2) = 2C_1$, whence

$$\chi^{\mathrm{red}}(\sigma, \alpha) = \frac{R'}{R} \cos^2 \alpha + \frac{R - R'}{2R} \sin^2 \alpha \tag{3.56}$$

This quantity has the property that isolates the temperature dependence of χ induced by the magnetic anisotropy. Besides, it embodies the angular dependence of the susceptibility. Figure 12 shows χ^{red} as a function of the angle between the anisotropy axis and the probing field (cf. Lin, 1961). As expected, the larger the $|\sigma|$, the more anisotropic the χ^{red} curves, becoming rather different from circles already for $|\sigma| \simeq 5$.

Figure 13 shows χ^{red} for the longitudinal and transverse components of the linear susceptibility (in this representation $\langle \chi^{\mathrm{red}} \rangle_{\mathrm{ran}}$ would take the constant value $\frac{1}{3}$). Both curves coincide at $\sigma = 0$, where the orientation of the field plays no role, taking the Langevin value $\frac{1}{3}$. It can also be seen that the maximum variation of χ^{red} with σ occurs when the probing field is parallel to the anisotropy axis. Note also that, *qualitatively*, the longitudinal- and the transverse-field cases interchange their roles when the sign of the anisotropy is

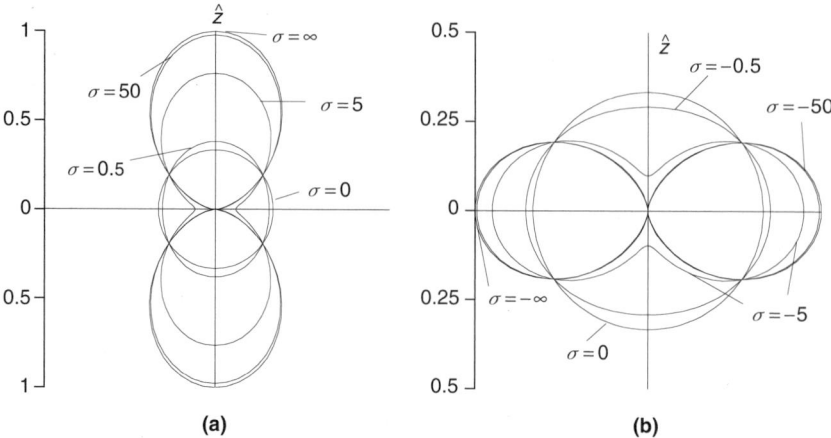

FIGURE 12. Polar plots showing the angular dependence of the reduced linear suscepti-bility χ^{red} [Eq. (3.56)] for various values of the dimensionless anisotropy parameter $\sigma = Kv/k_B T$: (*a*) easy-axis anisotropy; (*b*) easy-plane anisotropy.

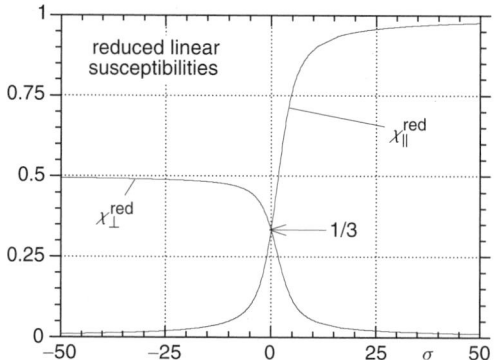

FIGURE 13. Reduced linear susceptibility (3.56) in the longitudinal, $\chi_{\parallel}^{\mathrm{red}}$, and transverse, $\chi_{\perp}^{\mathrm{red}}$, field cases versus the dimensionless anisotropy parameter $\sigma = Kv/k_B T$ (cf. Fig. 6).

reversed. This statement, which is supported by Fig. 12, is associated with the qualitatively "equivalent" magnetization behavior in the easy-axis and easy-plane anisotropy cases when the probing field points in the "easy-magnetization region" or in the "hard-magnetization region," regions that interchange them-selves when the sign of the anisotropy is changed.

3. Generalizations

a. Probing-Field Derivative of the Magnetization. The definition of the linear susceptibility as the coefficient of the linear term in the expansion of the magnetization in powers of the external field, of course agrees with that in terms of the field derivative of the magnetization at zero field, namely, $\chi = \mu_0[\partial\langle\vec{m}\cdot\hat{b}\rangle_e/\partial B]|_{B=0}$. This definition suggests the immediate generalization

$$\chi = \mu_0 \frac{\partial\langle\vec{m}\cdot\hat{b}\rangle}{\partial(\Delta B)}\bigg|_{\Delta B=0} \qquad (3.57)$$

where $\Delta\vec{B} = \Delta B\hat{b}$ is an external probing field ($\hat{b}$ stands now for the unit vector in the direction of the probing field). The absence of the subscript "e" in the thermal equilibrium averages is used to indicate that they are taken with respect to the total energy (system plus perturbation). The unperturbed system can already be subjected to a constant (bias) field $\vec{B}$ not necessarily collinear with $\Delta\vec{B}$.

Indeed, the calculation of the linear susceptibility can be carried out by starting from a total Hamiltonian $\mathcal{H}_T = \mathcal{H} - \vec{m}\cdot\Delta\vec{B}$, where the actual form of the unperturbed Hamiltonian $\mathcal{H}$ is not required. Let us first calculate

$$\frac{\partial\left\langle(\vec{m}\cdot\hat{b})^n\right\rangle}{\partial(\Delta B)} = \frac{\partial}{\partial(\Delta B)} \frac{\int d\Omega(\vec{m}\cdot\hat{b})^n e^{-\beta\mathcal{H}_T}}{\int d\Omega\, e^{-\beta\mathcal{H}_T}}$$

$$= \beta \frac{\mathcal{Z}\int d\Omega(\vec{m}\cdot\hat{b})^{n+1}e^{-\beta\mathcal{H}_T} - \int d\Omega(\vec{m}\cdot\hat{b})^n e^{-\beta\mathcal{H}_T}\int d\Omega(\vec{m}\cdot\hat{b})e^{-\beta\mathcal{H}_T}}{\mathcal{Z}^2}$$

where as usual $\int d\Omega(\cdot) = (1/2\pi)\int_{-1}^{1} d(\cos\vartheta)\int_0^{2\pi} d\varphi(\cdot)$. From this result we get the general relation

$$\frac{\partial\left\langle(\vec{m}\cdot\hat{b})^n\right\rangle}{\partial(\Delta B)} = \beta\left[\left\langle(\vec{m}\cdot\hat{b})^{n+1}\right\rangle - \left\langle(\vec{m}\cdot\hat{b})^n\right\rangle\langle\vec{m}\cdot\hat{b}\rangle\right] \qquad (3.58)$$

the $n = 1$ particular case of which merely reads

$$\frac{\partial\langle\vec{m}\cdot\hat{b}\rangle}{\partial(\Delta B)} = \beta\left[\left\langle(\vec{m}\cdot\hat{b})^2\right\rangle - \langle\vec{m}\cdot\hat{b}\rangle^2\right] \qquad (3.59)$$

and holds irrespective of the magnitude of $\Delta\vec{B}$. When this equation is evaluated at $\Delta B = 0$ and inserted in Eq. (3.57), one gets the celebrated expression for the linear susceptibility in terms of the statistics of the thermal equilibrium fluctuations of the magnetic moment in the absence of the probing field, namely

$$\chi = \frac{\mu_0}{k_B T}\left[\left\langle(\vec{m}\cdot\hat{b})^2\right\rangle_e - \langle\vec{m}\cdot\hat{b}\rangle_e^2\right] \qquad (3.60)$$

where $\langle\ \rangle_e$ denotes the equilibrium average in the absence of the perturbation.

The relation (3.60) is valid for *any* form of the Hamiltonian. When $\mathcal{H}$ is given by $\mathcal{H} = -(Kv/m^2)(\vec{m} \cdot \hat{n})^2$ [cf. Eq. (2.2)], the preceding averages in the absence of the probing field are in fact zero-field averages, which are directly related with the coefficients C_i by Eq. (2.42). Thus, by inserting

$$\langle \vec{m} \cdot \hat{b} \rangle_e|_{B=0} = 0, \qquad \left\langle (\vec{m} \cdot \hat{b})^2 \right\rangle_e|_{B=0} = m^2 2 C_1$$

into Eq. (3.60), one indeed recovers the expression (3.47) for χ.

b. Tensor Structure. The linear susceptibility is in fact a tensor defined by

$$\chi_{ij} = \mu_0 \frac{\partial \langle m_i \rangle}{\partial (\Delta B_j)}\bigg|_{\Delta B=0} \tag{3.61}$$

Note that the diagonal elements are given by Eq. (3.57) when $\hat{b}$ points along $\hat{x}$, $\hat{y}$, and $\hat{z}$. By a derivation analogous to that leading to Eq. (3.60), one arrives at the result

$$\chi_{ij} = \frac{\mu_0}{k_B T} \left[\langle m_i m_j \rangle_e - \langle m_i \rangle_e \langle m_j \rangle_e \right] \tag{3.62}$$

Because χ_{ij} is a symmetric second-rank tensor, it can be diagonalized by a suitable change of variables. Let us assume that this diagonalization has already been carried out. Then, if a probing field $\Delta \vec{B} = \Delta B \hat{b}$ is applied to the system, the projection of the average magnetic moment onto $\hat{b}$ is given in the linear response range by (we use $\mu_0 \langle m_i \rangle \simeq \mu_0 \langle m_i \rangle|_{\Delta B=0} + \Sigma_j \chi_{ij} \Delta B_j + \cdots$)

$$\mu_0 \langle \Delta \vec{m} \rangle \cdot \hat{b} \simeq (\chi_{xx} \cos^2 \alpha + \chi_{yy} \cos^2 \beta + \chi_{zz} \cos^2 \gamma) \Delta B \tag{3.63}$$

where (α, β, γ) are the direction cosines of $\hat{b}$ (in the coordinate system that diagonalizes χ_{ij}). The quantity into the brackets defines an *effective* linear susceptibility χ, which is in fact what we have been calling linear susceptibility throughout.

c. Average Linear Susceptibility for Randomly Distributed Anisotropy Axes revisited. On the basis of the expressions above, we can derive the result mentioned for the linear susceptibility of an ensemble of equivalent dipole moments whose Hamiltonian has inversion symmetry and their intrinsic axes are distributed at random (see also García-Palacios, Jönsson, and Svedlindh, 2000).

For an ensemble of independent dipole moments, the contribution of each dipole to χ is analogous to that occurring in Eq. (3.63), with (in principle) different direction cosines and diagonal elements χ_{ii} for each dipole. However, if these elements are all equal, we can write the total effective susceptibility as

$$\chi = \chi_{xx} \left\langle \cos^2 \alpha \right\rangle + \chi_{yy} \left\langle \cos^2 \beta \right\rangle + \chi_{zz} \left\langle \cos^2 \gamma \right\rangle \qquad (3.64)$$

where $\langle \rangle$ denotes now average over the ensemble of dipoles. Note that for the assumption about the equality of the tensor elements to hold, the dipoles must be equivalent (in the sense of having the same characteristic parameters), and the orientation of the intrinsic axes (which diagonalize the linear susceptibility tensor for each $\vec{m}$) with respect to the main reference frame must be irrelevant in determining the χ_{ii}; this excludes, for instance, the presence of an external (bias) field. Then, if those intrinsic axes are distributed at random, the effective linear susceptibility (3.64) reduces to

$$\langle \chi \rangle_{\text{ran}} = \frac{1}{3}(\chi_{xx} + \chi_{yy} + \chi_{zz}) = \frac{\mu_0}{3k_{\text{B}}T} \left\{ m^2 - \left[\langle m_x \rangle_{\text{e}}^2 + \langle m_y \rangle_{\text{e}}^2 + \langle m_z \rangle_{\text{e}}^2 \right] \right\}$$

where Eq. (3.62) has been used to express the χ_{ii}. Finally, if the Hamiltonian of each dipole has inversion symmetry ($\langle m_i^{2n+1} \rangle_{\text{e}} = 0$), one has $\langle m_i \rangle_{\text{e}} = 0$, $i = x, y, z$, so that $\langle \chi \rangle_{\text{ran}}$ further reduces to

$$\langle \chi \rangle_{\text{ran}} = \frac{\mu_0 m^2}{3k_{\text{B}}T} \qquad (3.65)$$

(note that the presence of a bias field could as well be excluded on the basis of the inversion symmetry condition). Equation (3.65) is the announced result: *For an ensemble of equivalent dipole moments whose Hamiltonian has inversion symmetry, the effective linear susceptibility is given by the Curie law when their intrinsic axes are distributed at random.* In particular this result is applicable to ensembles of spins with *arbitrary* anisotropy energy.

d. General Formula for any Axially Symmetric Hamiltonian. We now calculate the linear susceptibility of a magnetic moment with an arbitrary axially symmetric Hamiltonian. The corresponding equilibrium probability distribution of $z = m_z/m$ is given by [cf. Eq. (2.26)]

$$P_{\text{e},\parallel}(z) = \mathcal{Z}_{\parallel}^{-1} \exp[-\beta \mathcal{H}(z)], \qquad \mathcal{Z}_{\parallel} = \int_{-1}^{1} dz \exp[-\beta \mathcal{H}(z)] \qquad (3.66)$$

where we have assumed that the symmetry axis points along $\hat{z}$. In such a reference frame, the susceptibility tensor is diagonal and the diagonal elements are given by

$$\chi_{ii} = \frac{\mu_0}{k_B T} \left[\langle m_i^2 \rangle_e - \langle m_i \rangle_e^2 \right], \qquad i = x, y, z \qquad (3.67)$$

Besides, because of the axial symmetry of the Hamiltonian, the susceptibility tensor has only two independent elements: $\chi_{\parallel} = \chi_{zz}$ and $\chi_{\perp} = \chi_{xx} = \chi_{yy}$.

Let us introduce the averages of the Legendre polynomials $p_n(z)$

$$p_1(z) = z, \qquad\qquad p_2(z) = \tfrac{1}{2}(3z^2 - 1)$$

$$p_3(z) = \tfrac{1}{2}(5z^3 - 3z), \qquad p_4(z) = \tfrac{1}{8}(35z^4 - 30z^2 + 3), \ldots \qquad (3.68)$$

with respect to the equilibrium probability distribution $P_{e,\parallel}(z)$, namely

$$S_n \overset{\text{def}}{=} \langle p_n(z) \rangle_e = \int_{-1}^{1} dz\, p_n(z) P_{e,\parallel}(z) \qquad (3.69)$$

In terms of these quantities, we can write $\chi_{\parallel}$ and $\chi_{\perp}$ as

$$\chi_{\parallel} = \frac{\mu_0 m^2}{k_B T} \left(\frac{1 + 2S_2}{3} - S_1^2 \right), \qquad \chi_{\perp} = \frac{\mu_0 m^2}{k_B T} \frac{1 - S_2}{3} \qquad (3.70)$$

for which we have employed

$$\langle m_z \rangle_e = m S_1, \qquad \langle m_z^2 \rangle_e = \tfrac{1}{3} m^2 (1 + 2S_2)$$

$$\langle m_{x,y} \rangle_e = 0, \qquad \langle m_{x,y}^2 \rangle_e = \tfrac{1}{2} \left(m^2 - \langle m_z^2 \rangle_e \right)$$

Equations (3.70) are valid, for example, for *any* axially symmetric anisotropy potential in a longitudinal bias field. For the simplest uniaxial anisotropy in a longitudinal bias field

$$-\beta \mathcal{H} = \sigma z^2 + \xi z \qquad (3.71)$$

one can derive the following explicit expressions for S_1 and S_2

$$S_1 = \frac{e^{\sigma}}{\sigma \mathcal{Z}_{\parallel}} \sinh \xi - h \qquad (3.72)$$

$$S_2 = \frac{3}{2} \left[\frac{e^{\sigma}}{\sigma \mathcal{Z}_{\parallel}} (\cosh \xi - h \sinh \xi) + h^2 - \frac{1}{2\sigma} \right] - \frac{1}{2} \qquad (3.73)$$

where $h = B/B_k = \xi/2\sigma$ and $\mathcal{Z}_{\parallel}$ is given by Eq. (2.25).*

---

* The formula for $S_1 = \langle z \rangle_e$, is essentially Eq. (3.29) for the longitudinal magnetization. In order to derive the formula for S_2, we can take advantage of some previous results. Note first

In the isotropic ($K = 0$) case, the linear susceptibility is more easily obtained directly from the definition (3.69) of the S_n with help from Eqs. (2.46)–(2.51). On so doing, one obtains

$$\chi_{\parallel} = \frac{\mu_0 m^2}{k_B T} L', \qquad \chi_{\perp} = \frac{\mu_0 m^2}{k_B T} \frac{1}{\xi} L \qquad (3.74)$$

where $L(\xi)$ is the Langevin function. Note that, since $L(\xi) = \xi/3 + \cdots$ for low fields [Eq. (3.43)], both components of Eq. (3.74) merge on the Curie susceptibility $\chi = \mu_0 m^2 / 3k_B T$ as the bias field goes to zero.

For $B = 0$, the linear susceptibility is sometimes found written in a number of alternative forms. Note first that in this case one has $S_1 = 0$. Therefore, on introducing the notation $\tilde{S}_2 = S_2(\sigma, \xi)|_{\xi=0}$, one gets from Eq. (3.70) the following formulas

$$\chi_{\parallel} = \frac{\mu_0 m^2}{k_B T} \frac{1 + 2\tilde{S}_2}{3}, \qquad \chi_{\perp} = \frac{\mu_0 m^2}{k_B T} \frac{1 - \tilde{S}_2}{3} \qquad (3.75)$$

(The quantity $\tilde{S}_2$ is sometimes written as S or merely S_2.) In order to directly check Eqs. (3.75) against Eqs. (3.53) one only needs to use

$$\frac{R'}{R} = \frac{\int_{-1}^{1} dz\, z^2 \exp(\sigma z^2)}{\int_{-1}^{1} dz \exp(\sigma z^2)} = \langle z^2 \rangle_e |_{B=0} = \frac{1}{3}(1 + 2S_2)|_{B=0} = \frac{1}{3}(1 + 2\tilde{S}_2) \qquad (3.76)$$

Alternative expressions for χ at $B = 0$ can also be written in terms of Kummer functions. Thus, on introducing C_1 from Eq. (2.41) into Eq. (3.47), one directly gets (cf. Coffey, Crothers, Kalmykov and Waldron, 1995b)

$$\chi_{\parallel} = \frac{\mu_0 m^2}{3k_B T} \frac{M\left(\frac{3}{2}, \frac{5}{2}; \sigma\right)}{M\left(\frac{1}{2}, \frac{3}{2}; \sigma\right)}, \qquad \chi_{\perp} = \frac{\mu_0 m^2}{3k_B T} \frac{M\left(\frac{1}{2}, \frac{5}{2}; \sigma\right)}{M\left(\frac{1}{2}, \frac{3}{2}; \sigma\right)} \qquad (3.77)$$

---

that the thermodynamical energy in the longitudinal-field case can be written as

$$\mathcal{U}_{\parallel} = \langle -Kvz^2 - mBz \rangle_e = -Kv\left(\langle z^2 \rangle_e + 2h\langle z \rangle_e \right) = -Kv\left(\frac{1 + 2S_2}{3} + 2hS_1 \right)$$

Then, on using Eq. (3.19) for $\mathcal{U}_{\parallel}$, taking Eq. (3.72) into account, and recalling that $J = 2(\cosh \xi + h \sinh \xi)$, one gets

$$\frac{1 + 2S_2}{3} = -\frac{\mathcal{U}_{\parallel}}{Kv} - 2hS_1 = \frac{e^\sigma}{\sigma \mathcal{Z}_{\parallel}}(\cosh \xi - h \sinh \xi) + h^2 - \frac{1}{2\sigma}$$

from which Eq. (3.73) follows.   Q.E.D.

4. Approximate Formulas for the Linear Susceptibility

We now derive approximate formulas for χ with the aim of bypassing, when possible, the use of expressions involving nonelementary functions. The formulas obtained, based on weak- and strong-anisotropy expansions, reasonably compare with the exact results in the whole temperature range.

We find it convenient to rewrite first the exact expression (3.52) for χ as follows

$$\chi = \frac{\mu_0 m^2}{k_B T} \frac{1}{3} \left[1 + \frac{1}{2} \left(3\frac{R'}{R} - 1 \right) (3\cos^2\alpha - 1) \right] \tag{3.78}$$

where the factor multiplying $(3\cos^2\alpha - 1)$ is precisely $\tilde{S}_2$ [see Eq. (3.76)]. In order to derive approximate formulas for χ in the unbiased case, we use the approximate results for R'/R derived in Appendix A. We can also get most of the following results (up to second order) if we start from the expansions of M_B in powers of σ [Eq. (3.34)] and the asymptotic expansion (3.38).

a. Weak-Anisotropy Range. In order to obtain an approximate formula for χ valid in the $|\sigma| \ll 1$ range, we insert the approximate R'/R from Eq. (A.24) into Eq. (3.78), getting

$$\chi|_{|\sigma|\ll 1} \simeq \frac{\mu_0 m^2}{3k_B T} \left[1 + \left(\frac{2}{15}\sigma + \frac{4}{315}\sigma^2 - \frac{8}{4725}\sigma^3 \right) (3\cos^2\alpha - 1) \right] \tag{3.79}$$

This equation yields a good approximation of the exact χ for $|\sigma| \le 2$. Note that, as they should, when the anisotropy axes are distributed at random, the corrections to the leading (isotropic) result vanish at *all* orders.

b. Strong-Anisotropy Ranges. Similarly, to obtain approximate formulas for χ valid in the $|\sigma| \gg 1$ ranges, we use the corresponding approximate expressions for R'/R derived in Appendix A.

For $\sigma \ll -1$, we insert R'/R from Eq. (A.27) into Eq. (3.78), getting

$$\chi|_{\sigma\ll -1} \simeq \frac{\mu_0 m^2}{k_B T} \left[\frac{1}{2} \sin^2\alpha - \frac{1}{4\sigma} (3\cos^2\alpha - 1) \right] \tag{3.80}$$

An approximate formula for the extreme easy-axis case can be derived in a similar way. On substituting the $\sigma \gg 1$ result (A.29) for R'/R in Eq. (3.78), we obtain

$$\chi|_{\sigma\gg 1} \simeq \frac{\mu_0 m^2}{k_B T} \left[\cos^2\alpha - \left(\frac{1}{2\sigma} + \frac{1}{4\sigma^2} + \frac{5}{8\sigma^3} \right) (3\cos^2\alpha - 1) \right] \tag{3.81}$$

Again, when the anisotropy axes are distributed at random, all the corrections to the leading plane-rotator and Ising results vanish identically. These approximate formulas compare well with the corresponding exact results for $|\sigma| \geq 5$, so that, on complementing Eqs. (3.79)–(3.81), one can cover the entire σ range reasonably. This merely follows from the patching (shown in Fig. 34 of Appendix A) of the exact R'/R provided by the approximate formulas with which the approximate results (3.79)–(3.81) for χ have been constructed.

For future reference, we finally write the longitudinal and transverse components of χ for strong anisotropy to order $1/|\sigma|$, namely

$$\chi_\| \simeq \frac{\mu_0 m^2}{k_B T} - \frac{\mu_0 m^2}{K v}, \qquad \chi_\perp \simeq \frac{\mu_0 m^2}{2 K v}, \qquad K > 0 \qquad (3.82)$$

and

$$\chi_\| \simeq \frac{\mu_0 m^2}{2|K|v}, \qquad \chi_\perp \simeq \frac{\mu_0 m^2}{2 k_B T} - \frac{\mu_0 m^2}{4|K|v}, \qquad K < 0 \qquad (3.83)$$

(Note the qualitative interchange of the roles of $\chi_\|$ and $\chi_\perp$ with the transformation $K \to -K$.)

c. Formulas in the Presence of a Longitudinal Bias Field. We can also obtain high-barrier approximations of the exact equilibrium susceptibilities in the presence of a longitudinal bias field. Those equations, which will be valid for $h \ll 1$, can be obtained by starting from the approximate expression (2.28) for the partition function. Thus, on applying the relations [readily obtainable from Eqs. (3.66) and (3.69)]

$$S_1 = \frac{1}{\mathcal{Z}_\|} \frac{\partial \mathcal{Z}_\|}{\partial \xi}, \qquad \frac{1}{3}(1 + 2S_2) = \frac{1}{\mathcal{Z}_\|} \frac{\partial \mathcal{Z}_\|}{\partial \sigma} \qquad (3.84)$$

to the approximate $\mathcal{Z}_\|$ mentioned, one obtains the following expressions from Eqs. (3.70):

$$\chi_\| \simeq \frac{\mu_0 m^2}{k_B T} \frac{1}{(\cosh \xi - h \sinh \xi)^2}$$

$$\times \left\{ (1 - h^2) - \frac{1}{\sigma} + \frac{1}{8\sigma^2} \left[1 - \frac{1 + 6h^2 + h^4}{(1 - h^2)^2} \cosh(2\xi) \right. \right.$$

$$\left. \left. + \frac{4h(1 + h^2)}{(1 - h^2)^2} \sinh(2\xi) \right] \right\} \qquad (3.85)$$

$$\chi_\perp \simeq \frac{\mu_0 m^2}{k_B T} \frac{1}{2\sigma} \frac{(1 + h^2) \cosh \xi - 2h \sinh \xi}{(1 - h^2)(\cosh \xi - h \sinh \xi)} \qquad (3.86)$$

For $B = 0$, these formulas duly reduce to Eqs. (3.82). Finally, on taking formally the $K \to \infty$ limit in these expressions (i.e., $\sigma \to \infty$ and $h = \xi/2\sigma \to 0$), one gets the 'Ising-type' equilibrium susceptibilities in a longitudinal bias field [cf. Eq. (3.49)]

$$\chi_\| \simeq \frac{\mu_0 m^2}{k_\mathrm{B} T} \frac{1}{\cosh^2 \xi}, \qquad \chi_\perp \simeq 0 \qquad\qquad (3.87)$$

Equations (3.85)–(3.87), will be used in Section V.

5. Temperature Dependence of the Linear Susceptibility

Figure 14 displays the linear susceptibility in a longitudinal bias field. The longitudinal component decreases with increasing B for a given T, since $\chi_\|(T, B)$ is the slope of the longitudinal magnetization curve at B (see Fig. 9). Regarding the temperature dependence of $\chi_\|$, because $\chi_\|(T, B = 0)$ is the *initial* slope of $M_{B,\|}$, it always increases as the thermal agitation is reduced. In contrast, $\chi_\|(T, B \neq 0)$ has a maximum as a function of the temperature and tends to zero at low temperatures. This is also a result of $\chi_\|(T, B \neq 0)$ being the slope of $M_{B,\|}$ at $B \neq 0$. Indeed, at high temperatures ($\xi \ll 1$), $\chi_\|$ also increases with decreasing thermal agitation. However, at low temperatures ($\xi \gg 1$), the slope of $M_{B,\|}$ versus B decreases as T is lowered — "high-field" magnetization approaching a straight line due to the saturation of $\langle \vec{m} \rangle_\mathrm{e}$ (see Fig. 9). Therefore, in the intermediate temperature range, $\chi_\|(T, B \neq 0)$ exhibits a maximum at the temperature where the "shoulder" of the magnetization curve passes through B. Note finally that, for this maximum to exist, the anisotropy is secondary, whereas a nonzero bias field is essential. Indeed, the longitudinal component of Eq. (3.74) for an *isotropic* spin also exhibits a maximum in $\chi_\|$ versus T if $B \neq 0$.

Concerning the transverse susceptibility, it exhibits a broad maximum as a function of T even at $B = 0$, so it cannot be attributed to the presence of the bias field. This maximum is to be interpreted in terms of the anisotropy-induced crossover from the free-rotator regime at high T to the discrete-orientation regime as T is lowered. Indeed, at low temperatures the transverse probing field competes with the anisotropy energy in aligning the magnetic moments, which are concentrated close to the potential minima. Then, the increase of the thermal agitation permits $\vec{m}$ to (statistically) separate from the poles and the (transverse) response increases. However, if the temperature is further increased, $\vec{m}$ becomes progressively unfastened from the anisotropy and the transverse field competes mainly with the thermal agitation in aligning $\vec{m}$; the response then exhibits a maximum and decreases as T is increased. In this transverse probing-field case, is the anisotropy, not the bias field, the essential element for the appearance of the maximum in the response.

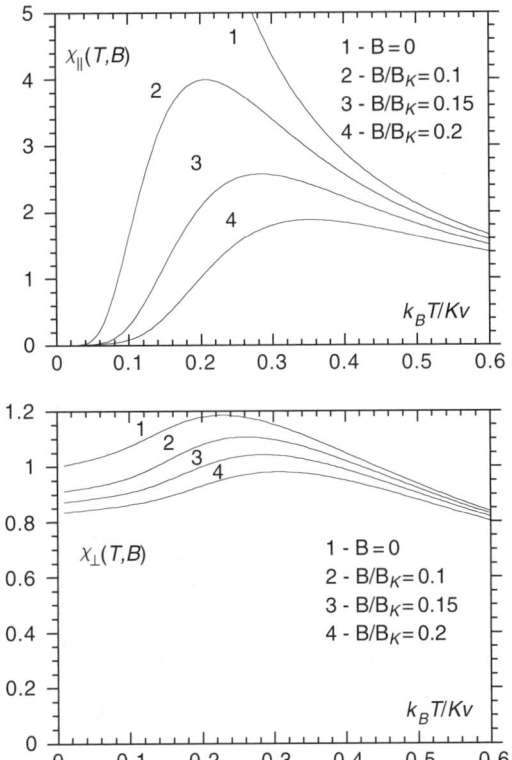

FIGURE 14. Longitudinal and transverse components of the linear susceptibility versus T in the unbiased case and in the presence of longitudinal bias fields [Eqs. (3.70)]. The anisotropy is assumed to be of easy-axis type ($K > 0$) and the susceptibilities are measured in units of $\mu_0 m/B_K = \mu_0 m^2/2Kv$ [the transverse equilibrium susceptibility at $T = 0$ in the unbiased case; see Eq. (3.82)].

Indeed, the transverse component of Eq. (3.74) for isotropic spins [i.e., $\chi_\perp = (\mu_0 m^2/k_B T)L/\xi$], starting from the nonzero value $\mu_0 m/B$ at $T = 0$, decreases *monotonically* with T in the whole temperature range [as $\chi_\perp \simeq (\mu_0 m/B)(1 - k_B T/mB)$ for $\xi \gg 1$, this decrease is linear at low T].

E. Nonlinear Susceptibilities

We shall now consider the nonlinear susceptibilities of classical spins with axially symmetric magnetic anisotropy. Part of the motivation to study the nonlinear susceptibilities is the suitability of these quantities in the study of collective phenomena in glassy systems, together with the glassy-like features exhibited by interacting magnetic nanoparticles (see, for example, Jonsson et al., 1995). Most of the following results were obtained by García-Palacios and Lázaro (1997), while the extension of the study to the (overdamped) dynamical case was done by Raĭkher and Stepanov (1997).

The nonlinear susceptibilities are defined as the coefficients of the nonlinear terms in the expansion of the magnetization in powers of the external field. Here, some of the parallel properties of these quantities will be illustrated with the first one of the series, χ_3. The basic expression for χ_3 can be obtained by comparing the H expansion of M_B (3.41) with its ξ expansion (3.42), to get

$$\chi_3 = \frac{\mu_0^3 m^4}{(k_B T)^3} 2(C_2 - C_1^2) \tag{3.88}$$

which involves the first two coefficients of the field expansion (2.35) of the partition function.

1. Nonlinear Susceptibilities: Particular Cases

Let us first write the expressions that emerge from Eq. (3.88) when one considers the various particular cases of the combination $2(C_2 - C_1^2)$ (see Table V).

a. Isotropic Case. For $\sigma \to 0$, one has $2(C_2 - C_1^2) = -\frac{1}{45}$, so that the Langevin χ_3 reads

$$\chi_{3,\text{Lan}} = -\frac{\mu_0^3 m^4}{45(k_B T)^3} \tag{3.89}$$

b. Ising Regime. For $\sigma \to \infty$, the combination of the C_i required reads $2(C_2 - C_1^2) = -\cos^4 \alpha/3$; accordingly, the Ising χ_3 is given by

$$\chi_{3,\text{Ising}} = -\frac{\mu_0^3 m^4 \cos^4 \alpha}{3(k_B T)^3} \tag{3.90}$$

which vanishes when the field points along a direction perpendicular to the anisotropy axis.

c. Plane-Rotator Regime. We have $2(C_2 - C_1^2) = -\sin^4 \alpha/16$ for $\sigma \to -\infty$, whence

$$\chi_{3,\text{rot}} = -\frac{\mu_0^3 m^4 \sin^4 \alpha}{16(k_B T)^3} \tag{3.91}$$

Here, the nonlinear susceptibility vanishes when the field points along the direction perpendicular to the plane of the rotator.

d. Longitudinal-Field Case. Finally, when the field is parallel to the anisotropy axis, one has $2(C_2 - C_1^2) = [R''/3R - (R'/R)^2]/2$, so that the corresponding nonlinear susceptibility reads

$$\chi_{3,\parallel} = \frac{\mu_0^3 m^4}{(k_B T)^3} \frac{1}{2} \left[\frac{1}{3} \frac{R''}{R} - \left(\frac{R'}{R} \right)^2 \right] \tag{3.92}$$

As occurs with the linear susceptibility, the magnetic anisotropy induces an additional dependence of χ_3 on T via the functions $R^{(l)}/R$, with the consequent departure from the T^{-3} dependences of the limit cases considered above.

2. Formulas for the Nonlinear Susceptibility

On introducing the complete expression for $2(C_2 - C_1^2)$ obtained from Eq. (2.67) into Eq. (3.88), we get the following general formula for χ_3:

$$\chi_3 = \frac{\mu_0^3 m^4}{(k_B T)^3} \left\{ \frac{1}{2} \left[\frac{1}{3} \frac{R''}{R} - \left(\frac{R'}{R} \right)^2 \right] \cos^4 \alpha \right.$$

$$+ \frac{1}{2} \left[\left(\frac{R'}{R} \right)^2 - \frac{R''}{R} \right] \cos^2 \alpha \sin^2 \alpha$$

$$\left. + \frac{1}{16} \left[-1 + 2\frac{R'}{R} - 2 \left(\frac{R'}{R} \right)^2 + \frac{R''}{R} \right] \sin^4 \alpha \right\} \tag{3.93}$$

This expression can alternatively be written in terms of the averages of the Legendre polynomials (3.69) evaluated at zero field (Raĭkher and Stepanov, 1997)

$$\chi_3 = \frac{\mu_0^3 m^4}{(k_B T)^3} \frac{1}{315} \left[\left(12\tilde{S}_4 - 70\tilde{S}_2^2 - 40\tilde{S}_2 - 7 \right) \cos^4 \alpha \right.$$

$$- 2 \left(18\tilde{S}_4 - 35\tilde{S}_2^2 + 10\tilde{S}_2 + 7 \right) \cos^2 \alpha \sin^2 \alpha$$

$$\left. + \frac{1}{2} \left(9\tilde{S}_4 - 35\tilde{S}_2^2 + 40\tilde{S}_2 - 14 \right) \sin^4 \alpha \right] \tag{3.94}$$

where $\tilde{S}_n = S_n(\sigma, \xi)|_{\xi=0}$. These formulas simplify notably when averaged over an ensemble of equivalent dipoles with a random distribution of anisotropy axes.

a. Average Nonlinear Susceptibility for Anisotropy Axes Distributed at Random. When the expressions (2.64) for the averages of the angular terms are introduced into Eq. (3.93), one gets the following formula for $\langle \chi_3 \rangle_{\text{ran}}$ [cf. Eq. (2.68)]

$$\langle \chi_3 \rangle_{\text{ran}} = \frac{\mu_0^3 m^4}{(k_B T)^3} \frac{1}{30} \left[2\frac{R'}{R} - 3\left(\frac{R'}{R}\right)^2 - 1 \right] \tag{3.95}$$

or, by using the relation $R'/R = (1 + 2\tilde{S}_2)/3$, the more compact form

$$\langle \chi_3 \rangle_{\text{ran}} = -\frac{\mu_0^3 m^4}{(k_B T)^3} \frac{1 + 2\tilde{S}_2^2}{45} \tag{3.96}$$

Note that, unlike $\langle \chi \rangle_{\text{ran}}$, which is given by the Curie law, χ_3 depends on the anisotropy energy even for anisotropy axes distributed at random. Indeed, we had already seen in Fig. 11 that, while one has $\langle M_B \rangle_{\text{ran}} \simeq mL(\xi)$ for low fields, as the field is increased $\langle M_B \rangle_{\text{ran}}$ bends downward more rapidly than the Langevin magnetization. Thus, not only $\langle \chi_3 \rangle_{\text{ran}} \neq \chi_{3,\text{Lan}}$, but $|\langle \chi_3 \rangle_{\text{ran}}| > |\chi_{3,\text{Lan}}|$ (by factors up to 3 and 1.5 at low T for $K > 0$ and $K < 0$, respectively).

  b. Reduced Nonlinear Susceptibility. In analogy with the reduced linear susceptibility (3.56), we can define a reduced nonlinear susceptibility isolating the anisotropy-induced temperature dependence of χ_3 as follows:

$$\chi_3^{\text{red}}(\sigma, \alpha) = \chi_3(\sigma, \alpha) \frac{(k_B T)^3}{\mu_0^3 m^4} = 2(C_2 - C_1^2) \tag{3.97}$$

Figure 15 displays $-\chi_3^{\text{red}}$ as a function of the angle between the anisotropy axis and the external field. It is shown that the χ_3^{red} curves become increasingly anisotropic as $|\sigma|$ increases, being quite different from circles already for $|\sigma| \simeq 1$. (The circles for the isotropic $-\chi_3^{\text{red}}|_{\sigma=0}$ correspond to the same radius $(\frac{1}{45})$, but they have different sizes in the plots since the maximum value of $-\chi_3^{\text{red}}$ is $\frac{1}{3}$ for $K > 0$ and $\frac{1}{16}$ for $K < 0$.)

  The upper panel of Fig. 16 shows χ_3^{red} vs. σ in the longitudinal and transverse field cases, as well as for anisotropy axes distributed at random. The three curves coincide at $\sigma = 0$, where the orientation of the magnetic field plays no role, taking the Langevin value $-\frac{1}{45}$. It is noticeable the large variation of χ_3^{red} with respect to σ for anisotropy axes parallel to the field. Note also that, although dramatically reduced, the anisotropy-induced temperature dependence of χ_3 is kept for anisotropy axes distributed at random. On the other hand, we can again remark that, *qualitatively*, the longitudinal and the transverse field cases interchange their rôles when the sign of the anisotropy is reversed (see also Fig. 15). For instance, for easy-plane anisotropy $\chi_{3,\parallel}$ rapidly vanishes as $|\sigma|$ departs from zero. The analogous result for easy-axis

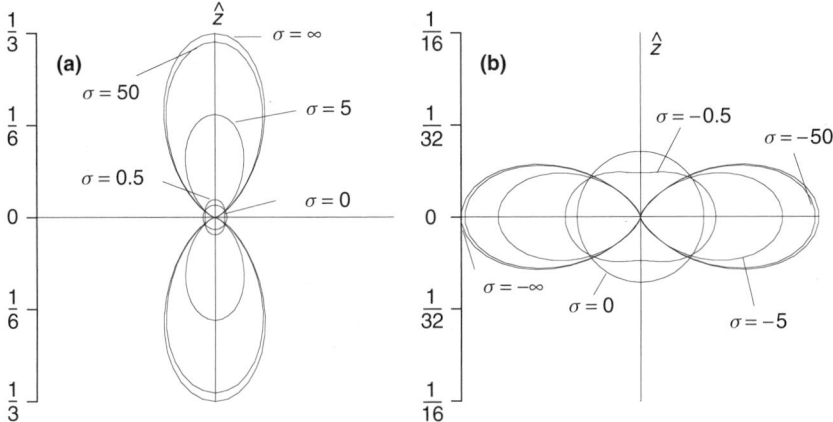

FIGURE 15. Polar plots showing the angular dependence of the reduced nonlinear susceptibility $-\chi_3^{\text{red}}$ [Eq. (3.97)] for various values of the dimensionless anisotropy parameter $\sigma = Kv/k_BT$. (a) easy-axis anisotropy; (b) easy-plane anisotropy.

anisotropy occurs in a transverse field; then $\chi_{3,\perp}$ rapidly decreases as σ departs from zero.

c. The Sign of the Nonlinear Susceptibility. As the nonlinear susceptibility is a measure of the initial departure of the magnetization from the linear field regime, and this departure usually consists of a bending downward, one is tempted to conclude that χ_3 is always a negative quantity. Indeed, Eq. (3.96) clearly shows that this is the case for $\langle \chi_3 \rangle_{\text{ran}}$ (in accordance with the downward bending of the corresponding magnetization in Fig. 11). However, this result is not general, as will be illustrated now with $\chi_{3,\perp}$. Let us calculate the low temperature ($\sigma \gg 1$) expression for $\chi_{3,\perp}$ by using the asymptotic methods of Appendix A:*

$$\chi_{3,\perp} = \frac{1}{16}\frac{\mu_0^3 m^4}{(k_BT)^3}\left[-1 + 2\frac{R'}{R} - 2\left(\frac{R'}{R}\right)^2 + \frac{R''}{R}\right] \simeq \frac{1}{16}\frac{\mu_0^3 m^4}{(k_BT)^3}\frac{1}{\sigma^4}$$

---

* As the first non-vanishing term in $\chi_{3,\perp}^{\text{red}}$ is of fourth order in $1/\sigma$ [see Eq. (A.30)], we need to compute one more coefficient b_i in the $\sigma \gg 1$ expansion of Appendix A. On doing this we get $b_4 = -\frac{37}{8}$, from which we obtain the fourth-order term of R'/R, and from this we can calculate the corresponding terms in $(R'/R)^2$ and R''/R.

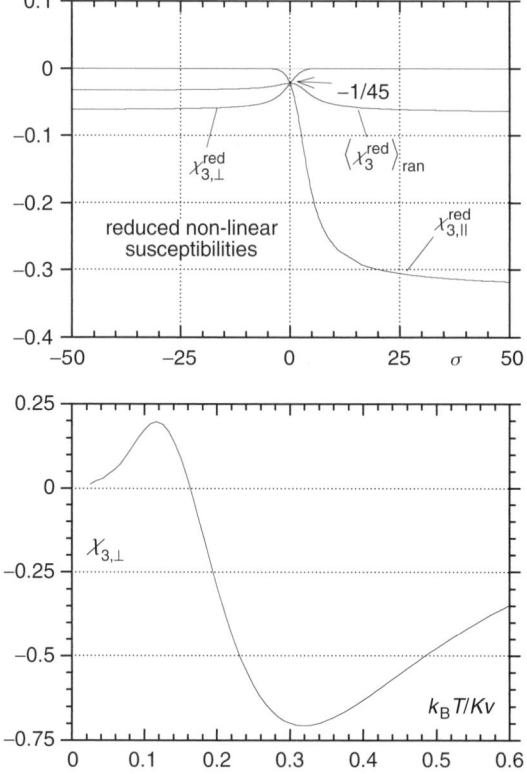

FIGURE 16. Upper panel: reduced nonlinear susceptibility (3.97) in the longitudinal, $\chi_{3,\parallel}^{\text{red}}$, and transverse, $\chi_{3,\perp}^{\text{red}}$, field cases, and for anisotropy axes distributed at random, $\langle\chi_3^{\text{red}}\rangle_{\text{ran}}$, versus the dimensionless anisotropy parameter $\sigma = Kv/k_BT$. Lower panel: temperature dependence of the transverse component of the nonlinear susceptibility [from Eq. (3.93)]. $\chi_{3,\perp}$ is measured in units of $m(\mu_0/B_K)^3$.

so that

$$\chi_{3,\perp} \simeq \frac{1}{2}m\left(\frac{\mu_0}{B_K}\right)^3\frac{k_BT}{Kv} \tag{3.98}$$

where $B_K = 2Kv/m$. Therefore, we see that, not only is $\chi_{3,\perp}$ positive at low temperatures, but it indeed increases linearly with T. At higher temperatures the expansion leading to Eq. (3.98) must break down and the corresponding corrections bring $\chi_{3,\perp}$ to the negative values that it must take at sufficiently high temperatures $[\chi_{3,\perp}|_{\sigma\ll1} \simeq \chi_{3,\text{Lan}} = -\mu_0^3m^4/45(k_BT)^3]$. Thus, from the knowledge of the limit temperature dependences ($\chi_{3,\perp} \propto T$ and $-1/T^3$), one concludes that $\chi_{3,\perp}$ must have two peaks and cross the temperature axis at a certain intermediate temperature. This is precisely what can be seen in the lower panel of Fig. 16, showing that $\chi_3 \leq 0$ is not a general result. As T decreases, $\chi_{3,\perp}$ has a negative minimum, increases, crosses zero, exhibits a secondary positive maximum, and eventually tends to zero from positive values at low temperatures. These are the typical features exhibited by the *dynamical* nonlinear susceptibility $\chi_3(\omega, T)$ (Raĭkher and Stepanov, 1997), but their occurrence in the *equilibrium* susceptibility is somewhat unexpected. This is another good example of the important effects of the magnetic anisotropy on the properties of superparamagnetic systems.

3. Generalizations

One can also derive the nonlinear susceptibility by means of the relation between the thermal equilibrium fluctuations of $\vec{m}$, in the absence of a probing field, and the actual magnetic response of the system, bypassing the explicit expansion of the magnetization in a series of powers of the field.

On inspecting the definition (3.41), one realizes that χ_3 can be obtained by differentiating the magnetization as $\chi_3 = \frac{1}{6}\mu_0^3 \partial^3 \langle \vec{m} \cdot \hat{b} \rangle_e / \partial B^3|_{B=0}$. This is directly generalized to

$$\chi_3 = \frac{1}{6}\mu_0^3 \left. \frac{\partial^3 \langle \vec{m} \cdot \hat{b} \rangle}{\partial(\Delta B)^3} \right|_{\Delta B=0} \tag{3.99}$$

where $\Delta \vec{B} = \Delta B \hat{b}$ is an external probing field and the averages are now taken with respect to the total energy of the system in the presence of $\Delta \vec{B}$.

On calculating the preceding third-order derivative by making repeated use of the Eq. (3.58), one arrives at the general result [cf. Eq. (3.60)]

$$\chi_3 = \frac{\mu_0^3}{(k_B T)^3} \frac{1}{6} \left[\left\langle (\vec{m} \cdot \hat{b})^4 \right\rangle_e - 4 \left\langle (\vec{m} \cdot \hat{b})^3 \right\rangle_e \langle \vec{m} \cdot \hat{b} \rangle_e - 3 \left\langle (\vec{m} \cdot \hat{b})^2 \right\rangle_e^2 \right.$$
$$\left. + 12 \left\langle (\vec{m} \cdot \hat{b})^2 \right\rangle_e \langle \vec{m} \cdot \hat{b} \rangle_e^2 - 6\langle \vec{m} \cdot \hat{b} \rangle_e^4 \right]$$

where the averages are finally taken in the absence of the probing field. Note, however, that if a bias field is applied, there is also a nonzero term in $(\Delta B)^2$, which defines the corresponding susceptibility χ_2 (see, for example, Raĭkher et al., 1997). Nevertheless, on assuming that no constant field is applied and noting that, consequently, the preceding averages at zero probing field are then zero-field averages, we can use $\langle (\vec{m} \cdot \hat{b})^{2n+1} \rangle_e|_{B=0} = 0$, to get

$$\chi_3 = \frac{\mu_0^3}{(k_B T)^3} \frac{1}{6} \left[\left\langle (\vec{m} \cdot \hat{b})^4 \right\rangle_e - 3 \left\langle (\vec{m} \cdot \hat{b})^2 \right\rangle_e^2 \right] \Bigg|_{B=0} \tag{3.100}$$

This relation between the nonlinear susceptibility and the thermal equilibrium fluctuations of the magnetic moment in zero field is valid for any form of the magnetic-anisotropy energy, provided this has inversion symmetry $\langle (\vec{m} \cdot \hat{b})^{2n+1} \rangle_e = 0$.

Finally, on returning to the uniaxial-anisotropy case and recalling that the zero-field averages of $(\vec{m} \cdot \hat{b})^{2i}$ are directly related with the coefficients C_i by Eq. (2.42), specifically

$$\left\langle (\vec{m} \cdot \hat{b})^2 \right\rangle_e \big|_{B=0} = m^2 2C_1, \qquad \left\langle (\vec{m} \cdot \hat{b})^4 \right\rangle_e \big|_{B=0} = m^4 12C_2$$

one gets

$$\frac{1}{6}\left[\left\langle (\vec{m} \cdot \hat{b})^4 \right\rangle_e - 3 \left\langle (\vec{m} \cdot \hat{b})^2 \right\rangle_e^2 \right]\bigg|_{B=0} = m^4 2(C_2 - C_1^2)$$

so that the expression (3.88) for χ_3 is reobtained.

4. Approximate Formulas for the Nonlinear Susceptibility

We now derive approximate expressions for χ_3, in an attempt to establish simple approximate expressions valid in wide temperature ranges. Again, in order to obtain the approximate formulas, we use the corresponding expressions for R'/R and R''/R derived in Appendix A. (We could also proceed from the weak- and strong-anisotropy formulas for M_B.) The approximate expressions for the combinations of the functions $R^{(l)}/R$ entering in the general formula (3.93) are given by Eqs. (A.25), (A.28), and (A.30).

a. Weak-Anisotropy Range. To obtain an approximate formula for χ_3 valid for weak anisotropy, we insert Eqs. (A.25) into Eq. (3.93), gather the terms with the same power of σ, and express the trigonometric factors in terms of $\cos^2 \alpha$ and $\cos^4 \alpha$ only, obtaining

$$\chi_3|_{|\sigma| \ll 1} \simeq -\frac{\mu_0^3 m^4}{45(k_B T)^3}\left[1 + \frac{8}{21}(3\cos^2 \alpha - 1)\sigma \right.$$

$$+ \frac{8}{105}(4\cos^4 \alpha - \cos^2 \alpha)\sigma^2$$

$$\left. + \frac{32}{10395}(21\cos^4 \alpha - 18\cos^2 \alpha + 4)\sigma^3 \right] \quad (3.101)$$

This equation is a good approximation of the exact χ_3 for $|\sigma| \leq 2$. Note that, in contrast to χ, only the *first* correction to the leading (isotropic) result vanishes when the anisotropy axes are distributed at random [recall Eq. (3.37)].

b. Strong-Anisotropy Ranges. Let us first consider the $\sigma \ll -1$ range. If we insert Eqs. (A.28) into Eq. (3.93) and gather the terms with the same power of $1/\sigma$, we obtain

$$\chi_3|_{\sigma \ll -1} \simeq -\frac{\mu_0^3 m^4 \sin^4 \alpha}{16(k_B T)^3}\left[1 + \frac{1}{\sigma} + (16\cot^2 \alpha - 1)\frac{1}{4\sigma^2}\right] \qquad (3.102)$$

This is the desired approximate formula for χ_3 valid in the easy-plane range. An approximate expression for $\sigma \gg 1$ can be obtained in a similar way. On inserting Eqs. (A.30) into Eq. (3.93) and gathering the terms with the same power of $1/\sigma$, we arrive at

$$\chi_3|_{\sigma \gg 1} \simeq -\frac{\mu_0^3 m^4 \cos^4 \alpha}{3(k_B T)^3}\left[1 - \frac{2}{\sigma} + (3\tan^2 \alpha - 1)\frac{1}{2\sigma^2} + (3\tan^2 \alpha - 4)\frac{1}{2\sigma^3}\right]$$
$$(3.103)$$

These strong-anisotropy equations match the corresponding exact results for $|\sigma| \geq 5$. In fact, with the combined use of Eqs. (3.101)–(3.103), one can almost cover the exact χ_3 in the whole temperature range. Again this arises directly from the reasonable patching shown in Appendix A of the exact R'/R and R''/R curves yielded by the approximate formulas employed.

5. Temperature Dependence of the Nonlinear Susceptibility

a. Theoretical results. We now study in more detail the temperature dependence of χ_3. Facing the subsequent particularization of the results to a number of systems of magnetic nanoparticles, we shall consider the occurrence of a distribution of particle volumes. We shall, however, take the anisotropy constant K and the spontaneous magnetization $M_s = m/v$ as fixed, that is, neither distribution in particle shape, nor size effects on M_s or K will be considered. Then, if the anisotropy axes of the particles with the same volume are distributed at random, one can write

$$\chi_3 = \int_0^\infty dv\, v^{-1} f(v)\langle \chi_3 \rangle_{\text{ran}}$$

where the factor v^{-1} occurs since $f(v)dv$ is taken as the fraction of the total volume occupied by particles with volumes in the interval $(v, v + dv)$.

In order to isolate the effect of the magnetic anisotropy on $\chi_3(T)$, we assume that M_s is independent of T. This condition, which is obeyed at temperatures well below the ordering temperature of the magnetic material constituting the particles, also yields temperature-independent anisotropy constants [this is apparent when the anisotropy is due to the magnetostatic self-energy; see Eq. (2.5)]. The computed quantity will be the dimensionless

$\tilde{\chi}_3 = \chi_3[K^3/(\mu_0^3 M_s^4)]$ and we shall employ a logarithmic-normal distribution for $f(v)$, namely

$$f(v) = \frac{1}{\sqrt{2\pi}\rho_v v} \exp\left\{ -\frac{[\ln(v/v_{\rm m})]^2}{2\rho_v^2} \right\}$$

where $v_{\rm m}$ is the *median* of the distribution and ρ_v is the standard deviation of $\ln(v)$.

Figure 17 displays χ_3 and the corresponding Ising and isotropic results vs. the temperature. As the influence of the anisotropy decreases with increasing T, χ_3 undergoes a smooth crossover from the low-temperature Ising regime to the high-temperature isotropic regime. For $\sigma_{\rm m} \gg 1$ ($\sigma_{\rm m} = Kv_{\rm m}/k_{\rm B}T$) and $|\sigma_{\rm m}| \ll 1$, the logarithmic slope $d\ln(-\chi_3)/d\ln(1/\sigma_{\rm m})$ tends to -3, indicating the limit T^{-3} dependences. However, logarithmic slopes lesser than -3 emerge in the transitional regime, where the departure of $\chi_3(T)$ from an inverse-temperature-cubed law is sizable. The rate of change of $\chi_3(T)$, moreover, increases as the anisotropy axes are aligned toward $\vec{B}$ (see Fig. 18). To illustrate, for a narrow volume distribution ($\rho_v = 0.25$), the maximum logarithmic slope changes from -3.53 for anisotropy axes distributed at random (see the lower panel of Fig. 17) to -3.98 for axes collinear with the field. Considering these significant deviations of $\chi_3(T)$ from a T^{-3} dependence, arguments discarding superparamagnetism based on this type of departure, such as those employed by Schiffer et al. (1995), should be carefully scrutinized.

Note finally that, when observed over limited temperature windows (e.g., those imposed by the unavoidably finite measurement time), an increase of the equilibrium $\chi_3(T)$ steeper than T^{-3} could resemble the high-temperature range of a quantity with a low-temperature divergence. This might misleadingly suggest the presence of appreciable interparticle interactions in the ensemble and, consequently, one could try a fit of the nonlinear susceptibility to, for example, $\chi_3(T) \propto (T - T_{\rm c})^{-3}$, obtaining false "critical" temperatures. If we do so with the $\chi_3(T)$ *theoretically* computed for the most diluted sample of Jonsson et al. (1995) over $100\,{\rm K} \leq T \leq 180\,{\rm K}$, we get the sizable value $T_{\rm c} \simeq 17.3\,{\rm K}$ (regression of the fit 0.99992). Note that, for $\omega/2\pi \sim 1\text{--}10^3$ Hz, effects associated with the finite measurement time appear at $T \lesssim 40\text{--}100$ K, below which one cannot measure the equilibrium χ_3.

b. Comparison with Experimental Data. Bitoh et al. (1993, 1995) measured the nonlinear dynamical susceptibility $\chi_3(\omega, T)$ for cobalt particles precipitated in a $Cu_{97}Co_3$ alloy. From the equilibrium (high-temperature) part of the χ_3 vs. T curve they obtained a mean logarithmic slope -3.17, whose departure from -3 was not accounted for.

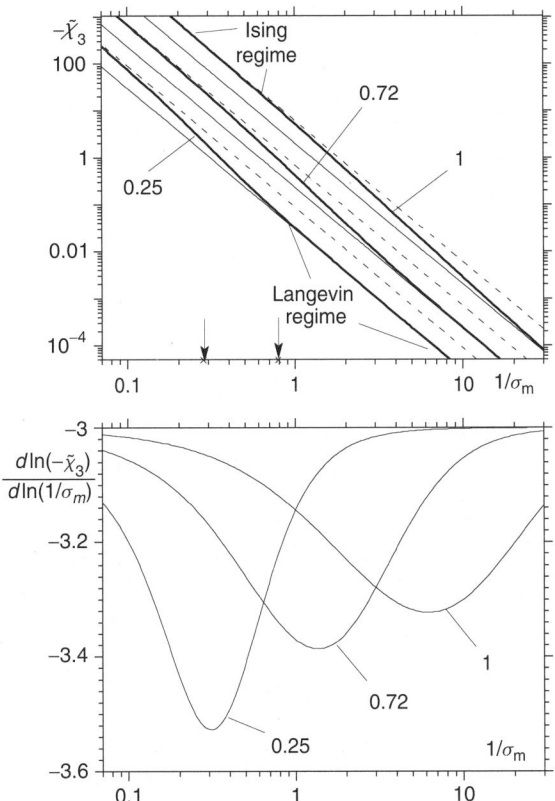

FIGURE 17. Upper panel: log–log plot of $-\tilde{\chi}_3$ versus $1/\sigma_m$ ($= k_B T/K v_m$) for a system with randomly distributed anisotropy axes. The straight lines correspond to the isotropic (thin solid) and Ising (dashed) nonlinear susceptibilities. The numbers mark the width ρ_v of the volume distribution. The mean slope of the $\rho_v = 0.72$ curve between the arrows is compared with the experiment of Bitoh et al. (1993) in the text. Lower panel: logarithmic slopes.

Their sample appears suitable to check the studied deviation of χ_3 from a T^{-3} law, since

1. The high Curie temperature of the particles ($\simeq 1400$ K) yields M_s feebly dependent on T in the range of the experiment (≤ 280 K).

2. The equilibrium linear susceptibility can be fitted to a Curie law with a mean logarithmic slope $\langle d \ln \chi / d \ln T \rangle = -1.01$, compatible with the absence of dipole–dipole interaction effects and anisotropy axes distributed at random.

Nevertheless, one could still argue that, as a result of finite size effects, the temperature dependence of the spontaneous magnetization of the cobalt

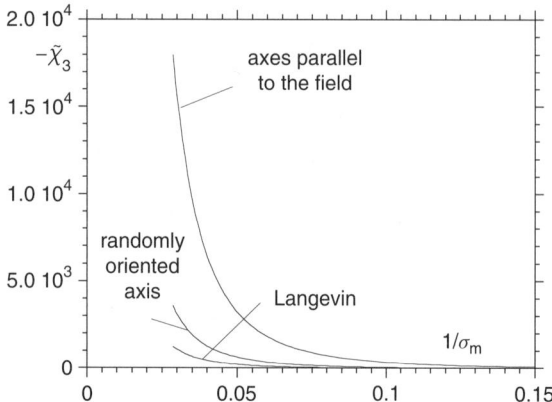

FIGURE 18. Effect of the alignment of the anisotropy axes toward $\vec{B}$ on the temperature dependence of the nonlinear susceptibility. The width of the volume distribution is $\rho_v = 0.25$.

particles could be larger than that of the bulk material, so that the measured temperature dependence of χ_3 could be attributed to such a phenomenon. However, the ascription of the extra $T^{-0.17}$ factor in $\chi_3(T)$ to $M_s(T)^4$ entails the occurrence of its square root in the Curie law [$\chi \propto M_s(T)^2$], yielding a total exponent $-(1 + 0.17/2) = -1.085$ for χ, which is not consistent with the measured one (-1.01).

Unfortunately, the high amplitude of the oscillating field employed in their experiment ($v_m M_s \Delta B/k_B \simeq 17$ K) might have induced nonlinear "saturation" effects on the measured susceptibilities at low temperatures, moving the volume distribution $f(v)$ that they derived from the $\chi(\omega, T)$ data, from the actual one. Even so, we have specialized the preceding calculation of $\chi_3(T)$ to the so-derived logarithmic normal $f(v)$. The temperature range of their experiment, in the dimensionless units $k_B T/K v_m$, is delimited in Fig. 17 by the arrows. Our calculation yields a mean logarithmic slope -3.25 that is within 3% of the experimentally determined value -3.17. One must anyway conclude that the sizable departure of the *theoretical* exponent from -3, renders mandatory the inclusion of anisotropy effects on the temperature dependence of χ_3 to achieve a complete understanding of this kind of experiments.

c. Proposed Experiments. In addition to searching for deviations of $\chi_3(T)$ from a T^{-3} law, we could measure the dependence of χ_3 on the angle between the anisotropy axis and the probing field in systems with oriented anisotropy axes. (Molecular magnetic clusters and textured frozen magnetic fluids are examples of systems with parallel axes where such experiments could be performed.) In a polar plot (see Fig. 15), $\chi_3(\alpha)$ will undergo an increasing deformation from a circle at high temperatures (isotropic χ_3) toward the characteristic two-looped shape of the Ising regime ($\chi_3|_{\text{Ising}} \propto \cos^4 \alpha$) as T decreases.

Other possible experiment could be to measure $\chi_3(T)$ in a magnetic fluid through the freezing point of the solvent, T_f. Recall that, due to the physical rotation of the particles in the fluid state, the magnetization is given by the Langevin law for each particle, irrespective of the anisotropy energy (Krueger, 1979). On the other hand, at temperatures below the freezing point, the anisotropy axes become immobilized; the magnetic anisotropy then takes reflection in the equilibrium quantities, and $\chi_3(T)$ would undergo a discontinuous change at T_f.* In contrast, if at T_f the anisotropy axes become immobilized in a random pattern, the linear equilibrium susceptibility would be continuous there [recall that $\langle\chi\rangle_{\text{ran}}$ does not depend on the anisotropy energy in a solid dispersion].

The relative size of the discontinuity in the nonlinear susceptibility $\Delta\chi_3/\chi_3$ at the freezing temperature is determined by the value of T_f in magnetic-anisotropy units, so that the size of the jump also depends on the anisotropy constants and the actual volume distribution. We have computed $\Delta\chi_3/\chi_3$ with the parameters of two magnetic fluids in the literature. First, for most diluted sample of Luo et al. (1991), $\Delta\chi_3/\chi_3$ would be small, because the freezing point of the carrier liquid is close to the isotropic regime. In contrast, for the most diluted sample of Jonsson et al. (1995), $\Delta\chi_3/\chi_3$ would be about 90%. Once more, if the anisotropy axes are frozen collinear with $\vec{B}$, this effect will be even more dramatic.[†]

IV. DYNAMICAL PROPERTIES: HEURISTIC APPROACH

A. Introduction

In this section we briefly consider a heuristic approach to the dynamics of classical magnetic moments in anisotropy potentials. We shall focus on the linear dynamical response, namely, the response of the system to a small-amplitude, oscillating or constant, magnetic field. The responses to both types of stimulus are related in a simple way, so that we merely employ the language of the linear dynamical response in the frequency domain — the linear dynamical susceptibility $\chi(\omega)$. This quantity, in addition to supplying valuable information about the intrinsic dynamics of the spins, is of relevance for general studies on

---

* This jump could be smeared out around T_f as a result of effects related with the immediacy of the critical point of the carrier liquid.

[†] However, for anisotropy axes *not* distributed at random, $\chi(T)$ would also exhibit a discontinuity at the freezing point of the magnetic fluid.

magnetic nanoparticle systems. For instance, under certain conditions $\chi(\omega)$ can be used to approximately determine the distribution of energy barriers (essentially particle volumes), occurring in assemblies of non-interacting magnetic nanoparticles (Shliomis and Stepanov, 1994). Besides, a rough estimate of the preexponential factor of the longitudinal relaxation time in the Arrhenius regime can also be derived from the $\chi(\omega)$ data.

The organization of this section is as follows. In Section IV.B various heuristic expressions that have been proposed to describe the linear dynamical response are discussed (they are compared with exact numerical results in Section V). In Section IV.C, the most general of those expressions is analyzed, illustrating how it can be used to get the energy-barrier distribution of magnetic nanoparticle ensembles. Finally, in Section IV.D some of the previous results will be illustrated with experiments performed on a frozen magnetic fluid of maghemite (γ-Fe$_2$O$_3$) nanoparticles. Part of the results of this section were presented by Svedlindh, Jonsson and García-Palacios (1997).

B.   Heuristic Treatment of the Linear Dynamical Response

Let us commence by considering the expression (3.54) for the linear *equilibrium* susceptibility in terms of its longitudinal and transverse components, namely

$$\chi = \chi_\| \cos^2 \alpha + \chi_\perp \sin^2 \alpha \qquad (4.1)$$

where α is the angle between the anisotropy axis and the probing field. The term $\chi_\| \cos^2 \alpha$ is proportional to the projection along the probing-field direction of the response of the magnetic moment to the longitudinal component (with respect to the anisotropy axis) of the probing field. Likewise, $\chi_\perp \sin^2 \alpha$ is proportional to the projection onto the probing field of the response of the spin to the transverse component of the field. As we know from Section III.D, averaging this equation with $\chi_\|$ and $\chi_\perp$ from Eq. (3.53), one gets $\langle \chi \rangle_{\text{ran}} = \mu_0 m^2 / 3 k_B T$. Consequently, in a noninteracting magnetic nanoparticle ensemble with anisotropy axes distributed at random, the linear *equilibrium* susceptibility *in the absence of an external bias field* is independent of the magnetic anisotropy (χ is then identical with that derived in a naive superparamagnetic model where the anisotropy is neglected). The main effect of the anisotropy is to introduce energy barriers that the spins need to overcome before equilibrium is reached, implying that the ensemble could, depending on the measurement time, display *magnetic relaxation*.

The relaxational mechanism consists of an orientational redistribution of the magnetic moments according to the conditions set by the magnetic anisotropy, temperature, and external field. The relaxation can be envisaged as a two-stage process: (1) the dipoles first redistribute inside the potential wells, with a characteristic time $\tau_\perp$ related with the inverse of the maximum precession frequency of the magnetic moments in the anisotropy field ($\sim 10^{-10} - 10^{-12}$ s); and then (2) the equilibration between the potential wells, which involves thermally activated rotations over the energy barriers, proceeds. This second mechanism can result in exceedingly slow magnetic relaxation since its characteristic time $\tau_{\parallel}$, which essentially follows an Arrhenius law [see Eq. (2.1)], ranges from picoseconds to geological timescales depending on the magnetic anisotropy, temperature, and external field.

A rigorous theoretical derivation of the linear *dynamical* susceptibility of classical spins in anisotropy potentials, as well as other dynamical quantities, is hindered by a number of mathematical difficulties (see Section V). Therefore, in order to describe the linear dynamical response of noninteracting magnetic nanoparticles, various simple expressions have been proposed in the literature. We shall mainly consider the expression suggested, on the basis of the two-stage relaxation process mentioned, by Shliomis and Stepanov (1993) to describe $\chi(\omega)$ at frequencies below the ferromagnetic-resonance frequency range. Besides, we shall show that this model contains as particular cases some models previously proposed.

In a study of magnetic fluids Shliomis and Stepanov suggested that $\chi(\omega)$ could be described as a sum of two independent Debye-type relaxation mechanisms: one for the response to the longitudinal component of the probing field and the other for the response to the transverse component (see also Raĭkher and Stepanov, 1997). The expression proposed can be generalized in order to describe the effect of a longitudinal *bias* field by merely writing

$$\chi_{\text{ShS}} = \frac{\chi_{\parallel}(T, B)}{1 + i\omega\tau_{\parallel}} \cos^2 \alpha + \frac{\chi_{\perp}(T, B)}{1 + i\omega\tau_{\perp}} \sin^2 \alpha \qquad (4.2)$$

where $\chi_{\parallel}$ and $\chi_{\perp}$ are the equilibrium susceptibilities (3.70).

Various expressions can be used for the characteristic times appearing in Eq. (4.2) (see Section V.C). However, for the purposes of this section it is sufficient to consider that in the high-barrier range $\tau_{\parallel}$ can be written in the Arrhenius form $\tau_{\parallel} = \tau_0 \exp(\Delta U/k_B T)$, where τ_0 is assumed to be a constant $\sim 10^{-10} - 10^{-12}$ s (i.e., we disregard the dependences of the preexponential factor on the temperature, external field, and the parameters of the particles in comparison with the dependences of the exponential term). Concerning the transverse relaxation time, for moderate frequencies (say, $\omega \lesssim 10^6$ Hz), the condition $\omega\tau_{\perp} \ll 1$ holds (Section V.C). One can then approximate $1/(1 + i\omega\tau_{\perp})$ by unity in Eq. (4.2), to get the *low-frequency* equation

$$\chi_{\text{ShS}}|_{\omega\tau_{\perp} \ll 1} \simeq \frac{\chi_{\parallel}}{1 + i\omega\tau_{\parallel}} \cos^2 \alpha + \chi_{\perp} \sin^2 \alpha \qquad (4.3)$$

The approximation used is equivalent to assume from the outset that the response to the transverse components of the probing field is instantaneous. In fact, very short measurement times, such as those obtained in neutron scattering or ferromagnetic resonance experiments, are required to probe the intra-potential-well dynamics (see Table I).

From now on we refer to Eq. (4.2), with the *exact* equilibrium susceptibilities, as the *Shliomis–Stepanov equation*. Further, the formula obtained when in the low-frequency Eq. (4.3) one uses the *high-barrier approximations* (3.85) and (3.86) for the equilibrium susceptibilities, will be called the *Gittleman–Abeles–Bozowski* equation (1974), since it properly generalizes their formula to $B \neq 0$ and an arbitrary anisotropy-axis orientation. Let us show this. On introducing Eqs. (3.85) and (3.86) evaluated at $B = 0$ [i.e., Eqs. (3.82)] into Eq. (4.3), one gets

$$\chi_{\mathrm{GAB}} \simeq \left[\frac{\mu_0 m^2}{k_\mathrm{B} T} \cos^2 \alpha + \frac{\mu_0 m^2}{Kv} \left(\frac{3}{2} \sin^2 \alpha - 1\right) + i\omega\tau_\| \frac{\mu_0 m^2}{2Kv} \sin^2 \alpha\right] \frac{1}{1 + i\omega\tau_\|} \tag{4.4}$$

which, when averaged over an ensemble with randomly distributed anisotropy axes (the second term in the square brackets then vanishes), reduces to the equation proposed by Gittleman et al. Finally, the expression obtained when one introduces the Ising-type Eqs. (3.87) into Eq. (4.2) [or Eq. (4.3)], namely

$$\chi_{\mathrm{Ising}} = \frac{\mu_0 m^2}{k_\mathrm{B} T} \frac{1}{\cosh^2 \xi} \frac{\cos^2 \alpha}{1 + i\omega\tau_\|} \tag{4.5}$$

will be called the *discrete-orientation* or *Ising dynamical susceptibility*.

C. Analysis of the Low-Frequency Shliomis–Stepanov Model

We now analyze the low-frequency Eq. (4.3) for an ensemble of noninteracting magnetic nanoparticles where there exists a distribution in particle parameters. If the distribution occurs mainly in one of the parameters, say, the volumes of the particles, and one assumes that the contribution of each particle to the linear susceptibility is given by an expression such as Eq. (4.3), one can write the linear susceptibility of the (unbiased) ensemble as

$$\chi(\omega, T) = \frac{\mu_0 M_s^2}{k_\mathrm{B} T} \frac{1}{K} \int_0^\infty dE\, f(E) E \left[\frac{R'}{R} \frac{\langle \cos^2 \alpha\rangle}{1 + i\omega\tau_\|} + \frac{R - R'}{2R} \langle \sin^2 \alpha\rangle\right] \tag{4.6}$$

In this equation the functions $R^{(l)}$ are evaluated at $\sigma = E/k_\mathrm{B} T$, $E = Kv$ (with K assumed equal for all particles) and $f(E)\,dE$ is the fraction of the total "magnetic" volume occupied by those particles with energy barriers in the interval $(E, E + dE)$. Note that the square of the magnetic moment has been written in terms of the spontaneous magnetization M_s as $m^2 = M_s^2 v^2$ and, since we are using the "occupied volume" representation of the distribution, one v is already incorporated into $f(E)$.

In Eq. (4.6), the orientational averages are taken with respect to the particles in $(E, E + dE)$ and could, in principle, depend on E. We shall not study this situation but merely consider that $\langle \cos^2 \alpha \rangle$ and $\langle \sin^2 \alpha \rangle$ are the same for each energy interval. One could also consider the cases where, due to finite size effects, M_s and K depend on v. Although this could be incorporated in the following considerations, we shall not take those dependences into account explicitly.

1. The Out-of-Phase Linear Dynamical Susceptibility and the Energy-Barrier Distribution

The out-of-phase component (imaginary part) of Eq. (4.6) reads

$$\chi''(\omega) = \frac{\mu_0 M_s^2 \langle \cos^2 \alpha \rangle}{k_B T} \frac{1}{K} \int_0^\infty dE \, f(E) E \frac{R'}{R} \frac{\omega \tau_\parallel}{1 + (\omega \tau_\parallel)^2} \tag{4.7}$$

to which the response to the transverse components of the probing field (with respect to each anisotropy axis) does not contribute due to the low-frequency assumption $\omega \lesssim 10^6$ Hz.

The term $\omega \tau_\parallel / [1 + (\omega \tau_\parallel)^2]$ in the integrand of Eq. (4.7), has a maximum at the energy barrier E_b for which $\omega \tau_\parallel = 1$ (see Fig. 19). On assuming a simple Arrhenius form for the relaxation time, $\tau_\parallel = \tau_0 \exp(E/k_B T)$, one finds $E_b = -k_B T \ln(\omega \tau_0)$, which explicitly depends on the temperature and the frequency. Besides, as a result of the *exponential* dependence assumed for $\tau_\parallel$, it follows from the definition of E_b that (1) $\tau_\parallel(E) \ll \tau_\parallel(E_b) = 1/\omega$, if $E < E_b$, whereas (2) $\tau_\parallel(E) \gg \tau_\parallel(E_b) = 1/\omega$, if $E > E_b$. By virtue of these properties, and considering that $1/\omega$ is the *measurement time* t_m in a dynamical experiment, E_b is called the *blocking barrier* (recall the considerations in Section I concerning the relative values of $\tau_\parallel$ and t_m). Similarly, one can define the corresponding *dimensionless blocking barrier* $\sigma_b = E_b/k_B T$, whence

$$E_b = -k_B T \ln(\omega \tau_0), \qquad \sigma_b = -\ln(\omega \tau_0) \tag{4.8}$$

For these two quantities one has, by definition, $\omega \tau_\parallel = 1$.

We shall not consider the finite height and width of the function $\omega \tau_\parallel / [1 + (\omega \tau_\parallel)^2]$, but we consider this function as a (unnormalized) Dirac delta centered at σ_b. This replacement works when the remainder terms in the integrand of the formula for $\chi''(\omega)$ (e.g., the energy-barrier distribution) change slowly enough in the interval about σ_b where $\omega \tau_\parallel / [1 + (\omega \tau_\parallel)^2]$ differs appreciably from zero. Concerning the term R'/R, when $\sigma_b \gtrsim 15-25$, its changes are not very large, because

$$\frac{d}{d\sigma}\left(\frac{R'}{R}\right) = \frac{R''}{R} - \left(\frac{R'}{R}\right)^2 \overset{\text{Eq. (A.31)}}{\simeq} \frac{1}{\sigma^2} \sim \frac{1}{400}, \qquad \text{for} \quad \sigma \sim 20$$

Values of $\sigma_b \gtrsim 15$–25 are typical for probing fields with $\omega \lesssim 10^3$ Hz. Under these conditions, $\omega\tau_\parallel/[1 + (\omega\tau_\parallel)^2]$ plays the role of a function proportional to a Dirac delta. In order to calculate the proportionality factor, one integrates that function over the entire energy range by means of the substitution $d\tau_\parallel = (\tau_\parallel/k_BT)\,dE$

$$\int_0^\infty dE\,\frac{\omega\tau_\parallel}{1 + (\omega\tau_\parallel)^2} = k_BT \int_{\tau_0}^\infty d\tau_\parallel\,\frac{\omega}{1 + (\omega\tau_\parallel)^2} \simeq \frac{\pi}{2}k_BT$$

where, on considering the low-frequency assumption ($\omega \lesssim 10^6$ Hz) and taking the tiny value of τ_0 ($\sim 10^{-10}$–10^{-12} s) into account, we have used the approximation $\arctan(\omega\tau_0) \lesssim \arctan(10^{-4}$–$10^{-6}) \simeq 0$. Therefore, when integrating functions slowly varying about $E_b = -k_BT\ln(\omega\tau_0)$, one can make use of the approximation

$$\frac{\omega\tau_\parallel}{1 + (\omega\tau_\parallel)^2} \simeq \frac{\pi}{2}k_BT\delta(E - E_b) \tag{4.9}$$

Thus, on calculating the integral in Eq. (4.7) by means of Eq. (4.9), one obtains (cf. Eq. (41) by Shliomis and Stepanov, 1994)

$$\chi''(\omega, T) = \frac{\pi}{2}\frac{\mu_0 M_s^2}{K}\langle\cos^2\alpha\rangle\frac{R'(\sigma_b)}{R(\sigma_b)}f(E_b)E_b \tag{4.10}$$

which directly relates the energy-barrier distribution and the out-of-phase linear dynamical susceptibility. Note that, even if we consider the weak temperature dependence of τ_0, this is further weakened when occurring inside the logarithm $\ln(\omega\tau_0)$, so that $\sigma_b = -\ln(\omega\tau_0)$ and thus the factor $R'(\sigma_b)/R(\sigma_b)$, are almost independent of T. Then, because $R'(\sigma_b)/R(\sigma_b)$ is also weakly dependent on ω, Eq. (4.10) shows that, approximately, all the dependence of χ'' on T and ω enters via the combination $E_b = -k_BT\ln(\omega\tau_0)$.[*] Therefore, if we plot χ'' versus $-k_BT\ln(\omega\tau_0)$, all the $\chi''(T)$ curves corresponding to different frequencies collapse onto a single "master" curve [proportional to

---

[*] We are also implicitly assuming that $d(M_s^2/K)/dT \simeq 0$. For instance, for the "shape" anisotropy of ellipsoids of revolution, M_s^2/K is in fact a geometric term [see Eq. (2.5)].

$f(E)E$ and with maximum at E_M]. Conversely, by fitting the frequency-dependent temperature of the maximum of $\chi''(T)$, denoted by $T_M(\omega)$, to the "Arrhenius law" $E_M = -k_B T_M(\omega)\ln(\omega\tau_0)$, one can get E_M and τ_0.

Note, however, that the parameter E_M, which is sometimes called "average energy barrier," is merely the maximum of the function $f(E)E$. Therefore, it is not necessarily related with a characteristic parameter of the energy-barrier distribution (incidentally, for the gamma and logarithmic-normal distributions E_M is equal to the *mean* and the *median* of the distribution, respectively).

2. The In-Phase Linear Dynamical Susceptibility

The in-phase component (real part) of Eq. (4.6) is given by

$$\chi'(\omega) = \frac{\mu_0 M_s^2}{k_B T}\frac{1}{K}\int_0^\infty dE\, f(E)E\left[\frac{R'}{R}\frac{\langle\cos^2\alpha\rangle}{1+(\omega\tau_\parallel)^2} + \frac{R-R'}{2R}\langle\sin^2\alpha\rangle\right]$$

$$(4.11)$$

where, because of the low-frequency assumption ($\omega\lesssim 10^6$ Hz), the response to the transverse components of the probing field contribute to $\chi'(\omega)$ with its thermal equilibrium value.

The term $1/[1+(\omega\tau_\parallel)^2]$ as a function of $\sigma = E/k_B T$ has the form of a smooth step about σ_b, whose width is of the order of the width of the peak of $\omega\tau_\parallel/[1+(\omega\tau_\parallel)^2]$ (see Fig. 19). However, when that term is under the integral sign and multiplied by functions that vary slowly around σ_b, we can

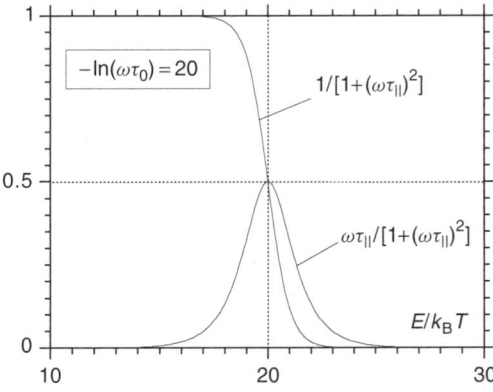

FIGURE 19. Real and imaginary parts of the Debye factor $1/(1+i\omega\tau_\parallel)$ versus the energy-barrier height $E/k_B T$. The relaxation time is given by the Arrhenius law $\tau_\parallel = \tau_0\exp(E/k_B T)$.

approximate $1/[1 + (\omega\tau_\parallel)^2]$ by a step function, namely

$$\frac{1}{1 + (\omega\tau_\parallel)^2} \simeq \begin{cases} 1 & \text{for} \quad E < E_\text{b} \\ 0 & \text{for} \quad E > E_\text{b} \end{cases} \tag{4.12}$$

Thus, on introducing Eq. (4.12) into Eq. (4.11) and rearranging the integration limits, one gets the approximate result

$$\chi'(\omega) = \frac{\mu_0 M_s^2}{k_\text{B} T} \frac{1}{K} \int_0^{E_\text{b}} dE\, f(E) E \left[\frac{R'}{R} \langle \cos^2 \alpha \rangle + \frac{R - R'}{2R} \langle \sin^2 \alpha \rangle \right]$$

$$+ \frac{\mu_0 M_s^2}{k_\text{B} T} \frac{1}{K} \int_{E_\text{b}}^\infty dE\, f(E) E \frac{R - R'}{2R} \langle \sin^2 \alpha \rangle \tag{4.13}$$

which can be interpreted as follows. Note first that only the particles with $E < E_\text{b}$, that is, those obeying $\tau_\parallel(E) \ll \tau_\parallel(E_\text{b}) = 1/\omega$, contribute to the first term. However, $1/\omega$ is the measurement time in a dynamical experiment, so that those particles are the *superparamagnetic* ones ($\tau_\parallel \ll 1/\omega$), and the first term is indeed their contribution to the linear *equilibrium* susceptibility. On the other hand, the particles with $E > E_\text{b}$, which are those contributing to the second term, satisfy $\tau_\parallel(E) \gg \tau_\parallel(E_\text{b}) = 1/\omega$, so that the overbarrier rotation process is not effective for them. These are the *blocked* particles, and contribute to $\chi'(\omega)$ via the fast rotations of their magnetic moments *inside* the potential wells toward the transverse components of the field. In fact, the second term in Eq. (4.13) is $\langle \sin^2 \alpha \rangle$ times the equilibrium transverse susceptibility of the blocked particles.

We finally note that, since $E_\text{b} = -k_\text{B} T \ln(\omega\tau_0)$, which appears in the integration limits of Eq. (4.13), the second term in this equation is small in comparison with the first one at sufficiently high temperatures. Then χ' is approximately equal to the equilibrium susceptibility, and, for anisotropy axes distributed at random, we can write

$$\langle \chi' \rangle_\text{ran}|_{\text{high } T} = \frac{\mu_0 M_s^2}{3k_\text{B} T} \frac{1}{K} \int_0^\infty dE\, f(E) E \equiv \frac{C}{T} \tag{4.14}$$

where C is the Curie constant.

3. The $\pi/2$ Law

We now explicitly derive, starting from the low-frequency Shliomis–Stepanov equation (4.6), a celebrated relation between $\partial\chi'/\partial \ln \omega$ and χ'' known as the $\pi/2$ *law*.

First, on rearranging the integration limits in Eq. (4.13), we can write $\chi'(\omega)$ as

$$\chi'(\omega) = \frac{\mu_0 M_s^2 \langle \cos^2 \alpha \rangle}{k_B T} \frac{1}{K} \int_0^{E_b} dE \, f(E) E \frac{R'}{R}$$

$$+ \frac{\mu_0 M_s^2 \langle \sin^2 \alpha \rangle}{k_B T} \frac{1}{K} \int_0^\infty dE \, f(E) E \frac{R - R'}{2R} \qquad (4.15)$$

where the last term, which is $\langle \sin^2 \alpha \rangle$ times the transverse equilibrium susceptibility of the *whole* ensemble, does not depend on ω. Then, on using $\partial E_b / \partial \ln \omega = -k_B T$ and the *Leibnitz formula*

$$\frac{d}{dx} \int_{g(x)}^{h(x)} dt \, F(x, t) = \{F[x, h(x)] h'(x) - F[x, g(x)] g'(x)\} + \int_{g(x)}^{h(x)} dt \frac{\partial}{\partial x} F(x, t)$$

$$(4.16)$$

one gets

$$\frac{\partial \chi'}{\partial \ln \omega} = -\frac{\mu_0 M_s^2}{K} \langle \cos^2 \alpha \rangle \frac{R'(\sigma_b)}{R(\sigma_b)} f(E_b) E_b \qquad (4.17)$$

Finally, on comparing this equation with Eq. (4.10), we obtain the desired relation between $\partial \chi' / \partial \ln \omega$ and χ'', namely

$$\chi'' = -\frac{\pi}{2} \frac{\partial \chi'}{\partial \ln \omega} \qquad (4.18)$$

For systems with a sufficiently wide distribution of relaxation times, the $\pi/2$ law is in fact a quite general result and independent of the dynamical model used, since it can then be derived from the Kramers–Krönig relations. These relations are merely based on general principles as the *linearity of the response*, and *causality* (i.e., the response at time t depends only on the values of the stimulus at times $t' < t$). For the sake of completeness, we shall repeat here one such derivation of the $\pi/2$ law by Böttcher and Bordewijk (1978, p. 58).

On writing one of the Kramers–Krönig relations in the form

$$\chi'(\omega) = \chi_S + \frac{2}{\pi} \int_0^\infty d\tilde{\omega} \frac{\tilde{\omega} \chi''(\tilde{\omega})}{\tilde{\omega}^2 - \omega^2} = \chi_S + \frac{2}{\pi} \int_{-\infty}^\infty d(\ln \tilde{\omega}) \frac{\tilde{\omega}^2 \chi''(\tilde{\omega})}{\tilde{\omega}^2 - \omega^2}$$

where χ_S is the adiabatic ($\omega \to \infty$) susceptibility ($\chi_\perp$ in our case), and approximating in the last integral the factor $\tilde{\omega}^2 / (\tilde{\omega}^2 - \omega^2)$ by a unit step function (with step at ω), one obtains

$$\chi'(\omega) \simeq \chi_S + \frac{2}{\pi} \int_{\ln \omega}^\infty d(\ln \tilde{\omega}) \chi''(\tilde{\omega})$$

Then, on differentiating this equation with respect to $\ln \omega$ by means of the Leibnitz formula (4.16), one finally gets the $\pi/2$ law.   Q.E.D.

The assumption of broad relaxation-time spectrum enters implicitly when approximating the factor $\tilde{\omega}^2/(\tilde{\omega}^2 - \omega^2)$ by a step function; the broad spectrum entails flat curves for $\chi''(\omega)$, so that the replacement mentioned does not introduce a significant error. This approximation is equivalent to the assumptions made above concerning the change of the functions appearing in the integrand of the equations for $\chi(\omega)$ in the range where the Debye factor has its maximum variation.

4. $\partial(T\chi')/\partial T$ and Its Relation with χ'' and the Energy-Barrier Distribution

Wohlfarth (1979), when studying spin glasses in the context of the superparamagnetic cluster model, proposed a method to obtain the energy-barrier distribution directly from the derivative $\partial(T\chi')/\partial T$. He considered a distribution of "blocking temperatures," which in our notation are $T_b = E_b/k_B$ ("blocking energies" in temperature units), disregarded the contribution of the blocked clusters, and wrote

$$\chi(T) \simeq \frac{C}{T} \int_0^T dT_b\, f(T_b) \qquad (4.19)$$

Here C is the Curie constant, and the susceptibility is the nonequilibrium susceptibility obtained in a dc experiment with a typical measurement time ~ 100 s. Then, by means of the inversion procedure [see the Leibnitz formula (4.16)]

$$f(T) = \frac{1}{C} \frac{\partial(T\chi)}{\partial T} \qquad (4.20)$$

he expressed the distribution of blocking temperatures in terms of the linear susceptibility.

Note that Eq. (4.19) can be considered as the particular case of Eq. (4.13) where the anisotropy axes are distributed at random (the term in the square brackets in the first integral then equals $\frac{1}{3}$) and the second integral (the $\chi_\perp$ of the blocked clusters) is neglected (Ising-type case). Besides, in order to establish this correspondence we must assume that his $f(T_b)$ incorporates the extra energy factor, that is, that $f(T_b) \propto f(E)E$.

Later on Lundgren, Svedlindh and Beckman (1981) derived a relation between χ'' and $\partial(T\chi')/\partial T$ for the following model

$$\chi(\omega) = \int_{\ln \tau_{min}}^{\ln \tau_{max}} d(\ln \tau) g(\tau) \chi(\tau) \frac{1}{1 + i\omega\tau} \qquad (4.21)$$

where $\chi(\tau)$ is the equilibrium susceptibility and $g(\tau)$ the distribution of relaxation times. They assumed $\chi(\tau) \propto 1/T$ and an Arrhenius dependence for τ, getting

$$\chi'' = -\frac{\pi}{2}\frac{1}{\ln(\omega\tau_0)}\frac{\partial(T\chi')}{\partial T} \qquad (4.22)$$

Because in the model (4.21), χ'' is also directly related with the distribution of relaxation times, Eq. (4.22) yields an inversion procedure analogous to that of Wohlfarth.

We now calculate $\partial(T\chi')/\partial T$ for the low-frequency Eq. (4.6). In this way, we shall take into account the effects of the finite width and depth of the anisotropy potential wells. Let us begin by taking the T derivative of $T\chi'$, with χ' given by Eq. (4.13) [or Eq. (4.15)]. Since the integrals in those equations also depend on T via the integration limits, the required T derivative can be taken by dint of the Leibnitz formula (4.16). On so doing, we get, after the rearrangement of the integration limits

$$
\begin{aligned}
\frac{\partial(T\chi')}{\partial T} &= -\ln(\omega\tau_0)\frac{\mu_0 M_s^2}{K}\left\langle \cos^2\alpha\right\rangle\frac{R'(\sigma_b)}{R(\sigma_b)}f(E_b)E_b \\
&\quad + \frac{\mu_0 M_s^2}{2K}\left\langle \sin^2\alpha\right\rangle\int_{E_b}^{\infty}\mathrm{d}E\, f(E)\left[\frac{R''}{R}-\left(\frac{R'}{R}\right)^2\right]\sigma^2 \\
&\quad + \frac{\mu_0 M_s^2}{2K}\left[3\left\langle \cos^2\alpha\right\rangle - 1\right]\int_0^{E_b}\mathrm{d}E\, f(E)\left[\frac{R''}{R}-\left(\frac{R'}{R}\right)^2\right]\sigma^2 \quad (4.23)
\end{aligned}
$$

where we have assumed that neither M_s nor K depend on T, and used $\partial E_b/\partial T = -k_B\ln(\omega\tau_0)$ as well as $(E/k_B)\partial\sigma/\partial T = -\sigma^2$.

Note that the first line on the right-hand side of Eq. (4.23) is directly related with the energy-barrier distribution. If the remainder terms were absent, this equation would give the inversion procedure of Wohlfarth (4.20). However, since the last two lines contain information about $f(E)$ in integral form, we see that the quantity $\partial(T\chi')/\partial T$ does not directly scan the energy-barrier distribution. Note in this connection that, unlike χ'' (or $\partial\chi'/\partial\ln\omega$), the quantity $[1/\ln(\omega\tau_0)]\partial(T\chi')/\partial T$ does not properly scale (in the sense of collapse onto a master curve) when represented against $-k_B T\ln(\omega\tau_0)$, due to the presence of the above-mentioned integral terms.

Next, on taking Eq. (4.10) into account we obtain the following relation between χ'' and $\partial(T\chi')/\partial T$

$$
\begin{aligned}
\chi'' = -\frac{\pi}{2}\frac{1}{\ln(\omega\tau_0)}\Bigg\{ &\frac{\partial(T\chi')}{\partial T} - \frac{\mu_0 M_s^2}{2K}\left\langle \sin^2\alpha\right\rangle\int_{E_b}^{\infty}\mathrm{d}E\, f(E)\left[\frac{R''}{R}-\left(\frac{R'}{R}\right)^2\right]\sigma^2 \\
&- \frac{\mu_0 M_s^2}{2K}\left[3\left\langle \cos^2\alpha\right\rangle - 1\right]\int_0^{E_b}\mathrm{d}E\, f(E)\left[\frac{R''}{R}-\left(\frac{R'}{R}\right)^2\right]\sigma^2\Bigg\}
\end{aligned}
$$

which is the counterpart of Eq. (4.22) in the low-frequency Shliomis–Stepanov model. Furthermore, since the angular factor in the last term on the right-hand side vanishes for anisotropy axes distributed at random, the preceding relation simplifies in that case to

$$\langle \chi'' \rangle_{ran} = -\frac{\pi}{2} \frac{1}{\ln(\omega\tau_0)} \left\{ \frac{\partial(T\langle\chi'\rangle_{ran})}{\partial T} - \frac{\mu_0 M_s^2}{3K} \int_{E_b}^{\infty} dE \, f(E) \left[\frac{R''}{R} - \left(\frac{R'}{R}\right)^2 \right] \sigma^2 \right\}$$

(4.24)

Finally, since $\sigma > \sigma_b \sim 20$–25, if $E > E_b$, one can replace $R''/R - (R'/R)^2$ in the integral term by its high-barrier approximation (A.31), to get

$$\langle \chi'' \rangle_{ran} = -\frac{\pi}{2} \frac{1}{\ln(\omega\tau_0)} \left\{ \frac{\partial}{\partial T} (T\langle\chi'\rangle_{ran}) - \frac{\mu_0 M_s^2}{3K} \int_{E_b}^{\infty} dE \, f(E) \right\} \qquad (4.25)$$

This is an interesting result; although the differences between χ'' and $\partial(T\chi')/\partial T$ are reduced after averaging for anisotropy axes distributed at random, some of them remain. These differences, and accordingly those of $\partial(T\chi')/\partial T$ with respect to the energy-barrier distribution, are again due to the presence of the second term on the right-hand side, which contains information about $f(E)$ in integral form. In addition, the lower the temperature, the larger the differences mentioned, because the lower integration limit in Eq. (4.25) decreases with T (recall that $E_b \propto k_B T$).

Note finally that, by using the high-barrier formula $\chi_\perp \simeq \mu_0 m^2 / 2Kv$ per particle [Eqs. (3.82)], the integral in Eq. (4.25) can alternatively be written in terms of the approximate transverse susceptibility of the blocked particles (at the temperature and frequency considered), namely

$$\langle \chi'' \rangle_{ran} = -\frac{\pi}{2} \frac{1}{\ln(\omega\tau_0)} \left\{ \frac{\partial}{\partial T} (T\langle\chi'\rangle_{ran}) - \frac{2}{3}\chi_{\perp,blo} \right\} \qquad (4.26)$$

Therefore, we find the T- and ω-dependent criterion (2/3) $\chi_{\perp,blo} \ll \partial(T\chi')/\partial T$, for the quantity $\partial(T\chi')/\partial T$ scanning the energy-barrier distribution as properly as χ'' (for anisotropy axes distributed at random only). Recall that no restriction of this type exists for the obtainment of the energy-barrier distribution from χ'' (or $\partial\chi'/\partial \ln \omega$). Note also that $\chi_{\perp,blo}$ is not only the transverse susceptibility of the blocked particles but, when multiplied by $\langle \sin^2 \alpha \rangle_{ran} = \frac{2}{3}$, is also their total contribution to the susceptibility, because the overbarrier relaxation mechanism is by definition "blocked" for those particles.

D. Comparison with Experiment

To conclude, we briefly illustrate some of the preceding results with experiments performed on a *frozen* magnetic fluid containing nanometric maghemite (γ-Fe$_2$O$_3$) particles.

The degree of dilution of the sample studied was ∼0.03% by volume, in order to avoid dipole–dipole interaction effects. This illustrates one advantage of the use of frozen magnetic fluids for fundamental studies on systems of magnetic nanoparticles — by simple dilution and subsequent freezing of the magnetic fluid, one can get a series of *solid* dispersions of nanoparticles where the strength of the interactions is tuned almost as desired. (This method also guarantees that all the samples have the same distribution in particle parameters.) Another advantage of these systems is that by means of the application of magnetic fields when freezing the samples, one can produce systems with different anisotropy-axis distributions. The sample considered here (Svedlindh et al., 1997) was frozen in zero field, so that a random distribution of the anisotropy axes is to be expected.

1. Comparison with the Ising-Type and Shliomis–Stepanov Models

Figure 20 displays the measured dynamical susceptibility as well as the Ising-type theoretical curves computed with the energy-barrier distribution derived from χ'' (left panels). While the calculated and experimental out-of-phase

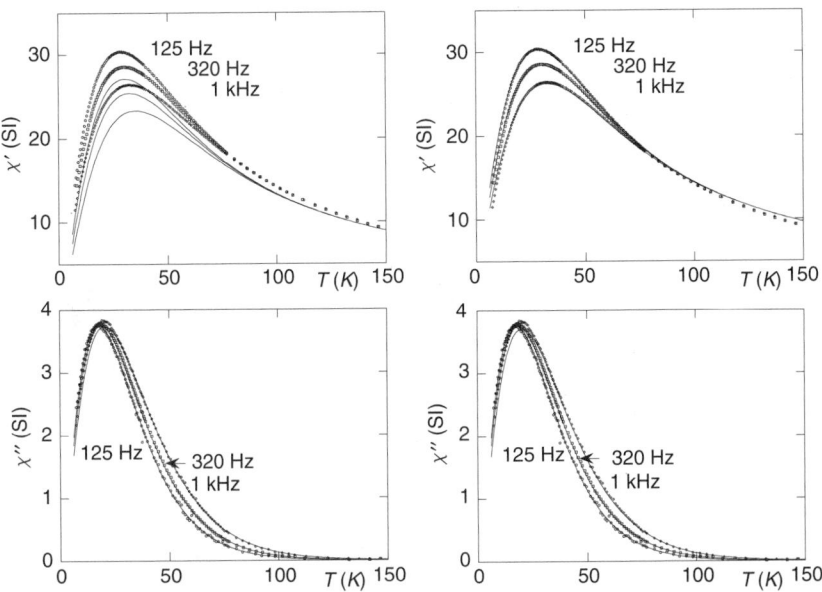

FIGURE 20. Temperature dependence of the in-phase (upper panels) and out-of-phase (lower panels) components of the dynamical susceptibility of a frozen magnetic fluid of maghemite particles. Left panels: solid lines computed with the Ising-type model where $\chi_\parallel = \mu_0 m^2 / k_B T$ and $\chi_\perp = 0$ (per particle). Right panels: solid lines computed with the low-frequency Shliomis–Stepanov equation (4.6).

susceptibilities compare to a high degree of precision (by construction), the matching of the in-phase curves is comparatively poor. One may guess that the reason for this poor matching is the absence of the transverse response in the model employed.* In order to check this hypothesis, Fig. 20 also displays the same experimental results together with the curves computed with the low-frequency Shliomis–Stepanov equation (right panels). One can see that the description of the experimental curves provided by this model has improved significantly.

2. Comparison of χ'' with $\partial(T\chi')/\partial T$

Equation (4.24) suggests that a joint plot of χ'' and $\partial(T\chi')/\partial T$ could be an alternative means to show the necessity of including the transverse response of the nanoparticles. When this contribution to the total response is negligible, those two curves should trace out the same energy-barrier dependence, whereas one would expect $(\pi/2\sigma_b)\partial(T\chi')/\partial T$ to be larger than χ'' otherwise. Moreover, one would also expect that, the lower the temperature, the larger the differences between the two curves, because the lower limit in the integral of Eq. (4.24) decreases with T. This is what is indeed observed in Fig. 21, giving further evidence of the necessity of including the transverse contribution to the total

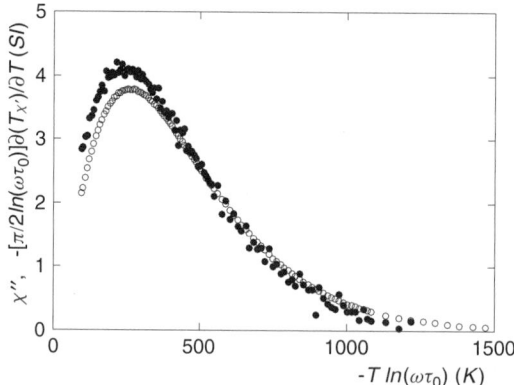

FIGURE 21. $\chi''(T)$ (hollow symbols) and $-[\pi/2\ln(\omega\tau_0)]\partial(T\chi')/\partial T$ (filled symbols) versus $-T\ln(\omega\tau_0)$ of a frozen magnetic fluid of maghemite particles at the frequency $\omega/2\pi = 320$ Hz.

---

* We use the phrase "take the transverse response into account" as an abbreviation for "take the finite width and depth of the anisotropy potential wells into account," since the lack of response to the transverse components of the probing field is perhaps the most characteristic feature of the Ising-type response.

response of the magnetic nanoparticle system considered here. The figure also confirms the point that $\partial(T\chi')/\partial T$ does not determine the energy-barrier distribution as accurately as χ'' does.

V. DYNAMICAL PROPERTIES: STOCHASTIC APPROACH

A. Introduction

In this section we study the dynamics of classical spins in the context of the *theory of stochastic processes*.

The rigorous theoretical study of the dynamics of classical spins in anisotropy potentials is hindered by certain mathematical difficulties. Therefore, in order to study the properties of these systems, numerical simulation techniques have also been used, being most of the studies that have been performed based on the Monte Carlo method. Although this method is a rigorous and efficient tool to compute thermal equilibrium quantities, the interpretation of the dynamical properties derived by means of Monte Carlo techniques, especially for non-Ising spins, is not free from criticism (Ettelaie and Moore, 1984; Binder and Stauffer, 1984). On the contrary, when using stochastic methods based on Fokker–Planck or Langevin equations, time does not merely label the sequential order of generated states when sampling the phase space, but is related with physical time.

For classical spins, the basic Langevin equation is the stochastic Landau–Lifshitz (–Gilbert) equation introduced by Brown (1963) (see also Kubo and Hashitsume, 1970). The *multiplicative* fluctuating terms occurring in this Langevin equation were treated in Brown's work, as well as in the subsequent theoretical developments, by means of the *Stratonovich stochastic calculus*. In this context, Brown constructed the celebrated Fokker–Planck (diffusion) equation for the time evolution of the *nonequilibrium* probability distribution of spin orientations.

In order to solve Brown's Fokker–Planck equation (a partial-differential equation of parabolic type) a number of techniques have been used, such as direct solution techniques (Rodé, Bertram and Fredkin, 1987) or more elaborate approaches involving continued fraction techniques or the numerical calculation of the eigenvalues and amplitudes of the relevant dynamical modes (Aharoni, 1964; Bessais, Ben Jaffel and Dormann, 1992; Coffey, Crothers, Kalmykov, Massawe and Waldron, 1994; Raĭkher and Stepanov, 1995*b*; Coffey, Crothers, Kalmykov and Waldron, 1995*a*).

An approach equivalent to solving the Fokker–Planck equation is to construct solutions of the underlying Langevin equation of the system. This *Langevin-dynamics* approach directly generates the stochastic trajectories of the variables of the system, from which averages can be computed. This is a relevant point since to solve the Fokker–Planck equation for multivariate systems, either numerically or analytically, is usually a formidable task. On the other hand, the Langevin-dynamics method requires an extensive computational effort and is less efficient than numerical methods especially suitable for noninteracting spins, such as those based on the Fokker–Planck equation mentioned in the previous paragraph. However, with a significant increase of the computational effort, the Langevin dynamics technique can also be used to study assemblies of interacting spins.

In this section we integrate the stochastic Landau–Lifshitz–Gilbert equation numerically in the context of the Stratonovich stochastic calculus. This is undertaken taking account of the underlying subtleties of the stochastic calculus as compared with the deterministic calculus. As the Langevin-dynamics method employed generates the selfsame stochastic trajectories of each individual magnetic moment, it provides much insight into the dynamics of the system. In addition, the theoretical study of single-particle phenomena is of special interest because dynamical measurements of *individual* magnetic *nano*particles have recently been performed (Wernsdorfer et al., 1997).

Concerning the response of an ensemble of classical spins (averaged quantities), the Langevin-dynamics method allows one to compute any desired quantity, such as hysteresis loops, field-cooled and zero-field-cooled magnetization curves, relaxation times, linear and nonlinear susceptibilities, thermal quantities, and, with appropriate relationships between lineshapes and correlation functions of the system, even spectroscopic quantities. We restrict our study to the linear dynamical response, which is chosen since it is a probe that enables one to examine the intrinsic dynamics of the system. In addition, because some relevant parameters of nanoparticle ensembles can be extracted from the analysis of the dynamical response data (see Section IV), an assessment of the accuracy of the heuristic equations employed in those analyses is necessary.

The organization of this section, which is an extended version of the results presented by García-Palacios and Lázaro (1998), is as follows. In order to provide the necessary background to undertake the study of the stochastic dynamics of classical spins, we begin in Section V.B with the study of the *deterministic* Landau–Lifshitz equation. Then, the Brown–Kubo–Hashitsume model for the stochastic dynamics of classical spins is discussed in Section V.C. The numerical method used to solve the stochastic Landau–Lifshitz (–Gilbert) equation is discussed in Section V.D. Finally, the results of the numerical integration of this Langevin equation are presented in Sections V.E and V.F. Specifically, Section V.E is devoted to the study of the trajectories of individual magnetic moments, while the dynamical response of the spin ensemble is studied in Section V.F.

B. Deterministic Dynamics of Classical Spins

We shall now study some aspects of the deterministic dynamics of classical magnetic moments.

1. The Gilbert and Landau–Lifshitz Equations

Let us start by considering the Gilbert equation of motion for a classical magnetic moment $\vec{m}$ (unpublished work, mentioned in Gilbert, 1955)

$$\frac{1}{\gamma}\frac{d\vec{m}}{dt} = \vec{m} \wedge \left[\vec{B}_{\text{eff}} - (\gamma m)^{-1}\lambda \frac{d\vec{m}}{dt} \right] \tag{5.1}$$

where γ is the magnetomechanical ratio and λ is a dimensionless damping coefficient. The *effective* field in this equation is given by

$$\vec{B}_{\text{eff}} = -\frac{\partial \mathcal{H}}{\partial \vec{m}} \tag{5.2}$$

where $\mathcal{H}$ is the Hamiltonian of $\vec{m}$ and $\partial/\partial\vec{m}$ stands for the gradient operator $[\partial f/\partial\vec{m} = (\partial f/\partial m_x)\hat{x} + (\partial f/\partial m_y)\hat{y} + (\partial f/\partial m_z)\hat{z}]$. For the justification of the occurrence of the expression (5.2) in the dynamical equations, the reader is referred to Section VI.B. Nevertheless, note that for $\mathcal{H} = -\vec{m} \cdot \vec{B}$ one indeed has $\vec{B}_{\text{eff}} = \vec{B}$, while in a more general situation $\vec{B}_{\text{eff}}$ incorporates the (deterministic) effects of the magnetic-anisotropy energy, the interaction with other spins, and other factors, on the dynamics of $\vec{m}$.

To illustrate, if the magnetic anisotropy is assumed to have the simplest axial symmetry (with symmetry axis $\hat{n}$) and $\vec{m}$ is subjected to an external constant field $\vec{B}$ and a low probing field $\Delta\vec{B}(t)$, the Hamiltonian reads [cf. Eq. (2.2)]

$$\mathcal{H}(\vec{m}, t) = -\vec{m} \cdot \left[\vec{B} + \Delta\vec{B}(t) \right] - \frac{Kv}{m^2}(\vec{m} \cdot \hat{n})^2 \tag{5.3}$$

In terms of $B_K = 2Kv/m$ [Eq. (2.6)], the effective field associated with this Hamiltonian can be written as

$$\vec{B}_{\text{eff}} = \vec{B} + \Delta\vec{B}(t) + \frac{B_K}{m}(\vec{m} \cdot \hat{n})\hat{n} \tag{5.4}$$

Note that the quantity $|B_K|$ is therefore the magnitude of the maximum *anisotropy field*

$$\vec{B}_{\text{a}} = \frac{B_K}{m}(\vec{m} \cdot \hat{n})\hat{n}$$

which occurs when $\vec{m} = \pm m\hat{n}$. The anisotropy field decreases as $\vec{m}$ approaches the equatorial region ($\vec{m} \perp \hat{n}$), where it vanishes. Recall finally that for easy-axis anisotropy in a longitudinal bias field ($\vec{B}\|\hat{n}$), the Hamiltonian has two minima at $\vec{m} = \pm m\hat{n}$ for $|B| < |B_K|$, with a potential barrier between than, whereas the upper (shallower) potential minimum disappears for $|B| \geq |B_K|$ (see Section II.B).

An equation of the Gilbert type can be cast into the archetypal Landau–Lifshitz form (1935) as follows. Take the vector product of $\vec{m}$ with both sides of Eq. (5.1)

$$\vec{m} \wedge \frac{\mathrm{d}\vec{m}}{\mathrm{d}t} = \gamma \vec{m} \wedge \left(\vec{m} \wedge \vec{B}_{\mathrm{eff}}\right) - \frac{\lambda}{m}\left[\vec{m}\left(\underbrace{\vec{m} \cdot \frac{\mathrm{d}\vec{m}}{\mathrm{d}t}}_{0}\right) - m^2\frac{\mathrm{d}\vec{m}}{\mathrm{d}t}\right]$$

where the triple vector product $\vec{m} \wedge [\vec{m} \wedge (\mathrm{d}\vec{m}/\mathrm{d}t)]$ has been expanded by using the rule

$$\vec{A} \wedge (\vec{B} \wedge \vec{C}) = \vec{B}(\vec{A} \cdot \vec{C}) - \vec{C}(\vec{A} \cdot \vec{B}) \qquad (5.5)$$

and $\vec{m} \cdot (\mathrm{d}\vec{m}/\mathrm{d}t) = 0$ (conservation of the magnitude of $\vec{m}$) follows from the starting equation (5.1). On introducing this result for $\vec{m} \wedge (\mathrm{d}\vec{m}/\mathrm{d}t)$ in the right-hand side of Eq. (5.1), passing $-\lambda^2 \mathrm{d}\vec{m}/\mathrm{d}t$ to the left-hand side, and introducing the "renormalized" magnetomechanical ratio $\tilde{\gamma} = \gamma/(1 + \lambda^2)$, one finally gets the desired Landau–Lifshitz form of the Gilbert equation

$$\frac{1}{\tilde{\gamma}}\frac{\mathrm{d}\vec{m}}{\mathrm{d}t} = \vec{m} \wedge \vec{B}_{\mathrm{eff}} - \frac{\lambda}{m}\vec{m} \wedge \left(\vec{m} \wedge \vec{B}_{\mathrm{eff}}\right) \qquad (5.6)$$

The celebrated Landau–Lifshitz relaxation (damping) term proportional to $-\vec{m} \wedge (\vec{m} \wedge \vec{B}_{\mathrm{eff}})$ drives $\vec{m}$ to the direction of $\vec{B}_{\mathrm{eff}}$, while λ measures the magnitude of the relaxation term relative to the precession term in the dynamical equation.

Conversely, one can start from Eq. (5.6) with $\tilde{\gamma}$ replaced by γ and then obtain its Gilbert equivalent equation. This is like Eq. (5.1) with γ replaced by a different "renormalized" magnetomechanical ratio: $\tilde{\gamma}' = \gamma \times (1 + \lambda^2)$.

There exists some controversy concerning which equation (Gilbert or Landau–Lifshitz) is more basic, or, equivalently, when one must use a renormalized γ. However, on recalling that both equations are anyway phenomenological ones, we can consider $\tilde{\gamma}$ (or $\tilde{\gamma}'$) to be a given constant for each magnetic moment. In addition, when $\lambda^2 \ll 1$ (weak damping), which is the common situation at least in bulk magnets, one has $\tilde{\gamma}' \simeq \tilde{\gamma} \simeq \gamma$, so that one does not need to worry about whether the magnetomechanical ratio occurring in a given formula is a bare or renormalized one.

Henceforth, we shall merely use the symbol γ in the dynamical quantities (as if we would have started from the Landau–Lifshitz equation). If one prefers to consider the Gilbert form as the commencing equation, one needs only to substitute $\gamma/(1 + \lambda^2)$ for γ in the corresponding formulas.

2. General Solution for Axially Symmetric Hamiltonians

We now investigate solutions of the deterministic Landau–Lifshitz equation

$$\frac{1}{\gamma}\frac{d\vec{m}}{dt} = \vec{m} \wedge \vec{B}_{\text{eff}} - \frac{\lambda}{m}\vec{m} \wedge \left(\vec{m} \wedge \vec{B}_{\text{eff}}\right) \tag{5.7}$$

restricting our attention to the case in which $\mathcal{H}(\vec{m})$ is axially symmetric. In this case, the effective field $\vec{B}_{\text{eff}}(\vec{m}) = -\partial\mathcal{H}/\partial\vec{m}$ is parallel to the symmetry axis, which can be chosen as the z axis, $\vec{B}_{\text{eff}} = B_{\text{eff}}(\vec{m})\hat{z}$. Then, on introducing the $\vec{m}$-dependent "frequency" $\omega_{\text{eff}}(\vec{m}) = \gamma B_{\text{eff}}(\vec{m})$, we can explicitly write the deterministic Landau–Lifshitz equation (5.7) as a system of coupled ordinary differential equations:

$$\frac{dm_x}{dt} = \omega_{\text{eff}}\left(m_y - \frac{\lambda}{m}m_x m_z\right)$$

$$\frac{dm_y}{dt} = \omega_{\text{eff}}\left(-m_x - \frac{\lambda}{m}m_y m_z\right)$$

$$\frac{dm_z}{dt} = \omega_{\text{eff}}\frac{\lambda}{m}\left(m^2 - m_z^2\right)$$

Next, on introducing spherical coordinates $m_z = m\cos\vartheta$ and $m_x + im_y = m\sin\vartheta\exp(-i\varphi)$ (we measure here the azimuthal angle clockwise), this system of differential equations can equivalently be written as

$$\frac{d\vartheta}{dt} = -\lambda\omega_{\text{eff}}\sin\vartheta \tag{5.8}$$

$$\frac{d\varphi}{dt} = -\frac{1}{\lambda\sin\vartheta}\frac{d\vartheta}{dt}, \qquad \left(\text{or}\quad \frac{d\varphi}{dt} = \omega_{\text{eff}}\right) \tag{5.9}$$

Equation (5.9) can readily be solved by separation of variables, to get

$$\varphi(\vartheta) - \varphi(\vartheta_0) = -\frac{1}{\lambda}\ln\left[\frac{\tan(\vartheta/2)}{\tan(\vartheta_0/2)}\right] \tag{5.10}$$

where $\int dx/\sin x = \ln[\tan(x/2)]$ has been used and $\vartheta_0 = \vartheta(t_0)$, where t_0 is the initial time. Concerning the equation (5.8) for ϑ, since $\omega_{\text{eff}} = \omega_{\text{eff}}(\vartheta)$, we can also separate the variables to obtain the following implicit expression for $\vartheta(t)$:

$$-\lambda(t - t_0) = \int_{\vartheta_0}^{\vartheta(t)} \frac{d\vartheta'}{\omega_{\text{eff}}(\vartheta') \sin \vartheta'} \tag{5.11}$$

Equations (5.10) and (5.11) are the solution of the deterministic Landau–Lifshitz equation (5.7) for *any* axially symmetric Hamiltonian $\mathcal{H}(\vartheta)$.

An important case is that in which $\lambda \ll 1$; this is known as the *weak-damping case*. Note first that Eq. (5.9) can also be written as $d\vartheta = -\lambda \sin \vartheta d\varphi$, which for weak damping yields $|d\vartheta| \ll |\sin \vartheta d\varphi|$. Then, the "displacement" of the tip of $\vec{m}$ along the polar direction $(\Delta \vartheta)$ in a time interval Δt is much smaller than the displacement along the tangential direction $(\sin \vartheta \Delta \varphi)$. It then makes sense to introduce a "position-dependent" frequency of rotation about $\hat{z}$, which is precisely given by $\omega_{\text{eff}} = \gamma B_{\text{eff}}$ [see the alternative form of Eq. (5.9)].

3. The Simplest Axially Symmetric Hamiltonian

Let us now specialize the preceding general solutions to the Hamiltonian obtained by the sum of the simplest axially symmetric anisotropy potential plus a longitudinal Zeeman term. Then [cf. Eq. (5.4)]

$$\vec{B}_{\text{eff}} = B\hat{z} + \frac{B_K}{m} m_z \hat{z} \tag{5.12}$$

and $\omega_{\text{eff}} = \gamma B_{\text{eff}}$ can be written as

$$\omega_{\text{eff}}(\vartheta) = \omega_B + \omega_K \cos \vartheta, \qquad \omega_B = \gamma B, \qquad \omega_K = \gamma B_K \tag{5.13}$$

The integral in the solution (5.11) is then given by

$$\int \frac{d\vartheta}{(\omega_B + \omega_K \cos \vartheta) \sin \vartheta} = \frac{1}{\omega_B + \omega_K} \ln \left[\tan \left(\frac{\vartheta}{2} \right) \right]$$

$$+ \frac{\omega_K}{\omega_B^2 - \omega_K^2} \ln \left[1 + \left(\frac{\omega_B - \omega_K}{\omega_B + \omega_K} \right) \tan^2 \left(\frac{\vartheta}{2} \right) \right]$$

$$+ \frac{\omega_K}{\omega_B^2 - \omega_K^2} \ln(\omega_B + \omega_K)$$

as can be checked by differentiation of the right-hand side. Therefore, from the general result (5.11) we get the still implicit solution

$$Ce^{-\lambda(\omega_B + \omega_K)t} = \tan \left(\frac{\vartheta}{2} \right) \left[1 + \left(\frac{\omega_B - \omega_K}{\omega_B + \omega_K} \right) \tan^2 \left(\frac{\vartheta}{2} \right) \right]^{\omega_K/(\omega_B - \omega_K)} \tag{5.14}$$

where the constant of integration C involves the terms evaluated at $t = t_0$.

4. Particular Cases

The implicit solution for $\vartheta(t)$ [Eq. (5.14)] turns into an explicit solution in various particular cases.

a. Dynamics in the Isotropic Case. Here $\omega_K = 0$, so that Eqs. (5.10) and (5.14) reduce to the celebrated results (see, for example, Chikazumi, 1978, Chapter 16)

$$\tan\left(\frac{\vartheta}{2}\right) = \tan\left(\frac{\vartheta_0}{2}\right) e^{-\lambda\omega_B(t-t_0)}, \qquad \varphi(t) - \varphi_0 = \omega_B(t - t_0)$$

Thus, the motion of $\vec{m}$ consists of a precession with frequency $\omega_B = \gamma B$ about $\hat{z}$ and a spiraling toward this axis with a characteristic time constant

$$\tau_B = \frac{1}{\lambda\omega_B} = \frac{1}{\lambda\gamma B} \tag{5.15}$$

Note that this is the characteristic decay time of $\tan(\vartheta/2)$; for $m_z = m\cos\vartheta$ in the vicinity of the minimum [$\tan(\vartheta/2) \simeq \vartheta/2$ and $\cos\vartheta \simeq 1 - \vartheta^2/2$], the characteristic time constant is $\tau_B/2$. Note also that, for $B < 0$, one has $\omega_B < 0$ and therefore $\lim_{t\to\infty} \tan(\vartheta/2) = \infty$, that is, $\vartheta \to \pi$ as $t \to \infty$, as it should.

b. Dynamics in the Zero-Field Case. Here $\omega_B = 0$, so that, by using $\tan\vartheta = 2\tan(\vartheta/2)/[1 - \tan^2(\vartheta/2)]$ in Eq. (5.14), one gets

$$\tan\vartheta = \tan\vartheta_0 \, e^{-\lambda\omega_K(t-t_0)} \tag{5.16}$$

Thus, the spiraling toward the minima has for $K > 0$ a characteristic time constant

$$\tau_K = \frac{1}{\lambda\omega_K} = \frac{1}{\lambda\gamma B_K} \tag{5.17}$$

or its absolute value if $K < 0$. In this easy-plane case one has $B_K, \omega_K < 0$, so that $\lim_{t\to\infty} \tan\vartheta = \infty$, that is, $\vartheta \to \pi/2$ as $t \to \infty$, and the magnetic moment eventually rests in the equatorial plane. This behavior against the change $B_K \to -B_K$ is different from the behavior against the transformation $B \to -B$ in the isotropic case (where $\vec{m}$ then falls into the $-\hat{z}$ minimum), and it is mathematically reflected by the occurrence of $\tan\vartheta$ in the solution of the unbiased case, whereas $\tan(\vartheta/2)$ appears in the solution of the isotropic case.

Note that for both signs of K, Eq. (5.16) yields $\vartheta(t) \in [0, \pi/2]$ if $\vartheta_0 \in [0, \pi/2]$ and $\vartheta(t) \in [\pi/2, \pi]$ when $\vartheta_0 \in [\pi/2, \pi]$. This reveals that, during the time evolution, $\vartheta(t)$ remains in the same hemisphere in which it was initially. For instance, $\vec{m}$ does not surmount the anisotropy-potential barrier when $K > 0$ (as it should in a deterministic damped dynamics), while for $K < 0$, $\vec{m}$ does not oscillate about (cross) the equatorial circle when spiraling towards the easy plane.

Concerning the azimuthal angle, by expressing $\tan(\vartheta/2)$ in terms of $\tan\vartheta$, one gets the following formula from Eq. (5.10)

$$\varphi(t) - \varphi_0 = \omega_K(t - t_0) - \frac{1}{\lambda}\ln\left[\frac{1 + \sec\vartheta_0}{1 \pm \sqrt{1 + \tan^2\vartheta_0\, e^{-2\lambda\omega_K(t-t_0)}}}\right]$$

where the plus sign corresponds to $\vartheta \in [0, \pi/2]$ and the minus sign to $\vartheta \in [\pi/2, \pi]$. From this equation it follows that the asymptotic $\lambda\omega_K(t - t_0) \gg 1$ behavior of the azimuthal angle is for $K > 0$

$$\Delta\varphi(t) \simeq \pm\omega_K(t - t_0)$$

which corresponds to a precession close to the bottom of the corresponding potential well with an angular velocity $\omega_K\hat{z}$ in the $z > 0$ well and $-\omega_K\hat{z}$ in the $z < 0$ well. For easy-plane anisotropy, one has $\tan(\vartheta/2)\overset{t\to\infty}{\to}1$, so that we find from Eq. (5.10) that the magnetic moment finally rests in the equatorial plane at $\varphi = \varphi(\vartheta_0) + \lambda^{-1}\ln[\tan(\vartheta_0/2)]$ (unless it starts at $\vartheta_0 = 0, \pi$, which are unstable equilibrium points).

c. Dynamics Close to the Potential Minima. The implicit solution (5.14) for $\vartheta(t)$ can also be explicitly written in the general case (both ω_K and ω_B different from zero) for the dynamics close to the potential minima (we consider the case $B_K > 0$ only). Let us initially assume $\vartheta \simeq 0$ [i.e., $\tan(\vartheta/2) \ll 1$]. Then, on retaining terms of order $\tan(\vartheta/2)$ in Eq. (5.14), we get $\tan(\vartheta/2) \simeq \tan(\vartheta_0/2)\exp[-\lambda(\omega_B + \omega_K)(t - t_0)]$ and $\varphi(t) - \varphi_0 \simeq (\omega_B + \omega_K)(t - t_0)$ by Eq. (5.10). However, within the same approximation ($\vartheta \ll 1$), we can replace the tangents by their arguments, getting

$$\vartheta(t) \simeq \vartheta_0 e^{-\lambda(\omega_B+\omega_K)(t-t_0)}, \qquad \varphi(t) - \varphi_0 \simeq (\omega_B + \omega_K)(t - t_0)$$

Thus, $\vec{m}$ precesses with frequency $\omega_B + \omega_K$ when spiraling toward the $\vartheta = 0$ potential minimum and the time constant of the decay of ϑ is $1/[\lambda(\omega_B + \omega_K)] = \tau_B\tau_K/(\tau_B + \tau_K)$. Note that the characteristic decay time of $m_z \propto \cos\vartheta \simeq 1 - \vartheta^2/2$, is a half of this result.

From the preceding equations we see that the approximation used ($\vartheta \ll 1$) is self-consistent if $\omega_B + \omega_K > 0$, that is, for any positive B and also for negative external fields of magnitude less than the anisotropy field $|B| < B_K$

(i.e., inasmuch as the $\vartheta = 0$ potential minimum exists; recall the discussion in Section II.B).

On the other hand, in the $\vartheta \simeq \pi$ case one has $\vartheta/2 \simeq \pi/2$ and, hence, $\tan(\vartheta/2) \gg 1$. Then, we can use $[(\omega_B - \omega_K)/(\omega_B + \omega_K)] \tan^2(\vartheta/2) \gg 1$ in Eq. (5.14) to get $\tan(\vartheta/2) \simeq \tan(\vartheta_0/2) \exp[\lambda(\omega_K - \omega_B)(t - t_0)]$, whence $\varphi(t) - \varphi_0 \simeq -(\omega_K - \omega_B)(t - t_0)$ by Eq. (5.10). However, when $\tan(\vartheta/2) \gg 1$, we can use the approximation $\tan\vartheta \simeq -2/\tan(\vartheta/2)$, so that on expanding $\tan\vartheta$ about $\vartheta = \pi$, we finally get

$$\vartheta(t) - \pi \simeq (\vartheta_0 - \pi)e^{-\lambda(\omega_K - \omega_B)(t-t_0)}, \qquad \varphi(t) - \varphi_0 \simeq -(\omega_K - \omega_B)(t - t_0)$$

Therefore, $\vec{m}$ precesses with frequency $\omega_K - \omega_B$ (about $-\hat{z}$) when spiraling toward the $\vartheta = \pi$ minimum, while ϑ decays with a characteristic time constant $1/[\lambda(\omega_K - \omega_B)] = \tau_B\tau_K/(\tau_B - \tau_K)$ (and m_z with a half of this value).

Note finally that the approximation used ($\pi - \vartheta \ll 1$) is self-consistent if $\omega_K - \omega_B > 0$, that is, for any negative B and also for positive B of magnitude less than the anisotropy field ($B < B_K$). Thus, in this case, and exhibiting a natural symmetry with the $\vartheta \simeq 0$ case, the motion is stable inasmuch as the $\vartheta = \pi$ minimum exists.

C.   Stochastic Dynamics of Classical Spins (Brown–Kubo–Hashitsume Model)

As a result of the interaction of a spin with the surrounding medium (phonons, conducting electrons, nuclear spins, etc.), its $T \neq 0$ dynamics is quite complicated. The complexity itself, however, permits an idealization of the phenomenon, by replacing the effect of the environment by a magnetic field randomly varying in time. Nevertheless, in order to describe the environmental effects properly and to attain a thermodynamically consistent description, the fluctuating terms must be supplemented with the analogue of a *relaxation* (damping or dissipative) term, to which they must be linked by *fluctuation–dissipation* relations.

We begin with a survey of how this general program is specialized to the study of the dynamics of classical magnetic moments. This was done by Brown (1963), in the context of the small-particle magnetism, and by Kubo and Hashitsume (1970), who studied generic classical spins. The subsequent developments based on each of these works have taken place separately in the literature. Nevertheless, both approaches are essentially equivalent and we shall present here a unified discussion of them.*

---

*Note that Kubo and Hashitsume say in their article that the main part of their work was done in the summer of 1963, so that both approaches are, in addition, contemporary.

1. Stochastic Dynamical (Langevin) Equations

In the Brown–Kubo–Hashitsume model the starting dynamical equation is the Gilbert equation (5.1), which already embodies a relaxation term, and the total field acting on $\vec{m}$ is the sum of the deterministic effective field $\vec{B}_{\mathrm{eff}}$ and a fluctuating or stochastic field $\vec{b}_{\mathrm{fl}}(t)$, namely

$$\frac{1}{\gamma}\frac{d\vec{m}}{dt} = \vec{m} \wedge \left[\vec{B}_{\mathrm{eff}} + \vec{b}_{\mathrm{fl}}(t) - (\gamma m)^{-1}\lambda\frac{d\vec{m}}{dt} \right] \qquad (5.18)$$

This is technically a nonlinear *stochastic differential equation*, and it is called the *stochastic Gilbert equation*. It suggest a heuristic analogy with the Langevin equation for ordinary Brownian motion since the "damping field" is proportional to minus the "velocity," $-(d\vec{m}/dt)$. However, the analogy ends here; in the dynamical equation for a Brownian particle [see, for example, Eq. (6.25)], a damping term proportional to minus the velocity enters in the Newton equation (i.e., in the equation for the acceleration), whereas $-(d\vec{m}/dt)$ enters in the equation for the "velocity" itself. Besides, the fluctuating terms enter in Eq. (5.18) in a *multiplicative* way.

As has been mentioned, the fluctuating field $\vec{b}_{\mathrm{fl}}(t)$ accounts for the effects of the interaction of $\vec{m}$ with the microscopic degrees of freedom (phonons, conducting electrons, nuclear spins, etc.), which cause fluctuations of the magnetic moment orientation. Those environmental degrees of freedom are *also* responsible for the damped precession of $\vec{m}$, since fluctuations and dissipation are related manifestations of one and the same interaction of the spin with its environment (see Section VI).

The customary assumptions about $\vec{b}_{\mathrm{fl}}(t)$ are that it is a Gaussian "stochastic process" with the following statistical properties

$$\langle b_{\mathrm{fl},k}(t)\rangle = 0, \qquad \langle b_{\mathrm{fl},k}(t)b_{\mathrm{fl},\ell}(t')\rangle = 2D\delta_{k\ell}\delta(t - t') \qquad (5.19)$$

(the first two *moments* determine a Gaussian process), where k and ℓ are Cartesian indices, the constant D measures the strength of the thermal fluctuations (assumed isotropic), and $\langle\ \rangle$ denotes an average taken over different *realizations* of the fluctuating field. (The constant D is determined on the grounds of statistical-mechanical considerations; see discussion below.) The Gaussian property of the fluctuations arises because they emerge from the interaction of $\vec{m}$ with a large number of microscopic degrees of freedom with equivalent statistical properties (central-limit theorem). On the other hand, the Dirac delta in the second Eq. (5.19) expresses that above certain temperature the autocorrelation time of $\vec{b}_{\mathrm{fl}}(t)$ (of microscopic scale) is much shorter than the rotational response time of the system ("white" noise), while the Krönecker delta expresses that the different components of $\vec{b}_{\text{fl}}(t)$ are assumed to be uncorrelated. Finally, it is also customarily assumed that the fluctuating fields acting on different magnetic moments are independent.

On starting from the stochastic Gilbert equation (5.18), the discussed transformation to the equivalent Landau–Lifshitz form yields (recall our convention for the magnetomechanical ratio)

$$\frac{1}{\gamma}\frac{d\vec{m}}{dt} = \vec{m} \wedge \left[\vec{B}_{\text{eff}} + \vec{b}_{\text{fl}}(t) \right] - \frac{\lambda}{m}\vec{m} \wedge \left\{ \vec{m} \wedge \left[\vec{B}_{\text{eff}} + \vec{b}_{\text{fl}}(t) \right] \right\} \qquad (5.20)$$

which will be called the *stochastic Landau–Lifshitz–Gilbert equation*. As will be shown later, the thermodynamical consistency of the approach entails that $|\vec{b}_{\text{fl}}| \sim \lambda^{1/2}$. Therefore, for weak damping ($\lambda \ll 1$), we can drop the fluctuating field from the relaxation term of Eq. (5.20), to arrive at

$$\frac{1}{\gamma}\frac{d\vec{m}}{dt} = \vec{m} \wedge \left[\vec{B}_{\text{eff}} + \vec{b}_{\text{fl}}(t) \right] - \frac{\lambda}{m}\vec{m} \wedge \left(\vec{m} \wedge \vec{B}_{\text{eff}} \right) \qquad (5.21)$$

This equation, which was in fact the equation studied by Kubo and Hashitsume (1970), will be called the *stochastic Landau–Lifshitz equation*, since in accordance with the spirit of its original deterministic counterpart, it describes weakly damped precession. Equation (5.21) is besides a Langevin equation more archetypal than Eq. (5.20), because the fluctuating and relaxation terms are not mixed.

Alternatively, one can bypass the reasoning employed to obtain Eq. (5.21) from Eq. (5.20), and consider the former as an independent stochastic model. We shall show that, when the condition of thermodynamical consistency is applied, the *average* properties derived both from Eqs. (5.20) and from (5.21) are completely equivalent.

Let us now discuss the multiplicative noise terms. Apparently, for a given D, Eqs. (5.20) or (5.21), supplemented by Eqs. (5.19), fully determine the dynamical problem under consideration. Nevertheless, because of the vector *products* of $\vec{m}$ and $\vec{b}_{\text{fl}}(t)$ occurring in those equations, the fluctuating field $\vec{b}_{\text{fl}}(t)$ enters in a *multiplicative* way. This fact gives rise to some formal problems because, for white multiplicative noise, any Langevin equation must be supplemented by an interpretation rule to properly define it (see, for example, van Kampen, 1981, p. 246).

Two dominant interpretations, which lead to either the Itô or the Stratonovich *stochastic calculus*, are usually considered, yielding different dynamical properties for the system. For instance, depending on the stochastic calculus used, disparate Fokker–Planck equations for the time evolution of the nonequilibrium probability distribution are obtained. The Itô calculus is commonly chosen on certain mathematical grounds, since rather general results of probability theory can then be employed. On the other hand, since the white noise is an idealization of physical noise with short autocorrelation time, the Stratonovich calculus is usually preferred in physical applications, since the associated results coincide with those obtained in the *formal* zero-correlation-time limit of fluctuations with finite autocorrelation time (see, for example, Risken, 1989). Concerning the spin-dynamics problem, both the seminal works of Brown (1963) and Kubo and Hashitsume (1970), as well as all the subsequent theoretical developments, are based, implicitly or explicitly, on the Stratonovich stochastic calculus.

2. *Fokker–Planck Equations*

We now consider the Fokker–Planck equation governing the time evolution of the nonequilibrium probability distribution of magnetic moment orientations. We need this equation to determine the amplitude of the fluctuating field in terms of the spin parameters and the temperature, so as to ensure the thermodynamical consistency of the approach. In addition, some analytic results can be obtained from the Fokker–Planck equation, which will be used as reference for the results obtained by numerical integration of the stochastic Landau–Lifshitz (–Gilbert) equation.

Brown (1963) derived the Fokker–Planck equation associated with the stochastic Landau–Lifshitz (–Gilbert) equation (5.20). By a different method and starting from the stochastic Landau–Lifshitz equation (5.21), Kubo and Hashitsume (1970) arrived at an equation for the probability distribution, which, when the autocorrelation times of $\vec{b}_{\text{fl}}(t)$ are much shorter than the precession period of $\vec{m}$, coincides with the Fokker–Planck equation of Brown in the absence of the anisotropy potential (they studied the case $\vec{B}_{\text{eff}} = \vec{B}$).[*]
We shall begin by giving a *unified* derivation of the Fokker–Planck equations associated with Eqs. (5.20) and (5.21).

a. Derivation of the Fokker–Planck Equations. Let us consider the general system of Langevin equations

$$\frac{\mathrm{d}y_i}{\mathrm{d}t} = A_i(\mathbf{y}, t) + \sum_k B_{ik}(\mathbf{y}, t)L_k(t) \tag{5.22}$$

where $i = 1, \ldots, n$, $\mathbf{y} = (y_1, \ldots, y_n)$ (the variables of the system), k runs over a given set of indices, and the "Langevin" sources $L_k(t)$ are independent Gaussian stochastic processes satisfying

$$\langle L_k(t) \rangle = 0, \qquad \langle L_k(t)L_\ell(t') \rangle = 2D\delta_{k\ell}\delta(t - t') \tag{5.23}$$

---

[*] For an alternative derivation starting from Eq. (5.21) see, for example, (Garanin, 1997).

When the functions $B_{ik}(\mathbf{y}, t)$ do depend on $\mathbf{y}$, the noise in the Langevin equations is termed *multiplicative*, whereas for $\partial B_{ik}/\partial y_j \equiv 0$ the noise is called *additive* (here the Itô and Stratonovich stochastic calculi coincide).

The time evolution of $P(\mathbf{y}, t)$, the nonequilibrium probability distribution of $\mathbf{y}$ at time t, is given by the Fokker–Planck equation (see, for example, Risken, 1989)

$$\frac{\partial P}{\partial t} = -\sum_i \frac{\partial}{\partial y_i}\left[\left(A_i + D\sum_{jk} B_{jk}\frac{\partial B_{ik}}{\partial y_j}\right)P\right] + \sum_{ij}\frac{\partial^2}{\partial y_i \partial y_j}\left[\left(D\sum_k B_{ik}B_{jk}\right)P\right]$$

(5.24)

where the Stratonovich calculus has been used to treat the (in general) multiplicative fluctuating terms in the Langevin equations (5.22) [when using the Itô calculus, the *noise-induced* drift coefficient $D\sum_{jk} B_{jk}(\partial B_{ik}/\partial y_j)$ is simply omitted]. On taking the y_j derivatives of the second term on the right-hand side (the diffusion term), one alternatively gets the Fokker–Planck equation in the form of a *continuity equation* for the probability distribution, namely

$$\frac{\partial P}{\partial t} = -\sum_i \frac{\partial}{\partial y_i}\left\{\left[A_i - D\sum_k B_{ik}\left(\sum_j \frac{\partial B_{jk}}{\partial y_j}\right) - D\sum_{jk} B_{ik}B_{jk}\frac{\partial}{\partial y_j}\right]P\right\}$$

(5.25)

where the term within the curly brackets defines the ith component of the current of probability $J_i(\mathbf{y}, t)$.

Next, on considering the *stochastic Landau–Lifshitz (–Gilbert) equation*, supplemented by the statistical properties (5.19), the following substitutions cast them into the form of the general system of Langevin equations (5.22): $(y_1, y_2, y_3) = (m_x, m_y, m_z)$, $L_k(t) = b_{\mathrm{fl},k}(t)$, and

$$A_i = \gamma\left[\vec{m} \wedge \vec{B}_{\mathrm{eff}} - \frac{\lambda}{m}\vec{m} \wedge \left(\vec{m} \wedge \vec{B}_{\mathrm{eff}}\right)\right]_i$$

(5.26)

$$B_{ik} = \gamma\left[\sum_j \varepsilon_{ijk}m_j + g\frac{\lambda}{m}\left(m^2\delta_{ik} - m_im_k\right)\right]$$

(5.27)

where ε_{ijk} is the antisymmetric unit tensor of rank three (Levi–Civita symbol)[*] and we have expanded the triple vector product $\vec{m} \wedge (\vec{m} \wedge \vec{b}_{\mathrm{fl}})$ by using the rule (5.5). The parameter g enables us to deal with both equations simultaneously; to obtain the stochastic Landau–Lifshitz–Gilbert equation (5.20), we

---

[*] This tensor is defined as the tensor antisymmetric in all three indices with $\varepsilon_{xyz} = 1$. Therefore, one can write the vector product of $\vec{A}$ and $\vec{B}$ as $(\vec{A} \wedge \vec{B})_i = \sum_{jk} \varepsilon_{ijk}A_jB_k$. In addition, one has the useful contraction property $\sum_k \varepsilon_{ijk}\varepsilon_{i'j'k} = \delta_{ii'}\delta_{jj'} - \delta_{ij'}\delta_{ji'}$.

put $g = 1$, whereas the stochastic Landau–Lifshitz equation (5.21) is recovered if $g = 0$, since in this case $\vec{b}_{\mathrm{fl}}(t)$ only enters in the precession term. Note that the B_{ik} depend on $\vec{m}$ in both cases; thus, *the noise terms in the stochastic Landau–Lifshitz (–Gilbert) equation are multiplicative.*

Next, on using $\partial m_i / \partial m_j = \delta_{ij}$, one first gets

$$\frac{\partial B_{ik}}{\partial m_j} = \gamma \left[\varepsilon_{ijk} - g\frac{\lambda}{m}(\delta_{ij}m_k + \delta_{kj}m_i) \right] \tag{5.28}$$

where the terms dependent on $m = \left(\sum_i m_i^2 \right)^{1/2}$ have not been differentiated because of the conservation of the magnitude of $\vec{m}$. (One can indeed check that differentiating those terms by using $\partial m / \partial m_j = m_j / m$ and repeating the following calculations, we arrive at the same final results.) Then, on taking $\varepsilon_{jjk} = 0$ into account one finds $\sum_j \partial B_{jk} / \partial m_j = -4g\gamma(\lambda/m)m_k$. From this result and Eq. (5.27) we get $\sum_k B_{ik}\left(\sum_j \partial B_{jk}/\partial m_j \right) = 0$, where we have used $\sum_{jk}\varepsilon_{ijk}m_jm_k = 0$ (due to the contraction of a symmetric tensor with an antisymmetric tensor) and $\sum_k(m^2\delta_{ik} - m_im_k)m_k = 0$. Therefore, he second term on the right-hand side of the general Fokker–Planck equation (5.25) vanishes identically in this case. In order to obtain the third term, we need to calculate first

$$\frac{1}{\gamma^2}\sum_k B_{ik}B_{jk}$$

$$= \sum_k \left[\sum_r \varepsilon_{irk}m_r + g\frac{\lambda}{m}\left(m^2\delta_{ik} - m_im_k \right) \right]\left[\sum_s \varepsilon_{jsk}m_s + g\frac{\lambda}{m}\left(m^2\delta_{jk} - m_jm_k \right) \right]$$

$$= \sum_{rs}(\delta_{ij}\delta_{rs} - \delta_{is}\delta_{rj})m_rm_s$$

$$+ g\frac{\lambda}{m}\left(m^2\sum_r (\varepsilon_{irj}m_r + \overbrace{\varepsilon_{jri}}^{-\varepsilon_{irj}}m_r) - m_j\underbrace{\sum_{kr}\varepsilon_{irk}m_rm_k}_{0} - m_i\underbrace{\sum_{ks}\varepsilon_{jsk}m_sm_k}_{0} \right)$$

with the underbrace under $\sum_r (\varepsilon_{irj}m_r + \varepsilon_{jri}m_r)$ labeled 0.

$$+ g\left(\frac{\lambda}{m} \right)^2\left[m^4\delta_{ij} - m^2(m_im_j + m_jm_i) + m_im_j\sum_k m_k^2 \right]$$

$$= \left(1 + g\lambda^2 \right)\left(m^2\delta_{ij} - m_im_j \right)$$

where we have taken into account that $g^2 = g$ and employed the above-mentioned contraction rule of ε_{ijk}. Then, on introducing the *Néel time*, namely

$$\frac{1}{\tau_N} = 2D\gamma^2 \left(1 + g\lambda^2\right) \tag{5.29}$$

which is the characteristic time of diffusion in the absence of potential (free-diffusion time; see discussion below), we get for the third term in Eq. (5.25)

$$-D\sum_{jk} B_{ik}B_{jk}\frac{\partial P}{\partial m_j} = \frac{1}{2\tau_N}\left[\vec{m} \wedge \left(\vec{m} \wedge \frac{\partial P}{\partial \vec{m}}\right)\right]_i \tag{5.30}$$

On introducing these results into Eq. (5.25) one finally arrives at the Fokker–Planck equation

$$\frac{\partial P}{\partial t} = -\frac{\partial}{\partial \vec{m}} \cdot \left[\gamma\vec{m} \wedge \vec{B}_{eff} - \gamma\frac{\lambda}{m}\vec{m} \wedge \left(\vec{m} \wedge \vec{B}_{eff}\right) + \frac{1}{2\tau_N}\vec{m} \wedge \left(\vec{m} \wedge \frac{\partial}{\partial \vec{m}}\right)\right] P \tag{5.31}$$

where $(\partial/\partial\vec{m})\cdot$ stands for the divergence operator $[(\partial/\partial\vec{m}) \cdot \vec{J} = \sum_i(\partial J_i/\partial m_i)]$. Thus, the Fokker–Planck equations associated with the stochastic Landau–Lifshitz–Gilbert equation (5.20) ($g = 1$) and the stochastic Landau–Lifshitz equation (5.21) ($g = 0$) are *both* given by Eq. (5.31); the only difference is the relation between the Néel time and the amplitude of the fluctuating field:

$$\frac{1}{\tau_N} = 2D\gamma^2 \left(1 + \lambda^2\right) \quad \text{(LLG)}, \qquad \frac{1}{\tau_N} = 2D\gamma^2 \quad \text{(LL)}$$

Equation (5.31) is equivalent to the Fokker–Planck equation derived by Brown (1963) [see also Eqs. (5.42)–(5.44) below].

b. Stationary Solution of the Fokker–Planck Equation and Comparison between the Stochastic Models. In order to ensure that the stationary properties of the system, derived from the Langevin equations (5.20) or (5.21), coincide with the correct thermal equilibrium properties, the Fokker–Planck equation associated with these Langevin equations is forced to have the Boltzmann distribution $P_e(\vec{m}) \propto \exp[-\beta\mathcal{H}(\vec{m})]$ as stationary solution.

To do so, note first that, by means of the definition $\vec{B}_{eff} = -\partial\mathcal{H}/\partial\vec{m}$, one can write $\partial P_e/\partial\vec{m}$ as

$$\frac{\partial P_e}{\partial \vec{m}} = \beta\vec{B}_{eff}P_e \tag{5.32}$$

From this result one can easily show that $\vec{m} \wedge \vec{B}_{\mathrm{eff}} P_{\mathrm{e}}$ is divergenceless (solenoidal).* Therefore, on taking these results into account when introducing the Boltzmann distribution in the Fokker–Planck equation (5.31), one gets

$$0 = \frac{\partial P_{\mathrm{e}}}{\partial t} = -\frac{\partial}{\partial \vec{m}} \cdot \left[-\gamma \frac{\lambda}{m} \vec{m} \wedge \left(\vec{m} \wedge \vec{B}_{\mathrm{eff}} \right) P_{\mathrm{e}} + \frac{\beta}{2\tau_{\mathrm{N}}} \vec{m} \wedge \left(\vec{m} \wedge \vec{B}_{\mathrm{eff}} \right) P_{\mathrm{e}} \right]$$

One then sees by inspection that, in order to have the Boltzmann distribution as stationary solution of the Fokker–Planck equation (5.31), it is sufficient to put

$$\gamma \frac{\lambda}{m} = \frac{\beta}{2\tau_{\mathrm{N}}} \qquad (5.33)$$

from which one gets the following expression for the Néel time

$$\tau_{\mathrm{N}} = \frac{1}{\lambda} \frac{m}{2\gamma k_{\mathrm{B}} T} \qquad (5.34)$$

Note that, since this result does not depend on the actual form of the Hamiltonian $\mathcal{H}$, it also holds for assemblies of interacting spins.

Therefore, as the thermodynamical consistency of the approach determines τ_{N} completely, we arrive at the important result that, once that the consistency condition is applied, *the Fokker–Planck equations associated with the stochastic Landau–Lifshitz (–Gilbert) and stochastic Landau–Lifshitz equations result to be identical.*[†] As τ_{N} is related with the amplitude D of the fluctuating field by different expressions [Eq. (5.29)], the only difference between the two stochastic models lies in the relation among D, λ, and T, namely

$$D = \frac{\lambda}{1 + g\lambda^2} \frac{k_{\mathrm{B}} T}{\gamma m} \qquad (5.35)$$

---

* This result follows from the general one

$$\frac{\partial}{\partial \vec{m}} \cdot (\vec{m} \wedge \vec{A}) = \sum_{i} \left(\sum_{jk} \varepsilon_{ijk} m_j \frac{\partial A_k}{\partial m_i} \right) = -\vec{m} \cdot \left(\frac{\partial}{\partial \vec{m}} \wedge \vec{A} \right)$$

when applied to $\vec{A} = B_{\mathrm{eff}} P_{\mathrm{e}}$, since $\vec{B}_{\mathrm{eff}} P_{\mathrm{e}}$ can be written as the gradient of a scalar by Eq. (5.32) and, thus, its rotational is zero.   Q.E.D.

[†] Since the stochastic Gilbert equation (5.18) is equivalent to the stochastic Landau–Lifshitz (–Gilbert) equation (5.20) with $\gamma \to \gamma/(1 + \lambda^2)$, the Fokker–Planck equation associated with the former is also given by Eq. (5.31) with τ_{N} from (5.34) after substituting $\gamma/(1 + \lambda^2)$ for γ. As $\tau_{\mathrm{N}}^{-1} \propto \gamma$, this gives a global timescale factor.

Let us write this result explicitly

$$D_{\mathrm{LLG}} = \frac{\lambda}{1+\lambda^2}\frac{k_{\mathrm{B}}T}{\gamma m}, \qquad D_{\mathrm{LL}} = \lambda\frac{k_{\mathrm{B}}T}{\gamma m}$$

so that we can compare with Brown's (1963) result. He wrote the right-hand side of the first of these equations as $(\eta/v)k_{\mathrm{B}}T$, since he began with the Gilbert equation $[\gamma \rightarrow \gamma/(1+\lambda^2)]$ and our $\lambda/\gamma m$ is equivalent to his η/v, where v is the volume of the nanoparticle.

The preceding Einstein-type relations between the amplitude of the thermal agitation field and the temperature, via the damping coefficient, ensure that the proper thermal equilibrium properties are obtained from the stochastic Landau–Lifshitz (–Gilbert) equation. They also ensure that the average dynamical properties associated with each one of these stochastic models are identical with each other (as those properties are determined by the same Fokker–Planck equation), even though the stochastic trajectories for a given realization of the fluctuating field $\vec{b}_{\mathrm{fl}}(t)$ are in principle different.

Later on we shall integrate the stochastic Landau–Lifshitz (–Gilbert) equation (5.20) numerically. Nevertheless, the considerations expressed above ensure that, if we integrate the stochastic Landau–Lifshitz equation (5.21) instead, we shall obtain the same results for the *averaged* quantities.

   c. Itô Case. It is to be noted that, since the relations (5.35) between the temperature and the amplitude of the fluctuating field [or equivalently Eq. (5.34)] are derived from Brown's Fokker–Planck equation (5.31), they *pertain to the Stratonovich stochastic calculus.* Indeed, after constructing the corresponding Fokker–Planck equation by using the Itô calculus, one finds that Eq. (5.34) does not ensure that the Boltzmann distribution is a solution of such an equation. Let us prove this.

   Let us first calculate the so-called *noise-induced* drift coefficient of the Fokker–Planck equation, namely $D\sum_{jk} B_{jk}(\partial B_{ik}/\partial y_j)$, which is the extra term accompanying A_i in Eq. (5.24). On introducing Eq. (5.27) for B_{ik} and the partial derivative (5.28) in the definition of the noise-induced drift, one finds

$$\frac{1}{\gamma^2}\sum_{jk} B_{jk}\frac{\partial B_{ik}}{\partial m_j} = \sum_{\ell j}\left(\overbrace{\sum_k \varepsilon_{j\ell k}\varepsilon_{ijk}}^{\delta_{ji}\delta_{\ell j}-\delta_{jj}\delta_{\ell i}}\right)m_\ell$$

$$- g\left(\frac{\lambda}{m}\right)^2\sum_{jk}\left(m^2\delta_{jk} - m_j m_k\right)\left(\delta_{ij}m_k + \delta_{kj}m_i\right)$$

$$= \sum_\ell (\delta_{i\ell} - 3\delta_{i\ell})m_\ell - g\left(\frac{\lambda}{m}\right)^2 m_i \sum_k \left(m^2\delta_{kk} - m_k m_k\right)$$

$$= -2\left(1 + g\lambda^2\right)m_i$$

where all the terms linear in λ have canceled out because of the contraction of symmetric tensors with antisymmetric ones. Therefore, on using the unified expression (5.29) for the Néel time, we can write the noise-induced drift coefficient as

$$D \sum_{jk} B_{jk} \frac{\partial B_{ik}}{\partial m_j} = -\frac{1}{\tau_N} m_i \qquad (5.36)$$

The Itô case of the Fokker–Planck equation is readily constructed by omitting the noise-induced drift coefficient in Eq. (5.24). As Eq. (5.36) shows, this term yields a contribution $-\tau_N^{-1} m_i P$ to the ith component of the current of probability J_i. However, in the Stratonovich case, that contribution was canceled by a term $\tau_N^{-1} m_i P$ originating from the second-order derivatives in the Fokker–Planck equation [this is a restatement of the vanishing of the second term on the right-hand side of the general Fokker–Planck equation (5.25) for the stochastic Landau–Lifshitz (–Gilbert) equation]. Anyway, the absence of the noise-induced contribution in the Itô equation yields a term $\tau_N^{-1} m_i P$ added to the Stratonovich J_i. Therefore, the Fokker–Planck equation associated with the stochastic Landau–Lifshitz (–Gilbert) equation, when this is interpreted in the Itô sense, can be written as [cf. Eq. (5.31)]

$$\frac{\partial P}{\partial t} = -\frac{\partial}{\partial \vec{m}} \cdot \left[\gamma \vec{m} \wedge \vec{B}_{\text{eff}} - \gamma \frac{\lambda}{m} \vec{m} \wedge \left(\vec{m} \wedge \vec{B}_{\text{eff}} \right) \right.$$
$$\left. + \frac{1}{\tau_N} \vec{m} + \frac{1}{2\tau_N} \vec{m} \wedge \left(\vec{m} \wedge \frac{\partial}{\partial \vec{m}} \right) \right] P \qquad (5.37)$$

Again, for the equilibrium distribution $\vec{m} \wedge \vec{B}_{\text{eff}} P_e$ is divergenceless and, if τ_N is given by Eq. (5.34), the second and fourth terms in the square brackets of Eq. (5.37) cancel each other (by construction). Therefore, the Itô Fokker–Planck equation yields, for $P = P_e$

$$0 = \frac{\partial}{\partial \vec{m}} \cdot (\vec{m} P_e) = \left(3 + \beta \vec{m} \cdot \vec{B}_{\text{eff}} \right) P_e \qquad \text{(Itô case)}$$

which is not necessarily satisfied by a general form of the Boltzmann distribution $P_e(\vec{m})$ (i.e., by a general form of the Hamiltonian). The simplest example of this inconsistency is provided by the dynamics in a constant potential. Then $\vec{B}_{\text{eff}} = 0$ and the equilibrium distribution — $P_e(\vec{m})$ uniform — is not a solution of the Itô case of the Fokker–Planck equation ($0 \neq 3P_e$). *Therefore, the stochastic Landau–Lifshitz (–Gilbert) equation, when interpreted in the Itô sense, does not yield the correct thermal equilibrium properties.*

We can give an even stronger argument against the interpretation of Eqs. (5.20) and (5.21) as Itô stochastic differential equations, based on the nonconservation of the magnitude of the magnetic moment. The deterministic counterpart of those equations [Eq. (5.6)] yields $0 = \vec{m} \cdot (d\vec{m}/dt) = \frac{1}{2}d(\vec{m}^2)/dt$, so that the magnitude of $\vec{m}$ is preserved during the time evolution. Nevertheless, when passing from ordinary to stochastic differential equations, specific rules of calculus (integration and differentiation) are required. In the context of the Stratonovich calculus, such rules are *formally* identical with the rules of the ordinary calculus. Therefore, $0 = \vec{m} \cdot (d\vec{m}/dt)$, which always follows from Eqs. (5.20) and (5.21), also entails $d(\vec{m}^2)/dt = 0$. However, when using the specific rules of differentiation of the Itô calculus one finds that $\vec{m} \cdot (d\vec{m}/dt) \neq \frac{1}{2}d(\vec{m}^2)/dt$ for those equations, which therefore *do not* conserve the magnitude of $\vec{m}$.*

   d. Fokker–Planck Equation in Spherical Coordinates. For future use, let us write the Fokker–Planck equation (5.31) in spherical coordinates, as was originally presented by Brown (1963).

First, on using $\gamma\lambda/m = \beta/2\tau_N$ [Eq. (5.33)], the Fokker–Planck equation (5.31) can be written as

$$\frac{\partial P}{\partial t} = -\frac{\partial}{\partial \vec{m}} \cdot \left\{ \gamma\vec{m} \wedge \vec{B}_{\text{eff}} - \frac{1}{2\tau_N}\vec{m} \wedge \left[\vec{m} \wedge \left(\beta\vec{B}_{\text{eff}} - \frac{\partial}{\partial \vec{m}} \right) \right] \right\} P \qquad (5.38)$$

Then, on introducing the dimensionless effective field $\vec{\xi}_{\text{eff}} = \beta m \vec{B}_{\text{eff}}$ and using once more the expression (5.33) for τ_N, we can cast Eq. (5.38) in the form

$$2\tau_N\frac{\partial P}{\partial t} = -\frac{\partial}{\partial \vec{m}} \cdot \left\{ \frac{1}{\lambda}\vec{m} \wedge \vec{\xi}_{\text{eff}} - \frac{1}{m}\vec{m} \wedge \left[\vec{m} \wedge \left(\vec{\xi}_{\text{eff}} - m\frac{\partial}{\partial \vec{m}} \right) \right] \right\} P \qquad (5.39)$$

On using now the formulas for the gradient and divergence operators in spherical coordinates, namely $(\vec{r} = \vec{m})$

$$\frac{\partial u}{\partial \vec{r}} = \hat{r}\frac{\partial u}{\partial r} + \hat{\vartheta}\frac{1}{r}\frac{\partial u}{\partial \vartheta} + \hat{\varphi}\frac{1}{r \sin \vartheta}\frac{\partial u}{\partial \varphi} \qquad (5.40)$$

$$\frac{\partial}{\partial \vec{r}} \cdot \vec{A} = \frac{1}{r^2}\frac{\partial}{\partial r}(r^2 A_r) + \frac{1}{r \sin \vartheta}\frac{\partial}{\partial \vartheta}(\sin \vartheta \, A_\vartheta) + \frac{1}{r \sin \vartheta}\frac{\partial A_\varphi}{\partial \varphi} \qquad (5.41)$$

---

*This can be demonstrated by using the Stratonovich *equivalents* of Eqs. (5.20) and (5.21) *when they are interpreted as Itô equations.* Those Stratonovich equivalent equations are obtained by augmenting the (now Itô) Eqs. (5.20) and (5.21) by $\tau_N^{-1}\vec{m}$, so that the stated result directly follows from the application of the ordinary rules of differentiation to the resulting equivalent equations.

along with the result $\vec{m} \wedge \vec{A} = m(-A_\varphi \hat{\vartheta} + A_\vartheta \hat{\varphi})$, one can write Eq. (5.39) in spherical coordinates as

$$2\tau_N \frac{\partial P}{\partial t} = -\frac{1}{\sin \vartheta} \left[\frac{\partial}{\partial \vartheta} (\sin \vartheta \tilde{J}_\vartheta) + \frac{\partial}{\partial \varphi} (\tilde{J}_\varphi) \right] \qquad (5.42)$$

where the spherical components of the *reduced* current of probability $[\tilde{J}_i = (2\tau_N/m)J_i]$ are given by

$$\tilde{J}_\vartheta = -\left[\frac{1}{k_B T} \left(\frac{\partial \mathcal{H}}{\partial \vartheta} - \frac{1}{\lambda} \frac{1}{\sin \vartheta} \frac{\partial \mathcal{H}}{\partial \varphi} \right) P + \frac{\partial P}{\partial \vartheta} \right] \qquad (5.43)$$

$$\tilde{J}_\varphi = -\left[\frac{1}{k_B T} \left(\frac{1}{\lambda} \frac{\partial \mathcal{H}}{\partial \vartheta} + \frac{1}{\sin \vartheta} \frac{\partial \mathcal{H}}{\partial \varphi} \right) P + \frac{1}{\sin \vartheta} \frac{\partial P}{\partial \varphi} \right] \qquad (5.44)$$

To get these expressions, we have also taken into account the definition (5.2) of the effective field in terms of the Hamiltonian $\mathcal{H}(\vec{m})$, which, together with Eq. (5.40), has allowed us to write the components of $\vec{\xi}_{\mathrm{eff}}$ as

$$\xi_{\mathrm{eff},\vartheta} = -\frac{1}{k_B T} \frac{\partial \mathcal{H}}{\partial \vartheta}, \qquad \xi_{\mathrm{eff},\varphi} = -\frac{1}{k_B T} \frac{1}{\sin \vartheta} \frac{\partial \mathcal{H}}{\partial \varphi} \qquad (5.45)$$

Finally, when Eqs. (5.43) and (5.44) are introduced in (5.42), Brown's Fokker–Planck equation emerges in its original form (1963).

e. The Axially Symmetric Fokker–Planck Equation as a Sturm–Liouville Problem. In an axially symmetric situation, that is, for $B_{\mathrm{eff},\varphi} = 0$ and $B_{\mathrm{eff},\vartheta} = B_{\mathrm{eff},\vartheta}(\vartheta)$, and restricting ourselves to solutions with axial symmetry $\partial P/\partial \varphi \equiv 0$ (the ones of interest when, for example, determining the steady-state solution in the presence of a longitudinal probing field), the Fokker–Planck equation (5.42) reduces to

$$2\tau_N \frac{\partial P}{\partial t} = -\frac{1}{\sin \vartheta} \frac{\partial}{\partial \vartheta} \left[\sin \vartheta \left(-\beta \frac{\partial \mathcal{H}}{\partial \vartheta} P - \frac{\partial P}{\partial \vartheta} \right) \right] \qquad (5.46)$$

Then, if we introduce the substitution $z = \cos \vartheta$ and use $(\partial f/\partial \vartheta) = -\sin \vartheta (\partial f/\partial z)$, the axially symmetric Fokker–Planck equation (5.46) can be written as

$$2\tau_N \frac{\partial P}{\partial t} = \frac{\partial}{\partial z} \left[\Omega(z) \left(\frac{\partial P}{\partial z} + \beta \mathcal{H}' P \right) \right] \qquad (5.47)$$

where we have used the shorthand $\Omega(z) = 1 - z^2$ and the prime denotes differentiation with respect to z. Note that, in this axially symmetric case, the precession terms [those multiplied by λ^{-1} in Eq. (5.39) or in Eqs. (5.43) and (5.44)] are absent from the Fokker–Planck equation. This entails that the effect of the damping parameter λ on the averaged quantities enters via the Néel time (5.34) only. Note that this no longer holds in nonaxially symmetric situations (e.g., in the presence of a transverse field).

The current of probability is defined by writing the Fokker–Planck equation (5.47) as a continuity equation for the probability distribution, namely, $2\tau_N(\partial P/\partial t) = -(\partial \tilde{J}_z/\partial z)$, whence

$$\tilde{J}_z = -\Omega(z)\left(\frac{\partial P}{\partial z} + \beta \mathcal{H}' P\right) \qquad (5.48)$$

Note that this expression can also be obtained from Eq. (5.43) for $\tilde{J}_\vartheta$, by using $\tilde{J}_z = -\tilde{J}_\vartheta \sin\vartheta$.

Now, by assuming a solution of Eq. (5.47) of the form $P(z,t) = T(t)F(z)$ (separation of variables), one gets $T(t) \propto \exp(-\Lambda t)$, while $F(z)$ then satisfies

$$\frac{d}{dz}\left\{\Omega(z)\,e^{-\beta\mathcal{H}(z)}\frac{d}{dz}\left[e^{\beta\mathcal{H}(z)}F(z)\right]\right\} = -(2\tau_N\Lambda)F(z) \qquad (5.49)$$

in which we have used the identity

$$e^{-\beta\mathcal{H}(z)}\frac{d}{dz}\left[e^{\beta\mathcal{H}(z)}F(z)\right] = \frac{dF}{dz} + \beta\mathcal{H}'F \qquad (5.50)$$

Further, on introducing the function $\phi(z) = e^{\beta\mathcal{H}(z)}F(z)$, Eq. (5.49) can equivalently be written as

$$\frac{d}{dz}\left[\Omega(z)\,e^{-\beta\mathcal{H}(z)}\frac{d\phi}{dz}\right] = -(2\tau_N\Lambda)\,e^{-\beta\mathcal{H}(z)}\phi(z) \qquad (5.51)$$

Therefore, to solve the Fokker–Planck equation in the axially symmetric case is transformed into the Sturm–Liouville problem of finding the eigenvalues Λ_k, and eigenfunctions $\phi_k(z)$ of Eq. (5.51).*

In order to prove that the problem defined by Eq. (5.51) is in fact a Sturm–Liouville problem, note first that $\Omega(z)e^{-\beta\mathcal{H}(z)} \neq 0$ inside the relevant interval $(-1, 1)$. The same holds for the function $e^{-\beta\mathcal{H}(z)}$ multiplying $\phi(z)$ on the right-hand side of Eq. (5.51). In addition, the differential operator on the left-hand side is *self-adjoint*, since, when expanding it, the coefficient of

---

* One can also define the more customary dimensionless eigenvalues by $\lambda_k = 2\tau_N\Lambda_k$.

$d\phi/dz$ is equal to the derivative of the coefficient of $d^2\phi/dz^2$. This completes the proof of that Eq. (5.51) defines a Sturm–Liouville problem. However, we must also check that the common boundary condition in Sturm–Liouville problems (see, for example, Arfken, 1985, p. 503)

$$p(z)\phi_1^*(z)\frac{d\phi_2}{dz}\bigg|_{z=-1} = p(z)\phi_1^*(z)\frac{d\phi_2}{dz}\bigg|_{z=1} \tag{5.52}$$

is satisfied. Here, $\phi_1(z)$ and $\phi_2(z)$ are two solutions of the differential equation being considered, ( )* stands for complex conjugation, and $p(z) = \Omega(z)e^{-\beta\mathcal{H}(z)}$ for the Sturm–Liouville problem (5.51). The proof is based on the fact that $\Omega(z)e^{-\beta\mathcal{H}(z)}d\phi/dz$ is proportional to J_z, as can be checked by using the definition (5.48) and the identity (5.50). The current of probability is tangent to the unit sphere, so that J_z must vanish at the poles. Thus $J_z|_{z=\pm1} = 0$, from which Eq. (5.52) follows, with the two sides of the equation equal to zero.

The property (5.52) is very important since from self-adjointness plus that boundary condition it follows the *Hermitian* character of the differential operator in the Sturm–Liouville problem. Hermitian operators have the following three important properties:

1. The eigenvalues Λ_k are real.
2. The eigenfunctions $\phi_k(z)$ are orthogonal with respect to a suitably chosen scalar product.
3. The eigenfunctions $\phi_k(z)$ [and therefore the $F_k(z)$] form a complete set.

Therefore, by using the completeness property, the general solution of the Fokker–Planck equation, $P(z,t)$, can be expanded in terms of the (basis) functions $F_k(z) = e^{-\beta\mathcal{H}(z)}\phi_k(z)$ as follows

$$P(z,t) = \mathcal{Z}^{-1}\exp[-\beta\mathcal{H}(z)] + \sum_{k\geq1}c_kF_k(z)\exp(-\Lambda_kt) \tag{5.53}$$

where $\mathcal{Z}^{-1}\exp[-\beta\mathcal{H}(z)]$ is the equilibrium ($t\rightarrow\infty$) solution (associated with the eigenvalue $\Lambda_0 = 0$) and the coefficients of the expansion c_k depend on the "initial conditions" (initial probability distribution).

In general, the eigenvalues and eigenfunctions of the Sturm–Liouville problem discussed above must be computed by means of numerical techniques. However, analytic results can be obtained for certain quantities without solving the full Sturm–Liouville problem (see Section V.C.4 below).

3. *Equations for Averages of the magnetic moment*

Let us now consider the dynamical equations for the averages of the magnetic moment with respect to the non-equilibrium probability distribution $P(\vec{m}, t)$.

(As these equations involve averaged quantities, they will be identical for the stochastic Landau–Lifshitz (–Gilbert) and stochastic Landau–Lifshitz models.)

The dynamical equations for the first two moments of a stochastic variable $y = (y_1, \ldots, y_n)$ that obeys the Fokker–Planck equation

$$\frac{\partial P}{\partial t} = -\sum_i \frac{\partial}{\partial y_i} \left[a_i^{(1)}(y, t)P \right] + \frac{1}{2} \sum_{ij} \frac{\partial^2}{\partial y_i\, \partial y_j} \left[a_{ij}^{(2)}(y, t)P \right]$$

are given by (see van Kampen, 1981, p. 130)

$$\frac{d}{dt} \langle y_i \rangle = \left\langle a_i^{(1)}(y, t) \right\rangle \tag{5.54}$$

$$\frac{d}{dt} \langle y_i y_j \rangle = \left\langle a_{ij}^{(2)}(y, t) \right\rangle + \left\langle y_i a_j^{(1)}(y, t) \right\rangle + \left\langle y_j a_i^{(1)}(y, t) \right\rangle \tag{5.55}$$

On comparing with the Fokker–Planck equation (5.24), taking Eqs. (5.26) and (5.27) into account, and using Eq. (5.36) for the noise-induced drift coefficient, we get the following expressions for the functions $a_i^{(1)}$ and $a_{ij}^{(2)}$ associated with the stochastic Landau–Lifshitz (–Gilbert) equation

$$a_i^{(1)}(\vec{m}, t) = \gamma \left[\vec{m} \wedge \vec{B}_{\text{eff}} - \frac{\lambda}{m} \vec{m} \wedge \left(\vec{m} \wedge \vec{B}_{\text{eff}} \right) \right]_i - \frac{1}{\tau_N} m_i$$

$$a_{ij}^{(2)}(\vec{m}, t) = \frac{1}{\tau_N} \left(m^2 \delta_{ij} - m_i m_j \right)$$

Thus, the dynamical equation for the first moment $\langle m_i \rangle(t) = \int_{|\vec{m}|=m} d^3\vec{m}\ m_i P(\vec{m}, t)$ reads

$$\frac{d}{dt} \langle \vec{m} \rangle = \gamma \left\langle \vec{m} \wedge \vec{B}_{\text{eff}} \right\rangle - \gamma \frac{\lambda}{m} \left\langle \vec{m} \wedge \left(\vec{m} \wedge \vec{B}_{\text{eff}} \right) \right\rangle - \frac{1}{\tau_N} \langle \vec{m} \rangle \tag{5.56}$$

where the results for the Cartesian components have been gathered in vector form. Note that the term $-\langle \vec{m} \rangle / \tau_N$ is analogous to the relaxation term in a Bloch-type equation (Garanin, Ishchenko and Panina, 1990). Analogously, for the second-order moments $\langle m_i m_j \rangle(t)$ one finds

$$\frac{d}{dt} \langle m_i m_j \rangle = -\frac{3}{2\tau_N} \left(\langle m_i m_j \rangle - \frac{1}{3} m^2 \delta_{ij} \right) + \gamma \left\langle m_i \left(\vec{m} \wedge \vec{B}_{\text{eff}} \right)_j \right\rangle$$

$$- \gamma \frac{\lambda}{m} \left\langle m_i \left[\vec{m} \wedge \left(\vec{m} \wedge \vec{B}_{\text{eff}} \right) \right]_j \right\rangle + i \leftrightarrow j \tag{5.57}$$

where $i \leftrightarrow j$ stands for the interchange in the entire previous expression of the subscripts i and j.

Equations (5.56) and (5.57) show that, for a general form of the Hamiltonian, no closed equation exists for the time evolution of the averages of the magnetic moment. For instance, even if $\vec{B}_{eff}$ does not depend on $\vec{m}$, for example, for $\vec{B}_{eff} = \vec{B}$, the Landau–Lifshitz-type relaxation term introduces $\langle m_i m_j \rangle(t)$ in the equation for $\langle m_i \rangle(t)$. Therefore, an additional differential equation for $\langle m_i m_j \rangle(t)$ is required [i.e., Eq. (5.57)]; however, that equation involves $\langle m_i m_j m_k \rangle(t)$, and so on. *The absence of closed dynamical equations for the averages is a major source of mathematical difficulties in the theoretical study of the dynamical properties of classical spins.*

We now discuss the *free-diffusion case*, in which the equations for the averages are not coupled and can, in addition, be explicitly solved. Here the Hamiltonian is constant (independent of $\vec{m}$), so that $\vec{B}_{eff} = 0$ and the equations for the first two moments reduce to

$$\frac{d}{dt}\langle m_i \rangle = -\frac{1}{\tau_N}\langle m_i \rangle, \qquad \frac{d}{dt}\langle m_i m_j \rangle = -\frac{3}{\tau_N}\left(\langle m_i m_j \rangle - \frac{1}{3}m^2 \delta_{ij} \right) \qquad (5.58)$$

Note that, because $\tau_N^{-1} \propto k_B T$ [Eq. (5.34)], this apparently academic case can be a reasonable approximation for sufficiently high temperatures, where the terms multiplied by τ_N^{-1} in Eqs. (5.56) (the above-mentioned Bloch-type term) and (5.57) dominate the remaining terms.

The solution for the first moment is

$$\langle m_i \rangle(t) = \langle m_i \rangle(t_0)e^{-(t-t_0)/\tau_N} \qquad (5.59)$$

which justifies calling the characteristic time constant τ_N the *free-diffusion* time. Similarly, on using $d\langle m_i m_j \rangle/dt = d(\langle m_i m_j \rangle - \frac{1}{3}m^2 \delta_{ij})/dt$, one gets for the second-order moments

$$\langle m_i m_j \rangle(t) = \frac{1}{3}m^2 \delta_{ij} + \left[\langle m_i m_j \rangle(t_0) - \frac{1}{3}m^2 \delta_{ij} \right] e^{-3(t-t_0)/\tau_N} \qquad (5.60)$$

For $(t - t_0) \gg \tau_N$, one therefore finds $\langle m_i \rangle(t) \to 0$ and $\langle m_i m_j \rangle(t) \to \frac{1}{3}m^2 \delta_{ij}$. Thus, the initial correlations between different components of the magnetic moment are lost for long times, while $\langle m_i^2 \rangle(t) \to \frac{1}{3}m^2, \forall i$ (random distribution of $\vec{m}$) as it should for the diffusion in a constant orientational potential or at very high temperatures. Note finally that these natural results are not obtained when one interprets the stochastic Landau–Lifshitz (–Gilbert) equation à la Itô.

4. Relaxation Times

We conclude our discussion of the Brown–Kubo–Hashitsume stochastic model by reviewing briefly various expressions for the relaxation times in the context of this model. For a comprehensive review including the comparison of different definitions and methods for the calculation of relaxation times, see (Coffey, 1998).

a. Longitudinal Relaxation Time. The longitudinal relaxation time is associated with the response to a field applied along the anisotropy axis. Therefore, the very definition of this quantity requires a previous discussion of the relaxation under such conditions.

Let us assume that the Hamiltonian $\mathcal{H}$ has uniaxial symmetry, so that the transformation discussed above of the Fokker–Planck equation into a Sturm–Liouville problem holds. Let us also assume that $\mathcal{H}$ contains, among other terms, a (longitudinal) Zeeman term $-\beta\mathcal{H}_{\text{Zeeman}} = \beta(m_z B) = z\xi$, where $\xi = \beta m B$ is the customary dimensionless magnetic field parameter. By averaging $m_z(t)$ with respect to the nonequilibrium probability distribution (5.53), the relaxation curve, after a sudden infinitesimal change at $t = 0$ on the applied field B by ΔB, reads

$$\langle m_z(\infty)\rangle - \langle m_z(t)\rangle = \mu_0^{-1}\Delta B\chi_{\parallel}\sum_{k\geq 1}a_k\exp(-\Lambda_k t) \qquad (5.61)$$

Here $\chi_{\parallel} = \mu_0\partial\langle m_z\rangle_e/\partial B$ is the longitudinal equilibrium susceptibility [$\langle\cdot\rangle_e$ denotes the thermal average in the absence of the perturbing field ΔB, i.e., with respect to the initial distribution $P_e = \mathcal{Z}_0^{-1}\exp(-\beta\mathcal{H}_0)$].* In Eq. (5.61) the Λ_k are the eigenvalues of the associated Sturm–Liouville problem and the a_k are the corresponding amplitudes, which are related with the constants c_k of Eq. (5.53) and also involve integrals of the form $\int_{-1}^{1}dzF_k(z)z$. Those amplitudes, by construction, obey the sum rule $\sum_{k\geq 1}a_k = 1$, as can be seen by considering that at $t = 0$ the system was in thermal equilibrium, so that $\mu_0\langle m_z(\infty) - m_z(0)\rangle = \Delta B\chi_{\parallel}$.

The eigenvalues are usually sorted in increasing order $0 = \Lambda_0 < \Lambda_1 \leq \Lambda_2\cdots$. The first nonvanishing eigenvalue, Λ_1, is commonly associated with the inter-potential-well dynamics, while the information about the intra-potential-well relaxation appears in the higher-order eigenvalues $\Lambda_k, k \geq 2$. In some cases, however, Λ_1 corresponds to a "long lived" mode and characterizes reasonably well the relaxation (except for its earliest stages).

---

* We omit the subscript $\parallel$ in the equilibrium distribution function and in the corresponding partition function.

On defining the longitudinal relaxation time as $\tau_\parallel = \Lambda_1^{-1}$, Brown (1963) derived the approximate results

$$\tau_\parallel \simeq \begin{cases} \tau_N \left[1 - \dfrac{2}{5}\sigma + \dfrac{48}{875}\sigma^2 \left(1 + \dfrac{175}{24}h^2 \right) \right]^{-1}, & \sigma \ll 1 \\[2ex] \tau_N \dfrac{\sqrt{\pi}}{2} \sigma^{-3/2} \dfrac{\exp[\sigma(1 + h^2)]}{(1 - h^2)(\cosh\xi - h\sinh\xi)}, & \sigma \gg 1 \end{cases} \qquad (5.62)$$

where $\sigma = Kv/k_BT$ is the dimensionless barrier-height parameter, $h = \xi/2\sigma$, and τ_N is the Néel time (5.34). To get these formulae Brown solved the Fokker–Planck equation perturbatively in the low potential-barrier case and with the use of the Kramers transition-state method in the high-barrier limit.

Cregg, Crothers and Wickstead (1994) *proposed* a simple expression for Λ_1 that is remarkably close to the exact Λ_1 in the entire σ range. It is essentially a formula that interpolates between the limiting results (5.62) of Brown and reads $(\tau_\parallel^{-1} = \Lambda_1)$

$$\tau_\parallel^{-1} \simeq \tau_N^{-1} \frac{1}{2} \left(1 - h^2 \right) \left(\frac{2}{\sqrt{\pi}} \frac{\sigma^{3/2}}{1 + 1/\sigma} + \sigma 2^{-\sigma} \right)$$

$$\times \left\{ \frac{1 - h}{\exp\left[\sigma(1 - h)^2 \right] - 1} + \frac{1 + h}{\exp[\sigma(1 + h)^2] - 1} \right\} \qquad (5.63)$$

Nevertheless, when the relaxation comprises different decay modes, a more useful characterization of the thermoactivation rate is provided by the *integral relaxation time* τ_{int}, which is in general defined as the area under the relaxation curve (normalized at $t = 0$) after a sudden infinitesimal change at $t = 0$ on the external control parameter. A general expression for the integral relaxation time associated with any one-dimensional Fokker–Planck equation was obtained by Jung and Risken (1985). Moro and Nordio (1985), in the context of thermoactivation phenomena in chemical-physics problems, also derived a similar expression.

In the context of the Brown–Kubo–Hashitsume model for classical spins, τ_{int} was calculated for systems with uniaxial anisotropy in a longitudinal magnetic field by Garanin et al. (1990). Here, the relaxing quantity is the average magnetic moment $\langle m_z(t)\rangle$, and the external control parameter is the magnetic field. Thus, the general definition reduces in this case to

$$\tau_{int} = \int_0^\infty dt\, \frac{\langle m_z(\infty)\rangle - \langle m_z(t)\rangle}{\langle m_z(\infty)\rangle - \langle m_z(0)\rangle} \qquad (5.64)$$

For a single exponential decay, specifically, $[\langle m_z(\infty)\rangle - \langle m_z(t)\rangle] \propto \exp(-t/\tau)$, this definition indeed yields $\tau_{int} = \tau$, whereas for a multiexponential decay,

τ_{int} is given by the average of the corresponding relaxation times weighted by the associated amplitudes. Indeed, when $\langle m_z(\infty) \rangle - \langle m_z(t) \rangle$ from Eq. (5.61) is substituted into the definition (5.64), τ_{int} emerges in the form

$$\tau_{int} = \frac{1}{\sum_{k \geq 1} a_k} \sum_{k \geq 1} a_k \int_0^\infty dt \exp(-\Lambda_k t) = \sum_{k \geq 1} a_k \Lambda_k^{-1} \qquad (5.65)$$

where we have used the sum rule $\sum_{k \geq 1} a_k = 1$.

In order to calculate τ_{int} without finding the eigenvalues and amplitudes of the Sturm–Liouville problem, Garanin et al. (1990) used the relation between τ_{int} and the low-frequency longitudinal susceptibility to get (see also Garanin, 1996, and Appendix B).

$$\tau_{int} = \frac{2\tau_N}{\partial \langle z \rangle_e / \partial \xi} \int_{-1}^1 \frac{dz}{1 - z^2} \frac{\Phi(z)^2}{P_e(z)} \qquad (5.66)$$

where the function $\Phi(z)$ is given by

$$\Phi(z) = \int_{-1}^z dz_1 P_e(z_1) \left[\langle z \rangle_e - z_1 \right] \qquad (5.67)$$

Equation (5.66), which is valid for *any* axially symmetric Hamiltonian, can readily be computed by means of the numerical integration of a double definite integral. Moreover, explicit expressions for $\Phi(z)$ can be derived for particular forms of the Hamiltonian (see Appendix B).

In the absence of a constant magnetic field (unbiased case), the integral relaxation time yields the results of Brown for Λ_1^{-1} in the appropriate limiting cases. However, τ_{int} depends on the whole set of eigenvalues Λ_k, and is in principle more informative than Λ_1. Indeed, in the presence of a bias field, the higher-order modes can make a substantial contribution to the relaxation, and Λ_1^{-1} can largely (exponentially) deviate from τ_{int} in the low-temperature region (Coffey, Crothers, Kalmykov and Waldron, 1995a; Garanin, 1996). Finally, in contrast to Λ_1, the integral relaxation time is, by definition, a directly mensurable quantity.

b. Transverse Relaxation Time. The formula usually employed for the transverse relaxation time is that yielded by the effective eigenvalue method (see, for example, Coffey, Kalmykov and Massawe, 1993)

$$\tau_\perp^{od} = 2\tau_N \frac{1 - S_2(\sigma, \xi)}{2 + S_2(\sigma, \xi)} \qquad (5.68)$$

where S_2 is the average of the second Legendre polynomial with respect to the thermal equilibrium distribution [Eq. (3.69)]. This equation, although valid for any axially symmetric Hamiltonian, does not take gyromagnetic effects into account [Eq. (5.68) holds only in the *overdamped* case, $\lambda \gg 1$].

On noting that from Eq. (3.76) one gets the relation $\tilde{S}_2 = \frac{1}{2}[3(R'/R) - 1]$ between $\tilde{S}_2 \equiv S_2(\sigma, \xi)|_{\xi=0}$ and R'/R, one sees that Eq. (5.68) reduces in the unbiased case to

$$\tau_\perp^{od}|_{\xi=0} = 2\tau_N \frac{1 - R'/R}{1 + R'/R} \tag{5.69}$$

Now, if we employ the small and large σ approximations for R'/R (Appendix A), we find

$$\frac{1 - R'/R}{1 + R'/R} \simeq \frac{1/\sigma}{2 - 1/\sigma} \quad (\sigma \gg 1), \qquad \frac{1 - R'/R}{1 + R'/R} \simeq \frac{2 - 4\sigma/15}{4 + 4\sigma/15} \quad (\sigma \ll 1)$$

whence one gets the limit behaviors of $\tau_\perp^{od}|_{\xi=0}$

$$\tau_\perp^{od}|_{\xi=0} \longrightarrow \begin{cases} 1/(\lambda\gamma B_K) & \text{as} \quad T \to 0 \\ 0 & \text{as} \quad T \to \infty \end{cases} \tag{5.70}$$

Thus, as it should, $\tau_\perp^{od}|_{\xi=0}$ goes to zero at high temperatures, whereas it tends to the constant deterministic result τ_K as $T \to 0$ [Eq. (5.17)].

Finally, it is shown in Fig. 22 that, in contrast to the longitudinal relaxation time, which may increase exponentially at low temperatures, $\tau_\perp^{od}|_{\xi=0}$ is

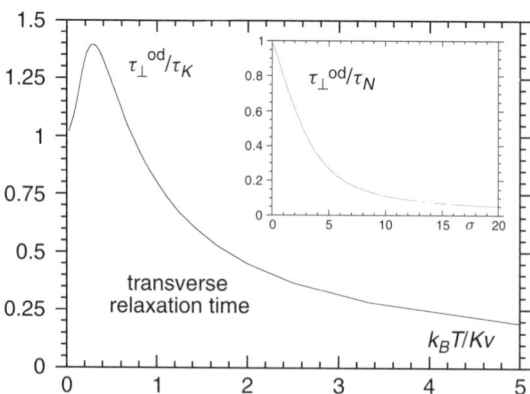

FIGURE 22. Transverse relaxation time $\tau_\perp^{od}|_{\xi=0}$ [in units of $\tau_K = 1/(\lambda\gamma B_K)$] versus the temperature. Inset: $\tau_\perp^{od}|_{\xi=0}/\tau_N$ versus $\sigma = Kv/k_B T$.

bounded from above. Indeed, from the graph one concludes that this transverse relaxation time is, at most, of the order of τ_K (specifically, $\tau_\perp^{od}|_{\xi=0} < 1.5\tau_K$).

On the other hand, we have mentioned that the expression (5.68) for the transverse relaxation time does not take the effects of the precession terms into account. In order to investigate the effects of these terms on the transverse response, Raĭkher and Shliomis (1975, 1994) studied the transverse dynamical susceptibility for $B = 0$, by a decoupling ansatz for the infinite system of differential equations for the averages of the magnetic moment [recall the discussion following Eqs. (5.56) and (5.57)]. They derived an expression for $\chi_\perp(\omega)$ and studied mainly the ferromagnetic-resonance frequency range. If one is interested in the low-frequency range, their $\chi_\perp(\omega)$ can be expanded in powers of ω, and then cast into the Debye-type form

$$\chi(\omega, T) \simeq \chi(T)(1 - i\omega\tau) \simeq \frac{\chi(T)}{1 + i\omega\tau}, \qquad \omega\tau \ll 1$$

where the last step is done with help from the binomial expansion $(1 + x)^\varepsilon = 1 + \varepsilon x + \cdots$. Then, the quantity multiplying $i\omega$ in the denominator defines an effective relaxation time useful in the *low*-frequency range, which is given by (see Appendix B)

$$\tau_\perp|_{\xi=0} = 2\tau_N \frac{1 - \tilde{S}_2}{2 + \tilde{S}_2} \frac{1}{1 + p(\sigma)/\lambda^2} \tag{5.71}$$

where

$$p(\sigma) = \frac{(3\tilde{S}_2)^2}{(2 + \tilde{S}_2)[2 + \tilde{S}_2(1 - 6/\sigma)]} \tag{5.72}$$

Note that, in the absence of the precession terms ($\lambda \to \infty$), Eq. (5.71) reduces to the unbiased case (5.69) of the overdamped result (5.68).

Finally, on considering that (we use $p(\sigma) \geq 0$)

$$\tau_\perp|_{\xi=0} = 2\tau_N \frac{1 - \tilde{S}_2}{2 + \tilde{S}_2} \frac{1}{1 + p(\sigma)/\lambda^2} \leq 2\tau_N \frac{1 - \tilde{S}_2}{2 + \tilde{S}_2} = \tau_\perp^{od}|_{\xi=0}$$

we also find that $\tau_\perp|_{\xi=0}$ is, at most, of the order of $\tau_K = (\lambda\gamma B_K)^{-1}$. For typical values of the quantities occurring in τ_K one has

$$\left.\begin{array}{l} \gamma = 1.76 \times 10^{11} \text{ T}^{-1} \text{ s}^{-1} \\ B_K \sim 50 \text{ mT} \\ \lambda \sim 0.01 - 1 \end{array}\right\} \longrightarrow \tau_K^{-1} \sim 10^8 - 10^{10} \text{ s}^{-1} \tag{5.73}$$

Thus, for moderate frequencies (say, $\omega < 10^6$ Hz) the condition $\omega\tau_\perp|_{\xi=0} \ll 1$ is obeyed, justifying (admittedly, nonrigorously) the step leading to the low-frequency Shliomis–Stepanov equation (4.3) in Section IV.

D. Numerical Method

In the remainder of this section we shall study the $T \neq 0$ dynamics by solving stochastic dynamical equations for classical spins numerically. To this end, we now discuss some topics related with the numerical integration of those Langevin equations.

1. Dimensionless Quantities

Let us first introduce a number of dimensionless quantities. The maximum anisotropy field $B_K = 2Kv/m$ provides a suitable reference magnetic-field scale that yields the dimensionless fields (in what follows we consider only easy-axis anisotropy $K > 0$)

$$\vec{h} = \frac{\vec{B}}{B_K}, \qquad \vec{h}_{\text{eff}} = \frac{\vec{B}_{\text{eff}}}{B_K}, \qquad \vec{h}_{\text{fl}}(t) = \frac{\vec{b}_{\text{fl}}(t)}{B_K} \qquad (5.74)$$

A suitable timescale is provided by τ_K, the deterministic relaxation time at $B = 0$ [Eq. (5.17)], which yields the dimensionless time

$$\bar{t} = \frac{t}{\tau_K}, \qquad \tau_K = \frac{1}{\lambda\gamma B_K} \qquad (5.75)$$

Note that in terms of τ_K and $\sigma = Kv/k_BT$, the Néel time (5.34) is merely given by

$$\tau_N = \sigma\tau_K \qquad (5.76)$$

2. Dimensionless Stochastic Landau–Lifshitz (–Gilbert) Equation

On using the dimensionless quantities introduced, the stochastic Landau–Lifshitz (–Gilbert) equation can be rewritten in a dimensionless form suitable for computation, namely

$$\frac{d\vec{e}}{d\bar{t}} = \frac{1}{\lambda}\vec{e} \wedge \left[\vec{h}_{\text{eff}} + \vec{h}_{\text{fl}}(\bar{t})\right] - \vec{e} \wedge \left\{\vec{e} \wedge \left[\vec{h}_{\text{eff}} + g\vec{h}_{\text{fl}}(\bar{t})\right]\right\} \qquad (5.77)$$

where $\vec{e} = \vec{m}/m$ is the unit vector in the direction of $\vec{m}$ and $g = 1$ for Eq. (5.20) while $g = 0$ for Eq. (5.21). The statistical properties of the dimensionless fluctuating field $\vec{h}_{\text{fl}}(\bar{t})$, which arise directly from those of $\vec{b}_{\text{fl}}(t)$ [Eqs. (5.19)], are

$$\langle h_{\text{fl},k}(\bar{t})\rangle = 0, \qquad \langle h_{\text{fl},k}(\bar{t})h_{\text{fl},\ell}(\bar{t}')\rangle = 2\overline{D}\delta_{k\ell}\delta(\bar{t} - \bar{t}') \qquad (5.78)$$

where [we use Eq. (5.35) for D and $\delta(t) = \delta(\bar{t}) \, d\bar{t}/dt = \delta(\bar{t})/\tau_K$] the dimensionless coefficient $\overline{D}$ is given by:

$$\overline{D} = \frac{D}{\tau_K B_K^2} = \frac{\lambda^2}{1 + g\lambda^2} \frac{k_B T}{m B_K} = \frac{\lambda^2}{1 + g\lambda^2} \frac{k_B T}{2Kv} \tag{5.79}$$

Let us finally cast the dimensionless Eq. (5.77) into the form of the general system of Langevin equations (5.22):

$$\frac{de_i}{d\bar{t}} = \overline{A}_i + \sum_k \overline{B}_{ik} h_{\mathrm{fl},k}(\bar{t}) \tag{5.80}$$

where k runs over x, y, z, and [cf. Eqs. (5.26) and (5.27)]

$$\overline{A}_i = \sum_k \left[\frac{1}{\lambda} \sum_j \varepsilon_{ijk} e_j + (\delta_{ik} - e_i e_k) \right] h_{\mathrm{eff},k} \tag{5.81}$$

$$\overline{B}_{ik} = \frac{1}{\lambda} \sum_j \varepsilon_{ijk} e_j + g(\delta_{ik} - e_i e_k) \tag{5.82}$$

3. Choice of the Numerical Scheme

As has been mentioned, the stochastic Landau–Lifshitz (–Gilbert) equation contains multiplicative white-noise terms [Eq. (5.27) or its dimensionless counterpart (5.82) clearly depend on the spin variables for both $g = 0$ and $g = 1$]. Together with difficulties at the level of definition, the occurrence of multiplicative white noise in a Langevin equation entails some technical problems as well. For instance, serious difficulties arise in developing high-order numerical integration schemes for this case (Kloeden and Platen, 1995). In general, the simple translation of a numerical scheme valid for deterministic differential equations does not necessarily yield a proper scheme in the stochastic case:

1. Depending on the original deterministic scheme chosen, its naive stochastic translation might converge to an Itô solution, to a Stratonovich solution, or to none of them.

2. Even if there exists proper convergence of the scheme chosen in the context of the stochastic calculus used, the *order of convergence* obtained is usually lower than that of the original deterministic scheme.

Let us consider the stochastic generalization of the deterministic Heun scheme, namely

$$y_i(t + \Delta t) = y_i(t) + \tfrac{1}{2} \left[A_i(\tilde{\mathbf{y}}, t + \Delta t) + A_i(\mathbf{y}, t) \right] \Delta t$$

$$+ \tfrac{1}{2} \sum_k \left[B_{ik}(\tilde{\mathbf{y}}, t + \Delta t) + B_{ik}(\mathbf{y}, t) \right] \Delta W_k \tag{5.83}$$

where Δt is the discretization time interval, $\mathbf{y} = \mathbf{y}(t)$, the $\tilde{y}_i$ are Euler-type supporting values

$$\tilde{y}_i = y_i(t) + A_i(\mathbf{y}, t)\Delta t + \sum_k B_{ik}(\mathbf{y}, t)\Delta W_k \qquad (5.84)$$

and the $\Delta W_k = \int_t^{t+\Delta t} dt' L_k(t')$ are *Gaussian* random numbers whose first two moments are

$$\langle \Delta W_k \rangle = 0, \quad \langle \Delta W_k \Delta W_\ell \rangle = (2D\Delta t)\delta_{k\ell} \qquad (5.85)$$

The Heun scheme converges *in quadratic mean* to the solution of the general system of stochastic differential equations (5.22) supplemented by Eqs. (5.23), *when interpreted in the sense of Stratonovich* (see, for example, Rümelin, 1982).

On the other hand, if one uses the Euler-type Eq. (5.84) as the numerical integration scheme [by identifying $y_i(t + \Delta t) = \tilde{y}_i$], the constructed trajectory *converges to the Itô solution* of the same system of equations (5.22). A proper Euler-type scheme in the context of the Stratonovich stochastic calculus is obtained when the deterministic drift in Eq. (5.84), A_i, is augmented by the noise-induced drift, namely

$$y_i(t + \Delta t) = y_i(t) + \left[A_i + D \sum_{jk} B_{jk} \frac{\partial B_{ik}}{\partial y_j} \right]_{(\mathbf{y},t)} \Delta t + \sum_k B_{ik}(\mathbf{y}, t)\Delta W_k$$

$$(5.86)$$

(for an alternative Euler-type algorithm for multiplicative noise, see Ramírez-Piscina, Sancho and Hernández-Machado, 1993). In order to use the scheme (5.86), one needs to calculate the corresponding noise-induced drift. This was already done, yielding Eq. (5.36), which can readily be adapted to the dimensionless Eq. (5.80):

$$\overline{D} \sum_{jk} \overline{B}_{jk} \frac{\partial \overline{B}_{ik}}{\partial e_j} = -\frac{1}{\tau_N/\tau_K} e_i = -\frac{k_B T}{K v} e_i$$

where Eq. (5.76) has been used to write the last equality.*

However, in order to choose the numerical scheme to undertake the integration of Eq. (5.77), it is convenient to determine first the *character* of the

---

* On recalling that in Eq. (5.77) the time is measured in units of τ_K, one realizes that the term $-(\tau_K/\tau_N)e_i$ corresponds to $-\langle m_i \rangle/\tau_N$ in the averaged dynamical equation (5.56). Indeed, by using $\langle \Delta W_k \rangle = 0$ for averaging Eq. (5.86) when particularized to the stochastic Landau–Lifshitz (–Gilbert) equation, one gets the discretized version of Eq. (5.56).

multiplicative noise in that equation. When the B_{ik} fulfill the relation

$$\sum_j B_{jk} \frac{\partial B_{i\ell}}{\partial y_j} = \sum_j B_{j\ell} \frac{\partial B_{ik}}{\partial y_j}, \qquad \forall i \qquad (5.87)$$

(i.e., *symmetry* with respect to the subscripts k and ℓ), the noise in the Langevin equations is said to be *commutative*. The condition of commutative noise is rather general and includes additive noise, $\partial B_{ik}/\partial y_j \equiv 0$, diagonal multiplicative noise, $B_{ik}(\mathbf{y}, t) = \delta_{ik} B_{ii}(y_i)$, and linear multiplicative noise, $B_{ik}(\mathbf{y}, t) = B_{ik}(t)y_i$ (see, for example, Kloeden and Platen, 1995, p. 348). In addition, when Eq. (5.87) is satisfied, the stochastic Heun scheme (5.83) has an order of convergence higher than the order of convergence of the Euler scheme (5.86) (see, for example, Rümelin, 1982).

Unfortunately, the noise in the stochastic Landau–Lifshitz (–Gilbert) equation is not only multiplicative but *noncommutative* as well. Indeed, on calculating the right-hand side of Eq. (5.87) with B_{ik} from Eq. (5.27), we find

$$\frac{1}{\gamma^2}\sum_j B_{j\ell}\frac{\partial B_{ik}}{\partial m_j} = -m_i \underbrace{\delta_{k\ell}}_{S} + \underbrace{m_k\delta_{i\ell}}_{ND}$$

$$+ g\frac{\lambda}{m}\left(\underbrace{m^2\varepsilon_{i\ell k}}_{A} - \underbrace{m_\ell \sum_j \varepsilon_{ijk}m_j - m_k\sum_r \varepsilon_{ir\ell}m_r}_{S} - \underbrace{m_i\sum_r \varepsilon_{kr\ell}m_r}_{A}\right)$$

$$- g\left(\frac{\lambda}{m}\right)^2\left(\underbrace{m^2 m_k\delta_{i\ell}}_{ND} + \underbrace{m^2 m_i\,\delta_{k\ell}}_{S} - \underbrace{2m_i\,m_k m_\ell}_{S}\right)$$

where S, A, and ND, indicate, respectively, symmetry, antisymmetry, and not-defined symmetry with respect to the subscripts k and ℓ. Therefore, owing to the presence of the last two types of terms, the commutative noise condition *is not* obeyed by either the stochastic Landau–Lifshitz (–Gilbert) or the stochastic Landau–Lifshitz equation.   Q.E.D.

For noncommutative noise the best order of convergence is attained (Rümelin, 1982) with the Heun scheme (5.83) but also with the simpler Euler algorithm (5.86) or the scheme of Ramírez-Piscina et al. (1993). Although the Heun scheme requires the evaluation of A_i and B_{ik} at two points per time step (at the initial and support ones), we have chosen it to integrate the stochastic Landau–Lifshitz (–Gilbert) equation. This is done because

1. The Heun scheme yields Stratonovich solutions of the stochastic differential equations naturally, without alterations to the drift term.

2. The deterministic part of the differential equations is treated with a second-order accuracy in Δt, which renders the Heun scheme numerically more stable than the Euler-type schemes.

We finally emphasize that, in order to integrate the stochastic Landau–Lifshitz (–Gilbert) equation numerically one *cannot* merely employ a bare Euler-type scheme such as (5.84), since this scheme yields Itô solutions of the differential equations. Even the stationary properties derived by means of such an approach would not coincide with the correct thermal equilibrium properties [recall the discussion following Eq. (5.37)].

4. Implementation

The integration of the stochastic Landau–Lifshitz (–Gilbert) equation is performed by starting from a given initial configuration, and updating recursively the state of the system, $\vec{m}(t) \rightarrow \vec{m}(t + \Delta t)$, by means of the set of finite-difference equations (5.83). This generates *stochastic trajectories* from which, when required, averages are directly computed. If one extrapolates the results obtained to zero discretization time interval Δt, the only error in the *averaged* quantities is a statistical error bar that can, in principle, be made arbitrarily small by averaging over a sufficiently large number of trajectories. We seldom carry out such $\Delta t \rightarrow 0$ limiting procedure, but we employ a discretization time interval small enough. Unless otherwise stated, the choice $\Delta t = 0.01\tau_K$ is made.

When computing average quantities, in order to minimize effects that are not caused by the application of the probing field $\Delta \vec{B}(t)$, the following *subtraction* method is used. Starting from the same initial configuration, the equations of motion are solved for two identical ensembles, one in the presence of $\Delta \vec{B}(t)$ and the other subjected to $-\Delta \vec{B}(t)$, and the time evolution analyzed is that of

$$M_{\text{sub}}(t) = \tfrac{1}{2} \left\{ \sum \vec{m} \left[\Delta \vec{B}(t) \right] - \sum \vec{m} \left[-\Delta \vec{B}(t) \right] \right\}$$

Moreover, we have found that this technique significantly diminishes the number of stochastic trajectories required to achieve convergence in the averaged results. In addition, the subtraction technique automatically eliminates the nonlinear terms *quadratic* in the probing field that could emerge.

Finally, the *Gaussian* random numbers required to simulate the ΔW_k entering in the integration schemes, are constructed from *uniformly* distributed random numbers by means of the Box–Muller algorithm. Thus, if r_1 and r_2 are random numbers uniformly distributed in the interval $(0,1)$ (as those pseudo–random numbers provided by a computer), the transformation

$$w_1 = \sqrt{-2\ln(r_1)}\cos(2\pi r_2)$$

$$w_2 = \sqrt{-2\ln(r_1)}\sin(2\pi r_2)$$

outputs w_1 and w_2, which are Gaussian-distributed random numbers of zero mean and variance unity (if one needs Gaussian numbers with variance σ^2, these are immediately obtained by multiplying the above w_i by σ). The generation of the random numbers is usually the slowest step in the recursive scheme. Therefore, when computing a quantity at various temperatures we generate all the trajectories at once, by using the same sequence of random numbers for the different temperatures.

E.   Stochastic Trajectories of Individual Spins

We shall now study the $T \neq 0$ dynamics of *individual* magnetic moments. To this end, we integrate the *stochastic* Landau–Lifshitz–Gilbert equation (5.20) numerically in the context of the Stratonovich calculus, by means of the stochastic generalization (5.83) of the Heun scheme. If one wishes to have a reference of the timescales involved, one can assume values such as those of Eq. (5.73), so that $\tau_K \sim 10^{-10} - 10^{-8}$ s.

1.   The Overbarrier Rotation Process

Figure 23 displays the projection of the trajectory of a magnetic moment with the simplest axially symmetric anisotropy onto selected planes. No magnetic field has been applied, so the graphs show the (in this sense) "intrinsic" dynamics.

The projection of $\vec{m}(t)$ onto a plane containing the anisotropy axis $\hat{n}$ (defining the $\hat{z}$ direction in Fig. 23) corresponds to a typical stochastic trajectory that starts close to one of the potential minima ($\vec{m} = m\hat{z}$) and, after some irregular rotations about it, reaches the potential-barrier (equatorial) region, where it wanders for a while, and eventually descends to the other potential minimum. Concerning the projection of this motion onto a plane perpendicular to the anisotropy axis, we have only shown the first stages, after the last potential-barrier crossing, of the damped precession of $\vec{m}$ about the anisotropy field when spiralling down to the bottom of the $m_z < 0$ potential well.

These graphs reveal the important role of the precession terms in the stochastic dynamics of the spin. Thus, the projection of $\vec{m}(t)$ onto the equatorial plane shows some of the typical irregular features of ordinary Brownian motion, although the rotary character is clearly exhibited. Concerning the projection of $\vec{m}(t)$ onto a plane containing the anisotropy axis, it can clearly be seen that crossing the potential barrier does not entail an immediate descent to the other potential minimum, but the precession terms together with an appropriate sequence of fluctuating fields can produce a rapid crossing back to the initial potential well.

For an ordinary, nongyromagnetic system, specifically, a mechanical system with inertia, the inertia guarantees that, unless the system reaches the top of the potential barrier with zero velocity, it will descend to the other potential

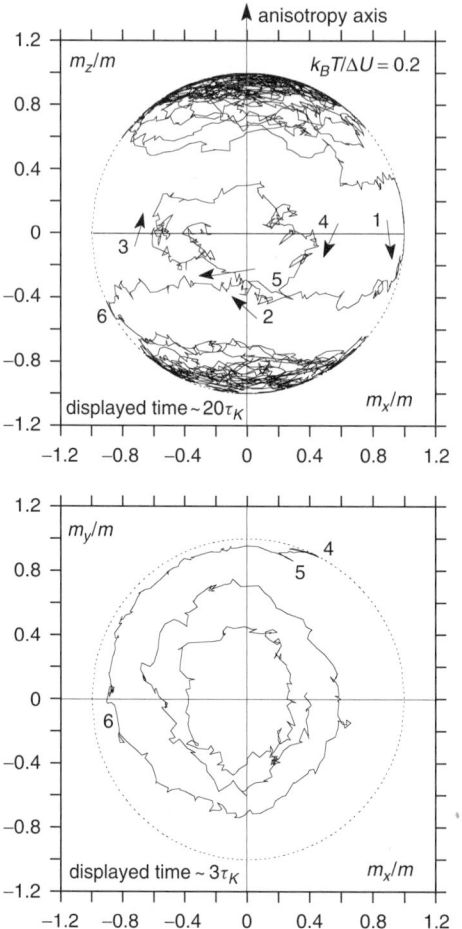

FIGURE 23. Two-dimensional projections of $\vec{m}(t)$, as determined by numerical integration of the stochastic Landau–Lifshitz (–Gilbert) equation (5.20). The magnetic-anisotropy energy is $-\Delta U(m_z/m)^2$, no external field has been applied, and the damping coefficient in the dynamical equation is $\lambda = 0.1$. Upper panel: projection onto a plane containing the anisotropy axis. Lower panel: projection onto a plane perpendicular to the anisotropy axis of the first stages of the damped precession down to the $\vec{m} = -m\hat{z}$ potential minimum, after the last potential-barrier crossing.

well with a large probability. Moreover, the forces, after the potential-barrier crossing, accelerate the system downward. In contrast, in the gyromagnetic case the dynamics is "noninertial" (the equations of motion are of first order in the time). Besides, the anisotropy field $\vec{B}_a = (B_K/m)m_z\hat{z}$ indeed drives $\vec{m}$ down to the bottom of the potential well, but this is effected via a damped precession about the anisotropy axis. Moreover, the effective precession "frequency" of this motion $\omega_{\text{eff}} \propto m_z$ is initially rather low because the anisotropy field is low near the top of the barrier ($m_z \simeq 0$). Consequently, in the beginning of the downward spiraling after a overbarrier crossing, the magnetic moment rotates (say, along a parallel of latitude) quite slowly not far from the top of the barrier, so that an appropriate sequence of fluctuations can drive it back to the initial potential well.

What is shown in Fig. 23 is precisely a multiple occurrence of this phenomenon; more than 10 barrier crossings can be identified in the overall excursion between the two minima. Besides, the magnetic moment might also had eventually fallen into the original potential well. As will be shown below, none of these processes is infrequent. The physical acumen of Brown (1959) is noteworthy since, on considering the gyromagnetic nature of the dynamics, he posed the possible occurrence of this kind of phenomena in his criticism of Néel's (1949) calculation of the relaxation time as the inverse of the rate of equatorial crossings of the magnetic moment.

2. The Effect of the Temperature

In order to assess the effect of the temperature on the dynamics of the spin, we have displayed in Fig. 24 some typical time evolutions of the projection of $\vec{m}$ onto the anisotropy axis.

As can be seen, at low temperatures (panel $k_{\text{B}}T/\Delta U = 0.12$), the dynamics consists merely of the rotations of the magnetic moment close to the bottom of the potential wells (intra-potential-well relaxation modes), while the overbarrier relaxation mechanism is "blocked." As the temperature is increased, the magnetic moment can carry out overbarrier rotations at the expense of the energy gained from the heat bath, and a number of them do occur during the displayed time interval (panels $k_{\text{B}}T/\Delta U = 0.18$ and 0.28). Finally, at higher temperatures (panel $k_{\text{B}}T/\Delta U = 0.4$), the spin effectuates a considerable number of overbarrier rotations during the observation time interval, exhibiting almost the thermal equilibrium distribution of orientations.

The curves of Fig. 24 resemble those of the experiments of Wernsdorfer et al. (1997) on individual magnetic nanoparticles (see Fig. 6 in that work). Furthermore, if the same trajectory is plotted with a larger sampling time interval, in order to mimic the finite resolution time of the measuring device, the resemblance is more apparent, since the curves then have less and wider angles (Fig. 25). (Recall that the strong dependence of the appearance of the time evolution curves on the sampling period is a typical feature of the stochastic dynamics.)

Note finally that in Fig. 24 a number of potential-barrier crossings followed by a rotation back to the original potential well can be identified (marked with small circles): one for $k_{\text{B}}T/\Delta U = 0.18$; three for $k_{\text{B}}T/\Delta U = 0.28$, where the

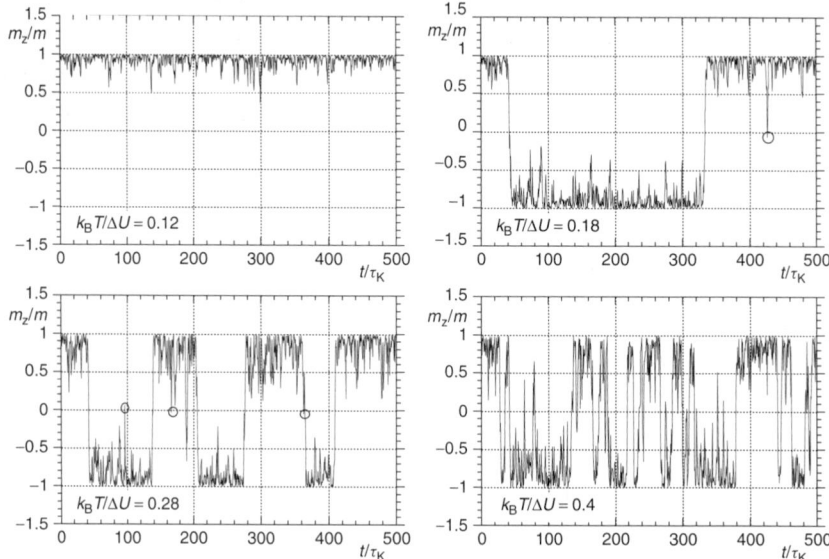

FIGURE 24. Projection onto the anisotropy axis of $\vec{m}(t)$, as determined by numerical integration of the stochastic Landau–Lifshitz–Gilbert equation (5.20), for various temperatures. The magnetic-anisotropy energy is $-\Delta U(m_z/m)^2$, $\vec{B} = 0$, and $\lambda = 0.1$. The small circles mark potential-barrier crossings followed by a backrotation to the initial potential well.

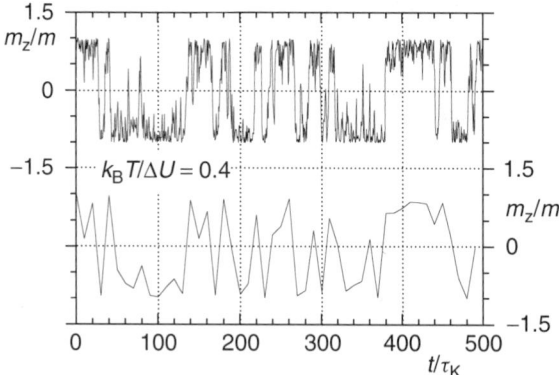

FIGURE 25. The same as in panel $k_B T / \Delta U = 0.4$ of Fig. 24, but the trajectory has also been plotted with a larger sampling time interval.

one occurring at $t/\tau_K \sim 360$ is a double crossing-back; and about seven for $k_B T/\Delta U = 0.4$ (not marked for the sake of clarity). It is also to be noted that an apparent single (or double) crossing-back can be multiple instead. Indeed, when the about 10 potential-barrier crossings of Fig. 23 are represented as m_z versus t, they seem to be a mere double crossing-back of the potential barrier.

3.   Projection of the Magnetic Moment onto the Probing-Field Direction

It is also illuminating to show the projection of the trajectories of individual spins onto the direction of a probing field $\Delta \vec{B}(t) = \Delta \vec{B} \cos(\omega t)$. Figure 26 shows this kind of trajectory in the intermediate temperature range.

The projection onto the anisotropy axis direction ($\Delta \vec{B} \| \hat{z}$) exhibits, as in the corresponding case of Fig. 24, a well-resolved bistability, and $\vec{m}$ "jumps" from one well to the other a number of times during a cycle of the probing field. Similar features are encountered when a longitudinal bias field is also applied; the main difference is that the upper potential well is less frequented by the spin. In contrast, the features of the stochastic trajectory obtained by projecting $\vec{m}(t)$ onto a direction perpendicular to the anisotropy axis ($\Delta \vec{B} \perp \hat{z}$)

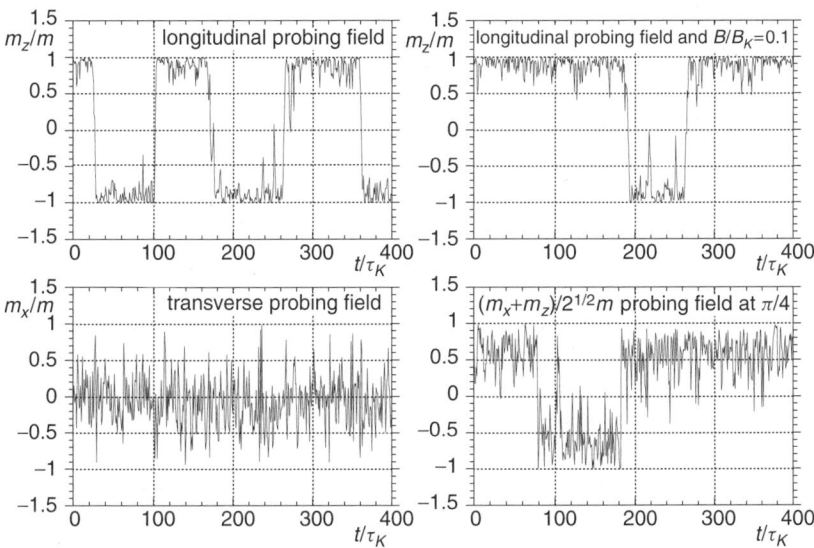

FIGURE 26.   Projection onto the direction of a probing field $\Delta \vec{B}(t) = \Delta \vec{B} \cos(\omega t)$ of $\vec{m}(t)$, as determined by numerical integration of the stochastic Landau–Lifshitz–Gilbert equation (5.20). The magnetic-anisotropy energy is $-\Delta U(m_z/m)^2$, and all the results are for $k_B T/\Delta U = 0.2$ and $\lambda = 0.1$. The displayed time interval corresponds to a complete cycle of the oscillating field ($\omega \tau_K/2\pi = 0.0025$). In the longitudinal probing-field case, results in the presence of a longitudinal bias field are also shown.

are markedly different (for example, this projection corresponds to plotting the trajectory of the upper panel of Fig. 23 as m_x vs. t). Here, the response is dominated by the fast ($\sim \tau_K$) intra-potential-well relaxation modes, and the transverse projection is a highly irregular sequence of sharp peaks. Finally, the projection of $\vec{m}(t)$ onto a $\Delta\vec{B}$ making an intermediate angle with the anisotropy axis ($\pi/4$ for the displayed curve), shows the magnetic bistability of the longitudinal projection, but the fast intra-potential-well motions are superimposed on it. This leads to a less well-resolved magnetic bistability.

Note finally that curves like those of Fig. 26 are the ones "analyzed" by the probing field in a dynamical "measurement". Recall also that the application of the oscillating field hardly changes the overall features of the curves from the free evolution ones. This is naturally so, since one applies a low enough field in order to probe the intrinsic dynamics of the system.

F. Dynamical Response of the Ensemble of Spins

Keeping Figs. 24–26 in mind, we shall undertake the study of the dynamical response of an ensemble of classical magnetic moments. As a suitable probe of the intrinsic dynamics of the system, we compute the linear dynamical susceptibility, $\chi(\omega)$, as a function of the temperature for various frequencies and orientations of an external probing field $\Delta\vec{B}(t) = \Delta\vec{B}\cos(\omega t)$.

We compute the dynamical response of ensembles of 1000 magnetic moments. We integrate numerically the stochastic Landau–Lifshitz (–Gilbert) equation of each spin by means of the stochastic Heun scheme (5.38), and analyze the time evolution of the total magnetic moment of the ensemble; the results for the dynamical susceptibility are typically averaged over 50–100 cycles of the oscillating field. In addition, in order to reduce the statistical error bars, we apply at each T the largest probing field without leaving the *equilibrium* linear response range (specifically, we scale the amplitude of the probing field with the temperature according to $m\Delta B = 0.3k_BT$).

The damping coefficient λ, the magnetic-anisotropy potential barrier $\Delta U = Kv$, and the magnitude of the magnetic moment m are assumed to be the same for each spin. For *noninteracting* entities the effects of a distribution in these parameters, as typically occurs in nanoparticle ensembles, could be taken into account by an appropriate rescaling and summation of the results thus obtained.

In all the figures that follow, the linear susceptibilities are measured in units of $\mu_0 m/B_K = \mu_0 m^2/2Kv$ [the transverse equilibrium susceptibility per spin at zero temperature in the absence of a bias field; see Eq. (3.82)]. Furthermore, when the statistical error bars of the numerical results are not shown, their size is, at most, equal to that of the plotted symbols. Finally, in order to have a reference of the discussed timescales, we can adopt the values of Eq. (5.73), so that $\tau_K^{-1} \sim 10^8$–10^{10} s^{-1} and the frequencies employed below ($\omega\tau_K/2\pi \sim 10^{-3} - 10^{-2}$) are then in the megahertz range.

1. Dynamical Response in the Absence of a Bias Field

We first study the response of the spin ensemble in the absence of a constant external field.

a. Longitudinal Response. Figure 27 displays the results for the longitudinal linear dynamical susceptibility vs. the temperature for an ensemble of spins with parallel anisotropy axes ($\Delta \vec{B} \| \hat{z}$). No bias field has been applied and a damping coefficient $\lambda = 0.1$ has been used.*

At low temperatures, the longitudinal relaxation time obeys the condition $\tau_\| \gg 2\pi/\omega$ [$t_m(\omega) = 2\pi/\omega$ is the *dynamical* measurement time]. Consequently, during a large number of cycles of the probing field, the probability of overbarrier rotations is almost zero. The response consists of the rotations of the magnetic moments close to the bottom of the potential wells (see the panel $k_B T / \Delta U = 0.12$ of Fig. 24), whose averaged (over the ensemble) projection onto the probing-field direction is quite small (but not zero; see the enlargement of the low-temperature range in Fig. 32). Moreover, as these intra-potential-well relaxation modes are very fast ($\sim \tau_K$), this small response is in phase with the probing field [see the low-T part of the phase shift $\phi = \arctan(\chi''/\chi')$ in the inset of Fig. 27].

As the temperature is increased, the spins can depart from the potential minima by means of the energy gained from the heat bath. Consequently, at a ω-dependent temperature ($k_B T / Kv \sim 0.1$–0.2 for the frequencies employed), it appears a small probability of surmounting the potential barrier during the observation time (this corresponds to the panel $k_B T / \Delta U = 0.18$ of Fig. 24). Accordingly, the averaged response starts to increase steeply with T. However, as the thermally activated response mechanism via overbarrier rotations is not efficient enough at these temperatures, the signal exhibits a considerable lag behind the probing field (see the inset of Fig. 27). This is also reflected by the occurrence of a sizable out-of-phase component of the response $\chi_\|''(T)$ (in fact, the response is mainly "out of phase").

At higher temperatures, the mechanism of overbarrier rotations becomes increasingly efficient (panel $k_B T / \Delta U = 0.28$ of Fig. 24). Consequently, after exhibiting a maximum, the phase shift starts to *decrease*, whereas the magnitude of the response still *increases* steeply with T (see the inset of Fig. 27). However, if the temperature is further increased, the very thermal agitation,

---

*Recall that, because of the axial symmetry considered, the effect of λ on the averaged quantities merely enters via the Néel time $\tau_N = \sigma \tau_K$ [see the discussion following Eq. (5.47)]. Because we measure the time in units of τ_K, the results presented for the *longitudinal* response are independent of the λ used.

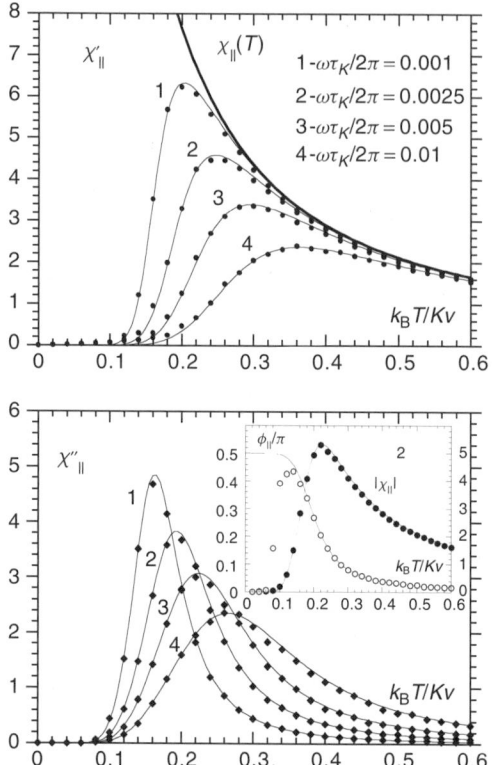

FIGURE 27. Longitudinal linear dynamical susceptibility $\chi_{\parallel}$ versus T in the absence of a bias field. The symbols represent the numerically computed $\chi_{\parallel}(\omega, T)$ and the thin solid lines represent Eq. (4.2) with $\tau_{\parallel}$ defined as integral relaxation time [Eq. (5.66)]. The heavy solid line in the upper panel indicates the thermal equilibrium susceptibility [Eq. (3.53)]. Inset (lower panel): modulus and phase shift $\phi = \arctan(\chi''/\chi')$ for $\omega\tau_K/2\pi = 0.0025$.

which up to these temperatures was responsible for the growth in the magnitude of the response, reaches a level that (1) efficiently produces overbarrier rotations, allowing the spins to approximately redistribute according to the instantaneous probing field, but, simultaneously, (2) sizably disturbs the alignment of the spins in the probing-field direction. Consequently, at a temperature above that of the phase maximum ($k_B T/Kv \sim 0.2\text{--}0.3$ for the frequencies considered), the magnitude of the response has a maximum and starts to decrease with increasing T. The frequency-dependent temperature at which this maximum occurs is called the *blocking temperature*.

Finally, at still higher temperatures ($k_B T/Kv \gtrsim 0.3\text{--}0.5$ for the frequencies considered) the inequality $\tau_{\parallel} \ll 2\pi/\omega$ holds. Then, in comparison with $\tau_{\parallel}^{-1}$, the rate of change of the probing field is quasistationary. Consequently, the spins can quickly redistribute according to the conditions set by the instantaneous probing field, being almost in the thermal equilibrium state associated with it (panel $k_B T/\Delta U = 0.4$ of Fig. 24). Then, the $\chi'_\parallel(T)$ curves corresponding to different frequencies sequentially superimpose on the linear equilibrium susceptibility, $\chi_\parallel(T)$, and, correspondingly, $\chi''_\parallel(T)$ goes to zero.

The occurrence of a frequency-dependent maximum in the dynamical response of a noisy nonlinear multistable system as a function of the noise intensity, is one of the features usually accompanying *stochastic resonance*. In this spin-dynamics case,[*] the maximum in the magnitude of the dynamical response as a function of T can be understood in terms of the quoted twofold role played by the temperature: (1) activating the dynamics of overbarrier rotations, enabling the spins to (statistically) follow the instantaneous field, but, (2) provoking the thermal misalignment of the spins from the driving-field direction. The maximum in the response as a function of T emerges as a result of the competition between these two effects.

b. Transverse Response. We now study the *transverse* dynamical response of an ensemble of magnetic moments with parallel anisotropy axes ($\Delta \vec{B} \perp \hat{z}$). Figure 28 displays the transverse dynamical susceptibility for various frequencies of the probing field (curves labeled 1; results in the presence of a bias field, to be discussed later, are also shown).

For this transverse probing-field geometry, the mechanism of inter-potential-well rotations plays a secondary dynamical role, since it pertains mainly to the components of the spins perpendicular to the probing field, whereas the response in the probing-field direction is the one analyzed. This consists of intra-potential-well rotations, which are very fast ($\sim \tau_K$) in comparison with $t_m = 2\pi/\omega$ (see the panel m_x vs. t of Fig. 26). Consequently, the dynamical susceptibilities obtained are close to the equilibrium susceptibility in the whole temperature range. Indeed, the $\chi'_\perp(T)$ curves corresponding to different frequencies are very close to one another (they visually coincide) and almost describe the equilibrium susceptibility $\chi_\perp(T)$ (heavy solid line), while the out-of-phase components $\chi''_\perp(T)$ are small. Furthermore, $\chi''_\perp$ is not only small in comparison with $\chi'_\perp$ but is also much smaller than the out-of-phase longitudinal susceptibility $\chi''_\parallel$ (cf. Fig. 27). Nevertheless, $\chi''_\perp$ provides interesting information concerning the dynamics of the spins, which will be discussed below.

---

[*] The specialization of the tools used when studying stochastic resonance to classical spins was done by Sadykov (1992), Raĭkher and Stepanov (1994, 1995*b*), and Pérez-Madrid and Rubí (1995).

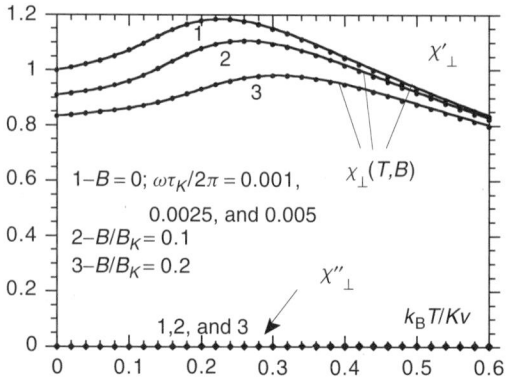

FIGURE 28. Transverse linear dynamical susceptibility $\chi_\perp$ versus T for the frequencies $\omega\tau_K/2\pi = 0.001$, 0.0025, and 0.005. The damping coefficient is $\lambda = 0.1$. Results in the unbiased case $(B = 0)$ and in the presence of the longitudinal bias fields $B/B_K = 0.1$ and 0.2 are shown (for $\omega\tau_K/2\pi = 0.005$ only). The heavy solid lines indicate the *equilibrium* susceptibilities [Eq. (3.70)]. $\chi'_\perp$ (circles) and $\chi''_\perp$ (rhombi) have intentionally been plotted on the same scale to show the relative smallness of the latter.

For the transverse response, the maximum of $\chi'_\perp$ versus T is due to the crossover from the free-rotator regime $(\sigma = Kv/k_B T \ll 1)$ to the discrete-orientation regime $(\sigma \gg 1)$, induced by the bistable magnetic-anisotropy potential. This is essentially a *thermal equilibrium* effect (see Section III.D), with a character markedly different from that of the *dynamical* maxima exhibited by the longitudinal susceptibility $\chi_\parallel(\omega, T)$.

c. Response for Anisotropy Axes Distributed at Random. Owing to the linearity of the response, when a distribution in anisotropy axis orientations occurs, $\chi(\omega)$ *in the absence of a bias field* is merely given by the weighted sum of the longitudinal and transverse dynamical responses (see the last panel of Fig. 26) and the weight factors are $\langle\cos^2\alpha\rangle$ and $\langle\sin^2\alpha\rangle$, respectively.*

The linear dynamical susceptibility for anisotropy axes distributed at random $(\langle\cos^2\alpha\rangle = \langle\sin^2\alpha\rangle/2 = \frac{1}{3})$ is displayed on Fig. 29. The out-of-phase component, $\langle\chi''\rangle_{ran}$, is overwhelmingly dominated by the responses to the components of the probing field *along* the different anisotropy axes, and it is almost $\frac{1}{3}\chi''_\parallel$ (cf. Fig. 27). On the other hand, the in-phase component, $\langle\chi'\rangle_{ran}$, is approximately $\frac{1}{3}\chi'_\parallel$ plus a nonuniform upward shift of magnitude $\frac{2}{3}\chi_\perp(T)$,

---

* α is the angle between the anisotropy axis and the probing field, and the angular brackets enclosing functions of α or susceptibilities, stand for average over the anisotropy axis distribution of an ensemble with the same parameters λ, $\Delta U = Kv$, and m.

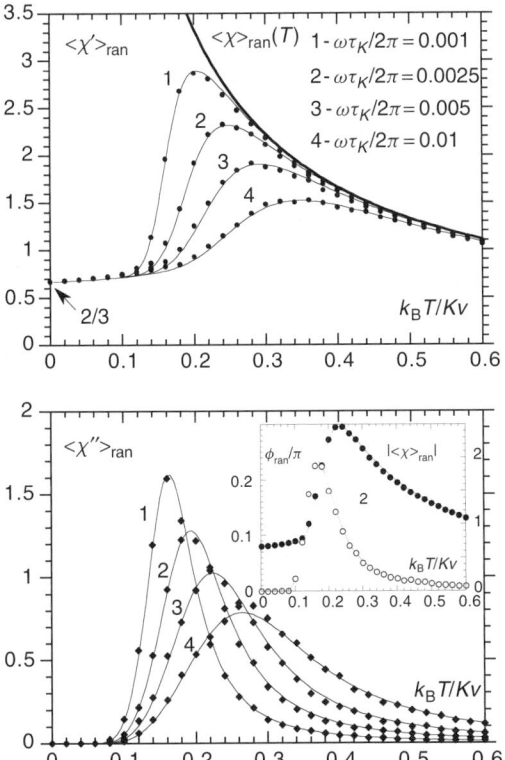

FIGURE 29. Linear dynamical susceptibility versus T for anisotropy axes distributed at random, $B = 0$, and $\lambda = 0.1$. The symbols represent the numerically computed $\langle\chi\rangle_{\mathrm{ran}}$ and the thin solid lines are Eq. (4.2) with $\tau_{\parallel}$ defined as integral relaxation time [Eq. (5.66)], and $\tau_{\perp}$ given by the effective transverse relaxation time (5.71). The heavy solid line in the upper panel indicates the thermal equilibrium susceptibility [Eq. (3.55)]. Inset (lower panel): modulus and phase shift $\phi = \arctan(\chi''/\chi')$ for $\omega\tau_K/2\pi = 0.0025$.

where $\chi_{\perp}(T)$ is the *equilibrium* transverse susceptibility. This occurs in such a way that (1) at high temperatures, the Curie law $\langle\chi\rangle_{\mathrm{ran}}|_{B=0} = \mu_0 m^2/3k_B T$ is obeyed (see Section III.D) and (2) at temperatures well below the blocking temperatures, the response consists mainly of the projection in the probing-field direction of the rotations of the magnetic moments close to the bottom of the potential wells toward the transverse components of the probing field $(\frac{2}{3}\chi_{\perp}|_{T\simeq0})$. Because of the short characteristic time of these intra-potential-well motions ($\sim\tau_K$; see Fig. 26), this low-temperature response is nearly instantaneous and in phase with the probing field (see the inset in the lower panel of Fig. 29).

We recall here that the relatively large value of the effective τ_0 ($\sim 10^{-8} - 10^{-7}$ s) in the Arrhenius law $\tau_\| \simeq \tau_0 \exp(\Delta U / k_B T)$, encountered in molecular magnetic clusters having high spin in their ground state, entails that experimental conditions with $\omega/2\pi \sim 10^3 - 10^4$ Hz already correspond to the frequency range considered here (the megahertz range if $\tau_K \sim 10^{-10} - 10^{-8}$ s). Indeed, these systems clearly exhibit the qualitative features of the linear dynamical susceptibility found at "high" (but below ferromagnetic resonance) frequencies: wide maxima in $\chi(\omega, T)$ versus T for only one potential barrier (relaxation time), sizable $\chi'(T)$ at temperatures well below the blocking temperatures, and flattening of the peak of $\chi''(T)$ with increasing ω (Barra et al., 1996; Gomes et al., 1998).

2. Dynamical Response in a Longitudinal Bias Field

We now study the effects of a constant magnetic field $\vec{B}$, applied along the common anisotropy axis direction of a spin ensemble with parallel anisotropy axes ($\vec{B} \| \hat{z}$).

a. Longitudinal Response. Figure 30 displays the longitudinal ($\Delta \vec{B} \| \hat{z} \| \vec{B}$) linear dynamical susceptibility of the system for various values of the bias field. The qualitative features of the susceptibility curves are similar to those encountered in the unbiased case ($B = 0$), and can be interpreted in terms of the same processes:

1. At low temperatures the response consists of the fast rotations of the magnetic moments close to the bottom of the potential wells, with blocking of the overbarrier relaxation mechanism.

2. As T is increased, the magnetic moments can appreciably depart from the potential minima by means of the energy gained from the heat bath, and the response starts to increase steeply with T (with a sizable lag behind the probing field).

3. If T is further increased, the system reaches the regime dominated by inter-potential-well rotations, exhibiting dynamical maxima first in the phase shift and then in the magnitude of the response.

4. In the high-temperature range, the magnetic moments are almost in the thermal-equilibrium state associated with the instantaneous probing field and, hence, $\chi'_\|(T, B)$ approaches to the linear equilibrium susceptibility while $\chi''_\|(T, B)$ tends to zero.

Thus, the dynamics is qualitatively similar to the dynamics in the unbiased case; the main difference is that the system now consists of bistable *nonsymmetric* entities (recall the panel $B/B_K = 0.1$ of Fig. 26).

We remark in passing that the simple idea that the application of a constant magnetic field reduces the potential barriers, so that the relaxation rate

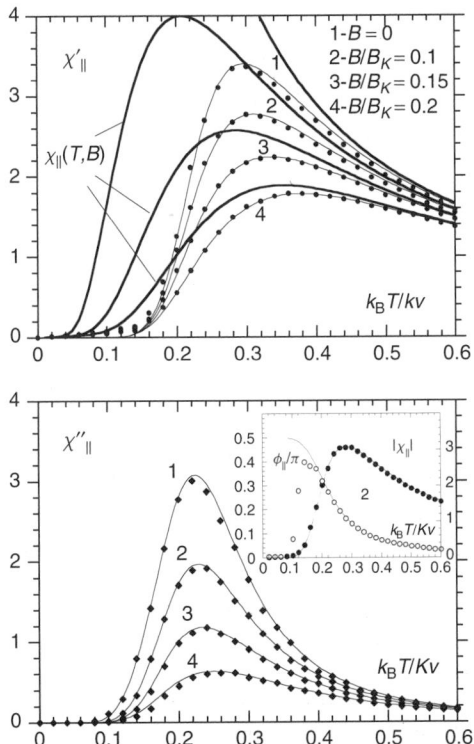

FIGURE 30. Longitudinal dynamical susceptibility $\chi_\parallel$ versus T, for $\lambda = 0.1$, $\omega\tau_K/2\pi = 0.005$, and various values of the longitudinal bias field. The symbols represent the numerically computed $\chi_\parallel(\omega, T, B)$ and the thin solid lines represent Eq. (4.2) with $\tau_\parallel$ defined as integral relaxation time [Eq. (5.66)]. The heavy solid lines in the upper panel indicate the corresponding equilibrium susceptibilities [Eq. (3.70)]. Inset (lower panel): modulus and phase shift $\phi = \arctan(\chi''/\chi')$ for $B/B_K = 0.1$.

increases and the blocking temperatures shift to lower temperatures, should be viewed with caution. The location of the maximum of the dynamical response depends not only on the potential-barrier heights but also on the form of the *equilibrium* response, which is markedly different from that of the unbiased case.* Indeed, for the frequencies and bias fields considered, the location of the maxima of $\chi_\parallel''(T)$ is not very sensitive to the bias field, while the maxima of $\chi_\parallel'(T)$ shift slightly to higher temperatures as B increases.

---

* In a bias field, because $\chi_\parallel(T, B)$ is the slope of the magnetization vs. field curve at B, instead of the initial slope of the unbiased case, the *equilibrium* response already exhibits a maximum as a function of T (see Section III.D).

b. Transverse Response. We finally consider the *transverse* dynamical response in the presence of a *longitudinal* bias field ($\Delta \vec{B} \perp \hat{z} \| \vec{B}$). Figure 28 also displays $\chi_\perp$ versus T for various values of the bias field at $\omega \tau_K / 2\pi = 0.005$ (curves labeled 2 and 3). The qualitative features of the response are again similar to those encountered in the unbiased case:

1. The mechanism of inter-potential-well rotations plays a secondary dynamical role, so the response is dominated by the fast intra-potential-well rotations.
2. The $\chi_\perp'(T, B)$ curves obtained are rather close to the corresponding equilibrium susceptibilities (heavy solid lines).
3. $\chi_\perp''(T, B)$ is small in comparison with both $\chi_\perp'(T, B)$ and $\chi_\|''(T, B)$.

3. Comparison with Different Analytic Expressions

We now compare the linear dynamical susceptibility, obtained by numerical integration of the stochastic Landau–Lifshitz (–Gilbert) equation, with the heuristic models discussed in Section IV.B and rigorous expressions. In this comparison *no adjustable parameter* will be employed.

We shall sometimes use the word *exact* when referring to the numerically computed quantities. Along with the feasible diminishing of the statistical error bars of the computed quantities by averaging over a sufficiently large number of trajectories, we also implicitly mean that the numerical results are *exact* in the context of the Brown–Kubo–Hashitsume stochastic model.

a. Longitudinal Response. Figure 31 shows the computed $\chi_\|(\omega)$ in the unbiased case and in the bias field $B/B_K = 0.1$. The results of the heuristic discrete-orientation equation (4.5), the Gittleman–Abeles–Bozowski model [Eq. (4.2) with the approximate Eq. (3.85)], and the Shliomis–Stepanov equation (4.2) are also shown. The longitudinal relaxation time $\tau_\|$, defined as the *integral relaxation time* τ_{int}, has been used in the three equations.

It is apparent that Eq. (4.5) fails to describe the numerical results; neither is the equilibrium (high-temperature) susceptibility properly described. Actually, the overall failure of this expression could be attributed mainly to the rough approximation used for its equilibrium part [Eq. (3.87)]. The probability that $\vec{m}$ makes a finite angle with the anisotropy axis is completely neglected in such a discrete-orientation equation.

Concerning the Gittleman–Abeles–Bozowski equation, it is more suitable than the discrete-orientation equation, especially for the matching of $\chi_\|''(T, B)$, although it fails to describe $\chi_\|'(T, B)$. Again, not even the equilibrium susceptibility is correctly described; the high-barrier approximation for $\chi_\|(T, B)$ occurring in this model [Eq. (3.85)], although better than the

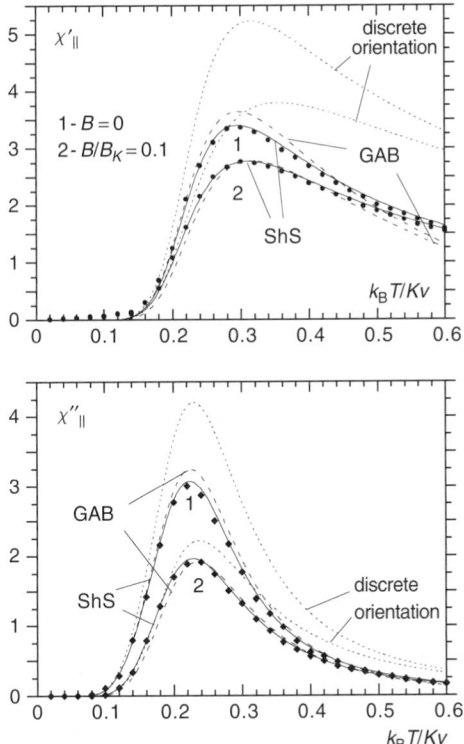

FIGURE 31. $\chi_\parallel$ versus T for $B = 0$ and $B/B_K = 0.1$ with $\omega\tau_K/2\pi = 0.005$ (symbols). The small dashing represents Eq. (4.5); the medium dashing Eq. (4.2) with the approximate Eq. (3.85); and the solid lines, Eq. (4.2). $\tau_\parallel$ defined as integral relaxation time [Eq. (5.66)], has been incorporated in the three equations.

discrete-orientation approximation, is still not sufficiently accurate at the relevant temperatures. Furthermore, for bias fields $B/B_K \gtrsim 0.15$, the departure of this model from the exact results becomes dramatic (results not shown).

In contrast, Eq. (4.2) approximates the numerical results reasonably. This is in agreement with the comparison carried out by Raĭkher et al. (1997) of the exact $\chi_\parallel(\omega)$ with what they called the "effective time approximation" [which is indeed equivalent to the use of the longitudinal component of Eq. (4.2) with $\tau_\parallel = \tau_{\text{int}}$]. Nevertheless, the exact analytic expression for $\chi_\parallel(\omega)$ comprises an infinite number of Debye-type relaxation mechanisms, namely (see Appendix B)

$$\chi_\parallel(\omega, T, B) = \chi_\parallel(T, B) \sum_{k=1}^{\infty} \frac{a_k(T, B)}{1 + i\omega/\Lambda_k(T, B)} \tag{5.88}$$

where a_k is the amplitude corresponding to the eigenvalue Λ_k of the Sturm–Liouville equation associated with the Fokker–Planck equation. (Recall that the first nonvanishing eigenvalue Λ_1 is associated with the inter-potential-well dynamics, whereas the higher-order eigenvalues, Λ_k, $k \geq 2$ are related with the intra-potential-well relaxation modes.) However, the mentioned agreement could be expected in the *unbiased* case since, as was shown numerically by Coffey et al. (1994): (1) $a_1(B = 0) \gg a_k(B = 0)$, $\forall k \geq 2$ and (2) $\Lambda_1^{-1}(B = 0) \simeq \tau_{\text{int}}(B = 0)$. Indeed, Coffey, Crothers, Kalmykov and Waldron (1995b) have shown that an expression equivalent to the longitudinal component of Eq. (4.2), together with the interpolation formula (5.63) for Λ_1^{-1}, well describes the longitudinal dynamical polarizability of the congeneric nematic liquid crystal with (unbiased) Meier–Saupe potential. (The *longitudinal* relaxation in this system is mathematically identical to that of classical magnetic moments.) On the other hand, in a constant longitudinal field the higher-order modes can make a substantial contribution *in the low-temperature region* ($\sigma \gg 1$), and then Λ_1^{-1} largely deviates from τ_{int} while $a_1 \gg a_k$ no longer holds (Coffey, Crothers, Kalmykov and Waldron, 1995a; Garanin, 1996). Nevertheless, for the frequencies employed here, the relevant dynamical phenomena occur in the range $\sigma \sim 2$–10, so that in the bias fields applied the conditions $a_1 \gg a_k$ and $\Lambda_1^{-1} \simeq \tau_{\text{int}}$ still hold approximately, and hence Eq. (4.2) describes the exact results reasonably.

However, one could expect, even for $B = 0$, a significant contribution of the intra-potential-well relaxation modes to the longitudinal response when the overbarrier dynamics is *blocked* at low T ($\omega/\Lambda_1 \gg 1$). Indeed, on scrutinizing Figs. 27 and 30, one sees that Eq. (4.2) predicts, both for $B = 0$ and $B \neq 0$, a smaller $\chi'_{\parallel}$ when departing from zero at temperatures well below the blocking temperatures than the exact $\chi'_{\parallel}$. In contrast, because the intra-potential-well modes are very fast ($\sim \tau_K$), their contribution to the out-of-phase susceptibility is comparatively smaller, so that $\chi''_{\parallel}$ is still described reasonably by the Debye-type term associated with the inter-potential-well dynamics ($\chi''_{\parallel} \simeq \chi_{\parallel}(\omega/\Lambda_1)/[1 + (\omega/\Lambda_1)^2]$).

These considerations are substantiated by comparing the numerical results with a remarkable asymptotic ($\sigma \gg 1$) expression for the longitudinal dynamical polarizability of the unbiased nematic liquid crystal derived by Storonkin (1985), namely

$$\chi_{\parallel} \simeq \frac{\mu_0 m^2}{k_B T} \left[\underbrace{\left(1 - \frac{1}{\sigma} - \frac{3}{4\sigma^2}\right) \frac{1}{1 + i\omega/\Lambda_1}}_{\text{inter-potential-well mode}} + \underbrace{\frac{1}{8\sigma^2} \left(\frac{1}{1 + i\omega/\Lambda_3} + \frac{1}{1 + i\omega/\Lambda_5} \right)}_{\text{intra-potential-well modes}} \right]$$

(5.89)

where [cf. Eq. (5.62) at $B = 0$]

$$\Lambda_1^{-1} \simeq \tau_N \frac{\sqrt{\pi}}{2} \sigma^{-3/2} \exp(\sigma) \left(1 + \frac{1}{\sigma} + \frac{7}{4\sigma^2}\right) \qquad (5.90)$$

$$\Lambda_3^{-1} \simeq \Lambda_5^{-1} \simeq \frac{1}{2} \frac{\tau_N}{\sigma} \left(1 + \frac{5}{2\sigma} + \frac{41}{4\sigma^2}\right) \qquad (5.91)$$

Note that $(\mu_0 m^2/k_B T)(1 - 1/\sigma - 3/4\sigma^2) \simeq \chi_\parallel(T) + \mathcal{O}(1/\sigma^2)$ [see Eqs. (3.53) and (A.29)], while the correction terms in Λ_1^{-1} agree with those derived by Brown (1979) (see also Coffey et al., 1994). Figure 32 shows that Eq. (5.89) remarkably describes the $B = 0$ numerical results at low temperatures. Note that, because $\Lambda_{3,5} \sim \tau_N/\sigma = \tau_K$ [Eq. (5.76)] and $\omega\tau_K \ll 1$ for the frequencies considered here, it follows that $1/(1 + i\omega/\Lambda_{3,5}) \simeq 1 - i\omega/\Lambda_{3,5}$. Therefore, since $(\mu_0 m^2/k_B T) \times (1/8\sigma^2) \propto k_B T$, the Storonkin formula (5.89) yields the low-temperature linear increase of $\chi_\parallel'$ with T due to the intra-potential-well relaxation modes, whereas their contribution to $\chi_\parallel''$ is reduced by the small factor $\omega/\Lambda_{3,5} \sim \omega\tau_K$.

The intra-potential-well relaxation modes take, in addition, a dramatic reflection in the phase shifts (Raĭkher and Stepanov, 1995b). As any expression of the form $\chi(\omega) = \chi/(1 + i\omega\tau)$ (Debye type), the *longitudinal* component of Eq. (4.2) yields a phase shift

$$\phi_\parallel = \arctan(\omega\tau_\parallel) \qquad (5.92)$$

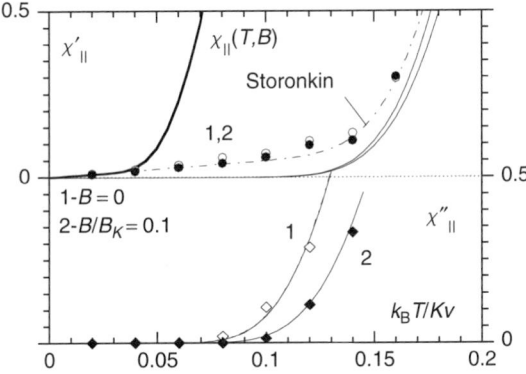

FIGURE 32. Detail of the low-temperature part of Fig. 31 showing the effect of the intra-potential-well relaxation modes. The heavy solid line indicates the equilibrium susceptibility for $B/B_K = 0.1$; the thin solid lines, Eq. (4.2); and the dashed-dotted lines, the asymptotic result (5.89) by Storonkin (for $B = 0$ only). The out-of-phase components of Eqs. (4.2) and (5.89) visually coincide.

which increases monotonically with decreasing T and, eventually, reaches $\pi/2$ since at low temperatures $\omega\tau_\| \gg 1$ (see the insets in the lower panels of Figs. 27 and 30). However, because the fast intra-potential-well relaxation modes yield an almost instantaneous contribution to the response, $\chi'_\|$ decreases with T less steeply than $\chi_\|/[1 + (\omega/\Lambda_1)^2]$ at low temperatures, whereas $\chi''_\|$ is still approximately given by $\chi_\|(\omega/\Lambda_1)/[1 + (\omega/\Lambda_1)^2]$. Consequently, the actual phase shift (insets of Figs. 27 and 30), also increases monotonically with decreasing T but, at a temperature close to that of the peak of $\chi''_\|(T)$, $\phi_\|(T)$ *exhibits a maximum* and then decreases to zero, since at low T the response is again "in phase" with the probing field because of the fast intra-potential-well modes. This behavior of the phase shift is qualitatively similar to that encountered in one-dimensional bistable systems (Morillo and Gómez-Ordóñez, 1993) and ascribed to the crossover from the "high-noise regime," dominated by intra-potential-well jumps, to the "low-noise regime," dominated by the fast intra-potential-well motions.

Concerning the phase behavior for noncollinear situations, we must bear in mind that the intra-potential-well motions make a relative contribution to the transverse response much larger than to the longitudinal response. Therefore, as the former contribution is somehow taken into account by Eq. (4.2), via the *equilibrium* transverse susceptibility, we find that, inasmuch as $\langle\cos^2\alpha\rangle$ departs from unity, the Shliomis–Stepanov equation describes the low-temperature phase shifts reasonably well (cf. the inset of Fig. 27 with that of Fig. 29). We finally remark that, because the intra-potential-well relaxation modes are very fast and, thus, $\chi''_\|$ is reasonably described by Eq. (4.2), while $\chi''_\perp$ is relatively small, the theoretical background of the methods of determination of the energy-barrier distribution of Section IV that are based on the use of the *out-of-phase* component of the low-frequency equation (4.3), result to be supported in the context of the Brown–Kubo–Hashitsume stochastic model.

b. Transverse Response. Figure 33 displays the corresponding comparison for $\chi_\perp(\omega)$ in the unbiased case for various values of the damping coefficient.* For the transverse relaxation time $\tau_\perp$, we have employed the effective relaxation time (5.71), which has been derived (Appendix B) from the low-frequency expansion of the equation for $\chi_\perp(\omega)$ of Raĭkher and Shliomis (1975, 1994).

---

* In the cases with larger damping coefficients, $\lambda = 0.5$ and 2, we have used a discretization time interval $\Delta t = 0.0025\tau_K$ in the numerical integration of the stochastic Landau–Lifshitz (–Gilbert) equation, instead of the value $\Delta t = 0.01\tau_K$ used in the rest of this section.

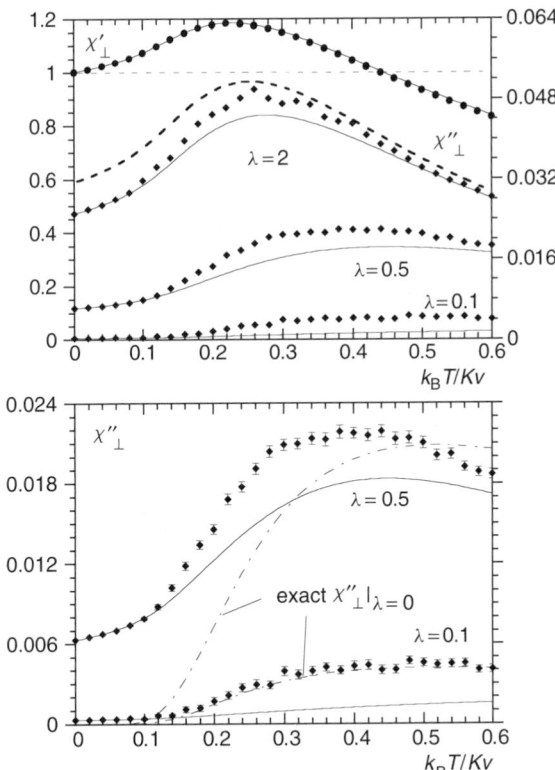

FIGURE 33. Upper panel: $\chi_\perp$ versus T for $B = 0$, $\omega\tau_K/2\pi = 0.005$, and various values of the damping coefficient λ. The circles represent $\chi'_\perp$, and the rhombi represent $\chi''_\perp$. The medium dashed line corresponds to the constant $\chi'_\perp$ given by Eq. (4.4) and the solid lines to Eq. (4.2) with $\tau_\perp$ given by the effective transverse relaxation time (5.71). The heavy dashed curve is $\chi''_\perp$ with $\tau_\perp$ given by the $\lambda \gg 1$ result (5.69). Lower panel: detail of $\chi''_\perp$ in the intermediate-to-weak damping range together with the exact zero-damping formula (5.93) (dashed–dotted lines).

For the transverse probing-field geometry, the discrete-orientation formula (4.5) predicts obviously an identically zero response, while the Gittleman–Abeles–Bozowski formula yields a constant $\chi'_\perp(T)$ and a zero $\chi''_\perp(T)$. In contrast, the exact $\chi'_\perp(T)$ is well described by Eq. (4.2), although, because $\omega\tau_\perp \ll 1$ holds in the considered frequency range, $\chi'_\perp(T)$ almost coincides with the equilibrium susceptibility $\chi_\perp(T)$. Concerning $\chi''_\perp(T)$, Eq. (4.2) with the effective expression (5.71) for $\tau_\perp$ matches the out-of-phase response only in the low-temperature range for the smallest damping coefficient considered ($\lambda = 0.1$). Nevertheless, Fig. 33 shows that, as the damping coefficient is enlarged, the matching between the numerical results and the simple Eq. (4.2) improves if one uses the effective $\tau_\perp$ proposed [Eq. (5.71)]. This constitutes an advance over the usual approach, where one employs the $\tau_\perp$ derived by the effective-eigenvalue method [Eq. (5.68)], which yields the heavy dashed curve of Fig. 33 *irrespective of* λ.

The above comparison is in agreement with that made by Kalmykov and Coffey (1997) of their numerical results, obtained by continued-fraction techniques, with the complete (but approximate) expression for $\chi_\perp(\omega)$ of Raĭkher and Shliomis (1975, 1994).* The failure of this expression for weak damping was explained in terms of the effects of the precession terms of the dynamical equation. When these terms dominate ($\lambda \ll 1$), due to the occurrence of a spread of the precession frequencies of $\bar{m}$ in the anisotropy field at intermediate temperatures (these frequencies are $\propto \gamma B_K m_z$), the response is not well described by a simple relaxation mechanism. Then, only at low temperatures, where the magnetic moments are concentrated close to the bottom of the potential wells (so the spread in precession frequencies is reduced), the exact results are well described by the $\chi_\perp(\omega)$ of Raĭkher and Shliomis.

The effects of the spread of the precession frequencies of $\bar{m}$ in the anisotropy field had already been investigated by Gekht (1983) and independently by Garanin et al. (1990). They derived the *exact* expression for $\chi_\perp''(\omega, T, B)$ in the $\lambda \to 0$ limit, which accounts for the effects of the spreading phenomenon discussed [the former author employed the deterministic (Hamiltonian) trajectories and the Kubo formula for the dynamical susceptibility, while the latter ones started from the Fokker–Planck equation]. Their formula can be written as

$$\chi_\perp''|_{\lambda=0} = \frac{\mu_0 m^2}{k_B T}\frac{\pi}{2}\frac{\tilde{\omega}}{(2|\sigma|)^3}\frac{(2\sigma)^2 - (\tilde{\omega} - \xi)^2}{\mathcal{Z}_\parallel}\exp\left(\frac{\tilde{\omega}^2 - \xi^2}{4\sigma}\right) \qquad (5.93)$$

where $\tilde{\omega} = \omega(m/\gamma k_B T)$, $\xi = mB/k_B T$, $\mathcal{Z}_\parallel$ is the longitudinal partition function (2.25), and $\chi_\perp''(\omega)$ is nonzero in the interval $(\tilde{\omega} - \xi)^2 \le (2\sigma)^2$. In order to compare the zero-damping formula (5.93) with the numerical results, we just write $\tilde{\omega} = \omega(2\lambda\tau_K\sigma)$, which for fixed $\omega\tau_K$ (as occurs in the plots) is a "function" of λ.

The lower panel of Fig. 33 shows that, for $\lambda = 0.5$, the dampingless Eq. (5.93) correctly gives the order of magnitude of the exact results at intermediate-to-high temperatures, while for $\lambda = 0.1$ a good agreement extending down to quite low temperatures can be seen. Since Eq. (5.93) is the exact $\lambda = 0$ result, this comparison indicates that, in the intermediate-to-weak damping regime, the contribution of the spread of the precession

---

* In the frequency range below the ferromagnetic resonance range, this formula is indistinguishable from the low-ω expansion used here.

frequencies of the magnetic moment to $\chi''_\perp(\omega)$ is sizable in comparison with the effects of the damping. Therefore, by omitting this zero-damping effect one could erroneously extract values of λ from the $\chi''_\perp(\omega)$ data that overestimate the actual λ and, for example, infer that the damping in superparamagnets is stronger than it is.

Another manifestation of the spreading effect was studied by Raĭkher and Stepanov (1995a). The contribution of the damping to the absorption line in intrinsic ferromagnetic resonance provokes a (unbounded) monotonic increase of the *linewidth* with the temperature, whereas the linewidths experimentally observed in certain magnetic nanoparticle systems are almost independent of the temperature (Hennion et al., 1994). However, the spread of precession frequencies in the anisotropy field also yields a contribution to the linewidths, which, in addition, saturates at high temperatures. Thus, the combination of both contributions leads to the appearance of an intermediate temperature regime, fairly wide for systems with low damping, in which the linewidth is quasiconstant.

VI. FOUNDATION OF THE STOCHASTIC DYNAMICAL EQUATIONS

A. Introduction

In this section we examine various topics related with the foundation of the Brown–Kubo–Hashitsume stochastic model and possible extensions of this model (García-Palacios, 1999).

1. Phenomenological Equations

The Brown–Kubo–Hashitsume model is phenomenological inasmuch as is constructed by augmenting known phenomenological equations (Gilbert or Landau–Lifshitz) by fluctuating fields. For subsequent reference, let us first rewrite the basic equations of this model (see Section V.C):

- *Stochastic Gilbert Equation*

$$\frac{1}{\gamma}\frac{d\vec{m}}{dt} = \vec{m} \wedge \left[\vec{B}_{\text{eff}} + \vec{b}_{\text{fl}}(t) - (\gamma m)^{-1}\lambda\frac{d\vec{m}}{dt} \right] \qquad (6.1)$$

This equation is equivalent to the stochastic Landau–Lifshitz–Gilbert equation (5.20), except for a "renormalization" of the magnetomechanical ratio.

- *Stochastic Landau–Lifshitz Equation*

$$\frac{1}{\gamma}\frac{d\vec{m}}{dt} = \vec{m} \wedge \left[\vec{B}_{\text{eff}} + \vec{b}_{\text{fl}}(t) \right] - \frac{\lambda}{m}\vec{m} \wedge \left(\vec{m} \wedge \vec{B}_{\text{eff}} \right) \qquad (6.2)$$

This equation may be regarded as the weak-damping case ($\lambda \ll 1$) of Eqs. (6.1) or (5.20), although it can be considered as an independent model as well. On the other hand, this is a Langevin equation more archetypal than those equations, in the sense that the fluctuating and relaxation (damping) terms are not mixed.

In these dynamical equations λ is a dimensionless damping coefficient and $\vec{B}_{\text{eff}} = -\partial \mathcal{H}/\partial \vec{m}$ is the (deterministic) effective field associated with the Hamiltonian of the spin $\mathcal{H}(\vec{m})$. This typically includes Zeeman and magnetic-anisotropy energy terms, for instance, for a spin with uniaxial anisotropy with symmetry axis $\vec{n}$

$$\mathcal{H} = -\vec{m} \cdot \vec{B} - \tfrac{1}{2}(B_K/m)(\vec{m} \cdot \vec{n})^2 \quad \Rightarrow \quad \vec{B}_{\text{eff}} = \vec{B} + \hat{K}\vec{m}$$

where $\hat{K}$ is a second-rank tensor with elements $K_{ij} = (B_K/m)n_i n_j$ [cf. Eq. (5.4)]. On the other hand, $\vec{b}_{\text{fl}}(t)$ is a fluctuating field, the statistical properties of which are

$$\langle b_{\text{fl},i}(t) \rangle = 0, \qquad \langle b_{\text{fl},i}(t)b_{\text{fl},j}(t') \rangle = \frac{2\lambda \delta_{ij}}{\gamma m} k_{\text{B}} T \delta(t - t') \qquad (6.3)$$

where we have taken into account that, if starting from the Gilbert equation, one must replace $\gamma \to \gamma/(1 + \lambda^2)$ in the results of Section V associated with the stochastic Landau–Lifshitz (–Gilbert) equation, so that D_{LLG} is then identical with D_{LL} [see Eq. (5.35)]. Finally, on introducing Eq. (5.33) into Eq. (5.38), the Fokker–Planck equation governing the time evolution of the nonequilibrium probability distribution of spin orientations, associated with the preceding Langevin equations, can be written as

$$\frac{\partial P}{\partial t} = -\frac{\partial}{\partial \vec{m}} \cdot \left\{ \gamma \vec{m} \wedge \vec{B}_{\text{eff}} P - \lambda \frac{\gamma}{m} \vec{m} \wedge \left[\vec{m} \wedge \left(\vec{B}_{\text{eff}} - k_{\text{B}} T \frac{\partial}{\partial \vec{m}} \right) P \right] \right\} \qquad (6.4)$$

where $(\partial/\partial \vec{m}) \cdot \vec{J} = \sum_i (\partial J_i/\partial m_i)$ and for the Gilbert case one must replace γ by $\gamma/(1 + \lambda^2)$.

The Brown–Kubo–Hashitsume stochastic model has been the basis of significant studies of the dynamics of classical magnetic moments. Nonetheless, there exist important microscopic relaxation mechanisms that cannot be accommodated in the context of this model, inasmuch as they do not produce a field-type perturbation on the spin ("field"-type fluctuations). An important example is the coupling of the spin to the lattice vibrations, which modulate the crystal field and the exchange and dipole–dipole interactions, and can produce fluctuations of the magnetic-anisotropy potential of the spin ("anisotropy"-type fluctuations).

In order to take this phenomenon into account, Garanin et al. (1990) generalized the preceding Langevin equations to $d\vec{m}/dt = \gamma\vec{m} \wedge [\vec{B}_{\text{eff}} + \vec{b}(t) + \hat{\kappa}(t)\vec{m}] - \vec{R}$. Here, $\vec{R}$ is a relaxation term to be determined and, in analogy with the expression $\vec{B}_{\text{eff}} = \vec{B} + \hat{K}\vec{m}$ for the effective field, $\vec{b}(t)$ is a stochastic *vector* that introduces the field-type part of the thermal fluctuations, while $\hat{\kappa}(t)$ is a stochastic *second-rank tensor*, so that $\hat{\kappa}(t)\vec{m}$ incorporates anisotropy-type fluctuations into the dynamical equation.

On assuming the correlation properties

$$\langle b_i(t)b_j(t')\rangle = \frac{2\lambda_{ij}}{\gamma m}k_{\text{B}}T\delta(t - t')$$

$$\langle b_i(t)\kappa_{jk}(t')\rangle = \frac{2\lambda_{i,jk}}{\gamma m}k_{\text{B}}T\delta(t - t') \qquad (6.5)$$

$$\langle \kappa_{ik}(t)\kappa_{j\ell}(t')\rangle = \frac{2\lambda_{ik,j\ell}}{\gamma m}k_{\text{B}}T\delta(t - t')$$

Garanin et al. constructed the associated Fokker–Planck equation

$$\frac{\partial P}{\partial t} = -\frac{\partial}{\partial\vec{m}}\cdot\left\{\gamma\vec{m}\wedge\vec{B}_{\text{eff}}P - \left[\vec{R} - \frac{\gamma}{m}k_{\text{B}}T\vec{m}\wedge\hat{G}\left(\vec{m}\wedge\frac{\partial}{\partial\vec{m}}\right)\right]P\right\} \qquad (6.6)$$

where $\hat{G}$ is a symmetric second-rank tensor related with the correlation coefficients of the fluctuating terms by

$$G_{ij} = \lambda_{ij} + \sum_k(\lambda_{i,jk} + \lambda_{j,ik})m_k + \sum_{k\ell}\lambda_{ik,j\ell}m_k m_\ell \qquad (6.7)$$

The relaxation term $\vec{R}$ was then determined by merely assuming that the Boltzmann distribution $P_{\text{e}}(\vec{m}) \propto \exp[-\mathcal{H}(\vec{m})/k_{\text{B}}T]$ is a stationary solution of the Fokker–Planck equation (6.6). This yields $\vec{R} = (\gamma/m)\vec{m}\wedge\hat{G}(\vec{m}\wedge\vec{B}_{\text{eff}})$, so the starting Langevin equation finally reads [cf. Eq. (6.2)]

$$\frac{1}{\gamma}\frac{d\vec{m}}{dt} = \vec{m}\wedge\left[\vec{B}_{\text{eff}} + \vec{b}(t) + \hat{\kappa}(t)\vec{m}\right] - \frac{1}{m}\vec{m}\wedge\hat{G}\left(\vec{m}\wedge\vec{B}_{\text{eff}}\right) \qquad (6.8)$$

For an arbitrary form of $\hat{G}$ the relaxation term in this equation deviates from the form proposed by Landau and Lifshitz (1935). Only for $G_{ij} = \lambda\delta_{ij}$, which, for instance, occurs when both the field-type and the anisotropy-type fluctuations are isotropic ($\lambda_{ij} \propto \delta_{ij}$ and $\lambda_{ik,j\ell} \propto \delta_{ij}\delta_{k\ell}$) and there is not interference between them ($\lambda_{i,jk} \equiv 0$), that archetypal relaxation term is recovered and the Fokker–Planck equation of Garanin et al. [Eq. (6.6)] reduces to Eq. (6.4).

2. Dynamical Approaches to the Phenomenological Equations

There have been several attempts to justify, starting from dynamical descriptions of a spin coupled to its surroundings, the phenomenological equations for the stochastic spin dynamics.

Smith and de Rozario (1976) considered a classical spin $\vec{m}$ coupled to a field $\vec{b}(\mathbf{P}, \mathbf{Q})$ depending on the canonical momenta and coordinates $(\mathbf{P}, \mathbf{Q})$ of the environment. They derived a master equation for $\vec{m}$ by "projecting out" the environment variables, which, when the modulation due to the surroundings is fast in comparison with the precession period of $\vec{m}$, reduces to the Fokker–Planck equation (6.4).

Seshadri and Lindenberg (1982) studied a test spin interacting through a Heisenberg-type Hamiltonian with an environment consisting of other spins. The interaction among the latter was treated by a mean-field approach, and a dynamical equation for the test spin was obtained to second order in the spin–environment coupling. The equation derived has the form of a generalized (i.e., containing "memory" terms) Langevin equation, whose fluctuating and relaxation terms naturally obey fluctuation–dissipation relations.

Jayannavar (1991) employed the *oscillator-bath* representation of the environment (Magalinskiĭ, 1959; Ullersma, 1966; Zwanzig, 1973; Caldeira and Leggett, 1983; Ford, Lewis and O'Connell, 1988), and assumed a coupling linear in both the spin variables and the oscillator coordinates (*bilinear coupling*). A generalized Langevin equation for the spin was derived, which, in the Markovian approach (no memory) and for isotropic fluctuations, formally reduces to the stochastic Gilbert equation (6.1). (A similar treatment was presented by Klik, 1992.) Equations of Landau–Lifshitz form, akin to those derived by Seshadri and Lindenberg, were also obtained in the weak-coupling regime.

Nevertheless, since spin–environment interactions linear in $\vec{m}$ produce a field-type perturbation on the spin (see discussion below), the treatments mentioned do not account for fluctuations of the magnetic anisotropy of the spin. In this section, in order to incorporate this phenomenon, we shall extend the bilinear coupling treatment of Jayannavar by considering general dependences of the spin-environment coupling on the spin variables. Furthermore, we also include interactions quadratic in the oscillator variables (the classical analogue of, for example, two-phonon or two-photon relaxation processes), which are essential at sufficiently high temperatures. Because the ordinary formalism of the environment of independent oscillators is not directly applicable when these quadratic couplings are included, we shall resort to a perturbational expansion in the spin–environment coupling, which is inspired on that of Cortés, West and Lindenberg (1985).

We shall obtain dynamical equations for the spin that have the structure of generalized Langevin equations with fluctuating terms $\vec{m} \wedge \vec{b}_{\mathrm{fl}}(\vec{m}, t)$ and concomitant relaxation terms. These will have the form of a vector product of $\vec{m}(t)$ with a memory integral, which includes $(\mathrm{d}\vec{m}/\mathrm{d}t)(t')$ or $(\vec{m} \wedge \vec{B}_{\mathrm{eff}})(t')$ for weak coupling, taken along the past history of the spin $(t' \leq t)$. In the Markovian approach, the equations derived will reduce to the form

$$\frac{\mathrm{d}\vec{m}}{\mathrm{d}t} = \gamma\vec{m} \wedge \left[\vec{B}_{\mathrm{eff}} + \vec{b}_{\mathrm{fl}}(\vec{m}, t)\right] - \vec{R}$$

where for couplings *linear* in the environmental variables the relaxation term is given by $\vec{R} = (1/m)\vec{m} \wedge \hat{\Lambda}^{(\mathrm{L})}(\mathrm{d}\vec{m}/\mathrm{d}t)$ or $\vec{R} = (\gamma/m)\vec{m} \wedge \hat{\Lambda}^{(\mathrm{L})}(\vec{m} \wedge \vec{B}_{\mathrm{eff}})$ for weak coupling, where $\hat{\Lambda}^{(\mathrm{L})}$ is a second-rank tensor depending on the structure of the coupling. In addition, when interactions *quadratic* in the environment variables are also taken into account, the relaxation term will depend explicitly on the temperature and, in the Markovian approach, $\vec{R}$ will take the form $\vec{R} = (\gamma/m)\vec{m} \wedge \hat{\Lambda}(\vec{m} \wedge \vec{B}_{\mathrm{eff}})$, with $\hat{\Lambda} = \hat{\Lambda}^{(\mathrm{L})} + k_{\mathrm{B}}T\hat{\Lambda}^{(\mathrm{Q})}$, where the additional tensor $\hat{\Lambda}^{(\mathrm{Q})}$ is determined by the quadratic portion of the coupling.

Since the fluctuating effective field $\vec{b}_{\mathrm{fl}}(\vec{m}, t)$ will depend in general on $\vec{m}$, it can incorporate fluctuations of the magnetic anisotropy of the spin. For instance, when the spin–environment interaction includes terms up to quadratic *in the spin variables*, $\vec{b}_{\mathrm{fl}}(\vec{m}, t)$ can be written as $\vec{b}(t) + \hat{\kappa}(t)\vec{m}$, and the correlation coefficients of the fluctuating terms are related to the tensors $\hat{\Lambda}$ by expressions identical with Eq. (6.7). In this way, the generalization of the classic Brown–Kubo–Hashitsume model effected by Garanin and co-workers will be formally obtained.

B.   Free Dynamics and Canonical Variables

The dynamical equation for an isolated classical spin with Hamiltonian $\mathcal{H}(\vec{m})$ is

$$\frac{1}{\gamma}\frac{\mathrm{d}\vec{m}}{\mathrm{d}t} = \vec{m} \wedge \vec{B}_{\mathrm{eff}}, \qquad \vec{B}_{\mathrm{eff}} = -\frac{\partial\mathcal{H}}{\partial\vec{m}} \tag{6.9}$$

By means of the formula (5.40) for the gradient operator in spherical coordinates, these vectorial equations, which merely express the precession of $\vec{m}$ about the instantaneous effective field, can be written as

$$\frac{\mathrm{d}\varphi}{\mathrm{d}t} = -\frac{\gamma}{m}\frac{1}{\sin\vartheta}\frac{\partial\mathcal{H}}{\partial\vartheta}, \qquad \frac{\mathrm{d}\vartheta}{\mathrm{d}t} = \frac{\gamma}{m}\frac{1}{\sin\vartheta}\frac{\partial\mathcal{H}}{\partial\varphi} \tag{6.10}$$

where φ and ϑ are, respectively, the azimuthal and polar angles of $\vec{m}$. Furthermore, these formulas are equivalent to the Hamiltonian equations

$$\frac{dq}{dt} = \frac{\partial \mathcal{H}}{\partial p}, \qquad \frac{dp}{dt} = -\frac{\partial \mathcal{H}}{\partial q}$$

with the conjugate canonical variables*

$$q = \varphi, \qquad p = \frac{m_z}{\gamma} \qquad\qquad (6.11)$$

In terms of the variables (6.11), the Cartesian components of the magnetic moment are given by

$$m_x = \sqrt{m^2 - (\gamma p)^2}\cos q, \quad m_y = \sqrt{m^2 - (\gamma p)^2}\sin q, \quad m_z = \gamma p \quad (6.12)$$

From these expressions for $m_i(p, q)$ and the definition of the Poisson bracket of two arbitrary dynamical variables

$$\{A, B\} \equiv \frac{\partial A}{\partial q}\frac{\partial B}{\partial p} - \frac{\partial A}{\partial p}\frac{\partial B}{\partial q}$$

one can readily obtain the customary Poisson bracket ("commutation") relations among the spin variables

$$\{m_i, m_j\} = \gamma \sum_k \varepsilon_{ijk} m_k$$

where ε_{ijk} is the Levi–Civita symbol.[†] In addition, on using the *chain rule* of the Poisson bracket, namely

$$\{f, g\} = \sum_{i,k} \frac{\partial f}{\partial x_i}\frac{\partial g}{\partial x_k}\{x_i, x_k\}, \qquad x_i = x_i(p, q)$$

---

* Gekht and Ignatchenko (1979) also discuss the Hamiltonian formulation of the dynamics of a classical spin. The alternative choice $\tilde{q} = m_z/\gamma$ and $\tilde{p} = -\varphi$ is equivalent to the one considered here through the *canonical* transformation $q = -\tilde{p}$ and $p = \tilde{q}$.

† To illustrate, from

$$\frac{\partial m_x}{\partial q} = -\left[m^2 - (\gamma p)^2\right]^{1/2}\sin q, \qquad \frac{\partial m_x}{\partial p} = -\gamma^2 p\left[m^2 - (\gamma p)^2\right]^{-1/2}\cos q$$

$$\frac{\partial m_y}{\partial q} = \left[m^2 - (\gamma p)^2\right]^{1/2}\cos q, \qquad \frac{\partial m_y}{\partial p} = -\gamma^2 p\left[m^2 - (\gamma p)^2\right]^{-1/2}\sin q$$

one gets $\{m_x, m_y\} = \gamma^2 p\sin^2 q + \gamma^2 p\cos^2 q = \gamma m_z$.   Q.E.D.

one gets the useful relation [cf. Eq. (13) by Smith and de Rozario (1976) and Eq. (1.4) by Gekht and Ignatchenko (1979)]

$$\{m_i, V(\vec{m})\} = -\gamma \left(\vec{m} \wedge \frac{\partial V}{\partial \vec{m}} \right)_i \qquad (6.13)$$

which is valid for any function of the spin variables $V(\vec{m})$.

Note finally that one can conversely *postulate* the relations $\{m_i, m_j\} = \gamma \sum_k \varepsilon_{ijk} m_k$ and then *derive* Eq. (6.9) starting from the basic Hamiltonian evolution equation $dm_i/dt = \{m_i, \mathcal{H}\}$ and using Eq. (6.13). This can be considered as a justification for the presence of the expression $\vec{B}_{\text{eff}} = -\partial \mathcal{H}/\partial \vec{m}$ in the dynamical equations for a classical spin.

C. Dynamical Equations for Couplings Linear in the Environment Variables

We shall now study a classical spin surrounded by an environment that can be represented by a set of independent classical harmonic oscillators. In spite of its academic appearance, those oscillators can correspond to the *normal modes* of an electromagnetic field, the lattice vibrations (in the harmonic approximation), or they can be an effective low-energy description of a more general surrounding medium (Caldeira and Leggett, 1983). We shall assume that the spin–environment interaction is *linear* in the coordinates of the oscillators but otherwise *arbitrary* in the spin variables. In this way, fluctuations of the magnetic anisotropy of the spin will be incorporated in the dynamical equations.

1. The Spin–Environment Hamiltonian

The total system consisting of the spin (the "system of interest") plus the oscillators representing the environment forms a *closed* dynamical system, which we describe by augmenting the isolated-spin Hamiltonian as follows:

$$\mathcal{H}_T = \mathcal{H}(\vec{m}) + \sum_\alpha \frac{1}{2} \left\{ P_\alpha^2 + \omega_\alpha^2 \left[Q_\alpha + \frac{\varepsilon}{\omega_\alpha^2} F_\alpha(\vec{m}) \right]^2 \right\}. \qquad (6.14)$$

Here, α is an oscillator index [e.g., the pair $(\vec{k}, s)$ formed by the wavevector and branch index of a normal mode of the environment] and the coupling terms $F_\alpha(\vec{m})$ are arbitrary functions of the spin variables (typically polynomials in $\vec{m}$). These terms may depend on the parameters of the oscillators ω_α, but not on their dynamical variables P_α, Q_α. On the other hand, for the sake of convenience in keeping track of the various orders, we have introduced a spin–environment coupling constant ε, which in the weak-coupling approximation will be considered small.

The terms proportional to F_α^2, which emerge when squaring $Q_\alpha + (\varepsilon/\omega_\alpha^2)F_\alpha$, are "counterterms" introduced to balance the coupling-induced renormalization of the Hamiltonian of the spin. The formalism takes as previously considered whether such a renormalization actually occurs for a given interaction (Caldeira and Leggett, 1983), so that $\mathcal{H}$ would already include it (whenever it exists). An advantage of this convention is that one deals with the experimentally accessible energy of the spin, instead of the "bare" one, which might be difficult to determine.

The introduction of *nonlinear* coupling terms $F_\alpha(\vec{m})$, as otherwise occur in various relevant situations ($F_\alpha \propto \sum m_k m_l$ for the magnetoelastic coupling of $\vec{m}$ to the lattice vibrations), will be essential to get fluctuations of the magnetic anisotropy of the spin. The starting Hamiltonian in the work of Jayannavar (1991) was similar to (6.14) but with a special type of *linear* $F_\alpha(\vec{m})$; the component m_i of the magnetic moment was coupled to the ith Cartesian component $Q_{\alpha,i}$ of certain three-dimensional oscillators. This specific *bilinear* interaction yielded, not only field-type fluctuations, but also uncorrelated ones. [Klik (1992) also considered couplings nonlinear in $\vec{m}$, but in that work the focus was on the existence of thermal equilibrium in the Markovian limit.]

2. Dynamical Equations: General Case

For the sake of simplicity in notation but also of generality, we cast the Hamiltonian (6.14) into the form

$$\mathcal{H}_\mathrm{T} = \mathcal{H}^{(\mathrm{m})}(p, q) + \sum_\alpha \tfrac{1}{2}\left(P_\alpha^2 + \omega_\alpha^2 Q_\alpha^2\right) + \varepsilon \sum_\alpha Q_\alpha F_\alpha(p, q) \qquad (6.15)$$

where q and p are the canonical coordinate and conjugate momentum of a system with Hamiltonian $\mathcal{H}(p, q)$ [in the spin-dynamics case p and q are given by Eqs. (6.11)], and the "modified" system Hamiltonian $\mathcal{H}^{(\mathrm{m})}$ augments $\mathcal{H}$ by the aforementioned counterterms

$$\mathcal{H}^{(\mathrm{m})} = \mathcal{H} + \frac{\varepsilon^2}{2} \sum_\alpha \frac{F_\alpha^2}{\omega_\alpha^2} \qquad (6.16)$$

The equation of motion for any dynamical variable C without explicit dependence on the time, $\partial C/\partial t \equiv 0$, is given by the basic Hamiltonian evolution equation

$$\frac{\mathrm{d}C}{\mathrm{d}t} = \{C, \mathcal{H}_\mathrm{T}\}$$

where the whole Poisson bracket is given by

$$\{A, B\} \equiv \frac{\partial A}{\partial q}\frac{\partial B}{\partial p} - \frac{\partial A}{\partial p}\frac{\partial B}{\partial q} + \sum_\alpha \frac{\partial A}{\partial Q_\alpha}\frac{\partial B}{\partial P_\alpha} - \frac{\partial A}{\partial P_\alpha}\frac{\partial B}{\partial Q_\alpha}$$

Therefore, the (coupled) equations of motion for *any dynamical variable* (*observable*) *of the system* $A(p,q)$ and the environment variables read ($C = A$, P_α, and Q_α)

$$\frac{dA}{dt} = \{A, \mathcal{H}^{(m)}\} + \varepsilon \sum_\alpha Q_\alpha \{A, F_\alpha\} \tag{6.17}$$

$$\frac{dQ_\alpha}{dt} = P_\alpha, \qquad \frac{dP_\alpha}{dt} = -\omega_\alpha^2 Q_\alpha - \varepsilon F_\alpha \tag{6.18}$$

The goal is to derive a dynamical equation for $A(p, q)$ involving the system variables only (*reduced* dynamical equation). Then, the corresponding equation for the spin will be obtained by replacing $A(p, q)$ in that equation by the Cartesian components of $\vec{m}$ [Eq. (6.12)].

On considering that in Eqs. (6.18) the term $-\varepsilon F_\alpha(t) = -\varepsilon F_\alpha[p(t), q(t)]$ plays the role of a time-dependent forcing on the oscillators, those equations can be explicitly integrated, yielding

$$Q_\alpha(t) = Q_\alpha^h(t) - \frac{\varepsilon}{\omega_\alpha} \int_{t_0}^t dt' \sin[\omega_\alpha(t - t')] F_\alpha(t') \tag{6.19}$$

where

$$Q_\alpha^h(t) = Q_\alpha(t_0) \cos[\omega_\alpha(t - t_0)] + \frac{P_\alpha(t_0)}{\omega_\alpha} \sin[\omega_\alpha(t - t_0)] \tag{6.20}$$

are the solutions of the *homogeneous* system of equations for the oscillators in the absence of the system–environment interaction (proper modes of the environment). Then, on integrating by parts in Eq. (6.19), one gets for the combination εQ_α that appears in Eq. (6.17)

$$\varepsilon Q_\alpha(t) = f_\alpha(t) - \left[\mathcal{K}_\alpha(t - t') F_\alpha(t')\right]_{t'=t_0}^{t'=t} + \int_{t_0}^t dt' \mathcal{K}_\alpha(t - t') \frac{dF_\alpha}{dt}(t') \tag{6.21}$$

where

$$f_\alpha(t) = \varepsilon Q_\alpha^h(t), \qquad \mathcal{K}_\alpha(\tau) = \frac{\varepsilon^2}{\omega_\alpha^2} \cos(\omega_\alpha \tau) \tag{6.22}$$

Next, in order to eliminate the environment variables from the equation for $A(p, q)$, one substitutes Eq. (6.21) back into Eq. (6.17). This yields a term $\sum_\alpha \{A, F_\alpha\} \mathcal{K}_\alpha(t - t_0) F_\alpha(t_0)$ that depends on the initial state of the system ($p(t_0), q(t_0)$) and produces a transient response that can be ignored in the long-time dynamics (we shall, however, return to this question below).* The parallel term $-\sum_\alpha \{A, F_\alpha\} \mathcal{K}_\alpha(0) F_\alpha(t)$, which is derivable from a Hamiltonian, is exactly balanced by the term emerging from the counterterms in $\{A, \mathcal{H}^{(m)}\}$. This can be readily shown by using $-\sum_\alpha \{A, F_\alpha\} \mathcal{K}_\alpha F_\alpha = \{A, -\frac{1}{2} \sum_\alpha \mathcal{K}_\alpha F_\alpha^2\}$, which follows from the *product rule* of the Poisson bracket

$$\{A, BC\} = \{A, B\}C + \{A, C\}B \qquad (6.23)$$

and then using $\mathcal{K}_\alpha(0) = \varepsilon^2/\omega_\alpha^2$ [see Eq. (6.22)].

Therefore, one is finally left with the *reduced* dynamical equation

$$\frac{dA}{dt} = \{A, \mathcal{H}\} + \sum_\alpha \{A, F_\alpha\} \left[f_\alpha(t) + \int_{t_0}^t dt' \mathcal{K}_\alpha(t - t') \frac{dF_\alpha}{dt}(t') \right] \qquad (6.24)$$

where the first term yields the free (conservative) time evolution of the system, while the second term incorporates the effects of the interaction of the system with its environment. The terms $f_\alpha(t)$ are customarily interpreted as *fluctuating* "forces" (or "fields"), whereas the integral term, which keeps in general memory of the previous history of the system, provides the *relaxation* due to the interaction with the surrounding medium.[†]

The origin of both types of terms can be traced back as follows. Recall that in Eq. (6.19) the time evolution of the oscillators has formally been written as if they were driven by (time-dependent) forces $-\varepsilon F_\alpha[p(t'), q(t')]$ depending on the dynamical state of the system. Therefore, $Q_\alpha(t)$ consists of the sum of the proper (free) mode $Q_\alpha^h(t)$ and the driven-type term, which naturally depends on the "forcing" (state of the system) at previous times. Then, the replacement of Q_α in the equation of the system variables by the driven-oscillator solution incorporates:

1. The time-dependent modulation due to the proper modes of the environment.

2. The "backreaction" on the system of its preceding action on the surrounding medium.

---

*In the ordinary independent oscillator model, one considers $F_\alpha(p, q) \propto q$ and the corresponding term can formally be removed from the dynamical equations by choosing the origin of the "coordinate frame" to lay at the "position" of the system at $t = t_0$, that is, $F_\alpha(t_0) \propto q(t_0) = 0$. However, this frame-dependent procedure cannot be employed if the system comprises different entities. In addition, in the spin-dynamics case with, for instance, $F_\alpha(\vec{m})$ linear in $\vec{m}$, one cannot set $\vec{m}(t_0) = \vec{0}$ due to the conservation of the magnitude of the spin.

[†] Note that without the integration by parts yielding Eq. (6.21), the Hamiltonian (renormalization) terms would occur inconveniently mixed in the integral term.

Thus, the formalism leads to a description in terms of a reduced number of dynamical variables at the expense of both explicitly time-dependent (fluctuating) terms and history-dependent (relaxation) terms (see Table VI).

Archetypal example: the Brownian particle. In order to particularize these general expressions to definite situations, the structure of the coupling terms F_α needs to be specified. For instance, on setting $F_\alpha(p, q) = -c_\alpha q$ (bilinear coupling), where the $c_\alpha = c_\alpha(\omega_\alpha)$ are coupling constants, and writing Eq. (6.24) for $A = q$ and $A = p$ with help from $\{p, B\} = -\partial B/\partial q$ and $\{q, B\} = \partial B/\partial p$, one gets the celebrated generalized Langevin equation for a "Brownian" particle (Zwanzig, 1973)

$$\frac{dq}{dt} = \frac{\partial \mathcal{H}}{\partial p}, \qquad \frac{dp}{dt} = -\frac{\partial \mathcal{H}}{\partial q} + f(t) - \int_{t_0}^{t} dt' \mathcal{K}(t - t')\frac{dq}{dt}(t') \qquad (6.25)$$

Here, $f(t) = \sum_\alpha c_\alpha f_\alpha(t)$ is the fluctuating force and $\mathcal{K}(\tau) = \sum_\alpha c_\alpha^2 \mathcal{K}_\alpha(\tau)$ is the memory kernel, and the relaxation term associated with it involves minus the velocity $-(dq/dt)(t')$ of the particle (*viscous damping*).

In general, when $\{A, F_\alpha\}$ in Eq. (6.24) is not constant, the fluctuating terms $f_\alpha(t)$ enter multiplying the system variables (*multiplicative* fluctuations). In this example, because $\{q, -c_\alpha q\} = 0$ and $\{p, -c_\alpha q\} = c_\alpha$, the fluctuations enter in an *additive* way.

3. Dynamical Equations: The Spin-Dynamics Case

Let us now particularize the preceding results to the dynamics of a classical spin. To this end, we introduce the coupling functions

$$F_\alpha(\vec{m}) = \sum_l c_\alpha^l V_l(\vec{m}) \qquad (6.26)$$

where l stands for a general index depending on the type of interaction, the coefficients c_α^l are spin–environment coupling constants, and the terms

TABLE VI
Terms Incorporating the Effects of the Interaction of the System with the Surrounding Medium
in the Reduced Dynamical Equation (6.24)

Term	Mechanism	Interpretation
$f_\alpha(t)$	Time-dependent modulation due to the proper modes of the environment	Fluctuating term
Integral term	Backreaction on the system of its preceding action on the environment	Relaxation term

$V_1(\vec{m})$ are certain functions of the spin variables. In order to motivate this expression, consider, for example, the magnetoelastic coupling of $\vec{m}$ to the lattice vibrations. The index 1 then stands for a pair of Cartesian indices (ij) and $V_1 \rightarrow V_{ij} = \sum_{k\ell} a_{ij,k\ell} m_k m_\ell$, where the $a_{ij,k\ell}$ are magnetoelastic coefficients.

In order to derive the reduced dynamical equation for the spin, we merely put $A = m_i$, $i = x,y,z$, in Eq. (6.24), and then use Eq. (6.13) to calculate the Poisson brackets required. On gathering the results thus obtained in vectorial form and using $\vec{B}_{\text{eff}} = -\partial \mathcal{H}/\partial \vec{m}$ and $dV_{1'}/dt = (\partial V_{1'}/\partial \vec{m}) \cdot (d\vec{m}/dt)$, we arrive at

$$\frac{1}{\gamma}\frac{d\vec{m}}{dt} = \vec{m} \wedge \left\{ \vec{B}_{\text{eff}} + \vec{b}_{\text{fl}}(\vec{m}, t) - \int_{t_0}^{t} dt'\, \hat{\Gamma}^{(\text{L})}(\vec{m}; t, t')\frac{d\vec{m}}{dt}(t') \right\} \qquad (6.27)$$

In this equation the *fluctuating magnetic field* is given by

$$\vec{b}_{\text{fl}}(\vec{m}, t) = -\sum_1 f_1(t)\frac{\partial V_1}{\partial \vec{m}} \qquad (6.28)$$

which involves the environmental proper modes via the fluctuating sources

$$f_1(t) = \varepsilon \sum_\alpha c_\alpha^1 Q_\alpha^{\text{h}}(t) \qquad (6.29)$$

Concerning the relaxation term in Eq. (6.27), the relaxation tensor reads*

$$\hat{\Gamma}^{(\text{L})}(\vec{m}; t, t') = \sum_{1,1'} \mathcal{K}_{11'}(t - t')\frac{\partial V_1}{\partial \vec{m}}(t)\frac{\partial V_{1'}}{\partial \vec{m}}(t') \qquad (6.30)$$

where the *memory kernel* is given by[†]

$$\mathcal{K}_{11'}(\tau) = \varepsilon^2 \sum_\alpha \frac{c_\alpha^1 c_\alpha^{1'}}{\omega_\alpha^2}\cos(\omega_\alpha \tau) \qquad (6.31)$$

Equation (6.27) contains $d\vec{m}/dt$ on its right-hand side, so it will be referred to as a *Gilbert-type* equation [cf. Eq. (6.1)]. For $\varepsilon \ll 1$, on replacing

---

* Although we omit the symbol of scalar product, the action of a dyadic $\vec{A}\vec{B}$ on a vector $\vec{C}$ is the standard one: $(\vec{A}\vec{B})\vec{C} \equiv \vec{A}(\vec{B} \cdot \vec{C})$.

[†] Note that $f_1(t) = \sum_\alpha c_\alpha^1 f_\alpha(t)$ and $\mathcal{K}_{11'}(\tau) = \sum_\alpha c_\alpha^1 c_\alpha^{1'}\mathcal{K}_\alpha(\tau)$, where $f_\alpha(t)$ and $\mathcal{K}_\alpha(\tau)$ are given by Eq. (6.22).

perturbatively that derivative by its conservative part, $d\vec{m}/dt \simeq \gamma\vec{m} \wedge \vec{B}_{\text{eff}}$, one gets the weak-coupling *Landau–Lifshitz-type* equation

$$\frac{1}{\gamma}\frac{d\vec{m}}{dt} = \vec{m} \wedge \left[\vec{B}_{\text{eff}} + \vec{b}_{\text{fl}}(\vec{m}, t)\right] - \vec{m} \wedge \left\{\int_{t_0}^{t} dt' \gamma\hat{\Gamma}^{(\text{L})}(\vec{m}; t, t') \left(\vec{m} \wedge \vec{B}_{\text{eff}}\right)(t')\right\}$$

(6.32)

which describes weakly damped precession.

For spin–environment interactions *linear* in the environment variables but that are otherwise *arbitrary* functions of $\vec{m}$, Eqs. (6.27) and (6.32) are the desired reduced dynamical equations for the spin. They have the structure of generalized Langevin equations with *fluctuating* terms $\vec{m} \wedge \vec{b}_{\text{fl}}(\vec{m}, t)$ (associated with the modulation by the proper modes of the environment) and history-dependent *relaxation* terms (corresponding to the backreaction on the spin of its previous action on the surrounding medium).

Note that $f_1(t)$ [Eq. (6.29)] is a sum of a large number of sinusoidal terms with different frequencies and phases; this can give to $f_1(t)$ the form of a highly irregular function of t that is expected for a fluctuating term. However, for a general form of the coupling functions $V_1(\vec{m})$, the term $\vec{b}_{\text{fl}}(\vec{m}, t)$ *cannot* be interpreted as a fluctuating *ordinary* field, since it may depend on $\vec{m}$, but it is rather a fluctuating *effective* field to be added to the deterministic effective field $\vec{B}_{\text{eff}} = -\partial\mathcal{H}/\partial\vec{m}$ [Eq. (6.9)]. This can be illustrated by phrasing the discussion in terms of the *fluctuating part* of the energy of the spin, namely [see Hamiltonian (6.15)], $\mathcal{H}_{\text{fl}} = \varepsilon\sum_{\alpha} Q_{\alpha}^{\text{h}}(t)F_{\alpha}(\vec{m})$. From this definition one first gets

$$\mathcal{H}_{\text{fl}}(\vec{m}, t) = \sum_{1} f_1(t)V_1(\vec{m}), \qquad \vec{b}_{\text{fl}}(\vec{m}, t) = -\frac{\partial\mathcal{H}_{\text{fl}}}{\partial\vec{m}}$$

(6.33)

so that $\vec{b}_{\text{fl}}$ can be derived from $\mathcal{H}_{\text{fl}}$ in the same way as $\vec{B}_{\text{eff}}$ is obtained from $\mathcal{H}$. Next, recall that the nonlinear part of $\mathcal{H}(\vec{m})$ carries the anisotropy-energy terms, for instance, $\mathcal{H} = -\vec{m} \cdot \vec{B} - \frac{1}{2}(B_K/m)(\vec{m} \cdot \vec{n})^2$ in a uniaxial crystal. Analogously, $\mathcal{H}_{\text{fl}}$ has the form $\mathcal{H}_{\text{fl}}(\vec{m}, t) = -\vec{m} \cdot \vec{b}_{\text{fl}}(t)$, with $\vec{b}_{\text{fl}}$ independent of $\vec{m}$, only for linear $V_1(\vec{m})$ (bilinear coupling), so that *the nonlinear part of $V_1(\vec{m})$ incorporates fluctuations of the magnetic anisotropy of the spin.*

To illustrate, if the spin–environment interaction includes terms up to quadratic in $\vec{m}$, one can write the coupling functions $V_1(\vec{m})$ as

$$V_1(\vec{m}) = \sum_{i} v_{1,i}m_i + \frac{1}{2}\sum_{ij} w_{1,ij}m_im_j$$

(6.34)

where the coupling constants $v_{1,i}$ and $w_{1,ij}$ incorporate the symmetry of the interaction. In this case, the fluctuating effective field (6.28) can be cast into the form [cf. Eq. (6.8)]

$$\vec{b}_{fl}(\vec{m}, t) = \vec{b}(t) + \hat{\kappa}(t)\vec{m} \tag{6.35}$$

with the following expressions for the fluctuating sources $\vec{b}(t)$ and $\hat{\kappa}(t)$ in terms of the coupling constants:

$$b_i(t) = -\sum_l f_l(t)v_{l,i}, \qquad \kappa_{ij}(t) = -\sum_l f_l(t)w_{l,ij}$$

Thus, $\vec{b}(t)$ is determined by the part of the coupling term $V_l(\vec{m})$ linear in the spin variables, while $\hat{\kappa}(t)$ is associated with the quadratic part.

As $\vec{b}(t)$ does not depend on $\vec{m}$, it can be interpreted as a fluctuating *ordinary* field. The fluctuations of $\hat{\kappa}(t)$, however, do not enter in this way, since they occur via $\sum_j \kappa_{ij}(t)m_j$, but they produce fluctuations of the magnetic-anisotropy potential of the spin, both of the direction of the anisotropy axes and of the magnitudes of the anisotropy constants. This is clearly perceived on considering that the fluctuating part of the energy of the spin (6.33) in this case takes the form

$$\mathcal{H}_{fl}(\vec{m}, t) = -\vec{m} \cdot \vec{b}(t) - \tfrac{1}{2}\vec{m} \cdot \hat{\kappa}(t)\vec{m}$$

This resembles the scenario encountered for a mechanical oscillator (Lindenberg and Seshadri, 1981), where the portion of the oscillator–environment coupling quadratic in the coordinate of the test oscillator yields, instead of a fluctuating force, a fluctuating contribution to its harmonic potential (*frequency-type* fluctuations). Finally, if $V_l(\vec{m})$ contains only nonlinear terms, such as those occurring in the magnetoelastic coupling mentioned ($V_l \propto \sum m_k m_\ell$), no field-type fluctuating terms emerge and only anisotropy-type fluctuations remain.

We remark in closing that, even for couplings linear in the spin variables, and hence for $\vec{b}_{fl}(t)$ independent of $\vec{m}$, the occurrence of the vector *product* $\vec{m} \wedge \vec{b}_{fl}$ in the dynamical equations entails that the fluctuating terms enter in a *multiplicative* way. This is at variance with the situation encountered in ordinary mechanical systems, where couplings linear in the system variables lead to additive fluctuations [see Eq. (6.25)], whereas multiplicative fluctuating terms emerge only for couplings nonlinear in the system variables (Lindenberg and Seshadri, 1981). To illustrate, for the mechanical oscillator mentioned above, the force- and frequency-type fluctuations provided by the first and second terms in $F_\alpha = -v_\alpha q - w_\alpha q^2$ are, respectively, additive and multiplicative, whereas in the gyromagnetic case the field-type fluctuations are already multiplicative. In the spin-dynamics case, in analogy with the results obtained for mechanical rigid rotators (Lindenberg, Mohanty and Seshadri, 1983), the multiplicative character of the fluctuations is a consequence of the Poisson bracket relations $\{m_i, m_j\} = \gamma \sum_k \varepsilon_{ijk} m_k$ for angular-momentum-type dynamical variables, which, even for F_α linear in $\vec{m}$, lead to nonconstant $\{A, F_\alpha\}$ in Eq. (6.24). In our derivation, this can straightly be traced back by virtue of the Poisson bracket formalism employed.

4. Statistical Properties of the Fluctuating Terms

In order to determine the statistical properties of the fluctuating sources $f_1(t)$, one usually assumes that the environment was in thermodynamical equilibrium at the *initial* time (recall that no statistical assumption has been explicitly introduced until this point). This initial state is customarily chosen in two different ways.

a. Decoupled Initial Conditions. The environment variables are distributed at $t = t_0$ according to the Boltzmann law associated with the environment Hamiltonian alone

$$P_e(\mathbf{P}(t_0), \mathbf{Q}(t_0)) \propto \exp\left[-\frac{\mathcal{H}_E(t_0)}{k_B T}\right] \qquad (6.36)$$

$$\mathcal{H}_E(t_0) = \sum_\alpha \tfrac{1}{2}\left[P_\alpha(t_0)^2 + \omega_\alpha^2 Q_\alpha(t_0)^2\right]$$

where $(\mathbf{P}, \mathbf{Q})$ stands for the set of canonical variables of the environment. The initial distribution is therefore Gaussian and one has for the first two moments of the environmental variables

$$\langle Q_\alpha(t_0)\rangle = 0, \qquad \langle P_\alpha(t_0)\rangle = 0$$

$$\langle Q_\alpha(t_0)Q_\beta(t_0)\rangle = \delta_{\alpha\beta}\frac{k_B T}{\omega_\alpha^2}, \qquad \langle Q_\alpha(t_0)P_\beta(t_0)\rangle = 0, \qquad \langle P_\alpha(t_0)P_\beta(t_0)\rangle = \delta_{\alpha\beta}k_B T$$

From these results one readily gets the averages of the proper modes over initial states of the environment (ensemble averages):

$$\langle Q_\alpha^h(t)\rangle = \underbrace{\langle Q_\alpha(t_0)\rangle}_{0} \cos[\omega_\alpha(t - t_0)] + \underbrace{\langle P_\alpha(t_0)\rangle}_{0} \frac{1}{\omega_\alpha}\sin[\omega_\alpha(t - t_0)]$$

$$\langle Q_\alpha^h(t)Q_\beta^h(t')\rangle = \underbrace{\langle Q_\alpha(t_0)Q_\beta(t_0)\rangle}_{\delta_{\alpha\beta}k_B T/\omega_\alpha^2} \cos[\omega_\alpha(t - t_0)]\cos[\omega_\beta(t' - t_0)]$$

$$+ \underbrace{\langle Q_\alpha(t_0)P_\beta(t_0)\rangle}_{0} \frac{1}{\omega_\beta}\cos[\omega_\alpha(t - t_0)]\sin[\omega_\beta(t' - t_0)]$$

$$+ \underbrace{\langle P_\alpha(t_0) Q_\beta(t_0) \rangle}_{0} \frac{1}{\omega_\alpha} \sin[\omega_\alpha(t - t_0)] \cos[\omega_\beta(t' - t_0)]$$

$$+ \underbrace{\langle P_\alpha(t_0) P_\beta(t_0) \rangle}_{\delta_{\alpha\beta} k_B T} \frac{1}{\omega_\alpha \omega_\beta} \sin[\omega_\alpha(t - t_0)] \sin[\omega_\beta(t' - t_0)]$$

$$= k_B T \frac{\delta_{\alpha\beta}}{\omega_\alpha^2} \{ \cos[\omega_\alpha(t - t_0)] \cos[\omega_\alpha(t' - t_0)]$$

$$+ \sin[\omega_\alpha(t - t_0)] \sin[\omega_\alpha(t' - t_0)] \}$$

so that

$$\langle Q_\alpha^h(t) \rangle = 0, \qquad \langle Q_\alpha^h(t) Q_\beta^h(t') \rangle = k_B T \frac{\delta_{\alpha\beta}}{\omega_\alpha^2} \cos[\omega_\alpha(t - t')] \qquad (6.37)$$

Thus, the fluctuating terms $f_1(t)$ [Eq. (6.29)] are Gaussian stochastic processes and the relevant averages over initial states of the environment are given by

$$\langle f_1(t) \rangle = 0 \qquad (6.38)$$

$$\langle f_1(t) f_{1'}(t') \rangle = k_B T \mathcal{K}_{11'}(t - t') \qquad (6.39)$$

Equation (6.39) relates the statistical time correlation of the fluctuating terms $f_1(t)$ with the relaxation memory kernels $\mathcal{K}_{11'}(\tau)$ occurring in the dynamical equations (*fluctuation–dissipation* relations). Short (long) correlation times of the fluctuating terms entail short-range (long-range) memory effects in the relaxation term, and vice versa. The emergence of this type of relations is not surprising in this context, since fluctuations and relaxation arise as different manifestations of the *same* interaction of the system with the surrounding medium.

b. Coupled Initial Conditions. The environment is assumed to be at $t = t_0$ in thermal equilibrium *in the presence of the system*, which is, however, taken as *fastened* in its initial state (see, for instance, Ford, Lewis and O'Connell, 1988). Therefore, the corresponding initial distribution of the environment variables is

$$P_e(\mathbf{P}(t_0), \mathbf{Q}(t_0)) \propto \exp\left[-\frac{\mathcal{H}_{SE}(t_0)}{k_B T} \right]$$

$$\mathcal{H}_{SE}(t_0) = \sum_\alpha \frac{1}{2} \left\{ P_\alpha(t_0)^2 + \omega_\alpha^2 \left[Q_\alpha(t_0) + \frac{\varepsilon}{\omega_\alpha^2} F_\alpha(t_0) \right]^2 \right\}$$

where the $F_\alpha(t_0)$ are taken as constants. In this case, the dropped terms depending on the initial state of the system $\mathcal{K}_\alpha(t - t_0)F_\alpha(t_0)$ [recall the remarks preceding Eq. (6.23)], which for $F_\alpha = \sum_l c_\alpha^l V_l$ lead to the terms $\sum_{l'} \mathcal{K}_{1l'}(t - t_0)V_{l'}(t_0)$, are not omitted but they are included into an alternative definition of the fluctuating sources, namely, $\tilde{f}_1(t) = f_1(t) + \sum_{l'} \mathcal{K}_{1l'}(t - t_0)V_{l'}(t_0)$. The statistical properties of these terms, as determined by the distribution introduced above, are given by expressions *identical* with Eqs. (6.38) and (6.39).

Notice that the recourse to the "process" of initial fastening (and subsequent releasing) of the system by an external agency can, to a certain extent, be circumvented on noting that the concomitant initial statistical properties of the environment are consistent with the notion of a timescale separation between the system and the surrounding medium; that is, the latter adjust rapidly to the state of the former (Lindenberg and West, 1984).

Note finally that the differences associated with assuming decoupled initial conditions or the more physically motivated coupled initial conditions diminish as long as the weak-coupling condition is met. Anyway, with both types of initial conditions, one obtains the *same* Langevin equation after a time, measured from t_0, of the order of the width of the memory kernels $\mathcal{K}_{1l'}(\tau)$, which is the characteristic time for the "transient" terms $\sum_{l'} \mathcal{K}_{1l'}(t - t_0)V_{l'}(t_0)$ to die out.

D. Dynamical Equations for Couplings Linear-Plus-Quadratic in the Environment Variables

The introduction of interactions nonlinear in the environment variables is mandatory when relaxation mechanisms involving more than one environmental normal mode (e.g., multiphonon or multiphoton processes) become relevant, as occurs at sufficiently high temperatures. When such nonlinear couplings are taken into account, one must resort to approximate methods to derive a reduced equation of motion for the spin. Here, we tackle the important weak-coupling case by a perturbational treatment.

1. The Spin–Environment Hamiltonian

Let us consider the following generalization of the Hamiltonian (6.14)

$$\mathcal{H}_T = \mathcal{H}(\vec{m}) + \sum_\alpha \frac{1}{2} \left\{ P_\alpha^2 + \omega_\alpha^2 \left[Q_\alpha + \frac{\varepsilon}{\omega_\alpha^2} F_\alpha(\vec{m}) \right]^2 \right\}$$
$$+ \frac{1}{2} \sum_{\alpha\beta} \left[\varepsilon Q_\alpha Q_\beta F_{\alpha\beta}(\vec{m}) + \frac{k_B T \varepsilon^2}{2\omega_\alpha^2 \omega_\beta^2} F_{\alpha\beta}(\vec{m})^2 \right] \quad (6.40)$$

where couplings quadratic in the coordinates of the oscillators representing the environment have been included. The part of this interaction depending on the spin variables is introduced via the functions $F_{\alpha\beta}$. Besides, embodying the additional counterterms (those proportional to $F_{\alpha\beta}^2$), the coupling-induced renormalization of the energy of the spin is balanced to order ε^2. This renormalization results to be explicitly dependent on the temperature for interactions nonlinear in the environment variables (see discussion below).

2. Dynamical Equations: General Case

Again, for the sake of simplicity and generality, we rewrite the Hamiltonian (6.40) as [cf. Eq. (6.15)]

$$
\mathcal{H}_T = \mathcal{H}^{(m)}(p, q) + \sum_{\alpha} \tfrac{1}{2} \left(P_\alpha^2 + \omega_\alpha^2 Q_\alpha^2 \right)
$$

$$
+ \varepsilon \left[\sum_{\alpha} Q_\alpha F_\alpha(p, q) + \tfrac{1}{2} \sum_{\alpha\beta} Q_\alpha Q_\beta F_{\alpha\beta}(p, q) \right] \qquad (6.41)
$$

where $\mathcal{H}^{(m)}$ augments the system Hamiltonian by the counterterms [cf. Eq. (6.16)]

$$
\mathcal{H}^{(m)} = \mathcal{H} + \frac{\varepsilon^2}{2} \left(\sum_{\alpha} \frac{F_\alpha^2}{\omega_\alpha^2} + k_B T \sum_{\alpha\beta} \frac{F_{\alpha\beta}^2}{2\omega_\alpha^2 \omega_\beta^2} \right) \qquad (6.42)
$$

The ordinary formalism of the environment of *independent* oscillators (Magalinskiĭ, 1959; Ullersma, 1966; Zwanzig, 1973; Caldeira and Leggett, 1983; Ford, Lewis and O'Connell, 1988) is not directly applicable when couplings nonlinear in the environment variables are included. For instance, $F_{\alpha\beta} Q_\alpha Q_\beta$ brings about an indirect interaction among the oscillators so that these are no longer independent. Because a reduced equation of motion for the system cannot easily be derived for an arbitrary strength of the coupling, we shall perform a perturbational treatment in the weak-coupling case by means of simple extensions of the treatment developed by Cortés et al. (1985).

In Appendix C the corresponding calculations are detailed for a class of Hamiltonians with quite general nonlinear couplings in both the system and the environment variables. The results obtained permit the incorporation of relaxation mechanisms involving any number of environmental normal modes into the dynamical equations of the system (under the weak-coupling condition mentioned). In the linear-plus-quadratic case considered here, we find the following *reduced* dynamical equation for any observable of the system $A(p, q)$ [cf. Eq. (6.24)]

$$\frac{dA}{dt} = \{A, \mathcal{H}\} + \sum_\alpha \{A, F_\alpha\} \left[f_\alpha(t) + \int_{t_0}^t dt' \mathcal{K}_\alpha(t - t') \frac{dF_\alpha}{dt}(t') \right]$$

$$+ \sum_{\alpha\beta} \{A, F_{\alpha\beta}\} \left[f_{\alpha\beta}(t) + \int_{t_0}^t dt' \mathcal{K}_{\alpha\beta}(t - t') \frac{dF_{\alpha\beta}}{dt}(t') \right] \qquad (6.43)$$

In this equation, the fluctuating terms $f_\alpha(t)$ and the corresponding kernels $\mathcal{K}_\alpha(\tau)$ are again given by Eqs. (6.22), whereas their counterparts for the quadratic portion of the coupling read

$$f_{\alpha\beta}(t) = \frac{\varepsilon}{2} Q_\alpha^h(t) Q_\beta^h(t) \qquad (6.44)$$

$$\mathcal{K}_{\alpha\beta}(\tau) = \frac{\varepsilon^2}{2} \frac{k_B T}{2\omega_\alpha^2 \omega_\beta^2} \left\{ \cos[(\omega_\alpha - \omega_\beta)\tau] + \cos[(\omega_\alpha + \omega_\beta)\tau] \right\} \qquad (6.45)$$

where the $Q_\alpha^h(t)$ are the environmental proper modes (6.20).

The treatment leading to Eq. (6.43) can be summarized in terms of the driven-oscillator picture discussed in Section VI.C. One part of the driving from the system now depends on the state of the oscillators [cf. Eqs. (6.18) with (C.3)]; this state is perturbatively replaced by the free evolution terms $Q_\alpha^h(t)$, and the backreaction on the system is averaged over initial states of the oscillators. This averaging yields the explicit dependence of the kernels $\mathcal{K}_{\alpha\beta}(\tau)$ on the temperature (and that of the associated counterterm $\frac{1}{2} \sum_{\alpha\beta} \mathcal{K}_{\alpha\beta}(0) F_{\alpha\beta}^2$).

3. Dynamical Equations: The Spin-Dynamics Case

In order to particularize the result (6.43) to the dynamics of a classical spin, the additional coupling functions $F_{\alpha\beta}$ are expressed as

$$F_{\alpha\beta}(\vec{m}) = \sum_q c_{\alpha\beta}^q V_q(\vec{m})$$

where the general index q is analogous to that introduced in the linear case [Eq. (6.26)], the coefficients $c_{\alpha\beta}^q$ are the spin–environment coupling constants for the quadratic part of the interaction, and the terms $V_q(\vec{m})$ are certain functions of the spin variables. To illustrate, for the coupling of $\vec{m}$ to the lattice vibrations including quadratic terms in the strain tensor ("two-phonon processes"), q stands for *two* pairs of Cartesian indices and, for example, $V_q \rightarrow V_{ij,k\ell} = \sum_{rs} b_{ijk\ell,rs} m_r m_s$, where the $b_{ijk\ell,rs}$ are second-order magnetoelastic coefficients.

Then, on merely replacing $A(p, q)$ in Eq. (6.43) by the Cartesian components of the magnetic moment and using Eq. (6.13) to calculate the corresponding Poisson brackets, one arrives at the following reduced equation of motion for $\vec{m}$ [cf. Eq. (6.32)]:

$$\frac{1}{\gamma}\frac{d\vec{m}}{dt} = \vec{m} \wedge \left[\vec{B}_{\text{eff}} + \vec{b}_{\text{fl}}(\vec{m}, t)\right]$$

$$- \vec{m} \wedge \left\{\int_{t_0}^{t} dt'\gamma \left[\hat{\Gamma}^{(\text{L})} + k_{\text{B}}T\hat{\Gamma}^{(\text{Q})}\right]_{(\vec{m};t,t')} \left(\vec{m} \wedge \vec{B}_{\text{eff}}\right)(t')\right\} \quad (6.46)$$

Here, the fluctuating effective field generalizes the expression (6.28) to

$$\vec{b}_{\text{fl}}(\vec{m}, t) = -\left[\sum_{\text{l}} f_{\text{l}}(t)\frac{\partial V_{\text{l}}}{\partial \vec{m}} + \sum_{\text{q}} f_{\text{q}}(t)\frac{\partial V_{\text{q}}}{\partial \vec{m}}\right] \quad (6.47)$$

where the $f_{\text{l}}(t)$ are given by Eq. (6.29) and the $f_{\text{q}}(t) = \sum_{\alpha\beta} c_{\alpha\beta}^{\text{q}} f_{\alpha\beta}(t)$ are additional fluctuating terms

$$f_{\text{q}}(t) = \frac{\varepsilon}{2}\sum_{\alpha\beta} c_{\alpha\beta}^{\text{q}} Q_{\alpha}^{\text{h}}(t)Q_{\beta}^{\text{h}}(t) \quad (6.48)$$

Concerning the relaxation terms, $\hat{\Gamma}^{(\text{L})}$ is again given by Eq. (6.30), while the part of the relaxation tensor associated with the quadratic part of the coupling reads

$$k_{\text{B}}T\hat{\Gamma}^{(\text{Q})}(\vec{m}; t, t') = \sum_{\text{q},\text{q}'} K_{\text{qq}'}(t - t')\frac{\partial V_{\text{q}}}{\partial \vec{m}}(t)\frac{\partial V_{\text{q}'}}{\partial \vec{m}}(t') \quad (6.49)$$

where the kernel is given by $K_{\text{qq}'}(\tau) = \sum_{\alpha\beta} c_{\alpha\beta}^{\text{q}} c_{\alpha\beta}^{\text{q}'} K_{\alpha\beta}(\tau)$ or, explicitly

$$K_{\text{qq}'}(\tau) = k_{\text{B}}T\frac{\varepsilon^2}{2}\sum_{\alpha\beta}\frac{c_{\alpha\beta}^{\text{q}} c_{\alpha\beta}^{\text{q}'}}{2\omega_{\alpha}^2\omega_{\beta}^2}\left\{\cos[(\omega_{\alpha} - \omega_{\beta})\tau] + \cos[(\omega_{\alpha} + \omega_{\beta})\tau]\right\} \quad (6.50)$$

Note that the equation (6.46) is of Landau–Lifshitz type since the derivative $d\vec{m}/dt$ that would appear in the relaxation term has been replaced, within the approximation used ($\varepsilon \ll 1$), by its free evolution part $d\vec{m}/dt \simeq \gamma\vec{m} \wedge \vec{B}_{\text{eff}}$ [see the remarks following Eq. (C.11)]. Note also that we have explicitly shown the temperature dependence of the relaxation term, which is caused by the quadratic portion of the coupling.

Equation (6.46) is the desired dynamical equation for the spin when its interaction with the environment is weak and embodies terms linear-plus-quadratic in the variables of the oscillators representing the environment. Note that, in the pictorial quantum-mechanical language, the term comprising $\cos(\omega_\alpha \tau)$ in the memory kernel (6.31) would correspond to a relaxation mechanism (transition) via the emission or absorption of a vibrational quantum of energy $\hbar\omega_\alpha$. Similarly, $\cos[(\omega_\alpha + \omega_\beta)\tau]$ in the kernel (6.50) would be associated with relaxation mechanisms with either the emission or the absorption of two vibrational quanta, whereas $\cos[(\omega_\alpha - \omega_\beta)\tau]$ would correspond to the absorption of one quantum and the emission of a second one (scattering processes).

Finally, the definition (6.33) of the fluctuating part of the energy of the spin can be generalized to

$$\mathcal{H}_{fl}(\vec{m}, t) = \sum_l f_l(t)V_l(\vec{m}) + \sum_q f_q(t)V_q(\vec{m}) \qquad (6.51)$$

whence $\vec{b}_{fl} = -\partial\mathcal{H}_{fl}/\partial\vec{m}$, in correspondence with $\vec{B}_{eff} = -\partial\mathcal{H}/\partial\vec{m}$. Remarks similar to those following Eq. (6.33) concerning the structure of $\mathcal{H}_{fl}(\vec{m}, t)$ for linear and nonlinear (in the spin variables) spin–environment interactions, and the corresponding nature of the fluctuations (field- and/or anisotropy-type), are in order here.

4. Statistical Properties of the Fluctuating Terms

The statistical properties of the $f_l(t)$, as determined by the initial distribution (6.36) of the environment variables (*decoupled initial conditions*), are given by Eqs. (6.38) and (6.39), whereas the statistical properties of the $f_q(t)$ and their cross-correlations read

$$\langle f_q(t) \rangle = 0 \qquad (6.52)$$

$$\langle f_l(t)f_q(t') \rangle = 0 \qquad (6.53)$$

$$\langle f_q(t)f_{q'}(t') \rangle = k_B T \mathcal{K}_{qq'}(t - t') \qquad (6.54)$$

In order to obtain Eq. (6.52) (i.e., centered fluctuating sources) as well as Eq. (6.54), we have assumed that $c_{\alpha\beta}^q \equiv 0$ for $\alpha = \beta$. If such a restriction is not applied, one has, for example, $\langle f_q(t) \rangle \neq 0$, which represents a nonvanishing average forcing of the spin. Note, however, that to retain those terms must cause no harm since, when the double sums over oscillators $\sum_{\alpha\beta}(\cdot)$ are transformed into double integrals for (quasi-) continuous distributions of oscillators, such $\alpha = \beta$ terms constitute a zero-measure set whose contribution can therefore be ignored.

The Gaussian property of the $f_q(t)$ can then be established on the basis that these terms are sums over a large number of contributions $c_{\alpha\beta}^q Q_\alpha^h(t) Q_\beta^h(t)$ with mean zero and equivalent statistical properties (central-limit theorem). On the other hand, Eq. (6.54) expresses that the fluctuating sources $f_q(t)$ and the relaxation memory kernels $\mathcal{K}_{\mathbf{qq'}}(\tau)$, associated with the quadratic portion of the coupling, also obey fluctuation–dissipation relations. In addition, the zero cross-correlations of Eq. (6.53) are also fluctuation–dissipation relations involving null kernels [see Eq. (C.17)].

We finally remark that on assuming *coupled initial conditions*, without modifying the definitions of the fluctuating terms, the corrections to Eqs. (6.38) and (6.52), and to the relations (6.39), (6.53), and (6.54) are, respectively, of order ε^2 and ε^3; these corrections are of order higher than the order of the terms retained in the weak-coupling approximation used (see Appendix C).

E. Markovian Regime and Phenomenological Equations

We now study the form that the dynamical equations derived exhibit in the absence of memory effects. Then, we consider some specific spin–environment interactions, formally obtaining the Langevin equations mentioned at the beginning of this section.

1. Markovian Regime

The Markovian regime arises when the relaxation memory kernels are sharply peaked at $\tau = 0$, the remainder terms in the memory integrals change slowly enough in the relevant range, and the kernels enclose a finite nonzero algebraic area. Under these conditions, one can replace the kernels by Dirac deltas and no memory effects occur.

a. Langevin Equations. Let us assume that the memory kernel (6.31) can be replaced by a Dirac delta

$$\mathcal{K}_{\text{ll'}}(\tau) = \frac{2\lambda_{\text{ll'}}}{\gamma m}\delta(\tau) \tag{6.55}$$

where the $\lambda_{\text{ll'}}$ are *damping coefficients* related with the strength and characteristics of the coupling (see Eq. (6.69) below). Then, on using $\int_0^{\infty} d\tau\, \delta(\tau)h(\tau) = h(0)/2$, Eq. (6.27) reduces to the *Gilbert-type* equation [cf. Eq. (6.1)]

$$\frac{1}{\gamma}\frac{d\vec{m}}{dt} = \vec{m} \wedge \left[\vec{B}_{\text{eff}} + \vec{b}_{\text{fl}}(\vec{m}, t) - (\gamma m)^{-1}\hat{\Lambda}^{(\text{L})}\frac{d\vec{m}}{dt}\right] \tag{6.56}$$

where $\hat{\Lambda}^{(\text{L})}(\vec{m})$ is a dimensionless second-rank tensor with elements

$$\Lambda_{ij}^{(\text{L})}(\vec{m}) = \sum_{\text{l,l'}} \lambda_{\text{ll'}} \frac{\partial V_{\text{l}}}{\partial m_i}\frac{\partial V_{\text{l'}}}{\partial m_j} \tag{6.57}$$

Likewise, on inserting Eq. (6.55) in the weak-coupling Eq. (6.32), we get the following *Landau–Lifshitz-type* equation [cf. Eq. (6.2)]:

$$\frac{1}{\gamma}\frac{d\vec{m}}{dt} = \vec{m} \wedge \left[\vec{B}_{\text{eff}} + \vec{b}_{\text{fl}}(\vec{m}, t)\right] - \frac{1}{m}\vec{m} \wedge \hat{\Lambda}^{(\text{L})}\left(\vec{m} \wedge \vec{B}_{\text{eff}}\right) \qquad (6.58)$$

Note that the tensor $\hat{\Lambda}^{(\text{L})}$, the precursor of which is the tensor $\hat{\Gamma}^{(\text{L})}$ [Eq. (6.30)] occurring in the memory integrals, is symmetrical since $\lambda_{\text{ll}'}$ is so [see Eq. (6.69), below].

On the other hand, the Markovian case of the dynamical equation for couplings linear-plus-quadratic in the environment coordinates emerges when the additional memory kernel can also be replaced by a Dirac delta, namely

$$\mathcal{K}_{\text{qq}'}(\tau) = \frac{2\lambda_{\text{qq}'}k_{\text{B}}T}{\gamma m}\delta(\tau) \qquad (6.59)$$

where we have explicitly shown the temperature dependence arising from the kernel (6.50). Under these conditions, Eq. (6.46) reduces to the *Landau–Lifshitz-type* equation

$$\frac{1}{\gamma}\frac{d\vec{m}}{dt} = \vec{m} \wedge \left[\vec{B}_{\text{eff}} + \vec{b}_{\text{fl}}(\vec{m}, t)\right] - \frac{1}{m}\vec{m} \wedge \hat{\Lambda}\left(\vec{m} \wedge \vec{B}_{\text{eff}}\right) \qquad (6.60)$$

where $\vec{b}_{\text{fl}}(\vec{m}, t)$ is now given by Eq. (6.47). In this equation the relaxation tensor

$$\hat{\Lambda} = \hat{\Lambda}^{(\text{L})} + k_{\text{B}}T\hat{\Lambda}^{(\text{Q})} \qquad (6.61)$$

where

$$\Lambda_{ij}^{(\text{Q})}(\vec{m}) = \sum_{\text{q,q}'}\lambda_{\text{qq}'}\frac{\partial V_{\text{q}}}{\partial m_i}\frac{\partial V_{\text{q}'}}{\partial m_j} \qquad (6.62)$$

introduces an explicit dependence on the temperature rooted in the quadratic portion of the coupling.

For a general form of the spin–environment interaction, due to the occurrence of the tensors $\hat{\Lambda}$, the structure of the relaxation terms in the above equations deviates from the forms proposed by Gilbert and Landau and Lifshitz. Such deviations can be produced by couplings nonlinear in $\vec{m}$, for which $\hat{\Lambda}_{ij}^{(\text{L})}$ and $\hat{\Lambda}_{ij}^{(\text{Q})}$ depend in general on the spin variables, but they also emerge when these tensors are independent of $\vec{m}$ (for example, for couplings linear in $\vec{m}$) but they are not proportional to δ_{ij}. The relaxation is then anisotropic because $-\vec{m} \wedge \hat{\Lambda}(\vec{m} \wedge \vec{B}_{\text{eff}})$ no longer points from $\vec{m}$ to the direction of $\vec{B}_{\text{eff}}$.

Finally, owing to the fluctuation–dissipation relations (6.39) and (6.54), the fluctuating terms corresponding to the Markovian memory kernels are delta-correlated in time. Consequently, the statistical properties of the fluctuating terms take the form

$$\langle f_1(t) \rangle = 0 \tag{6.63}$$

$$\langle f_1(t) f_{1'}(t') \rangle = \frac{2\lambda_{11'}}{\gamma m} k_B T \delta(t - t') \tag{6.64}$$

and

$$\langle f_q(t) \rangle = 0 \tag{6.65}$$

$$\langle f_1(t) f_q(t') \rangle = 0 \tag{6.66}$$

$$\langle f_q(t) f_{q'}(t') \rangle = \frac{2(\lambda_{qq'} k_B T)}{\gamma m} k_B T \delta(t - t') \tag{6.67}$$

Note the double occurrence of $k_B T$ in the last relation.

b. Damping Coefficients. On taking Eqs. (6.55) and (6.59) into account, one can calculate the damping coefficients from the area enclosed by the memory kernels, namely

$$\frac{\lambda_{11'}}{\gamma m} = \int_0^\infty d\tau \, \mathcal{K}_{11'}(\tau), \qquad \frac{\lambda_{qq'} k_B T}{\gamma m} = \int_0^\infty d\tau \, \mathcal{K}_{qq'}(\tau) \tag{6.68}$$

For the Markovian approximation to work, these areas must be (1) *finite* and (2) *different from zero*.

However, since it could be difficult to find the kernels exactly in some cases, it is convenient to have alternative means for calculating the areas required only. Thus, on inserting the definitions of the kernels (6.31) and (6.50) into the integrals given above and using $\int_0^\infty d\tau \cos(\omega\tau) = \pi\delta(\omega)$, we arrive at the following expressions for the damping coefficients in terms of the distribution of normal modes and spin–environment coupling constants:

$$\frac{\lambda_{11'}}{\gamma m} = \pi\varepsilon^2 \sum_\alpha \frac{c_\alpha^1 c_\alpha^{1'}}{\omega_\alpha^2} \delta(\omega_\alpha) \tag{6.69}$$

$$\frac{\lambda_{qq'}}{\gamma m} = \pi\frac{\varepsilon^2}{2} \sum_{\alpha\beta} \frac{c_{\alpha\beta}^q c_{\alpha\beta}^{q'}}{2\omega_\alpha^2 \omega_\beta^2} \left[\delta(\omega_\alpha - \omega_\beta) + \delta(\omega_\alpha + \omega_\beta) \right] \tag{6.70}$$

Note that the Dirac deltas in these formulas make sense under integral signs for (quasi-) continuous distributions of environmental modes. Recall in this connection that the coupling constants can depend on the frequencies of these normal modes.

c. Fokker–Planck Equations. The Markovian Langevin equations can be employed to construct the corresponding Fokker–Planck equations governing the time evolution of the nonequilibrium probability distribution of spin orientations $P(\vec{m}, t)$. On examining the statistical properties (6.64) and (6.67), one realizes that, to do so, we need to deal with Langevin equations where the noise terms *are not* statistically independent.

Let us then consider the general system of Langevin equations

$$\frac{dy_i}{dt} = A_i(\mathbf{y}, t) + \sum_k B_{ik}(\mathbf{y}, t)L_k(t) \tag{6.71}$$

where $i = 1, \ldots, n$, $\mathbf{y} = (y_1, \ldots, y_n)$, k runs over a given set of indices, and the Langevin sources $L_k(t)$ are Gaussian stochastic processes satisfying

$$\langle L_k(t) \rangle = 0, \qquad \langle L_k(t)L_\ell(t') \rangle = 2D_{k\ell}\delta(t - t') \tag{6.72}$$

Here, the constant (symmetric) matrix $D_{k\ell}$ accounts for the possible correlations among the $L_k(t)$ [cf. Eq. (5.23)].

The time evolution of $P(\mathbf{y}, t)$, the nonequilibrium probability distribution of $\mathbf{y}$ at time t, is given by the following generalization of the (Stratonovich) Fokker–Planck equation (5.24):

$$\frac{\partial P}{\partial t} = -\sum_i \frac{\partial}{\partial y_i}\left[\left(A_i + \sum_{jk\ell} B_{j\ell}D_{\ell k}\frac{\partial B_{ik}}{\partial y_j}\right)P\right]$$

$$+ \sum_{ij}\frac{\partial^2}{\partial y_i \partial y_j}\left[\left(\sum_{k\ell}B_{ik}D_{k\ell}B_{j\ell}\right)P\right]$$

As in Section V.C, we take the y_j derivatives of the diffusion term in order to cast the Fokker–Planck equation into the form of a continuity equation for the probability distribution

$$\frac{\partial P}{\partial t} = -\sum_i \frac{\partial}{\partial y_i}\left\{\left[A_i - \sum_{k\ell}B_{ik}D_{k\ell}\left(\sum_j \frac{\partial B_{j\ell}}{\partial y_j}\right) - \sum_{jk\ell}B_{ik}D_{k\ell}B_{j\ell}\frac{\partial}{\partial y_j}\right]P\right\} \tag{6.73}$$

Note that, for uncorrelated fluctuations, $D_{k\ell} = D\delta_{k\ell}$, these equations duly reduce to Eqs. (5.24) and (5.25).

Now, on considering the *Landau–Lifshitz-type equation* (6.58), supplemented by the statistical properties (6.63) and (6.64), the substitutions [cf. Eqs. (5.26) and (5.27)]

$$(k, \ell) = (\mathbf{l}, \mathbf{l'}), \qquad (y_1, y_2, y_3) = (m_x, m_y, m_z)$$

$$L_1(t) = f_1(t), \qquad D_{1l'} = \frac{\lambda_{1l'}}{\gamma m} k_B T$$

$$A_i = \gamma \left[\vec{m} \wedge \vec{B}_{\text{eff}} - \frac{1}{m} \vec{m} \wedge \hat{\Lambda}^{(L)} \left(\vec{m} \wedge \vec{B}_{\text{eff}} \right) \right]_i$$

$$B_{il} = -\gamma \sum_{rs} \varepsilon_{irs} m_r \frac{\partial V_1}{\partial m_s}$$

cast those equations into the form of the general system of Langevin equations (6.71) supplemented by Eqs. (6.72). Therefore, on using [cf. Eq. (5.28)]

$$\frac{\partial B_{il}}{\partial m_j} = -\gamma \left(\sum_s \varepsilon_{ijs} \frac{\partial V_1}{\partial m_s} + \sum_{rs} \varepsilon_{irs} m_r \frac{\partial^2 V_1}{\partial m_j \partial m_s} \right)$$

one finds that $\sum_j \partial B_{jl}/\partial m_j \equiv 0$, $\forall l$, due to the fact that $\varepsilon_{jjs} = 0$ and because of the vanishing of the contraction of symmetric tensors with antisymmetric ones. Consequently, the second term on the right-hand side of the general Fokker–Planck equation (6.73) also vanishes in this case. For the third term, by repeated use of $(\vec{J} \wedge \vec{J}')_i = \sum_{rs} \varepsilon_{irs} J_r J'_s$ and recalling the definition (6.57), we obtain

$$-\sum_{jll'} B_{il} D_{1l'} B_{jl'} \frac{\partial P}{\partial m_j} = \frac{\gamma}{m} k_B T \left[\vec{m} \wedge \hat{\Lambda}^{(L)} \left(\vec{m} \wedge \frac{\partial P}{\partial \vec{m}} \right) \right]_i$$

On introducing these results into Eq. (6.73) one eventually arrives at the Fokker–Planck equation [cf. Eqs. (6.4) and (6.6)]

$$\frac{\partial P}{\partial t} = -\frac{\partial}{\partial \vec{m}} \cdot \left\{ \gamma \vec{m} \wedge \vec{B}_{\text{eff}} P - \frac{\gamma}{m} \vec{m} \wedge \hat{\Lambda}^{(L)} \left[\vec{m} \wedge \left(\vec{B}_{\text{eff}} - k_B T \frac{\partial}{\partial \vec{m}} \right) P \right] \right\}$$
$$(6.74)$$

where $(\partial/\partial \vec{m}) \cdot \vec{J} = \sum_i (\partial J_i/\partial m_i)$. In addition, by means of similar considerations and allowing the index in the Langevin sources $L_k(t)$ to run also over the indices $\mathbf{q}$, the Landau–Lifshitz-type equation (6.60) leads to a Fokker–Planck equation analogous to the preceding one with $\hat{\Lambda}^{(L)}$ augmented to $\hat{\Lambda} = \hat{\Lambda}^{(L)} + k_B T \hat{\Lambda}^{(Q)}$, namely

$$\frac{\partial P}{\partial t} = -\frac{\partial}{\partial \vec{m}} \cdot \left\{ \gamma \vec{m} \wedge \vec{B}_{\text{eff}} P - \frac{\gamma}{m} \vec{m} \wedge \hat{\Lambda} \left[\vec{m} \wedge \left(\vec{B}_{\text{eff}} - k_B T \frac{\partial}{\partial \vec{m}} \right) P \right] \right\} \quad (6.75)$$

Concerning the stationary solution of these Fokker–Planck equations, one can use $\vec{B}_{\text{eff}} = -\partial \mathcal{H}/\partial \vec{m}$ and $(\partial/\partial \vec{m}) \cdot (\vec{m} \wedge \vec{B}_{\text{eff}} P_e) = 0$ (see Section V.C), to demonstrate that the Boltzmann distribution, $P_e(\vec{m}) \propto \exp[-\mathcal{H}(\vec{m})/k_B T]$, is indeed a stationary solution of Eqs. (6.74) and (6.75). This entails that under external stationary conditions $P(\vec{m}, t) \xrightarrow{t \to \infty} P_e(\vec{m})$, that is, the spin eventually reaches the thermal equilibrium distribution of orientations. Note that this is a consequence of the formalism employed, instead of a constraint imposed separately, as is done in the phenomenological approaches (see Section V.C).

Note nevertheless that we have only proved the thermal equilibration for Eqs. (6.58) and (6.60), namely, in the weak-coupling case. In this connection, it is to be recalled that, inasmuch as the spin–environment coupling Hamiltonians themselves are commonly obtained via perturbation theory (so they are "small" in some sense), the study of the arbitrary-coupling case of such Hamiltonians is mainly of an academic interest.

2. Brown–Kubo–Hashitsume Model

When the spin–environment interaction is linear in the spin variables, the Markovian equations obtained formally reduce to the equations occurring in the Brown–Kubo–Hashitsume model. To illustrate, let us consider the simpler case of couplings linear in the environment coordinates. Then, if the $V_l(\vec{m})$ are linear in $\vec{m}$, both the relaxation tensor $\hat{\Lambda}^{(L)}$ and the fluctuating field $\vec{b}_{\mathrm{fl}}$ are independent of $\vec{m}$ [see Eqs. (6.57) and (6.28), respectively]. From the statistical properties (6.63) and (6.64) of the fluctuating sources $f_l(t)$, one then gets [cf. Eqs. (6.3)]

$$\langle b_{\mathrm{fl},i}(t)\rangle = 0, \qquad \langle b_{\mathrm{fl},i}(t)b_{\mathrm{fl},j}(t')\rangle = \frac{2\Lambda_{ij}^{(L)}}{\gamma m}k_B T \delta(t - t') \qquad (6.76)$$

where the last result establishes the relation between the structure of the correlations among the components of $\vec{b}_{\mathrm{fl}}(t)$ and the form of the relaxation tensor $\hat{\Lambda}^{(L)}$.* The corresponding result by Jayannavar (1991) comprised an uncorrelated $\vec{b}_{\mathrm{fl}}(t)$ (a diagonal $\Lambda_{ij}^{(L)}$ in our formulation) due to the special bilinear interaction that he considered [recall the discussion following Eq. (6.14)].

Then, if the spin–environment interaction yields uncorrelated *and* isotropic fluctuations ($\Lambda_{ij}^{(L)} = \lambda \delta_{ij}$), one finds that (1) the statistical properties (6.76) reduce to (6.3); (2) the Langevin equations (6.56) and (6.58) reduce, respectively, to the stochastic Gilbert [Eq. (6.1)] and Landau–Lifshitz [Eq. (6.2)] equations; and (3) the Fokker–Planck equation (6.74) reduces to (6.4). Thus, the phenomenological Brown–Kubo–Hashitsume model is formally obtained.

Note that these results also hold when couplings quadratic in the environment variables are included [Eq. (6.60)], except that the relaxation terms

---

*Note that for $\vec{b}_{\mathrm{fl}}(\vec{m}, t)$ depending on $\vec{m}$, one cannot merely employ Eqs. (6.63) and (6.64) to derive the statistical properties of $\vec{b}_{\mathrm{fl}}(\vec{m}, t)$, since $\vec{m}(t)$ and $f_l(t)$ *are not* independent.

(effective damping coefficients) are then explicitly dependent on the temperature.

3. Garanin–Ishchenko–Panina Model

We shall now show that the weak-coupling Landau–Lifshitz-type equations (6.58) and (6.60), formally reduce to the Langevin equation (6.8) of Garanin et al., when the spin–environment interaction includes terms up to quadratic in the *spin variables*. In this case, the coupling functions V_1 and V_q can be written as the natural extension of Eq. (6.34):

$$V_1(\vec{m}) = \sum_i v_{1,i} m_i + \tfrac{1}{2} \sum_{ij} w_{1,ij} m_i m_j \tag{6.77}$$

$$V_q(\vec{m}) = \sum_i v_{q,i} m_i + \tfrac{1}{2} \sum_{ij} w_{q,ij} m_i m_j \tag{6.78}$$

where the $v_{1,i}$, $w_{1,ij}$, $v_{q,i}$, and $w_{q,ij}$ are coupling constants incorporating the symmetry of the interaction. As in Section VI.C, the fluctuating effective field (6.47) can be separated in an ordinary-field part and an anisotropy-field part

$$\vec{b}_{\text{fl}}(\vec{m}, t) = \vec{b}(t) + \hat{\kappa}(t)\vec{m} \tag{6.79}$$

while, in this case, the expressions for the fluctuating sources in terms of the coupling constants are generalized to

$$b_i(t) = -\left[\sum_1 f_1(t) v_{1,i} + \sum_q f_q(t) v_{q,i} \right]$$

$$\kappa_{ij}(t) = -\left[\sum_1 f_1(t) w_{1,ij} + \sum_q f_q(t) w_{q,ij} \right]$$

Naturally, the fluctuating part of the energy of the spin (6.51), which gives $\vec{b}_{\text{fl}} = -\partial \mathcal{H}_{\text{fl}}/\partial \vec{m}$, also takes in this case the form $\mathcal{H}_{\text{fl}} = -\vec{m} \cdot \vec{b}(t) - \tfrac{1}{2}\vec{m} \cdot \hat{\kappa}(t)\vec{m}$.

In the Markovian regime, the auto-correlations and cross-correlations of $\vec{b}(t)$ and $\hat{\kappa}(t)$ can be obtained by dint of Eqs. (6.64), (6.66), and (6.67). Such correlations can be cast into the form proposed by Garanin et al. [Eq. (6.5)], with the following expressions for the correlation coefficients

$$\lambda_{ij} = \sum_{1,1'} \lambda_{11'} v_{1,i} v_{1',j} + k_B T \sum_{q,q'} \lambda_{qq'} v_{q,i} v_{q',j}$$

$$\lambda_{i,jk} = \sum_{l,l'} \lambda_{ll'} v_{l,i} w_{l',jk} + k_B T \sum_{q,q'} \lambda_{qq'} v_{q,i} w_{q',jk} \tag{6.80}$$

$$\lambda_{ik,jl} = \sum_{l,l'} \lambda_{ll'} w_{l,ik} w_{l',jl} + k_B T \sum_{q,q'} \lambda_{qq'} w_{q,ik} w_{q',jl}$$

Concerning the relaxation term, the tensor $\hat{\Lambda} = \hat{\Lambda}^{(L)} + k_B T \hat{\Lambda}^{(Q)}$ [Eq. (6.61)] associated with the coupling functions (6.77) and (6.78), is given by

$$\Lambda_{ij} = \sum_{l,l'} \lambda_{ll'} \left(v_{l,i} + \sum_k w_{l,ik} m_k \right) \left(v_{l',j} + \sum_\ell w_{l',j\ell} m_\ell \right)$$

$$+ k_B T \sum_{q,q'} \lambda_{qq'} \left(v_{q,i} + \sum_k w_{q,ik} m_k \right) \left(v_{q',j} + \sum_\ell w_{q',j\ell} m_\ell \right)$$

However, this expression can be written in terms of the correlation coefficients (6.80) as

$$\Lambda_{ij} = \lambda_{ij} + \sum_k (\lambda_{i,jk} + \lambda_{j,ik}) m_k + \sum_{k\ell} \lambda_{ik,j\ell} m_k m_\ell \tag{6.81}$$

which is identical with the relation (6.7) between the tensor $\hat{G}$ in Eq. (6.8) and the correlation coefficients in Eq. (6.5).

Therefore, we find that when the spin–environment coupling includes terms up to quadratic in the spin variables, the structures of the fluctuating effective field $\vec{b}_{\text{fl}}(\vec{m}, t)$ and of the relaxation term $\vec{R} = (\gamma/m)\vec{m} \wedge \hat{\Lambda}(\vec{m} \wedge \vec{B}_{\text{eff}})$ in the Landau–Lifshitz-type equation (6.60), as well as the relation between them, are identical with the structures and mutual relations of the corresponding terms in the Langevin equation (6.8) of Garanin et al. (1990). Naturally, the Fokker–Planck equation (6.75) then reduces to Eq. (6.6).

F.   Discussion

Starting from a Hamiltonian description of a classical spin interacting with the surrounding medium, we have derived generalized Langevin equations, which, in the Markovian approach, reduce to known stochastic equations of motion for classical magnetic moments.

Note, however, that the presented derivation of the equations of Garanin et al. (1990) and, similarly, the previous derivations of the equations occurring in the Brown–Kubo–Hashitsume model (Smith and de Rozario, 1976; Seshadri and Lindenberg, 1982; Jayannavar, 1991; Klik, 1992), are formal in the sense that one must still investigate specific realizations of the spin-plus-environment whole system, and then prove that the assumptions employed (mainly that of Markovian behavior) are at least approximately met. A paradigmatic case in which the Markovian approach breaks down, is the case of the magnetoelastic coupling of the spin to the lattice vibrations (in two or three dimensions) *linear* in the corresponding normal modes (Garg and Kim, 1991). The associated memory kernel crosses zero, changes it sign, and tends to zero from negative values as $\tau \to \infty$, *enclosing a zero algebraic area*. One then gets identically zero $\lambda_{ll'}$ by Eq. (6.68) and hence a zero tensor $\hat{\Lambda}^{(L)}$ by Eq. (6.57). Therefore, on replacing such a kernel by a Dirac delta, one looses the relaxational effects associated with the portion of the coupling *linear* in the environment variables ("one-phonon processes"), which are dominant at sufficiently low temperatures.

On the other hand, we have considered the classical regime of the environment and the spin. A classical description of the environment is adequate, for example, for the coupling to low-frequency normal modes, while, for instance, the magnetic moment of a nanometric particle ($m \sim 10^3 - 10^5 \ \mu_B$) behaves, except for very low temperatures, as a classical spin. In addition, the equations derived might also serve as a limit description of the semiclassical dynamics of molecular magnetic clusters with high spin ($S \gtrsim 10$) in their ground state.

VII. SUMMARY AND CONCLUSIONS

To conclude, let us summarize the most important results presented in this chapter. Approximate and exact results for a number of *thermal equilibrium* quantities for noninteracting classical spins with a simple axially symmetric anisotropy potential have been derived and analyzed. The results obtained also apply to systems described as assemblies of classical dipole moments with Hamiltonians comprising a coupling term to an (electric or magnetic) external field plus an axially symmetric orientational potential. Concerning their application to superparamagnetic systems, it has been shown that the magnetic anisotropy plays a fundamental role in the thermal equilibrium properties of these systems and, consequently, that the approaches that ignore these effects, on the basis of a restrictive ascription of superparamagnetism to the temperature range where the anisotropy energy is smaller than the thermal energy, are inadequate.

In the study of the *dynamics* of individual magnetic moments by the Langevin dynamics approach, interesting phenomena in the overbarrier rotation process have been found, such as crossing-back and multiple crossing of the potential barrier, which can be explained in terms of the gyromagnetic nature of the system. The results for the linear dynamical susceptibility $\chi(\omega)$, obtained from the stochastic Landau–Lifshitz (–Gilbert) equation, have been compared with different analytical expressions used to model the relaxation of nanoparticle ensembles, assessing their accuracy. It has been found that, among a number of heuristic expressions for $\chi(\omega)$, only the simple formula proposed by Shliomis and Stepanov matches the coarse features of the susceptibility reasonably. On the other hand, we have investigated the effects of the intra-potential-well relaxation modes on the low-temperature longitudinal dynamical response, showing their relatively small reflection in the $\chi_{\parallel}(\omega, T)$ curves (remarkably small in $\chi_{\parallel}''$) and confirming their dramatic influence on the phase shifts. Concerning the transverse response, the sizable relative contribution to $\chi_{\perp}'(\omega)$ of the spread of the precession frequencies of the spins in the anisotropy field at intermediate-to-high temperatures, has been demonstrated by comparing the numerical results with the exact zero-damping expression for $\chi_{\perp}''(\omega)$. Taking this effect into account may be relevant to properly assess the strength of the damping in superparamagnetic systems.

Dynamical equations for a classical spin interacting with the surrounding medium have finally been derived by means of the formalism of the oscillator-bath environment. The customary bilinear coupling treatment has been extended to couplings that depend arbitrarily on the spin variables and that are linear or linear-plus-quadratic in the environment dynamical variables. The equations obtained have the structure of generalized Langevin equations, which, in the Markovian approach, formally reduce to known semiphenomenological equations of motion for classical magnetic moments. Specifically, the generalization of the stochastic Landau–Lifshitz equation that incorporates fluctuations of the magnetic anisotropy of the spin, has been obtained for spin–environment interactions including terms up to quadratic in the *spin* variables. Besides, the portion of the coupling quadratic in the *environment* variables introduces an explicit dependence of the effective damping coefficients on the temperature.

APPENDIX A. THE FUNCTIONS $R^{(l)}(\sigma)$

In this appendix we summarize some properties of the function $R(\sigma)$ and its derivatives:

$$R^{(l)}(\sigma) = \int_0^1 dz\, z^{2l} \exp(\sigma z^2), \qquad l = 0, 1, 2, \ldots$$

These functions, which were introduced by Raĭkher and Shliomis (1975), play an important role in the study of the equilibrium and dynamical properties of classical spins with the simplest axially symmetric anisotropy potential. We also obtain approximate expressions for the most familiar combinations of the type $R^{(l)}/R$, which will be valid in the ranges $|\sigma| \ll 1$ and $|\sigma| \gg 1$. These approximate formulas can be employed to derive the corresponding approximate expressions for a number of quantities.

1. Relations with Known Special Functions

The functions $R^{(l)}(\sigma)$ are related with certain special functions, including *the Kummer functions, error functions, and the Dawson integral.*

The definition of the confluent hypergeometric (Kummer) functions is (see Arfken, 1985, p. 753)

$$M(a, c; x) = \sum_{n=0}^{\infty} \frac{(a)_n}{(c)_n} \frac{x^n}{n!}, \qquad c \neq 0, -1, -2, \ldots \tag{A.1}$$

$$(a)_n = a(a+1)\cdots(a+n-1) = \frac{(a+n-1)!}{(a-1)!}, \qquad (a)_0 = 1$$

where $(a)_n$ is the Pochhammer symbol. For noninteger argument the factorial signs are to be interpreted as gamma functions $a! \overset{\text{def}}{=} \Gamma(a+1)$ with

$$\Gamma(z) = \int_0^{\infty} dt\, t^{z-1} e^{-t}, \qquad \Re(z) > 0 \tag{A.2}$$

where $\Re(\cdot)$ denotes real part. The relation between the functions $R^{(l)}(\sigma)$ and Kummer functions reads

$$R^{(l)}(\sigma) = \frac{M(l + \frac{1}{2}, l + \frac{3}{2}; \sigma)}{2l + 1}, \qquad l = 0, 1, 2, \ldots \tag{A.3}$$

On using $M(a, c; x = 0) = 1$ [see Eq. (A.1)], one gets from Eq. (A.3) as a corollary the derivatives of $R(\sigma)$ at the origin

$$R^{(l)}(0) = \frac{1}{2l + 1}, \qquad l = 0, 1, 2, \ldots. \tag{A.4}$$

The relations (A.3) can easily be derived from the following integral representation of the Kummer function

$$M(a, c; x) = \frac{2\,\Gamma(c)}{\Gamma(a)\Gamma(c-a)} \int_0^1 dz\, e^{x z^2} z^{2a-1} (1 - z^2)^{c-a-1}, \qquad \Re(c) > \Re(a) > 0 \tag{A.5}$$

which follows from the more familiar one (see Arfken, 1985, p. 754)

$$M(a, c; x) = \frac{\Gamma(c)}{\Gamma(a)\Gamma(c-a)} \int_0^1 dt\, e^{xt} t^{a-1} (1 - t)^{c-a-1}, \qquad \Re(c) > \Re(a) > 0 \tag{A.6}$$

by dint of the substitution $t = z^2$. Thus, for $a = l + \frac{1}{2}$ and $c = l + \frac{3}{2}$, one has $c - a = 1$, so that

$$\frac{2\Gamma(c)}{\Gamma(a)\Gamma(c-a)} = \frac{2\Gamma(l+\frac{3}{2})}{\Gamma(l+\frac{1}{2})\Gamma(1)} = 2l + 1$$

where $\Gamma(z+1) = z\Gamma(z)$ and $\Gamma(1) = 1$ have been employed. Then, on using $c - a - 1 = 0$ and $2a - 1 = 2l$, the right-hand side of Eq. (A.3) can be written by means of the integral representation (A.5) as

$$\frac{M(l+\frac{1}{2},l+\frac{3}{2};\sigma)}{2l+1} = \frac{1}{2l+1} \times (2l+1) \int_0^1 dz\, e^{\sigma z^2} z^{2l} \overset{\text{def}}{=} R^{(l)}(\sigma), \qquad \text{Q.E.D.}$$

On introducing the *error* functions of real and "imaginary" argument, namely

$$\text{erf}(x) = \sqrt{\frac{4}{\pi}} \int_0^x dt\, \exp(-t^2), \qquad \text{erfi}(x) = \sqrt{\frac{4}{\pi}} \int_0^x dt\, \exp(t^2) \qquad (A.7)$$

one can alternatively write $R(\sigma)$ as

$$R(\sigma) = \begin{cases} \sqrt{\dfrac{\pi}{4\sigma}}\, \text{erfi}(\sigma^{1/2}) & \text{for} \quad \sigma > 0 \\[3mm] \sqrt{\dfrac{\pi}{4|\sigma|}}\, \text{erf}(|\sigma|^{1/2}) & \text{for} \quad \sigma < 0 \end{cases} \qquad (A.8)$$

The less familiar $\text{erfi}(x)$ is directly related with the Dawson integral

$$D(x) = \exp(-x^2) \int_0^x dt\, \exp(t^2) \qquad (A.9)$$

which is a tabulated function also available in certain mathematical libraries of computers. Consequently, the first equation in (A.8) is essentially the known relation between $R(\sigma)$ and the Dawson integral (see Coffey, Cregg and Kalmykov, 1993, p. 368)

$$R(\sigma) = \frac{\exp(\sigma)}{\sqrt{\sigma}} D(\sqrt{\sigma}), \qquad \sigma > 0 \qquad (A.10)$$

which, as indicated, holds only for positive argument.

Proofs are as follows:

1. By means of the substitution $t = \sqrt{\pm\sigma}\, z$, where the upper and lower signs correspond, respectively, to $\sigma > 0$ and $\sigma < 0$, one finds

$$\sqrt{\frac{\pi}{4(\pm\sigma)}} \times \left\{ \begin{array}{l} \mathrm{erfi}(\sqrt{\sigma}) \\ \mathrm{erf}(\sqrt{-\sigma}) \end{array} \right. = \sqrt{\frac{\pi}{4(\pm\sigma)}} \sqrt{\frac{4}{\pi}} \int_0^{\sqrt{\pm\sigma}} dt \exp(\pm t^2)$$

$$= \underbrace{\sqrt{\frac{1}{(\pm\sigma)}} \sqrt{\pm\sigma} \int_0^1 dz \exp(\sigma z^2)}_{1}$$

from which Eqs. (A.8) follow.   Q.E.D.

2. Equations (A.7) and (A.9) immediately yield

$$\mathrm{erfi}(x) = \sqrt{\frac{4}{\pi}} \int_0^x dt \exp(t^2) = \sqrt{\frac{4}{\pi}} \exp(x^2) D(x)$$

from which one gets Eq. (A.10) through the already demonstrated Eq. (A.8).   Q.E.D.

2.   Recurrence Relations

The functions $R^{(l)}$ satisfy the following equivalent recurrence relations:

$$R^{(l+1)} = \frac{e^\sigma - (2l+1)R^{(l)}}{2\sigma}, \qquad R^{(l)} = \frac{e^\sigma - 2\sigma R^{(l+1)}}{2l+1} \tag{A.11}$$

which can readily be obtained by integrating by parts the definition of $R^{(l)}$. The $l = 0$ particular case of these relations is frequently employed. It can be written in the following equivalent forms

$$R' = \frac{e^\sigma - R}{2\sigma} \quad \Leftrightarrow \quad R = e^\sigma - 2\sigma R' \quad \Leftrightarrow \quad \frac{e^\sigma}{R} = 1 + 2\sigma\frac{R'}{R} \tag{A.12}$$

where the prime denotes derivative with respect to σ.

One can also derive recurrence relations among the combinations $R^{(l)}/R$, which occur in the expressions for a number of quantities (e.g., the linear and nonlinear susceptibilities). On dividing both sides of the first Eq. (A.11) by R and using Eq. (A.12) to eliminate $e^\sigma/(2\sigma R)$, one gets the following relation between quotients of the form $R^{(l)}/R$:

$$\frac{R^{(l+1)}}{R} = \frac{R^{(1)}}{R} + \frac{1}{2\sigma}\left[1 - (2l+1)\frac{R^{(l)}}{R}\right] \tag{A.13}$$

The following particular case

$$\frac{R''}{R} = \frac{R'}{R} - \frac{1}{2\sigma}\left(3\frac{R'}{R} - 1\right) \tag{A.14}$$

is especially useful. For instance, it can be employed to calculate R''/R from R'/R.

3.   Series Expansions

Series expansions for $R(\sigma)$ and its derivatives can easily be obtained from the corresponding expansions of the Kummer functions.

a.   Power Series

From the relations (A.3) between $R^{(l)}(\sigma)$ and Kummer functions, one can construct the Taylor expansion of the former through the power series (A.1) for the latter. For the quotient of Pochhammer symbols required, one gets

$$\frac{1}{2l+1}\frac{(l+\frac{1}{2})_n}{(l+\frac{3}{2})_n} = \frac{1}{2l+1}\frac{(l+n-\frac{1}{2})!/(l-\frac{1}{2})!}{(l+n+\frac{1}{2})!/(l+\frac{1}{2})!} = \frac{1}{2(l+n)+1}$$

from which we obtain the desired power series of $R^{(l)}(\sigma)$:

$$R^{(l)}(\sigma) = \sum_{n=0}^{\infty} \frac{1}{n!}\frac{\sigma^n}{2(l+n)+1}$$

b.   Asymptotic Formula for Large Positive Argument

For $x \gg 1$, the Kummer functions are approximately given by (see Arfken, 1985, p. 757)

$$M(a, c; x) = \frac{\Gamma(c)}{\Gamma(a)}\frac{e^x}{x^{c-a}}$$

$$\times \left[1 + \frac{(1-a)(c-a)}{x} + \frac{(1-a)(2-a)(c-a)(c-a+1)}{2x^2} + \cdots\right] \tag{A.15}$$

Then, on using the relations (A.3) and noting that in this case $1 - a = -(2l - 1)/2$ and $c - a = 1$, we obtain the following asymptotic expansion of $R^{(l)}(\sigma)$:

$$R^{(l)}(\sigma) = \frac{e^\sigma}{2\sigma}\left\{1 - \frac{(2l-1)}{2\sigma} + \frac{(2l-1)(2l-3)}{4\sigma^2} + \cdots\right\}, \qquad \sigma \gg 1 \tag{A.16}$$

This expansion generalizes for an arbitrary l the results derived by Raĭkher and Shliomis (1975) for $l = 0, 1, 2, 3$. Note finally that, one can use Eq. (A.16) to take the $\sigma \to \infty$ limit of the quotient $R^{(l)}/R$, getting

$$\frac{R^{(l)}}{R} \simeq \frac{1 - (2l-1)/2\sigma + (2l-1)(2l-3)/4\sigma^2 + \cdots}{1 + 1/2\sigma + 3/4\sigma^2 + \cdots} \xrightarrow{\sigma \to \infty} 1, \qquad \forall l \tag{A.17}$$

c. Asymptotic Formula for Large Negative Argument

Asymptotic expressions for $R^{(l)}(\sigma \ll -1)$ can be derived from the asymptotic expansion of the Kummer functions for large negative argument. The latter is easily obtained from the expansion (A.15) for large positive argument by dint of *Kummer's first formula* $M(a, c; x) = e^x M(c - a, c; -x)$ (see Arfken, 1985, p. 754)

$$M(a, c; x) \simeq \frac{\Gamma(c)}{\Gamma(c - a)} \frac{1}{(-x)^a}$$

$$\times \left[1 + \frac{(c - a - 1)a}{x} + \frac{(c - a - 2)(c - a - 1)a(a + 1)}{2x^2} + \cdots \right] \tag{A.18}$$

Then, taking once more the relations (A.3) into account, one obtains the approximate expression

$$R^{(l)}(\sigma) \simeq \frac{\pi^{1/2}}{2^{2l+1}} \frac{(2l)!}{l!} \frac{1}{(-\sigma)^{l+1/2}}, \qquad \sigma \ll -1 \tag{A.19}$$

for the derivation of which we have also employed the following useful result for the gamma function of half-odd-integer argument

$$\Gamma\left(l + \tfrac{1}{2}\right) = \frac{\pi^{1/2}}{2^{2l}} \frac{(2l)!}{l!} \tag{A.20}$$

Note that the next terms in the asymptotic expansion (A.19) vanish identically, since $c - a - 1 = 0$ in this case [see Eq. (A.18)]. Finally, for the quotient $R^{(l)}/R$ one gets the limit

$$\frac{R^{(l)}}{R} \simeq \frac{1}{2^{2l}} \frac{(2l)!}{l!} \frac{1}{(-\sigma)^l} \xrightarrow{\sigma \to -\infty} 0, \qquad \forall l \geq 1 \tag{A.21}$$

To conclude, as an exercise of consistency, one can obtain from the derived $|\sigma| \ll 1$ and $\sigma \gg 1$ expansions of $R(\sigma)$, via the relation (A.10), the known power and asymptotic series of the Dawson integral (see, for example, Coffey, Cregg and Kalmykov, 1993, p. 368):

$$D(x) = \begin{cases} x - \dfrac{2}{3}x^3 + \dfrac{4}{15}x^5 + \cdots, & x \ll 1 \\[2ex] \dfrac{1}{2x}\left(1 + \dfrac{1}{2x^2} + \cdots\right), & x \gg 1 \end{cases}$$

4.   Approximate Formulas for R'/R and R''/R

We now derive approximate expressions for R'/R valid in the $|\sigma| \ll 1$ and $|\sigma| \gg 1$ ranges. These formulas, along with the recurrence relations (A.13) between consecutive $R^{(l)}/R$, would provide approximate expressions for $R^{(l)}/R$ with $l \geq 2$. We shall explicitly give these approximate formulas for R''/R.

The following approximate expressions will be obtained by constructing approximate solutions of the differential equation that the function $G = R'/R$ satisfies, namely

$$\frac{dG}{d\sigma} = \frac{1}{2\sigma}(1 - 3G) + G(1 - G) \tag{A.22}$$

which can easily be derived from Eq. (A.14).

a.   Power Series

To obtain $G|_{|\sigma|\ll 1}$, we shall seek for a solution of the differential equation (A.22) in the form of a power series $G = \sum_{n=0}^{\infty} a_n \sigma^n$. Prior to do that, however, in order to remove the singularities in the coefficients in that differential equation, they are multiplied by 2σ, yielding the equivalent equation $2\sigma(dG/d\sigma) = (1 - 3G) + 2\sigma G(1 - G)$. This is a nonhomogeneous nonlinear differential equation, and these features will take reflection in the form of the constructed solution.

On inserting $G = \sum_{n=0}^{\infty} a_n \sigma^n$ into the preceding differential equation, redefining the summation indices in order to obtain the same exponent for σ under each summation symbol, and equating coefficients, one gets for the a_n:

$$a_0 = \tfrac{1}{3}, \qquad \left(n + \tfrac{3}{2}\right) a_n = a_{n-1} - \sum_{k=0}^{n-1} a_k a_{n-1-k}, \qquad \text{for} \quad n \geq 1$$

The fact that a_0 is not a free parameter results from the nonhomogeneous character of the differential equation. In addition, the preceding recurrence relation among the a_n shows that, as a consequence of the nonlinearity of the differential equation, the computation of each coefficient involves all the previous ones. Finally, on computing the first few coefficients, $G = R'/R$ emerges in the approximate form

$$G \simeq \frac{1}{3}\left(1 + \frac{4}{15}\sigma + \frac{8}{315}\sigma^2 - \frac{16}{4725}\sigma^3 - \frac{32}{31185}\sigma^4\right) \qquad (A.23)$$

We have carried out the expansion up to the fourth order in σ because some quantities are approximately obtained up to terms of order σ^3 and, for example, R''/R involves G' [see Eq. (A.26)].

The formulas required to derive some approximate expressions in the main text are

$$\frac{R'}{R} \simeq \frac{1}{3}\left(1 + \frac{4}{15}\sigma + \frac{8}{315}\sigma^2 - \frac{16}{4725}\sigma^3\right)$$

$$\left(\frac{R'}{R}\right)^2 \simeq \frac{1}{9}\left(1 + \frac{8}{15}\sigma + \frac{64}{525}\sigma^2 + \frac{32}{4725}\sigma^3\right) \qquad (A.24)$$

$$\frac{R''}{R} \simeq \frac{1}{5}\left(1 + \frac{8}{21}\sigma + \frac{16}{315}\sigma^2 - \frac{32}{10395}\sigma^3\right)$$

For instance, the combinations entering in the equations for the nonlinear susceptibility read

$$\frac{1}{2}\left[\frac{1}{3}\frac{R''}{R} - \left(\frac{R'}{R}\right)^2\right] \simeq -\frac{1}{45}\left(1 + \frac{16}{21}\sigma + \frac{8}{35}\sigma^2 + \frac{32}{1485}\sigma^3\right)$$

$$\frac{1}{2}\left[\left(\frac{R'}{R}\right)^2 - \frac{R''}{R}\right] \simeq -\frac{2}{45}\left(1 + \frac{4}{21}\sigma - \frac{4}{105}\sigma^2 - \frac{32}{2079}\sigma^3\right)$$

$$\frac{1}{16}\left[-1 + 2\frac{R'}{R} - 2\left(\frac{R'}{R}\right)^2 + \frac{R''}{R}\right] \simeq -\frac{1}{45}\left(1 - \frac{8}{21}\sigma + \frac{128}{10395}\sigma^3\right) \quad (A.25)$$

The expression for R''/R in Eq. (A.24) has been obtained from R'/R, through the relation

$$\frac{R''}{R} = G' + G^2 \qquad (A.26)$$

which is directly demonstrated by taking the derivative $G' = (R'/R)' = (R''/R) - (R'/R)^2$.

b. Asymptotic Formulas

We now derive approximate expressions for $G = R'/R$ valid in the $|\sigma| \gg 1$ ranges. To this end we make in Eq. (A.22) the substitution $\varrho = 1/\sigma$, which casts it into the form

$$-\varrho^2 \frac{dG}{d\varrho} = \frac{\varrho}{2}(1 - 3G) + G(1 - G)$$

Let us seek for solutions of this differential equation in the form of a series of powers of ϱ.* On inserting $G = \sum_{n=0}^{\infty} b_n \varrho^n$ into this equation, redefining conveniently the summation indices, and equating coefficients, one gets for the b_n:

$$b_0(1 - b_0) = 0, \qquad b_1 = \frac{1}{2}\frac{3b_0 - 1}{1 - 2b_0},$$

$$(1 - 2b_0)b_n = \left(\frac{5}{2} - n\right) b_{n-1} + \sum_{k=1}^{n-1} b_k b_{n-k}, \qquad \text{for} \quad n \geq 2$$

Again, the first coefficient is not a free parameter, and the recurrence relation involves all the coefficients preceding a given one.

As could be expected from the fact that we are searching for solutions in two different asymptotic ranges ($\sigma \to \pm\infty$), we obtain two different solutions. The one that corresponds to the choice $b_0 = 0$ (denoted G_1), when expressed in terms of the original variable $\sigma = 1/\varrho$, takes the simple form

$$G_1 = -\frac{1}{2\sigma}, \qquad \text{for} \quad b_0 = 0$$

since all the following terms vanish identically. [As can be readily seen, $G = -1/2\sigma$ is an exact solution of the original differential equation (A.22), although, since it diverges at $\sigma = 0$, it is not the selfsame R'/R.] The solution that corresponds to the choice $b_0 = 1$ (denoted G_2) is given by

$$G_2 = 1 - \frac{1}{\sigma} - \frac{1}{2\sigma^2} - \frac{5}{4\sigma^3} + \cdots, \qquad \text{for} \quad b_0 = 1$$

We must now ascribe each solution to one of the two asymptotic ranges. On recalling Eqs. (A.17) and (A.21), we immediately conclude that G_1 and G_2 correspond, respectively, to the $\sigma \ll -1$ and $\sigma \gg 1$ ranges. Note also that the $\sigma \ll -1$ result can *directly* be obtained from the asymptotic results (A.19).

We can now write the combinations of $R(\sigma)$ and its derivatives that are required in the main text to construct approximate formulas for various

---

*A similar method was employed by Raĭkher and Shliomis (1975) to derive the aforementioned asymptotic series for the first few $R^{(l)}(\sigma)$.

quantities. For $\sigma \ll -1$, these are

$$\frac{R'}{R} \simeq -\frac{1}{2\sigma}, \qquad \left(\frac{R'}{R}\right)^2 \simeq \frac{1}{4\sigma^2}, \qquad \frac{R''}{R} \simeq \frac{3}{4\sigma^2} \tag{A.27}$$

and the combinations

$$\frac{1}{2}\left[\frac{1}{3}\frac{R''}{R} - \left(\frac{R'}{R}\right)^2\right] \simeq 0$$

$$\frac{1}{2}\left[\left(\frac{R'}{R}\right)^2 - \frac{R''}{R}\right] \simeq -\frac{1}{4\sigma^2} \tag{A.28}$$

$$\frac{1}{16}\left[-1 + 2\frac{R'}{R} - 2\left(\frac{R'}{R}\right)^2 + \frac{R''}{R}\right] \simeq -\frac{1}{16}\left(1 + \frac{1}{\sigma} - \frac{1}{4\sigma^2}\right)$$

Similarly, for $\sigma \gg 1$ we have

$$\frac{R'}{R} \simeq 1 - \frac{1}{\sigma} - \frac{1}{2\sigma^2} - \frac{5}{4\sigma^3}$$

$$\left(\frac{R'}{R}\right)^2 \simeq 1 - \frac{2}{\sigma} - \frac{3}{2\sigma^3} \tag{A.29}$$

$$\frac{R''}{R} \simeq 1 - \frac{2}{\sigma} + \frac{1}{\sigma^2} - \frac{1}{2\sigma^3}$$

and their combinations

$$\frac{1}{2}\left[\frac{1}{3}\frac{R''}{R} - \left(\frac{R'}{R}\right)^2\right] \simeq -\frac{1}{3}\left(1 - \frac{2}{\sigma} - \frac{1}{2\sigma^2} - \frac{2}{\sigma^3}\right)$$

$$\frac{1}{2}\left[\left(\frac{R'}{R}\right)^2 - \frac{R''}{R}\right] \simeq -\left(\frac{1}{2\sigma^2} + \frac{1}{2\sigma^3}\right) \tag{A.30}$$

$$\frac{1}{16}\left[-1 + 2\frac{R'}{R} - 2\left(\frac{R'}{R}\right)^2 + \frac{R''}{R}\right] \simeq 0$$

Let us also write the leading terms in the $|\sigma| \ll 1$ and $|\sigma| \gg 1$ expansions of the combination $R''/R - (R'/R)^2$, which occurs in some expressions studied in Sections III and IV:

$$\frac{R''}{R} - \left(\frac{R'}{R}\right)^2 \simeq \begin{cases} \dfrac{1}{2\sigma^2} & \text{for} \quad \sigma \ll -1 \\[2mm] \dfrac{4}{45} & \text{for} \quad |\sigma| \ll 1 \\[2mm] \dfrac{1}{\sigma^2} & \text{for} \quad \sigma \gg 1 \end{cases} \qquad (A.31)$$

The appropriate combinations of Eqs. (A.24), (A.27), and (A.29) almost patch the corresponding exact curves over the entire σ range. This is shown in Fig. 34, where it can be seen that the use of the $|\sigma| \ll 1$ results, replaced at some point between $|\sigma| = 2$ and $|\sigma| = 4$ by the corresponding $|\sigma| \gg 1$ formulas, is a reasonable approximation of the exact functions.

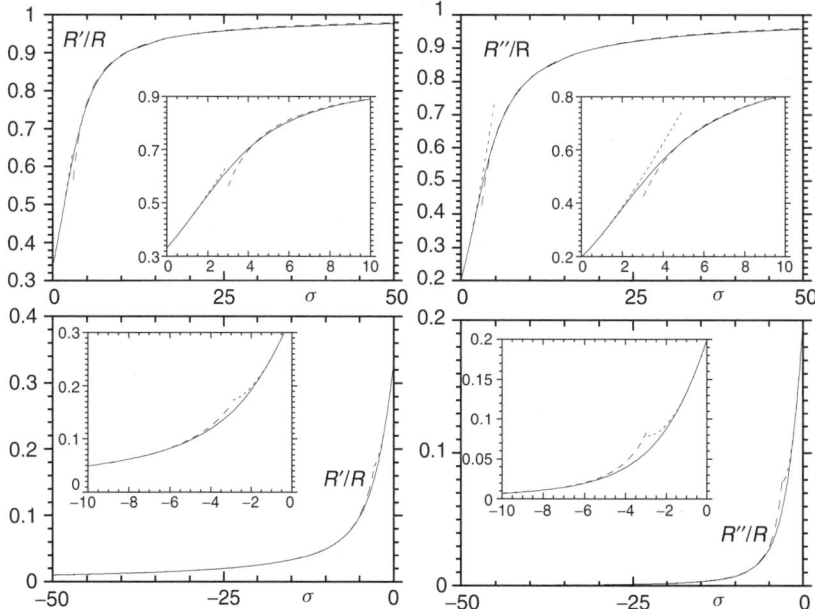

FIGURE 34.   The functions R'/R and R''/R together with their small and large σ approximations. The continuous lines represent the exact functions. Long dashes: large $|\sigma|$ approximations (A.27) and (A.29). Short dashes: small σ approximations (A.24). The insets show the details of the zones where the small σ approximation might be replaced by the corresponding large σ approximation, without a significant loss of accuracy.

APPENDIX B. DERIVATION OF THE FORMULAS FOR THE RELAXATION TIMES

In this appendix we give demonstrations of the formulas (5.66) and (5.71) for the relaxation times.

1. Integral Relaxation Time

a. The Integral Relaxation Time and the Low-Frequency Dynamical Susceptibility

The integral relaxation time defined by Eq. (5.64) is expressible in terms of the eigenvalues Λ_k and amplitudes a_k of the Sturm–Liouville problem associated with the axially symmetric Fokker–Planck equation [Eq. (5.65)]. In addition, τ_{int} can also be written in terms of the low-frequency dynamical susceptibility. To show this, let us first recall the general result from linear response theory

$$\chi(\omega) = \chi - i\omega \int_0^\infty dt \, e^{-i\omega t} \frac{\langle m(\infty) \rangle - \langle m(t) \rangle}{\Delta B} \tag{B.1}$$

where $\langle m(t) \rangle$ is the relaxing quantity, $\chi(\omega)$ its susceptibility counterpart, χ the equilibrium susceptibility, and ΔB the infinitesimal change in the external control parameter. On applying this result to $\langle m_z(\infty) \rangle - \langle m_z(t) \rangle$ from Eq. (5.61) and using the sum rule $\sum_{k \geq 1} a_k = 1$, one finds

$$\chi_\parallel(\omega) = \chi_\parallel \sum_{k \geq 1} a_k \left\{ 1 - i\omega \int_0^\infty dt \exp\left[-(i\omega + \Lambda_k)t \right] \right\}$$

from which it follows that

$$\chi_\parallel(\omega) = \chi_\parallel \sum_{k \geq 1} \frac{a_k}{1 + i\omega\Lambda_k^{-1}} \tag{B.2}$$

Thus, each exponential mode in the relaxation curve (5.61) gives a Debye-type factor in $\chi_\parallel(\omega)$ weighted by a_k and with characteristic time Λ_k^{-1}. Then, on expanding $\chi_\parallel(\omega)$ for low frequencies by dint of the binomial formula, one finally gets

$$\chi_\parallel(\omega) \simeq \chi_\parallel \sum_{k \geq 1} a_k \left(1 - i\omega\Lambda_k^{-1} + \cdots \right) = \chi_\parallel (1 - i\omega\tau_{\text{int}} + \cdots) \tag{B.3}$$

where we have again used $\sum_{k \geq 1} a_k = 1$ and taken Eq. (5.65) into account. Equation (B.3) demonstrates that the calculation of τ_{int} can effectively be reduced to the calculation of the low-frequency dynamical susceptibility.

b. Perturbational Solution of the Fokker–Planck Equation in the Presence of a Low Sinusoidal Field

In order to calculate $\chi_{\parallel}(\omega)$, one applies a low sinusoidal field, $\Delta\xi(t) = \Delta\xi \exp(i\omega t)$, where $\Delta\xi = m\Delta B/k_B T$, along the z (symmetry) axis and then calculates the solution of the axially symmetric Fokker–Planck equation (5.47) in such a situation. Since $-\beta\mathcal{H}(z, t) = -\beta\mathcal{H}_0(z) + z\Delta\xi(t)$, where $\mathcal{H}_0$ is the unperturbed Hamiltonian, we shall seek for a solution for the probability distribution in the stationary regime of the form

$$P(z, t) = P_e(z)[1 + q(z)\Delta\xi(t)] \tag{B.4}$$

where $P_e = \mathcal{Z}_0^{-1} \exp(-\beta\mathcal{H}_0)$ is the equilibrium probability distribution in the absence of the oscillating field.

On introducing the $P(z, t)$ from Eq. (B.4) into the Fokker–Planck equation (5.47), one gets, to first order in $\Delta\xi$ (linear response regime), the following second-order differential equation for $q(z)$

$$\left(-\beta\mathcal{H}_0' + \frac{d}{dz}\right)\left[\Omega(z)\frac{dq}{dz}\right] - i\omega 2\tau_N q = -\Omega(z)\beta\mathcal{H}_0' + \Omega' \tag{B.5}$$

where $\Omega(z) = 1 - z^2$ and the primes denote differentiation with respect to z. On the other hand, by taking the introduced form of $P(z, t)$ into account, the nonequilibrium average of the z component of magnetic moment can be written as

$$\langle m_z(t)\rangle = \int_{-1}^{1} dz\, P(z, t)m_z = m\langle z\rangle_e + m\,\Delta\xi(t)\int_{-1}^{1} dz\, P_e(z)q(z)z$$

where $\langle z\rangle_e = \int_{-1}^{1} dz\, P_e(z)z$ is the equilibrium average in the unperturbed case. Next, since $\Delta\xi(t) = (m\Delta B/k_B T)\exp(i\omega t)$, the dynamical susceptibility, which is the coefficient of $\mu_0^{-1}\Delta B e^{i\omega t}$ in $\langle m_z(t)\rangle$, is given by

$$\chi_{\parallel}(\omega) = \frac{\mu_0 m^2}{k_B T}\int_{-1}^{1} dz\, P_e(z)q(z)z \tag{B.6}$$

Comparison of this equation with Eq. (B.3) reveals that only the low-frequency part of $q(z)$ is required to calculate τ_{int}. This is important since Eq. (B.5) cannot be solved analytically in the general case. In contrast, it can be solved perturbatively for low ω because, for $\omega = 0$, only $q'(z)$ and $q''(z)$ occur in that equation. This enables one to lower the order of the differential equation (B.5) by introducing an auxiliary function $g(z) = q'(z)$, and solving successively the system of first-order differential equations for $q(z)$ and $g(z)$.

Let us accomplish this. First, one introduces the perturbational expansion

$$q(z) = q_0(z) - (i\omega)q_1(z) + (i\omega)^2 q_2(z) - \cdots$$

into Eq. (B.6), getting

$$\chi_\parallel(\omega) = \frac{\mu_0 m^2}{k_B T} \int_{-1}^{1} dz\, P_e(z) q_0(z) z - i\omega \frac{\mu_0 m^2}{k_B T} \int_{-1}^{1} dz\, P_e(z) q_1(z) z + \cdots$$

Then, on comparing this result with Eq. (B.3), one obtains the following integral representation of τ_{int}

$$\tau_{int} = \frac{1}{\partial\langle z\rangle_e/\partial\xi} \int_{-1}^{1} dz\, P_e(z) q_1(z) z \qquad (B.7)$$

where we have used $\chi_\parallel = (\mu_0 m^2/k_B T)\partial\langle z\rangle_e/\partial\xi$ [cf. Eqs. (3.59) and (3.60); we can differentiate with respect to B since this is parallel to the probing field ΔB]. Equation (B.7) shows that the calculation of τ_{int} effectively reduces to that of $q_1(z)$. In order to obtain this quantity, we introduce the preceding perturbational expansion of $q(z)$, along with $g_i \equiv dq_i/dz$, into Eq. (B.5), getting

$$\left(-\beta\mathcal{H}_0' + \frac{d}{dz}\right)\left\{\Omega(z)\left[g_0 - (i\omega)g_1 + (i\omega)^2 g_2 - \cdots\right]\right\}$$
$$- i\omega 2\tau_N[q_0 - (i\omega)q_1 + \cdots] = -\Omega(z)\beta\mathcal{H}_0' + \Omega' \qquad (B.8)$$

The zeroth-order equation has the thermal equilibrium solution

$$q_0 = z - \langle z\rangle_e \qquad (B.9)$$

as can be checked by using the definition of $q(z)$ and expanding the equilibrium probability distribution associated with $\beta\mathcal{H} = \beta\mathcal{H}_0 - z\Delta\xi$ [i.e., the $\omega = 0$ limit of $\beta\mathcal{H}(t)$] in powers of $\Delta\xi$.

The $(i\omega)$-order term of Eq. (B.8) reads

$$\left(-\beta\mathcal{H}_0' + \frac{d}{dz}\right)[\Omega(z)g_1] + 2\tau_N q_0 = 0$$

This differential equation can be integrated by quadratures, yielding

$$g_1(z) = \frac{2\tau_N}{\Omega(z)} \exp[\beta\mathcal{H}_0(z)][c_1 + \mathcal{Z}_0\Phi(z)] \qquad (B.10)$$

where $\mathcal{Z}_0$ is the (unperturbed) partition function and $\Phi(z)$ is given by

$$\Phi(z) = \int_{-1}^{z} dz_1 P_e(z_1) \underbrace{(\langle z \rangle_e - z_1)}_{-q_0(z_1)} \tag{B.11}$$

On using the condition $J_z|_{z=\pm 1} = 0$ (which follows from the tangency of the current of probability to the unit sphere) and $\Phi(-1) = \Phi(1) = 0$ (which immediately follow from the preceding definition), one gets $c_1 = 0$ for the integration constant. Consequently, $q_1(z) = \int^z dz_2 \, g_1(z_2)$ is given by

$$q_1(z) = c_2 + 2\tau_N \int_{-1}^{z} \frac{dz_2}{\Omega(z_2)} \frac{\Phi(z_2)}{P_e(z_2)} \equiv c_2 + \tilde{q}_1(z) \tag{B.12}$$

where we have written $\mathcal{Z}_0 \exp(\beta\mathcal{H}_0) = 1/P_e$. The new integration constant, c_2, can be obtained by solving the $(i\omega)^2$-order equation and imposing again the tangency condition of the current of probability at the boundaries. On doing so, one finds $c_2 = -\langle \tilde{q}_1 \rangle_e = -\int_{-1}^{1} dz \, P_e(z)\tilde{q}_1(z)$, where $\tilde{q}_1(z)$ is the integral term in Eq. (B.12).

<h3>c.   The Garanin–Ishchenko–Panina Formula</h3>

We can already do the integral involving $q_1(z)$ in the formula (B.7) for the integral relaxation time:

$$\int_{-1}^{1} dz \, P_e(z) q_1(z) z = \int_{-1}^{1} dz \, P_e(z) \left[\tilde{q}_1(z) - \langle \tilde{q}_1 \rangle_e \right] z$$

$$= \int_{-1}^{1} dz \, P_e(z)\tilde{q}_1(z) z - \langle z \rangle_e \underbrace{\int_{-1}^{1} dz \, P_e(z)\tilde{q}_1(z)}_{\langle \tilde{q}_1 \rangle_e}$$

$$= -\int_{-1}^{1} \underbrace{dz \, P_e(z)\big(\langle z \rangle_e - z\big)}_{d\Phi(z) \text{ by Eq. (B.11)}} \tilde{q}_1(z)$$

$$= -\underbrace{\left[\Phi(z)\tilde{q}_1(z) \right]_{-1}^{1}}_{0 \text{ by } \Phi(-1)=\Phi(1)=0} + \int_{-1}^{1} dz \, \Phi(z) \underbrace{\frac{2\tau_N}{\Omega(z)} \frac{\Phi(z)}{P_e(z)}}_{\tilde{q}_1'(z) \text{ by Eq. (B.12)}}$$

Then, on introducing this result into Eq. (B.7), one obtains

$$\tau_{\text{int}} = \frac{2\tau_N}{\partial \langle z \rangle_e / \partial \xi} \int_{-1}^{1} \frac{dz}{\Omega(z)} \frac{\Phi(z)^2}{P_e(z)} \tag{B.13}$$

whence, on recalling that $\Omega(z)$ is a shorthand for $1 - z^2$, one finally gets the result (5.66) of Garanin et al. (1990).

d. Explicit Expressions for $\Phi(z)$

Let us conclude with the calculation of explicit expressions for $\Phi(z)$ for particular forms of the Hamiltonian. Let us assume that $\mathcal{H}_0$ comprises the simplest uniaxial anisotropy term, plus a Zeeman term, i.e., $-\beta\mathcal{H}_0 = \sigma z^2 + \xi z$ [see Eq. (2.3)].

1. *Isotropic Case.* When $\sigma = 0$, the equilibrium probability distribution is given by Eq. (2.18). Thus, one of the contributions to $\Phi(z)$ is

$$\langle z \rangle_e \int_{-1}^{z} dz_1\, P_{e,\mathrm{Lan}}(z_1) = L(\xi)\frac{e^{\xi z} - e^{-\xi}}{2\sinh\xi}$$

where we have used $\langle z \rangle_e = L(\xi)$, where $L(\xi)$ is the Langevin function. The remainder contribution to $\Phi(z)$ is

$$-\int_{-1}^{z} dz_1\, P_{e,\mathrm{Lan}}(z_1) z_1 = -\frac{\xi}{2\sinh\xi}\frac{\partial}{\partial\xi}\int_{-1}^{z} dz_1\, \exp(\xi z_1)$$

$$= -\frac{e^{\xi z}}{2\sinh\xi}\left[\left(z - \frac{1}{3}\right) + e^{-\xi(1+z)}\left(1 + \frac{1}{\xi}\right)\right]$$

On adding these two contributions and recalling the definition (2.49) of the Langevin function, one finally gets the explicit result

$$\Phi_{\mathrm{Lan}}(z) = \frac{P_{e,\mathrm{Lan}}(z)}{\xi}\left[\coth\xi - z - \frac{\exp(-\xi z)}{\sinh\xi}\right] \tag{B.14}$$

2. *Zero-Field Case.* For $\xi = 0$, the equilibrium probability distribution is given by Eq. (2.21). Therefore, $\langle z \rangle_e = 0$ and

$$\Phi(z) = -\frac{1}{2R(\sigma)}\int_{-1}^{z} dz_1\, \exp\left(\sigma z_1^2\right) z_1$$

Then, on expressing the result of the integral in terms of the probability distribution (2.21), one gets [note that $P_{e,\mathrm{unb}}(-1) = P_{e,\mathrm{unb}}(1)$]

$$\Phi_{\mathrm{unb}}(z) = \frac{1}{2\sigma}[P_{e,\mathrm{unb}}(1) - P_{e,\mathrm{unb}}(z)] \tag{B.15}$$

2. Effective Transverse Relaxation Time

We now derive Eq. (5.71) for the effective transverse relaxation time by performing the low-frequency expansion of the formula for $\chi_\perp(\omega)$ of Raĭkher and Shliomis (1975, 1994).

a. The Raĭkher–Shliomis Formula for the Transverse Dynamical Susceptibility

The expression for $\chi_\perp(\omega)$ derived by these authors can be written as

$$\chi_\perp(\omega, T) = \chi_\perp(T)\frac{\lambda_a(\lambda_b + i\omega 2\tau_N) + \lambda_c}{(\lambda_1 + i\omega 2\tau_N)(\lambda_2 + i\omega 2\tau_N)} \tag{B.16}$$

where $\chi_\perp(T)$ is the equilibrium transverse susceptibility (3.53). The coefficients λ_a, λ_b, and λ_c are given, in terms of the functions $R^{(l)}(\sigma)$ [Eq. (2.33)] and the dimensionless damping coefficient λ in the Landau–Lifshitz equation, by

$$\lambda_a = \frac{R + R'}{R - R'}, \qquad \lambda_b = \frac{R - 3R' + 4R''}{R' - R''}, \qquad \lambda_c = \frac{2\sigma}{\lambda^2}\frac{3R' - R}{R - R'}$$

while λ_1 and λ_2 are the roots of the second-degree equation $x^2 - (\lambda_a + \lambda_b)x + (\lambda_a\lambda_b + \lambda_c) = 0$. Utilizing the fact that the roots x_1 and x_2 of $ax^2 + bx + c = 0$ obey $x_1 + x_2 = -(b/a)$ and $x_1 x_2 = c/a$, we can write the expression in the denominator of $\chi_\perp(\omega)$ in terms of λ_a, λ_b, and λ_c as

$$(\lambda_1 + i\omega 2\tau_N)(\lambda_2 + i\omega 2\tau_N) = (\lambda_a\lambda_b + \lambda_c) - 4\omega^2\tau_N^2 + i\omega 2\tau_N(\lambda_a + \lambda_b)$$

Accordingly, the transverse susceptibility (B.16) can equivalently be written as

$$\chi_\perp(\omega, T) = \chi_\perp(T)\frac{1 + i\omega 2\tau_N\dfrac{\lambda_a}{\lambda_a\lambda_b + \lambda_c}}{1 - 4\omega^2\tau_N^2\dfrac{1}{\lambda_a\lambda_b + \lambda_c} + i\omega 2\tau_N\dfrac{\lambda_a + \lambda_b}{\lambda_a\lambda_b + \lambda_c}} \tag{B.17}$$

b. Low-Frequency Expansion of $\chi_\perp(\omega)$ and Effective Transverse Relaxation Time

On expanding $\chi_\perp(\omega)$ from (B.17) in powers of $\omega\tau_N$ to first order, we get the simple result

$$\frac{\chi_\perp(\omega, T)}{\chi_\perp(T)} \simeq 1 - i\omega 2\tau_N\frac{\lambda_b}{\lambda_a\lambda_b + \lambda_c} \simeq \frac{1}{1 + i\omega 2\tau_N\dfrac{\lambda_b}{\lambda_a\lambda_b + \lambda_c}} \tag{B.18}$$

where the last approximate equality has been obtained by means of the binomial expansion $(1 + x)^\varepsilon = 1 + \varepsilon x + \cdots$. Therefore, in the low-frequency range $\chi_\perp(\omega)$ has a Debye-type form, so that the quantity multiplying $i\omega$ can be interpreted as an effective relaxation time, namely

$$\tau_\perp|_{\xi=0} = 2\tau_N \frac{1}{\lambda_a} \frac{1}{1 + \lambda_c/\lambda_a\lambda_b} \tag{B.19}$$

To conclude, with help from the results of Appendix A, let us write the coefficients λ_a, λ_b, and λ_c in terms of $\tilde{S}_2$ [the average of the second Legendre polynomial (3.73) at zero field]:

$$\lambda_a = \frac{2 + \tilde{S}_2}{1 - \tilde{S}_2}, \qquad \lambda_b = \frac{2\sigma}{3} \frac{2 + \tilde{S}_2(1 - 6/\sigma)}{\tilde{S}_2}, \qquad \lambda_c = \frac{1}{\lambda^2} \frac{6\sigma\tilde{S}_2}{1 - \tilde{S}_2}$$

Form these equations we get

$$\frac{1}{\lambda_a} = \frac{1 - \tilde{S}_2}{2 + \tilde{S}_2}, \qquad \frac{\lambda_c}{\lambda_a\lambda_b} = \frac{1}{\lambda^2} \frac{(3\tilde{S}_2)^2}{(2 + \tilde{S}_2)[2 + \tilde{S}_2(1 - 6/\sigma)]}$$

which, when inserted in Eq. (B.19), yield the effective transverse relaxation time (5.71).

Note finally that, as introduced, the effective transverse relaxation time is a sort of transverse *integral* relaxation time $\tau_{int,\perp}$ [cf. the first approximate equality of Eq. (B.18) with Eq. (B.3)]. However, its usefulness is questionable in the transverse-field case as the magnetization relaxation curve then comprises *oscillating* terms, so that the area under such a curve may largely overestimate the relaxation rate.

APPENDIX C. REDUCED EQUATIONS OF MOTION FOR NONLINEAR SYSTEM–ENVIRONMENT COUPLINGS

In this appendix we derive a reduced equation of motion for any dynamical variable $A(p, q)$ whose time evolution is determined by the Hamiltonian (6.41). This will be carried out by means of a perturbational expansion in the coupling parameter ε. Nevertheless, we first study the weak-coupling dynamics associated with a larger class of Hamiltonians of the form

$$\mathcal{H}_T = \mathcal{H}^{(m)}(p, q) + \sum_\alpha \tfrac{1}{2}(P_\alpha^2 + \omega_\alpha^2 Q_\alpha^2) + \varepsilon \sum_N \mathcal{B}^N(\mathbf{Q})F_N(p, q) \tag{C.1}$$

where the coupling terms $\mathcal{B}^N(\mathbf{Q})$ are *arbitrary* functions of the environment coordinates $\mathbf{Q}$ and N stands for a general index, which can run, for example, over single oscillator indices, pairs, triplets, and so on (α, $\alpha\beta$, $\alpha\beta\gamma$, ...). On the other hand, the modified system Hamiltonian $\mathcal{H}^{(m)}$ augments the system Hamiltonian $\mathcal{H}$ by appropriate counterterms, which will be determined below.

We shall first derive the reduced dynamical equations associated with the Hamiltonian (C.1), so that one could incorporate relaxation mechanisms involving any number of environmental normal modes into the dynamical equations of the system. This will be done by a perturbational treatment that is an extension of that developed by Cortés, West and Lindenberg (1985) to deal with a system–environment coupling *linear* in the system coordinate [the case $F_N(p, q) \propto q$ of the Hamiltonian (C.1)], but otherwise arbitrary in the environment coordinates.* Eventually, we shall particularize the results obtained to the Hamiltonian (6.41), which is recovered when

1. N runs only over single oscillator indices α and pairs $\alpha\beta$.
2. The corresponding coupling terms are $\mathcal{B}^{\alpha}(\mathbf{Q}) = Q_\alpha$ and $\mathcal{B}^{\alpha\beta}(\mathbf{Q}) = \frac{1}{2} Q_\alpha Q_\beta$.

The coupled dynamical equations for $A(p, q)$ and the environment variables associated with the Hamiltonian (C.1) are [cf. Eqs. (6.17) and (6.18)]

$$\frac{dA}{dt} = \{A, \mathcal{H}^{(m)}\} + \varepsilon \sum_N \mathcal{B}^N(\mathbf{Q})\{A, F_N\} \tag{C.2}$$

$$\frac{dQ_\alpha}{dt} = P_\alpha, \qquad \frac{dP_\alpha}{dt} = -\omega_\alpha^2 Q_\alpha - \varepsilon \sum_N \mathcal{B}_\alpha^N(\mathbf{Q}) F_N \tag{C.3}$$

where we have used the shorthand

$$\mathcal{B}_\alpha^N = \frac{\partial \mathcal{B}^N}{\partial Q_\alpha}$$

Equations (C.3) can *formally* be integrated, yielding an equation akin to Eq. (6.19) with $F_\alpha(t') \to \sum_N \mathcal{B}_\alpha^N[\mathbf{Q}(t')]F_N(t')$, namely

$$Q_\alpha(t) = Q_\alpha^{h}(t) - \frac{\varepsilon}{\omega_\alpha} \int_{t_0}^{t} dt' \, \sin[\omega_\alpha(t - t')] \sum_N \mathcal{B}_\alpha^N[\mathbf{Q}(t')]F_N(t')$$

where the $Q_\alpha^{h}(t)$ are the solutions (6.20) for the free oscillators and $F_N(t') = F_N[p(t'), q(t')]$. On integrating by parts in this equation, one gets [cf.

---

* Brun (1993) also treated rather general nonbilinear interactions by perturbation theory.

Eq. (6.21)]

$$Q_\alpha(t) = Q_\alpha^h(t) - \varepsilon \sum_N \left[D_\alpha^N(\mathbf{Q}; t, t') F_N(t') \right]_{t'=t_0}^{t'=t} + \varepsilon \int_{t_0}^t dt' \sum_N D_\alpha^N(\mathbf{Q}; t, t') \frac{dF_N}{dt}(t')$$

(C.4)

where we have introduced the indefinite integral

$$D_\alpha^N(\mathbf{Q}; t, t') = \frac{1}{\omega_\alpha} \int^{t'} dt'' \sin[\omega_\alpha(t - t'')] \mathcal{B}_\alpha^N[\mathbf{Q}(t'')]$$

(C.5)

Recall that writing $Q_\alpha(t)$ in the form (C.4) by an integration by parts allows one to separate the Hamiltonian (renormalization) and relaxational terms (see Section VI.C). However, Eq. (C.4) gives $Q_\alpha(t)$ in implicit form, since $Q_\alpha(t)$ also appears on the right-hand side via $\mathcal{B}_\alpha^N(\mathbf{Q})$. Thus, Eq. (C.4) is an explicit solution only in the linear $\mathcal{B}^N(\mathbf{Q})$ case of the Hamiltonian (6.15).

For weak system–environment interactions, we shall solve Eq. (C.4) for $Q_\alpha(t)$ perturbatively in ε. However, as pointed out by Cortés et al. (1985), in order to eventually obtain a thermodynamically consistent description, the expansion cannot be uniform in ε. If one keeps fluctuating terms up to order ε^k, the relaxation terms must be retained up to order ε^{2k}, in order to obtain proper fluctuation–dissipation relations [see, for example, Eqs. (6.29), (6.31), and (6.39)].

The ε expansion of $Q_\alpha(t)$ reads

$$Q_\alpha(t) = Q_\alpha^h(t) + \varepsilon \delta Q_\alpha(t) + \cdots$$

where $\varepsilon \delta Q_\alpha(t)$ is given by the second plus third terms on the right-hand side of Eq. (C.4) when $\mathbf{Q}^h$ (the zeroth-order term) is substituted for $\mathbf{Q}$ in $D_\alpha^N(\mathbf{Q}; t, t')$, namely

$$\varepsilon \delta Q_\alpha(t) = -\varepsilon \sum_N \left[D_\alpha^N(\mathbf{Q}^h; t, t') F_N(t') \right]_{t'=t_0}^{t'=t} + \varepsilon \int_{t_0}^t dt' \sum_N D_\alpha^N(\mathbf{Q}^h; t, t') \frac{dF_N}{dt}(t')$$

[i.e., we iterate Eq. (C.4) into itself]. The corresponding expansion of the coupling functions is given by

$$\varepsilon \mathcal{B}^N(\mathbf{Q}) = \varepsilon \mathcal{B}^N(\mathbf{Q}^h) + \varepsilon^2 \sum_\alpha \mathcal{B}_\alpha^N(\mathbf{Q}^h) \delta Q_\alpha + \cdots$$

(C.6)

which enters into Eq. (C.2). The term

$$f_N(t) \equiv \varepsilon \mathcal{B}^N[\mathbf{Q}^h(t)]$$

(C.7)

per analogy with $f_\alpha(t) = \varepsilon Q^{\mathrm{h}}_\alpha(t)$ [Eq. (6.22)], is interpreted as the lowest-order fluctuation. Following the program of Cortés et al. (1985), we retain fluctuations only to this order.*

Concerning the backreaction part, one first defines the quantities

$$\mathcal{K}^{N,M}(t, t') = \varepsilon^2 \left\langle \sum_\alpha \mathcal{B}^N_\alpha[\mathbf{Q}^{\mathrm{h}}(t)]D^M_\alpha(\mathbf{Q}^{\mathrm{h}}; t, t') \right\rangle \tag{C.8}$$

$$\delta\mathcal{K}^{N,M}(t, t') = \varepsilon^2 \sum_\alpha \mathcal{B}^N_\alpha[\mathbf{Q}^{\mathrm{h}}(t)]D^M_\alpha(\mathbf{Q}^{\mathrm{h}}; t, t') - \mathcal{K}^{N,M}(t, t') \tag{C.9}$$

so that the second term in the expansion (C.6) can exactly be decomposed as

$$\varepsilon^2 \sum_\alpha \mathcal{B}^N_\alpha(\mathbf{Q}^{\mathrm{h}})\delta Q_\alpha = -\left[\sum_M \left[\mathcal{K}^{N,M}(t, t') + \delta\mathcal{K}^{N,M}(t, t') \right] F_M(t') \right]^{t'=t}_{t'=t_0}$$

$$+ \int_{t_0}^t dt' \sum_M \left[\mathcal{K}^{N,M}(t, t') + \delta\mathcal{K}^{N,M}(t, t') \right] \frac{dF_M}{dt}(t')$$

Each kernel $\mathcal{K}^{N,M}$ gives a different type of contribution whereas the contribution of $\delta\mathcal{K}^{N,M}$ can be interpreted as fluctuations around the former. As these fluctuations are of order higher (ε^2) than the fluctuations that we are retaining in the present treatment, the terms $\delta\mathcal{K}^{N,M}$ will be omitted. On the other hand, the terms $\sum_M \mathcal{K}^{N,M}(t, t_0)F_M(t_0)$ in $\varepsilon^2 \sum_\alpha \mathcal{B}^N_\alpha \delta Q_\alpha$ will also be ignored as they are the generalization of those terms that give a transient in the response (see Section VI.C; recall, however, that they could be incorporated into an alternative definition of the fluctuating sources but, as they are of order ε^2, they would anyway be ignored). Finally, the parallel terms $-\sum_M \mathcal{K}^{N,M}(t, t)F_M(t)$ give the Hamiltonian contributions. In order to prove this, note first that, since $\mathcal{K}^{N,M}(t, t')$ contains equilibrium averages, it depends on $(t - t')$ and, hence, $\mathcal{K}^{N,M}(t, t)$ is independent of t. By the same reasoning, one can demonstrate the symmetry property $\mathcal{K}^{N,M} = \mathcal{K}^{M,N}$.[†] Then, by using the product rule of the Poisson bracket (6.23), one finds that the contribution originating from $-\sum_M \mathcal{K}^{N,M}(t, t)F_M(t)$ in the equation for $A(p, q)$ is

---

* To ensure $\langle f_N(t) \rangle = 0$, where the angular brackets denote average over initial states of the oscillators, one could assume that, for instance, at least one coordinate enters in $\mathcal{B}^N(\mathbf{Q})$ an odd number of times. Nevertheless, as discussed after Eqs. (6.52)–(6.54), such a restriction is not actually needed when the frequency spectrum of the oscillators is sufficiently dense.

† We shall anyway verify explicitly these two results for the Hamiltonian (6.41).

given by

$$-\sum_{NM} \mathcal{K}^{N,M}(0)\{A, F_N\}F_M = \left\{A, -\tfrac{1}{2}\sum_{NM}\mathcal{K}^{N,M}(0)F_NF_M\right\}$$

which is indeed derivable from a (time-independent) Hamiltonian. This term is associated with the coupling-induced renormalization of the energy of the system and is balanced by the counterterms incorporated in $\mathcal{H}^{(m)}$, now explicitly identified as [cf. Eq. (6.16)]

$$\mathcal{H}^{(m)} = \mathcal{H} + \tfrac{1}{2}\sum_{NM}\mathcal{K}^{N,M}(0)F_NF_M \qquad (C.10)$$

On collecting the terms whose retention has thus far been argued and introducing them into Eq. (C.2), one finally gets the (approximate) reduced equation of motion for any dynamical variable $A(p, q)$ [cf. Eq. (6.24)]:

$$\frac{dA}{dt} = \{A, \mathcal{H}\} + \sum_N \{A, F_N\}\left[f_N(t) + \int_{t_0}^{t} dt' \sum_M \mathcal{K}^{N,M}(t - t')\frac{dF_M}{dt}(t')\right]$$
$$(C.11)$$

In addition, within the approximation used (fluctuating and relaxation terms to orders ε and ε^2, respectively), one can replace dF_M/dt in the memory integral by its conservative part $dF_M/dt \simeq \{F_M, \mathcal{H}\}$. Finally, one can establish fluctuation–dissipation relations by means of arguments parallel to those presented by Cortés et al. (1985).

To conclude, we shall particularize these results to the linear-plus-quadratic couplings of the Hamiltonian (6.41). As has been mentioned, this is recovered when N runs over single oscillator indices α, with $\mathcal{B}^\alpha = Q_\alpha$, and pairs $\alpha\beta$, with $\mathcal{B}^{\alpha\beta} = \tfrac{1}{2}Q_\alpha Q_\beta$. Then, the fluctuating terms $f_N(t) = \varepsilon\mathcal{B}^N[\mathbf{Q}^h(t)]$ are given by $f_\alpha(t) = \varepsilon Q_\alpha^h(t)$ [Eq. (6.22)] and $f_{\alpha\beta}(t) = (\varepsilon/2)Q_\alpha^h(t)Q_\beta^h(t)$ [Eq. (6.44)]. On the other hand, by inserting the derivatives

$$\mathcal{B}_\gamma^\alpha = \frac{\partial \mathcal{B}^\alpha}{\partial Q_\gamma} = \delta_{\alpha\gamma} \qquad (C.12)$$

$$\mathcal{B}_\gamma^{\alpha\beta} = \frac{\partial \mathcal{B}^{\alpha\beta}}{\partial Q_\gamma} = \frac{1}{2}(\delta_{\alpha\gamma}Q_\beta + \delta_{\beta\gamma}Q_\alpha) \qquad (C.13)$$

into Eq. (C.5), the functions $D_\gamma^N(\mathbf{Q}; t, t')$ emerge in the form ($N = \alpha, \alpha\beta$)

$$D_\gamma^\alpha(\mathbf{Q}; t, t') = \frac{\delta_{\alpha\gamma}}{\omega_\alpha^2}\cos[\omega_\alpha(t - t')] \qquad (C.14)$$

$$D_\gamma^{\alpha\beta}(\mathbf{Q}; t, t') = \frac{1}{\omega_\gamma} \int^{t'} dt'' \sin\left[\omega_\gamma(t - t'')\right] \frac{1}{2} \left[\delta_{\alpha\gamma}Q_\beta(t'') + \delta_{\beta\gamma}Q_\alpha(t'')\right]$$

$$(C.15)$$

Therefore, on taking the averages in Eq. (C.8) with respect to the distribution (6.36) (decoupled initial conditions) by means of Eqs. (6.37), we get for the kernels $\mathcal{K}^{N,M}$ (see proofs below)

$$\mathcal{K}^{\alpha,\beta}(\tau) = \delta_{\alpha\beta} \frac{\varepsilon^2}{\omega_\alpha^2} \cos(\omega_\alpha \tau) \tag{C.16}$$

$$\mathcal{K}^{\alpha,\beta\gamma}(\tau) = \mathcal{K}^{\alpha\beta,\gamma}(\tau) = 0 \tag{C.17}$$

$$\mathcal{K}^{\alpha\beta,\gamma\delta}(\tau) = \frac{1}{2}(\delta_{\alpha\gamma}\delta_{\beta\delta} + \delta_{\alpha\delta}\delta_{\beta\gamma})$$

$$\times \frac{\varepsilon^2}{2} \frac{k_B T}{2\omega_\alpha^2\omega_\beta^2} \left\{\cos[(\omega_\alpha - \omega_\beta)\tau] + \cos[(\omega_\alpha + \omega_\beta)\tau]\right\} \tag{C.18}$$

These kernels satisfy the properties mentioned above; they depend on $\tau = t - t'$ and are symmetric with respect to the indices separated by commas, which correspond to the general indices N, M.

On introducing all these results in Eq. (C.11), the resulting dynamical equation for $A(p, q)$ is given by Eq. (6.43). For the sake of simplicity, we have introduced in that equation the kernels $\mathcal{K}_\alpha(\tau)$ and $\mathcal{K}_{\alpha\beta}(\tau)$, which are defined in terms of the above kernels by

$$\mathcal{K}^{\alpha,\beta}(\tau) = \delta_{\alpha\beta}\mathcal{K}_\alpha(\tau)$$

$$\mathcal{K}^{\alpha\beta,\gamma\delta}(\tau) = \tfrac{1}{2}(\delta_{\alpha\gamma}\delta_{\beta\delta} + \delta_{\alpha\delta}\delta_{\beta\gamma})\mathcal{K}_{\alpha\beta}(\tau)$$

Besides, on explicitly writing the counterterm of Eq. (C.10) in this linear-plus-quadratic case, one arrives at Eq. (6.42).

Note finally that, since $\mathcal{B}_\gamma^\alpha(\mathbf{Q}^h)D_\gamma^\beta(\mathbf{Q}^h; t, t')$ does not depend on $\mathbf{Q}^h$, the kernel $\mathcal{K}_\alpha(\tau)$ is not affected by the averaging procedure, whereas this renders $\mathcal{K}_{\alpha\beta}(\tau)$ explicitly dependent on the temperature. In this connection, we remark that the modifications of this last kernel obtained when one assumes coupled initial conditions, begin at order ε^3.

Derivation of the Kernels

1. *Derivation of $\mathcal{K}^{\alpha,\beta}(\tau)$.* From Eqs. (C.12) and (C.14) and the general definition $\mathcal{K}^{N,M}(t, t') = \varepsilon^2 \langle \sum_\rho \mathcal{B}_\rho^N D_\rho^M \rangle$, one gets

$$\mathcal{K}^{\alpha,\beta}(t,t') = \varepsilon^2 \left\langle \sum_\rho \mathcal{B}_\rho^\alpha D_\rho^\beta \right\rangle$$

$$= \varepsilon^2 \left\langle \sum_\rho \delta_{\alpha\rho} \frac{\delta_{\beta\rho}}{\omega_\beta^2} \cos\left[\omega_\beta(t-t')\right] \right\rangle = \delta_{\alpha\beta} \frac{\varepsilon^2}{\omega_\alpha^2} \cos[\omega_\alpha(t-t')]$$

where the average has played no role.   Q.E.D.

2. *Derivation of $\mathcal{K}^{\alpha,\beta\gamma}(\tau)$.* From Eqs. (C.12) and (C.15) we obtain

$$\langle \mathcal{B}_\rho^\alpha D_\rho^{\beta\gamma} \rangle = \frac{\delta_{\alpha\rho}}{\omega_\rho} \int^{t'} dt'' \sin\left[\omega_\rho(t-t'')\right] \frac{1}{2} [\delta_{\beta\rho} \underbrace{\langle Q_\gamma^h(t'') \rangle}_{0} + \delta_{\gamma\rho} \underbrace{\langle Q_\beta^h(t'') \rangle}_{0}] = 0$$

where Eqs. (6.37) have been employed. Therefore, from this result and the general definition (C.8) it follows that $\mathcal{K}^{\alpha,\beta\gamma}(t,t') = 0$.   Q.E.D.

3. *Derivation of $\mathcal{K}^{\alpha\beta,\gamma}(\tau)$.* The average of the product of Eqs. (C.13) and (C.14) evaluated at $\mathbf{Q}^h$ is zero as well. Indeed

$$\langle \mathcal{B}_\rho^{\alpha\beta} D_\rho^\gamma \rangle = \frac{1}{2} [\delta_{\alpha\rho} \underbrace{\langle Q_\beta^h(t) \rangle}_{0} + \delta_{\beta\rho} \underbrace{\langle Q_\alpha^h(t) \rangle}_{0}] \frac{\delta_{\gamma\rho}}{\omega_\gamma^2} \cos[\omega_\gamma(t-t')] = 0$$

whence one gets the stated result $\mathcal{K}^{\alpha\beta,\gamma}(t,t') = 0$.   Q.E.D.

4. *Derivation of $\mathcal{K}^{\alpha\beta,\gamma\delta}(\tau)$.* Finally, for the average of the product of Eqs. (C.13) and (C.15) evaluated at $\mathbf{Q}^h$, one has

$$\langle \mathcal{B}_\rho^{\alpha\beta} D_\rho^{\gamma\delta} \rangle = \left\langle \frac{1}{2} [\delta_{\alpha\rho} Q_\beta^h(t) + \delta_{\beta\rho} Q_\alpha^h(t)] \right.$$

$$\left. \times \frac{1}{\omega_\rho} \int^{t'} dt'' \sin[\omega_\rho(t-t'')] \frac{1}{2} [\delta_{\gamma\rho} Q_\delta^h(t'') + \delta_{\delta\rho} Q_\gamma^h(t'')] \right\rangle$$

Therefore, we need to calculate the following average

$$\langle [\delta_{\alpha\rho} Q_\beta^h(t) + \delta_{\beta\rho} Q_\alpha^h(t)][\delta_{\gamma\rho} Q_\delta^h(t'') + \delta_{\delta\rho} Q_\gamma^h(t'')] \rangle$$

$$= k_B T \left\{ \delta_{\alpha\rho} \delta_{\gamma\rho} \frac{\delta_{\beta\delta}}{\omega_\beta^2} \cos[\omega_\beta(t-t'')] + \delta_{\alpha\rho} \delta_{\delta\rho} \frac{\delta_{\beta\gamma}}{\omega_\beta^2} \cos[\omega_\beta(t-t'')] \right.$$

$$\left. + \delta_{\beta\rho} \delta_{\gamma\rho} \frac{\delta_{\alpha\delta}}{\omega_\alpha^2} \cos[\omega_\alpha(t-t'')] + \delta_{\beta\rho} \delta_{\delta\rho} \frac{\delta_{\alpha\gamma}}{\omega_\alpha^2} \cos[\omega_\alpha(t-t'')] \right\}$$

$$= \frac{k_{\mathrm{B}}T}{\omega_\alpha^2 \omega_\beta^2} \Big\{ \delta_{\alpha\rho}(\delta_{\gamma\rho}\delta_{\beta\delta} + \delta_{\delta\rho}\delta_{\beta\gamma})\omega_\alpha^2 \cos[\omega_\beta(t - t'')]$$

$$+ \delta_{\beta\rho}(\delta_{\gamma\rho}\delta_{\alpha\delta} + \delta_{\delta\rho}\delta_{\alpha\gamma})\omega_\beta^2 \cos[\omega_\alpha(t - t'')] \Big\}$$

where we have used Eqs. (6.37). Next, on multiplying this expression by $\sin[\omega_\rho(t - t'')]/\omega_\rho$, and summing over ρ, we obtain

$$\sum_\rho \big\langle [\delta_{\alpha\rho}Q_\beta^{\mathrm{h}}(t) + \delta_{\beta\rho}Q_\alpha^{\mathrm{h}}(t)][\delta_{\gamma\rho}Q_\delta^{\mathrm{h}}(t'') + \delta_{\delta\rho}Q_\gamma^{\mathrm{h}}(t'')] \big\rangle \frac{\sin[\omega_\rho(t-t'')]}{\omega_\rho}$$

$$= \frac{k_{\mathrm{B}}T}{\omega_\alpha^2 \omega_\beta^2}(\delta_{\alpha\gamma}\delta_{\beta\delta} + \delta_{\alpha\delta}\delta_{\beta\gamma})$$

$$\times \big\{ \omega_\alpha \cos[\omega_\beta(t-t'')]\sin[\omega_\alpha(t - t'')] + \omega_\beta \cos[\omega_\alpha(t-t'')]\sin[\omega_\beta(t-t'')] \big\}$$

Then, on taking into account that $(d/dt'')\{\cos[\omega_\alpha(t - t'')]\cos[\omega_\beta(t - t'')]\}$ is equal to the term within the curly brackets shown above when calculating the integral occurring in $\mathcal{K}^{\alpha\beta,\gamma\delta}(t, t') = \varepsilon^2 \langle \sum_\rho \mathcal{B}_\rho^{\alpha\beta} D_\rho^{\gamma\delta} \rangle$, we arrive at

$$\mathcal{K}^{\alpha\beta,\gamma\delta}(t, t') = \frac{1}{2}(\delta_{\alpha\gamma}\delta_{\beta\delta} + \delta_{\alpha\delta}\delta_{\beta\gamma})\frac{\varepsilon^2}{2}\frac{k_{\mathrm{B}}T}{\omega_\alpha^2 \omega_\beta^2} \cos[\omega_\alpha(t - t')]\cos[\omega_\beta(t - t')]$$

whence one immediately obtains Eq. (C.18).   Q.E.D.

ACKNOWLEDGMENTS

Helpful discussions with Prof. W. T. Coffey, Dr. D. A. Garanin, Dr. F. Falo, Prof. J. M. Sancho, Dr. P. Svedlindh, Dr. F. J. Lázaro (who supervised the PhD thesis on which this work has been based), and Prof. Yu. P. Raĭkher are gratefully acknowledged. I also wish to express my gratitude to the colleagues of *Instituto de Ciencia de Materiales de Aragón* (Consejo Superior de Investigaciones Científicas–Universidad de Zaragoza) and *Department of Materials Science, Division of Solid State Physics* (Uppsala University) with which I have worked during these years. This work has partially been supported by Diputación General de Aragón and the Swedish Natural Science Research Council (NFR).

REFERENCES

1. A. Aharoni, *Phys. Rev.* **135**, A447–A449 (1964).

2. G. Arfken, *Mathematical Methods for Physicists*, 3rd ed. Academic Press, Boston, 1985.

3. A.-L. Barra, P. Debrunner, D. Gatteschi, C. E. Schultz, and R. Sessoli, *Europhys. Lett.* **35**, 133–138 (1996).

4. L. Bessais, L. Ben Jaffel, and J. L. Dormann, *Phys. Rev. B* **45**, 7805–7815 (1992).

5. K. Binder and D. Stauffer, in *Applications of the Monte Carlo Method in Statistical Physics*, (K. Binder, ed.), p. 1, Springer, Berlin, 1984.

6. T. Bitoh, K. Ohba, M. Takamatsu, T. Shirane, and S. Chikazawa, *J. Phys. Soc. Jpn.* **62**, 2583–2586 (1993).

7. T. Bitoh, K. Ohba, M. Takamatsu, T. Shirane, and S. Chikazawa, *J. Phys. Soc. Jpn.* **64**, 1311–1319 (1995).

8. C. J. F. Böttcher and P. Bordewijk, *Theory of Electric Polarization* Vol. II, Elsevier, Amsterdam, 1978.

9. W. F. Brown, Jr., *J. Appl. Phys.* **30**, 130s–133s (1959).

10. W. F. Brown, Jr., *Phys. Rev.* **130**, 1677–1686 (1963).

11. W. F. Brown, Jr., *IEEE Trans. Magn.* **MAG-15**, 1196–1208 (1979).

12. W. F. Brown, Jr. and A. H. Morrish, *Phys. Rev.* **105**, 1198–1201 (1957).

13. T. A. Brun, *Phys. Rev. D* **47**, 3383–3393 (1993).

14. A. O. Caldeira and A. J. Leggett, *Ann. Phys.* **149**, 374–456 (1983).

15. R. W. Chantrell, N. Y. Ayoub, and J. Popplewell, *J. Magn. Magn. Mater.* **53**, 199–207 (1985).

16. S. Chikazumi, *Physics of Magnetism*, Krieger, New York, 1978.

17. W. T. Coffey, *Adv. Chem. Phys.* **103**, 259–333 (1998).

18. W. T. Coffey, P. J. Cregg, and Yu. P. Kalmykov, *Adv. Chem. Phys.* **83**, 263–464 (1993).

19. W. T. Coffey, D. S. F. Crothers, Yu. P. Kalmykov, E. S. Massawe, and J. T. Waldron, *Phys. Rev. E* **49**, 1869–1882 (1994).

20. W. T. Coffey, D. S. F. Crothers, Yu. P. Kalmykov, and J. T. Waldron, *Phys. Rev. B* **51**, 15947–15956 (1995*a*).

21. W. T. Coffey, D. S. F. Crothers, Yu. P. Kalmykov, and J. T. Waldron, *Physica A* **213**, 551–575 (1995*b*).

22. W. T. Coffey, Yu. P. Kalmykov, and E. S. Massawe, *Adv. Chem. Phys.* **85**, 667–792 (1993).

23. E. Cortés, B. J. West, and K. Lindenberg, *J. Chem. Phys.* **82**, 2708–2717 (1985).

24. P. J. Cregg and L. Bessais, *J. Magn. Magn. Mater.* **202**, 554–564 (1999*a*).

25. P. J. Cregg and L. Bessais, *J. Magn. Magn. Mater.* **203**, 265–267 (1999*b*).

26. P. J. Cregg, D. S. F. Crothers, and A. W. Wickstead, *J. Appl. Phys.* **76**, 4900–4902 (1994).

27. R. Ettelaie and M. A. Moore, *J. Phys. A* **17**, 3505–3520 (1984).

28. G. W. Ford, J. T. Lewis, and R. F. O'Connell, *Phys. Rev. A* **37**, 4419–4428 (1988).

29. D. A. Garanin, *Phys. Rev. E* **54**, 3250–3256 (1996).

30. D. A. Garanin, *Phys. Rev. B* **55**, 3050–3057 (1997).

31. D. A. Garanin, V. V. Ishchenko, and L. V. Panina, *Theor. Math. Phys. (USSR)* **82**, 169–179 (1990), [*Teor. Mat. Fiz.* **82**, 242 (1990)].

32. J. L. García-Palacios, *Eur. Phys. J. B* **11**, 293–308 (1999).

33. J. L. García-Palacios, P. Jönsson, and P. Svedlindh, *Phys. Rev. B* **61** (2000).

34. J. L. García-Palacios and F. J. Lázaro, *Phys. Rev. B* **55**, 1006–1010 (1997).

35. J. L. García-Palacios and F. J. Lázaro, *Phys. Rev. B* **58**, 14937–14958 (1998).

36. A. Garg and G.-H. Kim, *Phys. Rev. B* **43**, 712–718 (1991).

37. R. S. Gekht, *Phys. Met. Metallogr. (USSR)* **55**, 12–17 (1983), [*Fiz. Met. Metalloved.* **55**, 225 (1983)].

38. R. S. Gekht and V. A. Ignatchenko, *Sov. Phys. JETP* **49**, 84–89 (1979), [*Zh. Eksp. Teor. Fiz.* **76**, 164 (1979)].

39. T. L. Gilbert, *Phys. Rev.* **100**, 1243 (1995).

40. J. I. Gittleman, B. Abeles, and S. Bozowski *Phys. Rev. B* **9**, 3891–3897 (1974).

41. A. M. Gomes, M. A. Novak, R. Sessoli, A. Caneschi, and D. Gatteschi, *Phys. Rev. B* **57**, 5021–5024 (1998).

42. M. Hennion, C. Bellouard, I. Mirebeau, J. L. Dormann, and M. Nogues, *Europhys. Lett.* **25**, 43–48 (1994).

43. I. S. Jacobs and C. P. Bean, in *Magnetism*, (G. T. Rado and H. Suhl, eds), Vol. III, pp. 271–350, Academic Press, New York–London, 1963.

44. A. M. Jayannavar, *Z. Phys. B* **82**, 153–156 (1991).

45. T. Jonsson, J. Mattsson, C. Djurberg, F. A. Khan, P. Nordblad, and P. Svedlindh, *Phys. Rev. Lett.* **75**, 4138–4141 (1995).

46. P. Jung and H. Risken, *Z. Phys. B* **59**, 469–481 (1985).

47. Yu. P. Kalmykov and W. T. Coffey, *Phys. Rev. B* **56**, 3325–3337 (1997).

48. I. Klik, *J. Stat. Phys.* **66**, 635–645 (1992).

49. P. E. Kloeden and E. Platen, *Numerical Solution of Stochastic Differential Equations*, 2nd corr. printing, Springer, Berlin, 1995.

50. D. A. Krueger, *J. Appl. Phys.* **50**, 8169–8171 (1979).

51. R. Kubo and N. Hashitsume, *Prog. Theor. Phys. Suppl.* **46**, 210–220 (1970).

52. L. D. Landau and E. M. Lifshitz, *Z Phys. Sowjet.* **8**, 153–169 (1935).

53. L. D. Landau and E. M. Lifshitz, *Statistical Physics* (Part 1), 3rd ed., Pergamon Press, Oxford, 1980.

54. L. D. Landau and E. M. Lifshitz, *Electrodynamics of Continuous Media*, 2nd ed., Pergamon Press, Oxford, 1984.

55. C. J. Lin, *J. Appl. Phys.* **32**, 233s–234s (1961).

56. K. Lindenberg, U. Mohanty, and V. Seshadri, *Physica A* **119**, 1–16 (1983).

57. K. Lindenberg and V. Seshadri, *Physica A* **109**, 483–499 (1981).

58. K. Lindenberg and B. J. West, *Phys. Rev. A* **30**, 568–582 (1984).

59. L. Lundgren, P. Svedlindh, and O. Beckman, *J. Magn. Magn. Mater.* **25**, 33–38 (1981).

60. W. Luo, S. R. Nagel, T. F. Rosenbaum, and R. E. Rosensweig, *Phys. Rev. Lett.* **67**, 2721–2724 (1991).

61. V. B. Magalinskiĭ, *Sov. Phys. JETP* **9**, 1381–1382 (1959), [*Zh. Eksp. Teor. Fiz.* **36**, 1942 (1959)].

62. A. Maiti and L. M. Falicov, *Phys. Rev. B* **48**, 13596–13601 (1993).

63. M. Morillo and J. Gómez-Ordóñez, *Phys. Rev. Lett.* **71**, 9–11 (1993).

64. G. Moro and P. L. Nordio, *Mol. Phys.* **56**, 255–269 (1985).

65. S. Mørup, *J. Magn. Magn. Mater.* **37**, 39–50 (1983).

66. L. Néel, *Ann. Geophys.* **5**, 99–136 (1949).

67. A. Pérez-Madrid and J. M. Rubí, *Phys. Rev. E* **51**, 4159–4164 (1995).

68. Yu. L. Raĭkher and M. I. Shliomis, *Sov. Phys. JETP* **40**, 526–532 (1975), [*Zh. Eksp. Teor. Fiz.* **67**, 1060 (1974)].

69. Yu. L. Raĭkher and M. I. Shliomis, *Adv. Chem. Phys.* **87**, 595–751 (1994).

70. Yu. L. Raĭkher and V. I. Stepanov, *J. Phys.: Condens. Matter* **6**, 4137–4145 (1994).

71. Yu. L. Raĭkher and V. I. Stepanov, *Phys. Rev. B* **51**, 16428–16431 (1995a).

72. Yu. L. Raĭkher and V. I. Stepanov, *Phys. Rev. B* **52**, 3493–3498 (1995b).

73. Yu. L. Raĭkher and V. I. Stepanov, *Phys. Rev. B* **55**, 15005–15017 (1997).

74. Yu. L. Raĭkher, V. I. Stepanov, A. N. Grigorenko, and P. I. Nikitin, *Phys. Rev. E* **56**, 6400–6409 (1997).

75. L. Ramírez-Piscina, J. M. Sancho, and A. Hernández-Machado, *Phys. Rev. B* **48**, 125–131 (1993).

76. H. Risken, *The Fokker-Planck Equation*, 2nd ed., Springer, Berlin 1989.

77. D. Rodé, H. N. Bertram, and D. R. Fredkin, *IEEE Trans. Magn.* **MAG-23**, 2224–2226 (1987).

78. W. Rümelin, *SIAM (Soc. Ind. Appl. Math.) J. Numer. Anal.* **19**, 604–613 (1982).

79. E. K. Sadykov, *J. Phys.: Condens. Matter* **4**, 3295–3298 (1992).

80. P. Schiffer, A. P. Ramirez, D. A. Huse, P. L. Gammel, U. Yaron, D. J. Bishop, and A. J. Valentino, *Phys. Rev. Lett.* **74**, 2379–2382 (1995).

81. V. Seshadri and K. Lindenberg, *Physica A* **115**, 501–518 (1982).

82. V. V. Shcherbakova, *Izvestiya, Earth Phys.* **14**, 308–309 (1978).

83. M. I. Shliomis and V. I. Stepanov, *J. Magn. Magn. Mater.* **122**, 176–181 (1993).

84. M. I. Shliomis and V. I. Stepanov, *Adv. Chem. Phys.* **87**, 1–30 (1994).

85. D. A. Smith and F. A. de Rozario, *J. Magn. Magn. Mater.* **3**, 219–233 (1976).

86. B. A. Storonkin, *Sov. Phys. Crystallogr.* **30**, 489–493 (1985), [*Kristallografiya* **30**, 841 (1985)].

87. P. Svedlindh, T. Jonsson, and J. L. García-Palacios, *J. Magn. Magn. Mater.* **169**, 323–334 (1997).

88. P. Ullersma, *Physica* **32**, 27, 56, 74, 90–96 (1966).

89. N. G. van Kampen, *Stochastic Processes in Physics and Chemistry*, North-Holland, Amsterdam, 1981.

90. W. Wernsdorfer, E. Bonet Orozco, K. Hasselbach, A. Benoit, B. Barbara, N. Demoncy, A. Loiseau, H. Pascard, and D. Mailly, *Phys. Rev. Lett.* **78**, 1791–1794 (1997).

91. F. G. West, *J. Appl. Phys.* **32**, 249s–250s. (1961).

92. H. D. Williams, K. O'Grady, M. El-Hilo, and R. W. Chantrell, *J. Magn. Magn. Mater.* **122**, 129–133 (1993).

93. E. P. Wohlfarth, *Phys. Lett.* **70A**, 489–491 (1979).

94. R. Zwanzig, *J. Stat. Phys.* **9**, 215–220 (1973).

RELAXATION TIMES FOR SINGLE-DOMAIN FERROMAGNETIC PARTICLES

E. E. C. KENNEDY

Department of Applied Mathematics and Theoretical Physics, The Queen's University of Belfast, Belfast, Co. Antrim, Northern Ireland

CONTENTS

Advances in Chemical Physics, Volume 112, Edited by I. Prigogine and Stuart A. Rice
ISBN 0-471-38002-4. © 2000 John Wiley & Sons, Inc.

I. INTRODUCTION

A. Purpose of This Review

It is the purpose of this review to present a number of methods for the calculation of various relaxation times for single-domain ferromagnetic particles. A procedure for calculating the relaxation time exactly from the matrix form of the Fokker–Planck equation, which describes the distribution of representative points on the surface of a unit sphere, is presented. The calculations allow for the external magnetic field to be applied at an oblique angle to the easy axis of magnetization, and they also allow for the effect of the dimensionless damping factor a to be included. A variety of asymptotic formulas are then presented, which describe various situations, in particular, low damping (LD) and intermediate to high damping (IHD), and use both axially symmetric and nonaxially symmetric potentials. The exact data extracted from the matrix form of the Fokker–Planck equation are then compared with the asymptotic formulas in order to determine their accuracy and ranges of validity. The applicability of the asymptotes to various experimental situations is also discussed.

The work discussed in this review is of great importance in the development of fundamental theories of magnetism, and in modeling magnetic materials, experiments and computerized simulations, and it is also of great importance in the area of paleomagnetism. Such work can also provide insight into some remarkable technological improvements in the area of information storage, since thermal instability limits are determined by the superparamagnetic effects of the media.

B. Superparamagnetic Effect

All ferromagnetic materials contain microscopic regions called *domains*, within which all magnetic moments are aligned. These domains have volumes of approximately 10^{-12}–10^{-8} m^3 and contain 10^{17}–10^{21} atoms [1]. The boundaries between the various domains having different orientations are called

FIGURE 1. Unmagnetized sample of a ferromagnetic material.

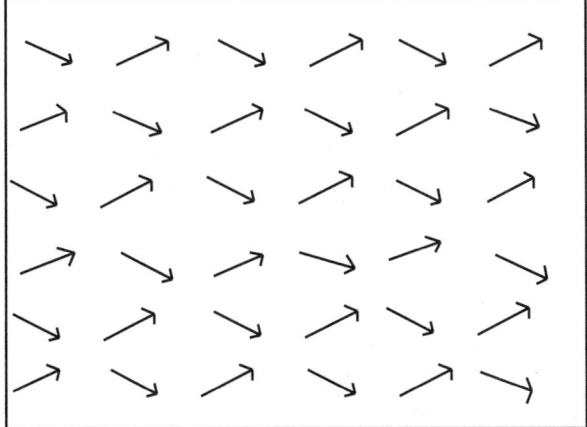

FIGURE 2. Magnetized sample of a ferromagnetic material.

domain walls. In an unmagnetized sample, the domains are randomly oriented so that the total magnetic moment is zero (see Fig. 1). However, when the sample is placed in an external magnetic field, the domains tend to align with the field, which results in a magnetized sample (see Fig. 2). When the external field is removed, the sample may retain a net magnetization in the direction of the original field. It is possible, however, for the alignment of the domains to be reoriented as a result of thermal agitation, although at ordinary temperatures, thermal agitation is not sufficiently high to disrupt the preferred orientation of the magnetic moments.

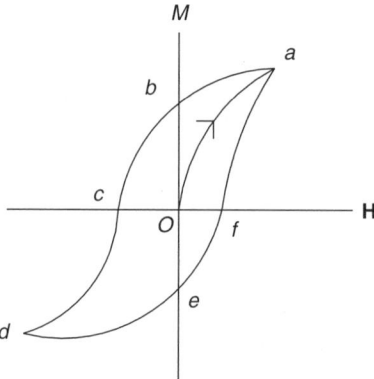

FIGURE 3. Magnetic hysterisis curve.

In order to explain the behavior described above, let us consider an unmagnetized sample of a ferromagnetic material as illustrated by Fig. 1, hence we suppose that at the point O on the hysterisis curve (Fig. 3), the domains are randomly oriented (i.e., there is no externally applied field at this point). As the external field **H** increases, the domains are almost all aligned at a. At this point, the sample is approaching saturation (i.e., almost all the domains are aligned in the same direction). Suppose then that the applied field is suddenly reduced to zero, then the magnetization curve (or hysterisis curve) follows path ab as shown in Fig. 3. However, we note that although the externally applied field is now zero, the sample remains magnetized (point b), and this is due to the alignment of a large number of the domains. This magnetic state is known as *remanant* [2] *magnetization*. If the external field **H** is then applied in the reverse direction, the domains reorient until the sample is again unmagnetized at point c, where the magnetization M is zero. If the external field is increased further in this direction, then the domains reorient themselves until the sample becomes magnetized in the opposite direction, approaching saturation at point d. Again reversing the direction of the field causes the magnetization curve to follow the path def; if the field is increased sufficiently, the curve will return to point a, where the sample will again have its maximum magnetization. The effect described here is called *magnetic hysterisis*, which shows that the magnetization of a ferromagnetic substance depends not only on the actual substance but also on the strength of the externally applied field.

Having described the behavior exhibited by ferromagnetic materials in the presence of an externally applied magnetic field, it is possible to highlight the technological importance of ferromagnetic materials since all magnetic recording media rely on the properties of small ferromagnetic particles. Now a ferromagnetic particle below a certain critical size (order of 150 Å in radius) consists of a single magnetic domain (this is discussed in more detail in the following section), and is in a uniform state of magnetization for any applied field. Such a particle is called a *single-domain ferromagnetic particle*. The magnetic storage industry pursue increased signal-to-noise ratios, and so the search for smaller ferromagnetic particles continues. However, difficulties arise where, at certain temperatures, ferromagnetic particles of this size exhibit behavior that inhibits the technological advancement of magnetic recording. The problem has been referred to as the *superparamagnetic limit*. [3].

The problem can be described in real terms as follows. We can magnetize a sample of a ferromagnetic material and the magnetic domains will align parallel to the applied field; we can also demagnetize it as described by the hysterisis curve in Fig. 3, but we would not expect the alignment of the domains to jump spontaneously and reverse their direction or return to a zero state of magnetic moment. However, in principle, any apparently stable magnetic state of a tape or magnet is only one of many local minima of the free energy [4]; thermal agitation can cause spontaneous jumps from one magnetic state to another. The apparent stability is due to the fact that the tape or magnet cannot get from one state to another without passing over an energy barrier (this aspect is discussed in further detail in the following section), which is very large in comparison with the thermal energy kT (where $k =$ Boltzmann's constant, and $T =$ absolute temperature). Hence, one would be forgiven for assuming that it was virtually impossible for the magnetic state of the magnet to change, since the probability per unit time of a jump over such a barrier is so small that the mean time that we would have to wait for such an event to happen would far exceed our own mean lifetime [4]. However, we are considering the *total* free energy of the system, and this undergoes a change during the hypothetical transition; moreover, this energy decreases as the volume of the system decreases. Thus for fine ferromagnetic particles at temperatures that are not too low, the barrier height becomes comparable with kT and spontaneous jumps from one state to another now become not only a strong possibility but also of great technological importance.

In the reverse case [i.e., when the barrier height is very small in comparison with kT (even smaller ferromagnetic particles)], a ferromagnetic particle behaves like a paramagnetic atom. A paramagnetic substance, like a ferromagnetic substance, involves the presence of atoms with permanent magnetic dipole moments where the dipoles interact weakly with one another, and are randomly oriented in the absence of an externally applied magnetic field. However, when the substance is subjected to an externally applied magnetic field, its dipoles tend to align with the field, but the alignment process must compete with the effects of thermal motion, and so this tends to randomize the dipole orientations; hence above a certain critical temperature the substance exhibits paramagnetic behavior. Let us suppose that the interaction between such particles is negligible; then the magnetization as a function of the applied field exhibits no hysterisis (and is determined by a Langevin function). This phenomenon is called *superparamagnetism* [3].

C. Domain Structure

Study of ferromagnetic materials and their properties began in the early twentieth century. In 1907, Weiss [5] proposed that a ferromagnetic material in the absence of an external magnetic field consists of elementary regions (called *Weiss* [5] *domains*) where each is magnetized almost to saturation in some (random) direction. His theories included the assumption that these regions coincided with the minute crystals of which the material was composed. However, his assumption was disputed in 1930 by Frenkel and Dorfman [6], who realized that the spontaneous magnetization was present in single crystals of the material also; hence they arrived at the conclusion that even a single crystal of a ferromagnetic substance consists of a number of "elementary magnets". At this time, it was thought that in the absence of an external magnetic field, the demagnetized state was the stable state in a large crystal of a ferromagnetic material. However, contrary to the assumption of Weiss [5], that the so-called domains are randomly oriented, Landau and Lifshitz [7] discussed the fact that the domains are oriented in such a way that the magnetic flux circuit lies entirely within the specimen. This internal flux closure indicates a high degree of ordering in the configuration of the domains, and so they suggested that the domains are actually made up of geometrically ordered elementary layers, and separated by domain walls (Bloch [2] walls) as in Fig. 4.

The existence of domains, the formation of the domains into elementary layers, and their permanent magnetization are all due to the minimization [2] of the total free energy of the system. In a finite sample of a ferromagnetic material, the total free energy comprises both surface and volume energies. The surface of the boundary between the domains (separated by the Bloch [2] walls) represents the surface energy, while the magnetic field is the

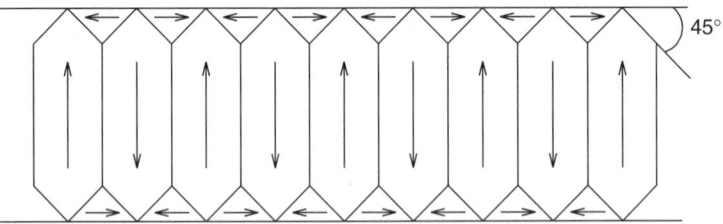

FIGURE 4.   Geometrically ordered layered structure of the domains.

volume energy. As the dimensions of the sample are reduced, the surface energies become more significant than the volume energies (since the volume is decreasing). When very small dimensions are reached, there will come a point where it is more favorable energetically to eliminate the domain boundaries so that the specimen becomes one domain and acts as a permanent magnet, the result of which is called a *single-domain ferromagnetic particle*. This phenomenon was first recognized by Frenkel and Dorfman [6], who calculated that these elementary regions of spontaneous magnetization must contain at least 10^{12} atoms. Experimental evidence of permanent magnetization in small ferromagnetic particles was first presented by Elmore [8] in 1938.

The dimension of a sample below which it was possible for a single domain to exist was calculated by Kittel [9] in 1946 as having a radius of approximately 150 Å. This was verified experimentally by Bean and Livingston [10], who found single-domain ferromagnetic particles with radii of 100 Å; also Frenkel and Dorfman [6] predicted from energy considerations that particles of the order of 100 Å should have single-ferromagnetic domains. Montgomery [11,12] in 1932 attempted to verify these predictions for nickel colloids, but was unsuccessful because the nickel oxidized before any measurements could be made. He did point out, however, that if the colloid particles were to consist of such small permanent magnets, then they should behave as the molecules of a classic paramagnetic gas, obeying the Langevin [13] equation. Also, in 1938, Elmore [8] found that a colloidal suspension of single-domain magnetic iron oxide particles exhibits paramagnetic behavior similar to Langevin paramagnetism. Thus, if we have a suspension of equally sized, isotropic (meaning that there is no variation in their properties with direction), single-domain ferromagnetic particles in a nonmagnetic medium, then we can express the magnetization M_0 of the suspension when in thermodynamic equilibrium with a constant applied field H by the Langevin equation where

$$M_0 = cM_s L(\xi), \qquad L(\xi) = \coth(\xi) - \frac{1}{\xi}, \qquad \xi = \frac{\mu H}{kT} \qquad (1.1)$$

where c is the volume concentration of the particles (this is assumed to be small enough to neglect particle interaction), M_s is the spontaneous magnetization of the ferromagnetic material, μ is the magnetic moment (magnetization × volume) of an individual particle, H is the magnitude of the applied magnetic field, k is Boltzmann's constant, and T is the absolute temperature of the suspension.

In practice, however, there exists a distribution in the size of the particle moments. If we are to compare the thermal equilibrium magnetization of a suspension of isotropic single-domain ferromagnetic particles with the Langevin treatment of atomic paramagnetism, then it should be noted that this idea differs from the Langevin approach in that the magnetic moment μ is not that of a single atom, but rather of a single-domain ferromagnetic particle that may contain more than 10^5 atoms that are ferromagnetically coupled by exchange forces.

Another consideration is that the single-domain particles are not completely isotropic in their properties, but will have anisotropic contributions to their total energy associated with the external shape of the particle, imposed stresses, or the crystalline structure itself. Figure 5 describes the shape of an isotropic particle and a particle with uniaxial [2] shape anisotropy.

Until now, we have discussed only the magnetization properties of an assembly of identical single-domain particles in thermal equilibrium with an applied field; however, the permanent-magnet properties of stable single-domain particles are remarkably different and are associated with the particles that are not in a state of thermodynamic equilibrium. We wish to consider the conditions under which an assembly of single-domain particles can achieve equilibrium.

A sufficiently fine, internally homogeneous ferromagnetic particle has a domain structure that may be characterized by a uniform vector magnetization $\mathbf{M} = M_s \mathbf{r}$. The quantity M_s, termed the *spontaneous magnetization*, is determined by the material and temperature. The unit vector $\mathbf{r}$, called the

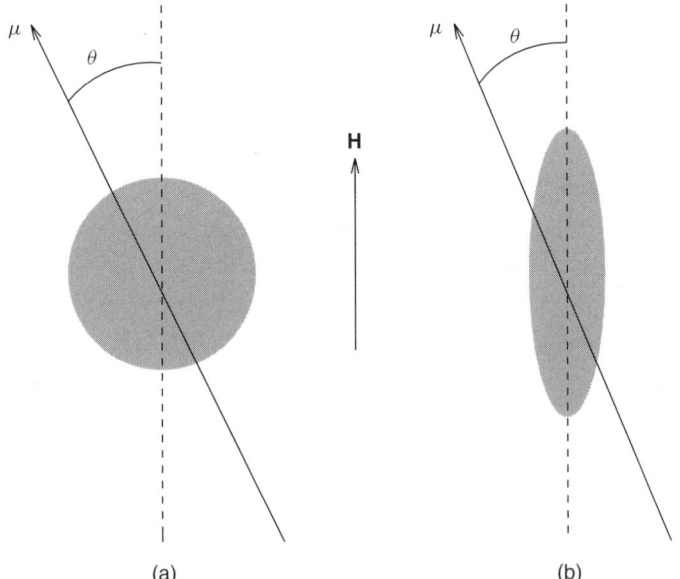

(a)   (b)

FIGURE 5. (*a*) Isotropic single-domain particle; (*b*) single-domain particle with uniaxial shape anisotropy.

magnetization orientation, is determined by the crystalline anisotropy, shape anisotropy, and external magnetic fields. In addition to the externally applied field, other considerations may include interparticle interactions. The externally applied field is assumed to be a constant uniform field $\mathbf{H} = H\mathbf{h}$ of magnitude H and direction $\mathbf{h}$. In such a system of particles, the behavior of the magnetization orientation $\mathbf{r}$ is treated as a Markov process, and the resulting probability distribution $W(\mathbf{r}, t)$ is called the *distribution of magnetization orientations*.

In thermal equilibrium, the system is treated as a Gibbs canonical ensemble, and from statistical mechanics, the equilibrium distribution is given by

$$W(\mathbf{r}) = A_0 e^{-\beta V(\mathbf{r})} \tag{1.2}$$

where $\beta = v/kT$, v is the magnetic particle volume, k is Boltzmann's constant, and T is the absolute temperature, A_0 is a normalization constant, and V is the Gibbs energy per unit volume of the system. The stable magnetization orientations are then the local minima of the free energy. We suppose that V is expressed in spherical polar-angle coordinates (as illustrated by Fig. 6), and V is said to be nonaxially symmetric when there is an explicit dependence on the azimuthal angle φ. However, V is said to be axially symmetric when there is no azimuthal (or φ) dependence. One of the most significant nonaxially symmetric cases is when V arises from uniaxial [2] anisotropy in the presence of a constant uniform external magnetic field at an oblique angle $\psi \neq 0, \pi$ to the easy axis, then

$$V = K(1 - \mathbf{r} \cdot \mathbf{e}_3^2) - HM_s\mathbf{r} \cdot \mathbf{h} \tag{1.3}$$

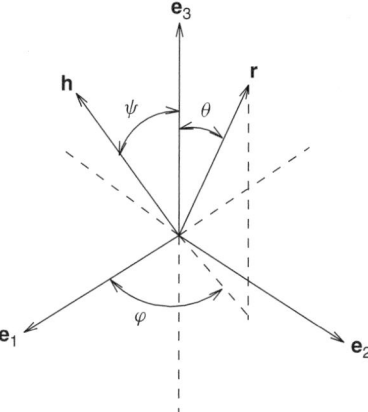

FIGURE 6. External field and magnetization orientations in terms of spherical polar coordinates.

where K is the anisotropy energy per unit volume and $\mathbf{e}_3$ denotes the direction of the uniaxial (i.e., the easy axis of magnetization) axis where

$$1 - \mathbf{r} \cdot \mathbf{e}_3^2 = 1 - \cos^2 \theta = \sin^2 \theta$$

and

$$\mathbf{r} \cdot \mathbf{h} = (\sin \theta \cos \varphi \mathbf{e}_1 + \sin \theta \sin \varphi \mathbf{e}_2 + \cos \theta \mathbf{e}_3)(\sin \psi \mathbf{e}_1 + \cos \psi \mathbf{e}_3)$$
$$= \sin \psi \sin \theta \cos \varphi + \cos \psi \cos \theta \qquad (1.4)$$

so that Eq. (1.3) becomes

$$V = K \sin^2 \theta - H M_s \sin \psi \sin \theta \cos \varphi - H M_s \cos \psi \cos \theta \qquad (1.5)$$

For a longitudinally applied field (i.e., a field applied parallel to the easy axis of magnetization, $\psi = 0$), Eq. (1.5) becomes

$$V = K \sin^2 \theta - H M_s \cos \theta \qquad (1.6)$$

For a transversely applied field (i.e., $\psi = \pi/2$), Eq. (1.5) may be written as

$$V = K \sin^2 \theta - H M_s \sin \theta \cos \varphi \qquad (1.7)$$

If K and H are sufficiently large, then the dependence on the azimuthal angle φ may be ignored (results of numerical computations in Ref. 14 provide evidence for the basis of such an assumption). Equation (1.7) can be approximated by

$$V = K \sin^2 \theta - H M_s \sin \theta \qquad (1.8)$$

Those cases presented by Eqs. (1.6) and (1.8) are termed *axially symmetric*, while all cases represented by Eq. (1.5) are said to be *nonaxially symmetric*. We suppose that V_1 and V_2 denote the values of V at local minima (stable magnetization orientations) that are separated by a local maximum, where the value of V is denoted by V_0 (see Fig. 7). Then, we say that the stable magnetization orientations are separated by energy barriers having barrier heights

$$v(V_0 - V_i), \qquad i = 1, 2 \qquad (1.9)$$

D. Relaxation Mechanisms

For a system that is not in thermal equilibrium, the evolution, or relaxation of the nonequilibrium distribution $W(\mathbf{r}, t)$ (where $\mathbf{r}$ is the magnetization orientation vector and is determined by the uniaxial shape anisotropy as shown

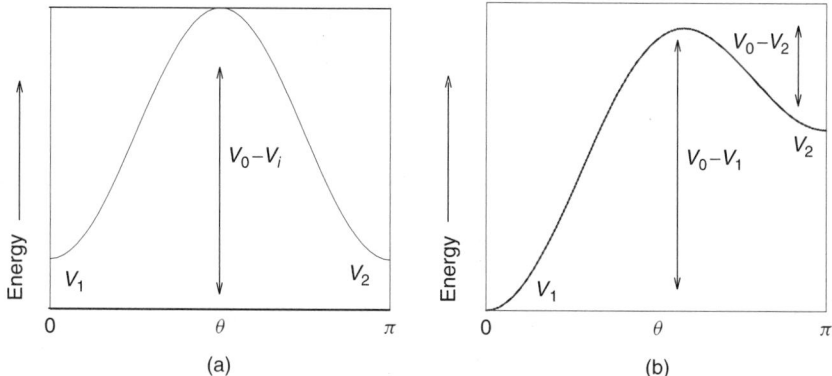

FIGURE 7. (*a*) Plot of V [as given in Eq. (1.6)] versus θ for a uniaxial anisotropy particle without field; (*b*) with a small field applied along $\theta = 0$.

in Fig. 5) into the equilibrium distribution $A_0 e^{-\beta V(\mathbf{r})}$ can be expressed as a sum of decaying exponentials. When the relaxation is dominated by a single exponential $e^{-t/\tau}$, then the corresponding time constant τ is referred to as the *relaxation time*. We consider two relaxation mechanisms:

- *Debye [15] Relaxation.* One method of approaching equilibrium in such a system is to physically rotate the individual particles, which is possible only if the particles are in a liquid medium as described by Elmore [8]. Using the mechanism of physical rotation, the factor determining the rate of approach to equilibrium is the viscosity of the liquid in which the particle suspension is held. The mechanism described here is known as *Debye* [15] *relaxation,* and has been used to treat the dielectric properties of solutions containing molecules with permanent electric dipole moments.

- *Néel [16] Relaxation.* Néel relaxation involves spontaneous Brownian-type rotations or jumps in the magnetization orientation within the particle due to thermal fluctuations, and occurs when the particles are of the order of 100–150 Å, so that the barrier heights in Eq. (1.9) are comparable to the thermal energy kT. The rate of approach to equilibrium will have an inverse exponential dependence on the ratio of the barrier height to thermal energy $\beta(V_0 - V_i)$.

However, for our purposes, it is assumed that the carrier is in a solid state throughout, and so the Debye [15] relaxation mechanism can be ignored. The magnitudes of the barrier heights give rise to the following cases:

1. *Ferromagnetic behavior* refers to the case when the barrier heights are very large in comparison with kT. When the temperature is not too low,

fine ferromagnetic particles exhibit this behavior, which accounts for the apparent stability of magnetization in many magnetic storage devices.

2. *Superparamagnetic behavior*, or thermal instability of magnetization, refers to the case of very small particles, where the barrier heights are small in comparison with kT. Thermal agitation causes continual changes in the magnetization orientation so that the system is constantly in thermal equilibrium. In the presence of a magnetic field, the system behaves like an ensemble of paramagnetic atoms displaying no hysterisis.

3. *Magnetic aftereffect behavior* refers to a narrow range of intermediate barrier heights. The system neither remains in a single state for a long time, nor attains thermal equilibrium in a short time so that the relaxation time is of the order of experimental times ($\approx 10^2$ seconds) [17].

E. The Discrete-Orientation Model

The behavior of the magnetization orientations may be described by three different models; the first of these is the discrete-orientation model. When the energy barriers are large in comparison with kT, but not so large as to completely prevent changes in the magnetization orientation **r** occurring, then we may assume that **r** is restricted to the stable orientations along the local minima of the free energy. The time behavior of **r** is treated as a discrete Markov process, and the distribution W is replaced by n_i, namely, the number of particles in the ith orientation. For a large number $n = \Sigma n_i$ of non-interacting particles, n_i changes with time in accordance with the master equation

$$\dot{n}_i = \sum_{i \neq j} (v_{j,i} n_j - v_{i,j} n_i) \tag{1.10}$$

where $v_{i,j}$ is the transition probability from orientation i to orientation j, that is, the probability of a particle in orientation i undergoing a transition to orientation j; $v_{i,j}$ will depend on the absolute temperature T, the anisotropy constant K, and the external field **H**.

For two orientations, as is the case for a uniaxial crystal with constant uniform external magnetic field, we let 1 refer to the positive orientation, and 2 the negative orientation so that Eq. (1.10) reduces to

$$\dot{n}_1 = -\dot{n}_2 = v_{2,1} n_2 - v_{1,2} n_1 \tag{1.11}$$

Setting $n_2 = n - n_1$ gives

$$\dot{n}_1 = -\dot{n}_2 = v_{2,1} n - (v_{1,2} + v_{2,1}) n_1 \tag{1.12}$$

so that n_1 approaches its final value according to the term $e^{-(v_{1,2}+v_{2,1})t}$, that is, with the time constant or relaxation (Néel [16] time)

$$\tau = (v_{1,2} + v_{2,1})^{-1} \tag{1.13}$$

It is normally assumed that

$$v_{i,j} = v_{i,j}^0 e^{-\beta(V_0 - V_i)} \tag{1.14}$$

where the $v_{i,j}^0$ vary slowly with temperature and so may be considered constant. When the barrier heights are equal, Eq. (1.13), using the assumption in Eq. (1.14), becomes

$$\tau = C e^{\beta(V_0 - V_1)} \tag{1.15}$$

where C is a constant.

F. The Néel Model

The Néel [16] model was originally applied to a system having uniaxial anisotropy that is initially in thermal equilibrium in the presence of a large magnetic field, which is then switched off at $t = 0$. There are two basic assumptions:

1. *Rigid-Coupling Assumption.* The spins of individual magnetic moments within the particle that contribute to the magnetization orientation of the particle are assumed to be rigidly coupled, and hence synchronous rotation of these spins occurs when the magnetization orientation is reversed.
2. *Discrete-Orientation Assumption.* When the energy barriers that oppose such reversals are large in comparison to the thermal energy kT, then the magnetization orientation, when treated as a stochastic variable, is assumed to adopt only a finite set of possible orientations, namely, the minima of the free energy [17].

Néel [16] then proposed the following formula for the relaxation time based on the discrete orientation approximation in Eq. (1.15):

$$\tau = \tau_N e^{\beta K} \tag{1.16}$$

The term τ_N is called the Néel [16] relaxation time and is given by

$$\tau_N = \frac{M_s}{3K\gamma |\lambda_s| \sqrt{\frac{\pi}{2\beta G}}} \tag{1.17}$$

where γ is the gyromagnetic ratio, λ_s is the empirical magnetization constant, and G is Young's modulus. τ_N is usually taken to be of the order of 10^{-9} s.

Néel's [16] discovery of the concept of thermal fluctuations in single-domain particles was initiated by the study of paleomagnetism since many rocks contain small amounts of magnetic material such as titanium iron oxides. By measuring the intensity and direction of remanent magnetization in many different types of rock, it was thought that such data would provide information on the strength and direction of the earth's magnetic field at the time the rock was formed. In 1956 Blackett [18] proposed that

> The detailed study of the natural magnetization in rocks is likely to allow us to trace back to the beginning of geological time, both the history of the earth's magnetic field, and the motion of the continental masses relative to one another and to the geographical pole.

However, the ability of the magnetic rocks to remember an earlier magnetic field depends on their ability to exist in thermal equilibrium with the earth's field at an early stage in their formation, and to later be frozen in a state of magnetization stable against later changes in the direction and strength of the field. In sedimentary rocks, the small grains of magnetic material can align themselves with the direction of the earth's field by physical rotation of the particles (in principle, Debye [15] relaxation), while the sediment is still wet. Later when the sediment becomes hard rock, the relaxation mechanism is lost and the magnetization is stable.

However, in igneous rocks, the material becomes magnetic by cooling through its Curie point. The rock therefore is already solid, and so physical rotation of the particles cannot occur. However, at high temperatures, energy barriers can be small in comparison with kT, and so single-domain particles may be able to reach thermal equilibrium by Néel [16] relaxation (rotation of the moment with respect to the particle) and will become stable when cooled through their "blocking temperatures."

When dealing with individual particles, it was found that the Néel [16] relaxation mechanism does not occur below a certain critical temperature, namely, the blocking temperature for particles with radii >25 Å. However, Bean and Livingston [10] suggested in their 1959 paper for particles <25 Å that it could be possible for their magnetization to tunnel quantally from one orientation to another down to a temperature of absolute zero. Experimental evidence for this idea was provided by Weil [19], who observed that particles of this dimension failed to reach their blocking temperature down to a temperature of around 1 K. However, the idea of quantum tunneling of magnetization is still under investigation.

G. Brown's Model

In his 1963 paper, Brown [20] criticized Neel's work on two counts:

1. The Néel model does not allow for particles spending time in interme-diate orientations before jumping, nor does it allow for jumps back to the initial stable orientation.
2. Néel's calculations did not allow the system to be treated as a gyromag-netic one.

In order to set up equations of motion describing the dynamic behavior of ferromagnetic particles, we begin by considering an uncompensated electron in a ferromagnetic particle. Such an electron possesses a magnetic moment μ and an angular momentum ω that arise from the motion of its charge e and mass m, respectively. The two are related by the equation

$$\mu = \gamma\omega \qquad (1.18)$$

where for spin electrons, $\gamma = -e/mc$ is the gyromagnetic ratio. In the presence of an effective magnetic field $\mathbf{H}_{eff}$, a torque given by $\mu \times \mathbf{H}_{eff}$ acts on each electron. Differentiating Eq. (1.18) gives the simple gyromagnetic equation

$$\dot{\mu} = \gamma\mu \times \mathbf{H}_{eff} \qquad (1.19)$$

since the above-mentioned torque ($\mu \times \mathbf{H}_{eff}$) equals the rate of change of angular momentum of the electron ($\dot{\omega}$). This simple gyromagnetic equation is the most basic equation describing the dynamic behavior of the magnetization of a single-domain ferromagnetic particle.

The first phenomenological equation of motion describing the average behavior of the magnetization vector $\mathbf{r}$ was put forward by Landau and Lifshitz [7] in 1935. If we have a single-domain ferromagnetic particle comprising many atoms, and having one easy axis of magnetization, then all such magnetic moments are assumed to be aligned so that

$$\dot{\mathbf{r}} = \gamma\mathbf{r} \times \mathbf{H}_{eff} \qquad (1.20)$$

(where the vector $\mathbf{r}$ represents the magnetization orientation). This equation represents the undamped precession of $\mathbf{r}$ about the axis of $\mathbf{H}_{eff}$. However, the observable behavior of the magnetization of a single-domain ferromagnetic particle is that of the alignment of $\mathbf{r}$ with $\mathbf{H}_{eff}$. With the aim of including this fact in the mathematical analysis, Landau and Lifshitz [7] introduced a second term into the equation, the tendency of which is to align $\mathbf{r}$ with $\mathbf{H}_{eff}$. Gilbert [21] then in 1955 proposed that an effective damping field term be introduced into the equation, thus giving

$$\dot{\mathbf{r}} = \gamma\mathbf{r} \times (\mathbf{H}_{eff} - \eta M_s\dot{\mathbf{r}}) \qquad (1.21)$$

where η is a phenomenological damping constant and M_s is the spontaneous magnetization. If we take the scalar product of this equation with $\mathbf{r}$, then we arrive at the expression

$$\dot{\mathbf{r}} \cdot \mathbf{r} = \gamma(\mathbf{r} \times \mathbf{H}_{\text{eff}}) \cdot \mathbf{r} - \eta\gamma M_s(\mathbf{r} \times \dot{\mathbf{r}}) \cdot \mathbf{r} = 0 \qquad (1.22)$$

If we then transpose η in Eq. (1.22) to the left-hand side, and take the vector product with $\mathbf{r}$, we obtain

$$\dot{\mathbf{r}} \times \mathbf{r} + \gamma\eta M_s(\mathbf{r} \times \dot{\mathbf{r}}) \times \mathbf{r} = \gamma(\mathbf{r} \times \mathbf{H})_{\text{eff}} \times \mathbf{r} \qquad (1.23)$$

By expanding the vector triple product on the left hand side of Eq. (1.23), and applying Eq. (1.22), we obtain

$$(\mathbf{r} \times \dot{\mathbf{r}}) \times \mathbf{r} = -(\dot{\mathbf{r}} \cdot \mathbf{r})\mathbf{r} + \dot{\mathbf{r}} = \dot{\mathbf{r}} \qquad (1.24)$$

Using Eq. (1.24), Eq. (1.23) becomes

$$\dot{\mathbf{r}} \times \mathbf{r} = -\gamma\eta M_s\dot{\mathbf{r}} + \gamma(\mathbf{r} \times \mathbf{H})_{\text{eff}} \times \mathbf{r} \qquad (1.25)$$

Then Gilbert's [21] equation (1.21) can be rewritten as

$$\dot{\mathbf{r}} \times \mathbf{r} = (\gamma\eta M_s)^{-1}(\dot{\mathbf{r}} - \gamma\mathbf{r} \times \mathbf{H}_{\text{eff}}). \qquad (1.26)$$

By equating the left-hand sides of Eqs. (1.25) and (1.26), we obtain

$$(\gamma\eta M_s)^{-1}(\dot{\mathbf{r}} - \gamma\mathbf{r} \times \mathbf{H}_{\text{eff}}) + \gamma\eta M_s\dot{\mathbf{r}} = \gamma(\mathbf{r} \times \mathbf{H})_{\text{eff}} \times \mathbf{r} \qquad (1.27)$$

and solving for $\dot{\mathbf{r}}$ provides the explicit form of Gilbert's [21] equation

$$\dot{\mathbf{r}} = \frac{b}{a}M_s\mathbf{r} \times \mathbf{H}_{\text{eff}} + bM_s(\mathbf{r} \times \mathbf{H}_{\text{eff}}) \times \mathbf{r} \qquad (1.28)$$

where

$$b = \frac{\gamma a}{(1 + a^2)M_s} \qquad (1.29)$$

and

$$a = \eta\gamma M_s \qquad (1.30)$$

Using the assumption $a^2 \cong 0$, Eq. (1.28) reduces to the earlier Landau–Lifshitz [7] equation (1.20). The term containing b/a will be referred to as the gyromagnetic term from this point forward. The term containing b will be referred to as the *alignment term*. The effective magnetic field can be expressed in terms of the free energy per unit volume V [17] by

$$\mathbf{H}_{\text{eff}} = -\frac{1}{M_s}\frac{\partial V}{\partial \mathbf{r}} \tag{1.31}$$

However, in 1963, Brown [20] extended the equations of motion (initiated by Landau and Lifshitz [7]) to describe not only the average but also the specific behavior of the magnetization of a single-domain ferromagnetic particle. By supposing that the effective damping field ($\eta M_s \dot{\mathbf{r}}$) describes only the statistical average of the rapidly fluctuating random forces due to thermal agitation, Brown [20] suggested that for an individual particle, the expression must be augmented by a random field term $\mathbf{h}(t)$, so that the expression becomes $\eta M_s \dot{\mathbf{r}} + \mathbf{h}(t)$, where

$$\langle \mathbf{h}(t) \rangle = 0, \qquad \langle h_i(t_1)h_j(t_2) \rangle = 2\eta\beta^{-1}\delta_{i,j}\delta(t_1 - t_2) \tag{1.32}$$

where the braces $\langle \# \rangle$ denote the statistical average over a large number of particles that have all started with the same orientation in configuration space and the $h_i(t)$ denote the rectangular components of $\mathbf{h}(t)$. It is assumed that the components obey the theorem of Isserlis (see Ref. 22)

$$\langle h(t_1) \cdots h(t_{2n+1}) \rangle = 0$$

$$\langle h(t_1) \cdots h(t_{2n}) \rangle = \sum \prod_{k_i < k_j} \langle h(t_{k_i})h(t_{k_j}) \rangle \tag{1.33}$$

where the sum is over all such products, each of which is formed by selecting n pairs from $2n$ time points. For example, if $n = 2$ [17] we have

$$\langle h(t_1)h(t_2)h(t_3)h(t_4) \rangle = \langle h(t_1)h(t_2)h(t_3)h(t_4) \rangle + \langle h(t_1)h(t_3)h(t_2)h(t_4) \rangle$$

$$+ \langle h(t_1)h(t_4)h(t_2)h(t_3) \rangle \tag{1.34}$$

Brown [4] was then able to derive the Fokker–Planck equation for the distribution of magnetization orientations using the methods of Wang and Uhlenbeck [23] on the basis of the assumption that we have just made. He also presented an alternative approach based on an argument of Einstein [24] by substituting Eq. (1.31) for the effective magnetic field into the explicit form of Gilbert's [21] equation so that we have

$$\dot{\mathbf{r}} = -\frac{b}{a}\mathbf{r} \times \frac{\partial V}{\partial \mathbf{r}} - b\left(\mathbf{r} \times \frac{\partial V}{\partial \mathbf{r}}\right) \times \mathbf{r} \tag{1.35}$$

Since $\partial V/\partial \mathbf{r}$ is perpendicular to $\mathbf{r}$, we have

$$\mathbf{r}\frac{\partial V}{\partial \mathbf{r}} = 0 \tag{1.36}$$

If we expand the vector triple product on the right-hand side of Eq. (1.35), namely

$$\left(\mathbf{r} \times \frac{\partial V}{\partial \mathbf{r}}\right) \times \mathbf{r} = \frac{\partial V}{\partial \mathbf{r}} - \left(\mathbf{r}\frac{\partial V}{\partial \mathbf{r}}\right)\mathbf{r} = \frac{\partial V}{\partial \mathbf{r}} \tag{1.37}$$

then Eq. (1.35) becomes

$$\dot{\mathbf{r}} = -\frac{b}{a}\mathbf{r} \times \frac{\partial V}{\partial \mathbf{r}} - b\frac{\partial V}{\partial \mathbf{r}} \tag{1.38}$$

Brown [4] based his work on the assumption that the individual magnetization orientations in a system of magnetic particles can be treated as a current of representative points moving around the surface of a unit sphere having a number density of $W\,(\mathbf{r}, t)$, and current density $\mathbf{J}\,(\mathbf{r}, t)$. Since such representative points can be neither created nor destroyed, then W and $\mathbf{J}$ satisfy the continuity equation

$$\dot{W} = -\left(\frac{\partial}{\partial \mathbf{r}}\right)\mathbf{J} \tag{1.39}$$

The representative points that are concentrated around the stable magnetization orientations are then dispersed by the influence of the random thermal forces. This may be represented mathematically by the inclusion of a diffusion term of the form $-\kappa'(\partial W/\partial \mathbf{r})$, where $\kappa' > 0$ is a constant at a given temperature. The current density $\mathbf{J}$ may then be written

$$\begin{aligned}
\mathbf{J} &= W\dot{\mathbf{r}} - \kappa'\frac{\partial W}{\partial \mathbf{r}} \\
&= -\frac{b}{a}W\mathbf{r} \times \frac{\partial V}{\partial \mathbf{r}} - bW\frac{\partial V}{\partial \mathbf{r}} - k'\frac{\partial W}{\partial \mathbf{r}}
\end{aligned} \tag{1.40}$$

Using this notation for the current density $\mathbf{J}$, Eq. (1.39) then becomes

$$\dot{W} = \frac{b}{a}\frac{\partial}{\partial \mathbf{r}}\left(W\mathbf{r} \times \frac{\partial V}{\partial \mathbf{r}}\right) + b\frac{\partial}{\partial \mathbf{r}}\left(W\frac{\partial V}{\partial \mathbf{r}}\right) + k'\frac{\partial^2 W}{\partial \mathbf{r}^2} \tag{1.41}$$

Using the product rule, the derivative in the gyroscopic term becomes

$$\frac{\partial}{\partial \mathbf{r}}\left(W\mathbf{r} \times \frac{\partial V}{\partial \mathbf{r}}\right) = \frac{\partial W}{\partial \mathbf{r}}\left(\mathbf{r} \times \frac{\partial V}{\partial \mathbf{r}}\right) + W\frac{\partial}{\partial \mathbf{r}}\left(\mathbf{r} \times \frac{\partial V}{\partial \mathbf{r}}\right) \tag{1.42}$$

The first term in Eq. (1.41) can be rewritten as

$$\frac{b}{a}\mathbf{r}\left(\frac{\partial V}{\partial \mathbf{r}} \times \frac{\partial W}{\partial \mathbf{r}}\right) \tag{1.43}$$

If we again apply the scalar triple product formula to the alignment term in Eq. (1.42), we obtain

$$\frac{\partial}{\partial \mathbf{r}}\left(\mathbf{r} \times \frac{\partial V}{\partial \mathbf{r}}\right) = \frac{\partial V}{\partial \mathbf{r}}\left(\frac{\partial}{\partial \mathbf{r}} \times \mathbf{r}\right) - \mathbf{r}\left(\frac{\partial}{\partial \mathbf{r}} \times \frac{\partial V}{\partial \mathbf{r}}\right) = 0 \tag{1.44}$$

Equation (1.41) may then be rewritten as

$$\dot{W} = \frac{b}{a}\mathbf{r}\left(\frac{\partial V}{\partial \mathbf{r}} \times \frac{\partial W}{\partial \mathbf{r}}\right) + b\frac{\partial}{\partial \mathbf{r}}\left(W\frac{\partial V}{\partial \mathbf{r}}\right) + k'\frac{\partial^2 W}{\partial \mathbf{r}^2} \tag{1.45}$$

This is the Fokker–Planck equation for the distribution of magnetization orientations. In thermal equilibrium, $\dot{W} = 0$, and so W reduces to the equilibrium distribution

$$W_0 = Ae^{-\beta V} \tag{1.46}$$

where $\beta = v/kT$ is the ratio of the magnetic particle volume to thermal energy. If we substitute this into Eq. (1.45), we obtain

$$-\frac{\beta b}{a}e^{-\beta V}\mathbf{r}\left(\frac{\partial V}{\partial \mathbf{r}} \times \frac{\partial V}{\partial \mathbf{r}}\right) + (b - \beta k')\frac{\partial}{\partial \mathbf{r}}\left(e^{-\beta V}\frac{\partial V}{\partial \mathbf{r}}\right) = 0 \tag{1.47}$$

The gyroscopic term vanishes, and since the prefactor in the final term is nonzero, then

$$k' = \frac{b}{\beta} \tag{1.48}$$

and so Eq. (1.45) becomes

$$\dot{W} = \frac{b}{a}\mathbf{r}\left(\frac{\partial V}{\partial \mathbf{r}} \times \frac{\partial W}{\partial \mathbf{r}}\right) + b\frac{\partial}{\partial \mathbf{r}}\left(W\frac{\partial V}{\partial \mathbf{r}}\right) + b\beta^{-1}\frac{\partial^2 W}{\partial \mathbf{r}^2} \tag{1.49}$$

It is more convenient to express this in terms of the Fokker–Planck operator L_{FP}, namely

$$\dot{W} = L_{\mathrm{FP}}W \tag{1.50}$$

where

$$L_{\mathrm{FP}}W = \frac{b}{a}\mathbf{r}\left(\frac{\partial V}{\partial \mathbf{r}} \times \frac{\partial W}{\partial \mathbf{r}}\right) + b\frac{\partial}{\partial \mathbf{r}}\left(W\frac{\partial V}{\partial \mathbf{r}}\right) + b\beta^{-1}\frac{\partial^2 W}{\partial \mathbf{r}^2} \tag{1.51}$$

is the Fokker–Planck operator. The gradient, divergence, and Laplacian in spherical polar coordinates r, θ, φ are

$$\frac{\partial A}{\partial \mathbf{r}} = \frac{\partial A}{\partial \theta}\bar{\mathbf{e}}_1 + \frac{1}{\sin\theta}\frac{\partial A}{\partial \varphi}\bar{\mathbf{e}}_2 \tag{1.52}$$

$$\frac{\partial \mathbf{A}}{\partial \mathbf{r}} = \frac{1}{\sin\theta}\left(\frac{\partial}{\partial \theta}(\sin\theta A_1) + \frac{\partial A_2}{\partial \varphi}\right) \tag{1.53}$$

$$\frac{\partial^2 A}{\partial \mathbf{r}^2} = \nabla^2 A = \frac{1}{\sin\theta}\frac{\partial}{\partial \theta}\left(\sin\theta\frac{\partial A}{\partial \theta}\right) + \frac{1}{\sin^2\theta}\frac{\partial^2 A}{\partial \varphi^2} \tag{1.54}$$

where $\mathbf{A} = A_1\bar{\mathbf{e}}_1 + A_2\bar{\mathbf{e}}_2$, and $\bar{\mathbf{e}}_1, \bar{\mathbf{e}}_2$ are orthogonal vectors on the surface of the unit sphere (refer to Appendix A). The gyroscopic term then becomes

$$\frac{b}{a}\mathbf{r}\cdot\left(\frac{\partial V}{\partial \mathbf{r}} \times \frac{\partial W}{\partial \mathbf{r}}\right) = \frac{b}{a}\mathbf{r}\cdot\begin{vmatrix} \bar{\mathbf{e}}_1 & \bar{\mathbf{e}}_2 & \bar{\mathbf{e}}_3 \\ \dfrac{\partial V}{\partial \theta} & \dfrac{1}{\sin\theta}\dfrac{\partial V}{\partial \varphi} & 0 \\ \dfrac{\partial W}{\partial \theta} & \dfrac{1}{\sin\theta}\dfrac{\partial W}{\partial \varphi} & 0 \end{vmatrix}$$

$$= \frac{b}{a\sin\theta}\left(\frac{\partial V}{\partial \theta}\frac{\partial W}{\partial \varphi} - \frac{\partial V}{\partial \varphi}\frac{\partial W}{\partial \theta}\right)\mathbf{r}\cdot\bar{\mathbf{e}}_3$$

$$= \frac{b}{a\sin\theta}\left(\frac{\partial V}{\partial \theta}\frac{\partial W}{\partial \varphi} - \frac{\partial V}{\partial \varphi}\frac{\partial W}{\partial \theta}\right) \tag{1.55}$$

and the alignment term becomes

$$b\frac{\partial}{\partial \mathbf{r}}\left(W\frac{\partial V}{\partial \mathbf{r}}\right) = b\frac{\partial}{\partial \mathbf{r}}\left(W\frac{\partial V}{\partial \theta}\bar{\mathbf{e}}_1 + \frac{W}{\sin\theta}\frac{\partial V}{\partial \varphi}\bar{\mathbf{e}}_2\right)$$

$$= \frac{b}{\sin\theta}\frac{\partial}{\partial \theta}\left(\sin\theta W\frac{\partial V}{\partial \theta}\right) + \frac{b}{\sin^2\theta}\frac{\partial}{\partial \varphi}\left(W\frac{\partial V}{\partial \varphi}\right)$$

$$= bW\nabla^2 V + b\left(\frac{\partial V}{\partial \theta}\frac{\partial W}{\partial \theta} + \frac{1}{\sin^2\theta}\frac{\partial V}{\partial \varphi}\frac{\partial W}{\partial \varphi}\right) \tag{1.56}$$

Using Eqs. (1.55) and (1.56), the Fokker–Planck equation in Eq. (1.51) is then

$$L_{FP}W = \beta^{-1}b\nabla^2 W + bW\nabla^2 V + b\left(\frac{\partial V}{\partial\theta}\frac{\partial W}{\partial\theta} + \frac{1}{\sin^2\theta}\frac{\partial V}{\partial\varphi}\frac{\partial W}{\partial\varphi}\right)$$

$$+ \frac{b}{a\sin\theta}\left(\frac{\partial V}{\partial\theta}\frac{\partial W}{\partial\varphi} - \frac{\partial V}{\partial\varphi}\frac{\partial W}{\partial\theta}\right) \tag{1.57}$$

and so Eq. (1.49) becomes

$$\dot{W} = \beta^{-1}b\nabla^2 W + bW\nabla^2 V + b\left(\frac{\partial V}{\partial\theta}\frac{\partial W}{\partial\theta} + \frac{1}{\sin^2\theta}\frac{\partial V}{\partial\varphi}\frac{\partial W}{\partial\varphi}\right)$$

$$+ \frac{b}{a\sin\theta}\left(\frac{\partial V}{\partial\theta}\frac{\partial W}{\partial\varphi} - \frac{\partial V}{\partial\varphi}\frac{\partial W}{\partial\theta}\right) \tag{1.58}$$

It is assumed that the Fokker–Planck equation, which may now be written in the form

$$\frac{\partial W}{\partial t} = L_{FP}W \tag{1.59}$$

has a solution of the form

$$W(\mathbf{r}, t) = T(t)F(\mathbf{r}) \tag{1.60}$$

It can be shown that the general solution to the Fokker–Planck equation may be represented by

$$W(\mathbf{r}, t) = W_0(\mathbf{r}) + \sum_{n=1}^{\infty} A_n F_n(\mathbf{r})e^{-p_n t} \tag{1.61}$$

The terms $F_n(\mathbf{r})$ satisfy Eq. (1.61) with $\partial/\partial t$ replaced by $-p_n$; p_n are the eigenvalues and F_n the corresponding eigenfunctions, which are determined by the requirement of single-valuedness and finiteness. The equilibrium solution $W_0 = Ae^{-\beta V}$ is the eigenfunction corresponding to the eigenvalue $p_0 = 0$ and $F_0(\mathbf{r}) = e^{-\beta V(\mathbf{r})}$, which is obtained by the requirements of statistical equilibrium and $\beta = v/kT$. The value of A_0 is determined by the normalisation of $\int W d\Omega$, where $d\Omega$ is the solid angle $\sin\theta\, d\theta\, d\varphi$, and A_n, $n \geq 1$ are determined by the initial values of W.

We suppose that the external magnetic field is represented by $H + H_1$ when $t < 0$ and simply by H when $t > 0$ where H_1 is a perturbing field such that $H_1 \ll H$, which is switched off at $t = 0$, and switched on at $t = -\infty$.

The system is assumed to be in thermal equilibrium at $t = 0, \infty$ so that the distribution of magnetization orientations may be given by

$$W(\mathbf{r}, 0) = W_1(\mathbf{r}) = A_0' e^{-\beta V(\mathbf{r}) + \xi_1 \mathbf{r} \cdot \mathbf{h}}$$

$$W(\mathbf{r}, \infty) = W_0(\mathbf{r}) = A_0 e^{-\beta V(\mathbf{r})} \tag{1.62}$$

where $\xi_1 = \beta H_1 / M_s$. The corresponding solution of the Fokker–Planck equation is then referred to as the *aftereffect solution*, and may be represented as

$$W(\mathbf{r}, t) = e^{-\beta V(\mathbf{r})}[1 + \xi_1 \alpha(t)] + O(\xi_1^2) \tag{1.63}$$

where $\alpha(t)$ decays from an initial value $\alpha(0) = \mathbf{r} \cdot \mathbf{h}$ to a final value $\alpha(\infty) = 0$. The stochastic variable U then has an expectation value

$$\langle U \rangle = \frac{\int U W d\Omega}{\int W d\Omega} \tag{1.64}$$

and $d\Omega$ denotes the element of solid angle, and the integration is over the unit sphere. We allow

$$\langle U \rangle_1 = \frac{\int U e^{-\beta V(\mathbf{r}) + \xi_1 \mathbf{r} \cdot \mathbf{h}} d\Omega}{\int e^{-\beta V(\mathbf{r}) + \xi_1 \mathbf{r} \cdot \mathbf{h}} d\Omega}$$

$$\langle U \rangle_0 = \frac{\int U e^{-\beta V(\mathbf{r})} d\Omega}{\int e^{-\beta V(\mathbf{r})} d\Omega} \tag{1.65}$$

to denote the expectation values in the presence and absence of the perturbing field, respectively. The equilibrium expectation value $\langle \mathbf{r} \cdot \mathbf{h} \rangle_1$ may be expressed in terms of $\langle \mathbf{r} \cdot \mathbf{h} \rangle_0$ by applying the linear approximation in which terms of $O(\xi_1^2)$ are ignored, and so

$$\langle \mathbf{r} \cdot \mathbf{h} \rangle_1 = \frac{\int \mathbf{r} \cdot \mathbf{h} e^{-\beta V + \xi_1 \mathbf{r} \cdot \mathbf{h}} d\Omega}{\int e^{-\beta V + \xi_1 \mathbf{r} \cdot \mathbf{h}} d\Omega} \approx \frac{\int \mathbf{r} \cdot \mathbf{h} e^{-\beta V}(1 + \xi_1 \mathbf{r} \cdot \mathbf{h}) d\Omega}{\int e^{-\beta V}(1 + \xi_1 \mathbf{r} \cdot \mathbf{h}) d\Omega}$$

$$= \frac{\langle \mathbf{r} \cdot \mathbf{h} \rangle_0 + \xi_1 \langle \mathbf{r} \cdot \mathbf{h}^2 \rangle_0}{1 + \xi_1 \langle \mathbf{r} \cdot \mathbf{h} \rangle_0} \approx (1 - \xi_1 \langle \mathbf{r} \cdot \mathbf{h} \rangle_0)(\langle \mathbf{r} \cdot \mathbf{h} \rangle_0 + \xi_1 \langle \mathbf{r} \cdot \mathbf{h} \rangle_0)$$

$$\approx \langle \mathbf{r} \cdot \mathbf{h} \rangle_0 + \xi_1(\langle \mathbf{r} \cdot \mathbf{h}^2 \rangle_0 - \langle \mathbf{r} \cdot \mathbf{h} \rangle_0^2) \tag{1.66}$$

This is often referred to as the *static case*, in that it avoids any considerations concerning the time dependence. The dynamic case is accomplished by using the formal solution described by Eq. (1.63)

$$\langle \mathbf{r} \cdot \mathbf{h} \rangle \approx \frac{\int \mathbf{r} \cdot \mathbf{h} e^{-\beta V[1+\xi_1 \alpha(t)]} d\Omega}{\int e^{-\beta V}[1 + \xi_1 \alpha(t)] d\Omega} = \frac{\langle \mathbf{r} \cdot \mathbf{h} \rangle_0 + \xi_1 \langle \alpha(t) \mathbf{r} \cdot \mathbf{h} \rangle_0}{1 + \xi_1 \langle \alpha(t) \rangle_0}$$

$$\approx (\langle \mathbf{r} \cdot \mathbf{h} \rangle_0 + \xi_1 \langle \alpha(t) \mathbf{r} \cdot \mathbf{h} \rangle_0)(1 - \xi_1 \langle \alpha(t) \rangle_0)$$

$$\approx \langle \mathbf{r} \cdot \mathbf{h} \rangle_0 + \xi_1 (\langle \alpha(t) \mathbf{r} \cdot \mathbf{h} \rangle_0 - \langle \alpha(t) \rangle_0 \langle \mathbf{r} \cdot \mathbf{h} \rangle_0) \qquad (1.67)$$

The aftereffect function is then

$$f(t) = \langle \mathbf{r} \cdot \mathbf{h} \rangle - \langle \mathbf{r} \cdot \mathbf{h} \rangle_0 = \xi_1 (\langle \alpha(t) \alpha(0) \rangle_0 - \langle \alpha(t) \rangle_0 \langle \alpha(0) \rangle_0)$$

$$= \xi_1 \mathrm{Cov}(\alpha(t), \alpha(0)) \qquad (1.68)$$

In particular

$$f(0) = \langle \mathbf{r} \cdot \mathbf{h} \rangle_1 - \langle \mathbf{r} \cdot \mathbf{h} \rangle_0 = \xi_1 (\langle \alpha^2(0) \rangle_0 - \langle \alpha(0) \rangle_0^2)$$

$$= \xi_1 \mathrm{Var}(\alpha(0)) \qquad (1.69)$$

The autocorrelation function is then defined as

$$C(t) = \frac{\mathrm{Cov}(\alpha(t), \alpha(0))}{\mathrm{Var}(\alpha(0))} = \frac{f(t)}{f(0)} \qquad (1.70)$$

and the correlation time is the area under the curve of the autocorrelation function

$$T = \int_0^\infty C(t) dt \qquad (1.71)$$

Assuming that Eq. (1.68) has the time dependence

$$f(t) = \sum_{n=0}^\infty a_n e^{-p_n t} \qquad (1.72)$$

Eqs. (1.70) and (1.71) then give

$$T = \frac{\sum_{n=0}^\infty a_n / p_n}{\sum_{n=0}^\infty a_n} \qquad (1.73)$$

If, in addition to Eq. (1.72), the time dependence of Eq. (1.68) is assumed to be

$$f(t) = f(0) e^{-p_{\mathrm{eff}} t} \qquad (1.74)$$

then p_{eff} is called the *effective eigenvalue*. Differentiation gives

$$p_{\text{eff}} = -\frac{\dot{f}(0)}{f(0)} = \frac{\sum_{n=0}^{\infty} a_n}{\sum_{n=0}^{\infty} p_n a_n} \tag{1.75}$$

If it is assumed that the smallest nonvanishing eigenvalue satisfies $p_1 \ll p_n$, $n \geq 2$, and the corresponding amplitude $A_1 \gg A_n$, $n \geq 2$, then

$$T, \tau_{\text{eff}} \approx \tau = \frac{1}{p_1} \tag{1.76}$$

The variable τ corresponds to the time constant associated with the longest-lived mode in Eq. (1.61) and will be referred to as the *relaxation time*. Suppose that the vectors $\mathbf{r}$ and $\mathbf{h}$ are represented in terms of orthogonal unit vectors $\{\mathbf{e}_1, \mathbf{e}_2, \mathbf{e}_3\}$, where $\mathbf{e}_3$ is parallel to the easy axis; then [cf. Eq. (1.4)]

$$\langle \mathbf{r} \cdot \mathbf{h} \rangle = \langle (\sin\theta\cos\varphi\mathbf{e}_1 + \sin\theta\sin\varphi\mathbf{e}_2 + \cos\theta\mathbf{e}_3)(\sin\psi\mathbf{e}_1 + \cos\psi\mathbf{e}_3) \rangle$$

$$= \cos\psi\langle\cos\theta\rangle + \sin\psi\langle\sin\theta\cos\varphi\rangle \tag{1.77}$$

As well as determining the relaxation time τ, it is also important to determine the dependence of the correlation time T_c on ψ. The values of the correlation time in the special limiting cases where $\psi = 0$, $\pi/2$ are denoted by $T_{\|}$ and $T_{\perp}$, respectively. These are addressed further in the following sections, and are referred to as the *longitudinal* and *transverse* correlation times (or relaxation times in the case of τ). The next section addresses the particular case of axial symmetry.

H. Solutions for Axial Symmetry

For the moment, only the case of axial symmetry is considered, and some of the results that have been presented by previous researchers are discussed. When the external field is applied parallel to the easy axis, then it is assumed that both the Gibbs free energy per unit volume V and the distribution of magnetization orientations W are axially symmetric [i.e., V and W are both independent of φ and $(\partial V/\partial\varphi) = 0$, $(\partial W/\partial\varphi) = 0$]. In fact, initial expressions for the relaxation rate τ^{-1} of the magnetization of a uniaxially anisotropic single-domain ferromagnetic particle were first presented by Néel [16] for the case where the Gibbs free energy is given by

$$V(\theta) = K \sin^2\theta \tag{1.78}$$

and also for the case when

$$V(\theta) = K \sin^2\theta - HM_s \cos\theta \tag{1.79}$$

When V and W are axially symmetric, the Fokker–Planck equation in Eq. (1.59) reduces to

$$\frac{\partial W}{\partial t} = \frac{b}{\sin\theta}\frac{\partial}{\partial\theta}\left[\sin\theta\left(\frac{\partial V}{\partial\theta}W + \beta^{-1}\frac{\partial W}{\partial\theta}\right)\right] \tag{1.80}$$

and so the gyromagnetic term is dropped from the equation. The current density represented by Eq. (1.41) when considering the longitudinal component only, reduces to

$$
\begin{aligned}
J_\theta &= -bW\frac{\partial V}{\partial\theta} - k'\frac{\partial W}{\partial\theta} \\
&= -b\left(\frac{\partial V}{\partial\theta}W + \beta^{-1}\frac{\partial W}{\partial\theta}\right)
\end{aligned}
\tag{1.81}
$$

since $k' = b/\beta$.

If we take as an independent variable $x = \cos\theta$, then the general solution to the simplified form of the Fokker–Planck Eq. (1.80) may be written as

$$W(x, t) = A_0 + \sum_{n=1}^{\infty} A_n F_n(x)e^{-p_n t} \tag{1.82}$$

with

$$W_0 = A_0 e^{-\beta V} \tag{1.83}$$

where F_n are the eigenfunctions and A_n the corresponding amplitudes. Equation (1.80) then takes the form

$$\frac{\partial W}{\partial t} = \frac{\partial}{\partial x}\left[(1 - x^2)\left(\frac{\partial V}{\partial x}W + \beta^{-1}\frac{\partial W}{\partial x}\right)\right] \tag{1.84}$$

When $\partial W/\partial t = 0$, this last equation can be integrated directly, and the differential equation satisfied by F_n may be expressed as the Sturm–Liouville problem

$$\frac{d}{dx}\left[(1 - x^2)e^{-\beta V}\frac{d}{dx}(e^{\beta V}F_n)\right] + \lambda_n F_n = 0 \tag{1.85}$$

where $\lambda_n = \beta p_n/b$ are determined by the requirement that F_n must be finite at $x = \pm 1$. The lowest eigenvalue is $\lambda = 0$, and it corresponds to the solution $F_0 = $ constant. It can be stated, then, that Brown's [4] Sturm–Liouville equation (1.85) approximately describes the behavior of the fluctuating magnetic moment of a single-domain ferromagnetic particle with uniaxial anisotropy, having a field applied parallel to the easiest axis of magnetization. λ_n provides a description of the relaxation time τ, since from Eq. (1.82), the individual decay rates p_n of the distribution function are given by $\lambda_n/2\tau_n$. Brown [4] therefore formulated a procedure for obtaining τ that is valid for all barrier heights ($\sigma = \beta K$), but did not actually calculate the range of eigenvalues of the equation for cases other than $\sigma = 0$. Instead he based his results on the assumption that

$$\tau = \frac{2\tau_n}{\lambda_1} \tag{1.86}$$

and obtained approximations for λ_1 in the limit of low and high energy barriers.

I. Brown's Low-Energy-Barrier Approximation for Axial Symmetry

Brown [4] devised a low-energy-barrier approximation for situations where the ratio of the barrier height to thermal energy is small, and when the Gibbs free energy per unit volume is as described by Eq. (1.79) (i.e., for a longitudinally applied field). In his low-energy-barrier approximation, Brown therefore used the assumption that $\beta V \ll 1$. In the limit $\beta \to 0$, Eq. (1.85) becomes Legendre's equation with eigenvalues $\lambda_n = n(n+1)$, $n = 0, 1, 2, 3, \ldots$. The eigenvalue 0 corresponds to the equilibrium solution (simply $W = $ constant), and the other eigenvalues determine the reciprocal time constants p_n, which are the integral multiples of b/β. Using the methods of perturbation theory, Brown derived the following expression for λ_1 as the first few terms in a series in a small parameter σ

$$\lambda_1 \approx 2 - \frac{4}{5}\sigma + \frac{96}{875}\sigma^2 + \frac{4}{5}\sigma^2 h^2 \tag{1.87}$$

where $h = HM_s/2K$. The relaxation time is then calculated from

$$\tau_\parallel = (\beta/b)\lambda_1 \approx \frac{\beta}{b}\left(2 - \frac{4}{5}\sigma + \frac{96}{875}\sigma^2 + \frac{4}{5}\sigma^2 h^2\right)^{-1} \tag{1.88}$$

However, this applies to the cases of least interest [i.e., when the ratio of the barrier height to thermal energy is small (low temperatures)]; therefore those more interesting cases (for higher barrier heights) are discussed in more detail in the following sections.

J. Brown's High-Energy-Barrier Approximation

For higher temperatures, Brown [4] devised a high-energy-barrier approximation and imposed the condition that $h < 1$ in order to ensure that the free energy (Gibb's free energy per unit volume) has a bistable potential. The procedure described here is based on a variation of the Kramers [25] transition-state theory. It differs slightly from the calculations of Brown [4] in that it allows for the stationary points to be situated at points other than the poles. For simplicity, Brown considered the case of two minima for $V(\theta)$ at 0 and π that are separated by a maximum at $\theta = \theta_0$. However, for the moment we suppose that the free energy $V(\theta)$ has minima at θ_i, $i = 1, 2$ and a maximum at θ_0 such that $\theta_1 < \theta_0 < \theta_2$. The corresponding values of the free energy per unit volume are V_i, $i = 0, 1, 2$ (see Fig. 7), and so the high-energy-barrier assumption can be expressed as $\beta(V_0 - V_i) \gg 1$. The Taylor series of V about θ_i truncated at the $(\theta - \theta_i)^2$ term is

$$V = V_i + \frac{k_i}{2!}(\theta - \theta_i)^2, \qquad k_i = \left(\frac{d^2 V}{d\theta^2}\right)_{\theta = \theta_i} \tag{1.89}$$

We suppose that quasiequilibrium has been obtained separately within the intervals $R_1 = (0, \alpha_1)$ and $R_2 = (\alpha_2, \pi)$, where α_1 and α_2 are chosen so that

$$0 \le \theta_1 < \alpha_1 < \theta_0 < \alpha_2 < \theta_2 \le \pi \tag{1.90}$$

and

$$e^{-\beta V_0} \ll e^{-\beta V(\alpha_i)} \ll e^{-\beta V_i} \tag{1.91}$$

Only a very small fraction of the particles will not have orientations within the intervals R_1 and R_2, specifically

$$W = W_i e^{-\beta(V - V_i)}, \qquad i = 1, 2 \tag{1.92}$$

where $W_i = W(\theta_i)$. The number of representative points in R_i is

$$n_i = 2\pi W i \int_{\theta \in R_i} e^{-\beta(V - V_i)} \sin \theta \, d\theta \cong 2\pi W_i I_i \tag{1.93}$$

and from Eq. (1.89)

$$I_i = \int_{\theta \in R_i} e^{-(\beta k_i/2)(\theta - \theta_i)^2} \sin \theta \, d\theta \tag{1.94}$$

The integral I_i is then evaluated for two separate limiting cases:

- When $\theta_i \to 0, \pi$, then $\sin \theta$ is replaced by θ, and the integration is extended over $[0, \infty]$ so that

$$I_i \to \int_0^\infty e^{-(\beta k_i/2)(\theta - \theta_i)^2} \theta \, d\theta \tag{1.95}$$

- When θ_i is far enough from the corresponding pole to ensure that $e^{-\beta k_i \theta_i^2/2}$, $e^{-\beta k_i(\pi-\theta_i^2)}/2 \gg 1$ $\sin\theta$ is replaced by $\sin\theta_i$. The integration is then extended over $[-\infty, \infty]$ so that

$$I_i \to \sin\theta_i \int_{-\infty}^{\infty} e^{-(\beta k_i/2)(\theta-\theta_i)^2} \mathrm{d}\theta = \sin\theta_i \sqrt{\frac{2\pi}{\beta k_i}} \qquad (1.96)$$

We use a divergenceless current density (which is independent of θ) to represent the net flow of particles from the overpopulated toward the underpopulated minimum, since in the interval (α_1, α_2), W is small but not too small to prevent the flow of particles. The total current is given by $\dot{n}_1 = -\dot{n}_2 = -2\pi \sin\theta J_\theta$; then, from Eq. (1.81)

$$\frac{\partial W}{\partial\theta} + \beta\frac{\partial V}{\partial\theta}W = \frac{\dot{n}_1}{2\pi b \sin\theta} \qquad (1.97)$$

If we then integrate Eq. (1.97) over (θ_1, θ_2), and multiply it by $e^{\beta V}$, it becomes

$$\left[W e^{\beta V}\right]_{\theta_1}^{\theta_2} = \frac{\beta\dot{n}_1}{2\pi b}\int_{\theta_1}^{\theta_2}\frac{e^{\beta V}}{\sin\theta}\mathrm{d}\theta \qquad (1.98)$$

The integral in Eq. (1.98) is evaluated by replacing V with Eq. (1.89), replacing $\sin\theta$ with $\sin\theta_0$, and extending the integration over $(-\infty, \infty)$ so that we have

$$\left[W e^{\beta V}\right]_{\theta_1}^{\theta_2} \cong \frac{\dot{n}_1 e^{\beta V_0}}{2\pi b \sin\theta_0}\sqrt{\frac{-2\pi\beta}{k_0}} \qquad (1.99)$$

From Eq. (1.92) then

$$\left[W e^{\beta V}\right]_{\theta_1}^{\theta_2} = \frac{1}{2\pi}\left(\frac{n_2 e^{\beta V_2}}{I_2} - \frac{n_1 e^{\beta V_1}}{I_1}\right) \qquad (1.100)$$

thus

$$\dot{n}_1 = b\sin\theta_0\sqrt{\frac{-k_0}{2\pi b}}\left(\frac{n_2 e^{-\beta(V_0-V_2)}}{I_2} - \frac{n_1 e^{-\beta(V_0-V_1)}}{I_1}\right) \qquad (1.101)$$

which is analogous to the discrete-orientation approximation [Eq. (1.11)] [17], namely

$$\dot{n}_1 = -\dot{n}_2 = v_{2,1}n_2 - v_{1,2}n_1 \qquad (1.102)$$

and so the transition probabilities are

$$
\begin{aligned}
v_{i,j} &= b \sin \theta_0 I_i^{-1} \sqrt{\frac{-k_0}{2\pi b}} e^{-\beta(V_0 - V_i)} \\
&= b k_i \sin \theta_0 \sqrt{\frac{-\beta k_0}{2\pi e^{-\beta(V_0 - V_i)}}} \qquad \theta_i = 0, \pi \\
&= \left(\frac{b \sin \theta_0}{2\pi \sin \theta_i} \right) \sqrt{-k_0 k_i} \; e^{-\beta(V_0 - V_i)} \qquad \theta_i \neq 0, \pi
\end{aligned}
\qquad (1.103)
$$

The relaxation time is then calculated by the formula

$$
\tau = (v_{1,2} + v_{2,1})^{-1} \qquad (1.104)
$$

which is the high-energy-barrier approximation formula. The procedures described above can then be applied particularly to the case of axial symmetry. If we have a particle possesing uniaxial anisotropy and having a field applied parallel to the easy axis of magnetization (longitudinally applied field at an angle $\psi = 0$), then the Gibbs free energy per unit volume is represented by Eq. (1.6), namely

$$
V = K(\sin^2 \theta - 2h \cos \theta) \qquad (1.105)
$$

The stationary points of this potential are determined by the solutions to the equation $V' = 0$:

$$
V' = 2K \sin \theta \cos \theta + 2Kh \sin \theta = 2K \sin \theta (\cos \theta + h) = 0 \qquad (1.106)
$$

The solutions to these equations are then given by $\sin \theta = 0$, and $\cos \theta = -h$, so then the stationary points are

$$
\theta_0 = \cos^{-1}(-h), \qquad \theta_1 = 0, \qquad \theta_2 = \pi \qquad (1.107)
$$

so that the condition $\theta_1 < \theta_0 < \theta_2$ is satisfied. In order to determine which of these are the local minima and maxima, it is necessary to evaluate V'' at the stationary points, where

$$
V'' = -2K \sin^2 \theta + 2K \cos^2 \theta + 2Kh \cos \theta \qquad (1.108)
$$

The values of V'' at the stationary points give

$$
k_0 = -2K(1 - h^2), \qquad k_1 = 2K(1 + h), \qquad k_2 = 2K(1 - h) \qquad (1.109)
$$

Since $k_0 < 0$, then θ_0 is the local maximum, and $k_1, k_2 > 0$, so θ_1 and θ_2 are the local minima. The values of V at the stationary points provide the ratios of the barrier heights to thermal energy, namely, $\beta(V_0 - V_i)$, since

$$V_0 = K(1 - h^2) + 2Kh^2 = K(1 + h^2) \tag{1.110}$$

$$V_1 = -2Kh \tag{1.111}$$

$$V_2 = 2Kh \tag{1.112}$$

and so

$$\beta(V_0 - V_1) = \sigma(1 + h)^2 \tag{1.113}$$

$$\beta(V_0 - V_2) = \sigma(1 - h)^2 \tag{1.114}$$

By noting that $\sin\theta_0 = \sqrt{1 - \cos^2\theta_0} = \sqrt{1 - h^2}$ and using Eqs. (1.110)–(1.114), the transition probabilities can be expressed as

$$\begin{matrix} \nu_{1,2} \\ \nu_{2,1} \end{matrix} = 2b\beta^{-1}\sigma^{3/2}\pi^{-1/2}(1 - h^2)(1 \pm h)e^{-\sigma(1\pm h)^2} \tag{1.115}$$

The longitudinal relaxation time then is given by

$$\tau_\parallel = \frac{\beta\sigma^{-3/2}\sqrt{\pi}}{2b(1 - h^2)[(1 + h)e^{-\sigma(1+h)^2} + (1 - h)e^{-\sigma(1-h)^2}]} \tag{1.116}$$

On the other hand, if we have a transversely applied field, specifically, one that is applied at an angle $\psi = \pi/2$ to the easy axis of magnetization, then the Gibbs free energy per unit volume is represented by Eq. (1.8), namely

$$V = K\sin\theta(\sin\theta - 2h) \tag{1.117}$$

Then the stationary points are given by $\cos\theta = 0$ and $\sin\theta = h$ and so $\theta_0 = \pi/2$, $\theta_1 = \sin^{-1}h$, $\theta_2 = \pi - \sin^{-1}h$ so that they comply with the condition $\theta_1 < \theta_0 < \theta_2$. In order to determine the local minima and maxima, we must evaluate V at the stationary points, where

$$V'' = 2K\cos^2\theta - 2K\sin\theta + 2Kh\sin\theta \tag{1.118}$$

By nothing that

$$\sin(\pi - \sin^{-1}h) = \sin\pi\cos(\sin^{-1}h) - \cos\pi\sin(\sin^{-1}h)$$

$$= \sin\pi\sqrt{1 - h^2} - h\cos\pi$$

$$= h \tag{1.119}$$

then the values of V'' at the stationary points are given by

$$k_0 = -2K(1-h), \qquad k_1 = k_2 = 2K(1-h^2) \qquad (1.120)$$

Since $k_0 < 0$, then θ_0 is the local maximum, and since $k_1, k_2 > 0$, then θ_1 and θ_2 are the local minima. If we evaluate V at the stationary points, we obtain

$$V_0 = K - 2Kh = K(1-2h) \qquad (1.121)$$

$$V_1 = V_2 = -Kh^2 \qquad (1.122)$$

and so the ratio of the barrier height to thermal energy can be expressed as

$$\beta(V_0 - V_1) = \beta(V_0 - V_2) = \sigma(1-h)^2 \qquad (1.123)$$

For the case of a transversely applied field, Eq. (1.115) becomes

$$\nu_{1,2} = \nu_{2,1} = \frac{b\sigma}{\pi\beta h}\sqrt{(1-h)(1-h)^2}\,e^{-\sigma(1-h)^2} \qquad (1.124)$$

and so the longest relaxation time in a transverse field is given by

$$\tau_\perp = \frac{\pi\beta h}{2b\sigma}(1-h)^{-1/2}(1-h^2)^{-1/2}e^{\sigma(1-h)^2} \qquad (1.125)$$

Brown [20] also derived Eq. (1.125) for the case when the Gibbs free energy per unit volume is represented as $V = -K\cos^2\theta$ (or $-Kx^2$) using the method of approximate minimization. This method has the advantage of avoiding the assumption of a divergenceless current density in (θ_1, θ_2). For a zero applied field ($h = 0$), then

$$\nu_{1,2} = \nu_{2,1} = 2b\beta^{-1}\sigma^{3/2}\pi^{-1/2}e^{-\sigma} \qquad (1.126)$$

and so in this case

$$\tau = \left(\frac{\beta\sqrt{\pi}}{4b}\right)\sigma^{-3/2}e^{\sigma} \qquad (1.127)$$

This equation is subject to the conditions $\lambda_n \geq n(n+1)e^{-\sigma}$ and $\lambda_1 \geq 2e^{-\sigma}$ so that $\sigma \geq (\pi/4)^{1/3} = 0.92$, hence for $\sigma < 0.92$, the value of the preceding expression for τ will be too large. Brown [20] himself did not present any exact solutions to his Sturm–Liouville equation [Eq. (1.85)] for any cases other than $h = 0$, but instead used his approximation formulas to estimate λ_1 for higher values of h. The first exact results to the equation were presented by Aharoni [27] in 1969. He expanded the solution to Eq. (1.85) as a series of Legendre polynomials:

$$F_n(x) = \sum_{n=0}^{\infty} a_n P_n(x) \tag{1.128}$$

By substituting this in Eq. (1.85) and using the differential and recurrence formulas of the Legendre polynomials (refer to Appendix B), Brown deduced that Eq. (1.128) is a solution to the Sturm–Liouville equation, provided that for every $m \geq 0$

$$\left(\frac{\lambda - m(m+1)}{2\sigma} + \frac{m(m+1)}{(2m-1)(2m+3)} \right) a_m - h \frac{m(m-1)}{2m-1} a_{m-1}$$

$$+ \frac{(m+1)(m+2)(m+3)}{(2m+3)(2m+5)} a_{m+2} + h \frac{(m+1)(m+2)}{(2m+3)} a_{m+1}$$

$$- \frac{m(m-1)(m-2)}{(2m-1)(2m-3)} a_{m-2} = 0 \tag{1.129}$$

For the case of $h = 0$, Brown separated the even and odd terms in Eq. (1.129), and replaced the resulting three-term recursion formula by a continued fraction. When $h \neq 0$, then the five-term recursion formula was used, and operating on the knowledge that the coefficients of a_m in Eq. (1.129) should vanish so that (1.129) will have a nonvanishing solution for the a_m values, then the determinant of the coefficients of a_m up to a certain order n was equated to zero. Then the smallest nonvanishing root was computed. As n increased, the smallest nonvanishing root was computed again, until the effect of the increase in n was negligible. Figure 8 provides the values of the eigenvalue λ_1 as computed by Aharoni [26,27] as a function of the barrier-height parameter σ, and the field h for particles with uniaxial anisotropy K in a magnetic field H at temperature T.

In 1974, further developments of Brown's [4] approach were made by the group of Shliomis [28], who showed how Brown's [4] work could be adapted to describe the dynamic behavior of suspensions of single-domain ferromagnetic particles in fluids (i.e., ferrofluids) [29]. They [28] obtained λ_1 exactly for the case when V is axially symmetric and is given by

$$V(\theta) = -K \cos^2 \theta \tag{1.130}$$

However, the underlying difference from previous cases was that the distribution function W may be φ-dependent. They [28] used a separation-of-variables technique and applied it to the resulting Fokker–Planck equation,

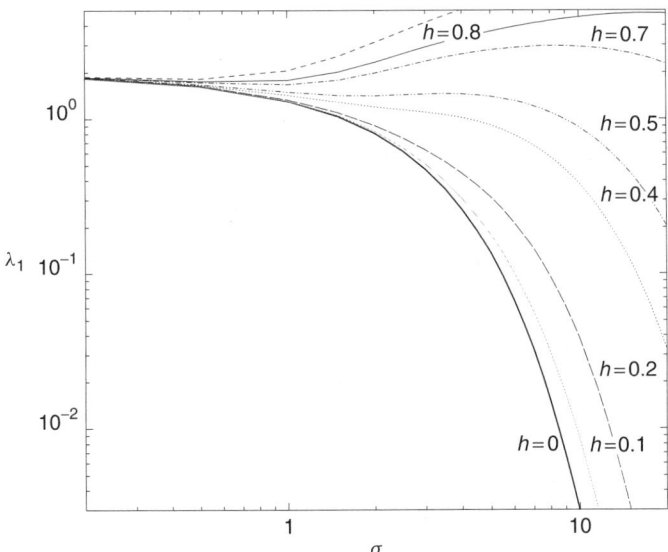

FIGURE 8. The eigenvalue λ_1 plotted as function of the barrier height parameter σ for various of h (1969 results of Aharoni).

an eigenvalue equation that was subjected to Galerkin's (see Ref. 30) numerical methods to retrieve Aharoni's [27] result. They [28] also obtained an approximate expression for τ for the case when $h = 0$, namely

$$\tau = \tau_\| = \tau_{\mathrm{n}} \left[1 - \sigma + \sigma \frac{F''(\theta)}{F'(\theta)} \right]^{-1} \qquad (1.131)$$

where

$$F(\sigma) = \int_0^1 e^{\sigma x^2} \, dx, \qquad x = \cos\theta \qquad (1.132)$$

where $'$ denotes differentiation with respect to θ, and $\tau_\|$ is the longitudinal relaxation rate.

Prior to the work of Coffey et al. [22] in 1993, only one set of results for $h \neq 0$ was available, and those were the results of Aharoni [27]. More recently, Coffey et al. [31] obtained exact results for λ_1 from Brown's [4] axially symmetric Fokker–Planck equation for the potential $V(\theta) = -K\cos^2\theta$ and expanded the distribution function W in terms of Legendre polynomials with time-dependent coefficients, and then solved the characteristic equation of the resulting set of differential recurrence relations to a suitable degree of accuracy. The results of the calculations following the work of Coffey et al. [31] are presented in the following sections.

II. CHARACTERISTIC TIMES FOR NONAXIALLY SYMMETRIC POTENTIALS

A. Magnetic Response Functions

Having discussed the case of axial symmetry in the introductory section, where the Gibbs free energy per unit volume was as described by Eqs. (1.6) and (1.8), we now present cases where the external field is applied at an oblique angle to the easy axis and the Gibbs free per unit volume is as described by Eq. (1.5). The purpose of these calculations being that in an actual sample of a ferromagnetic material, the easy direction of the magnetization is in an arbitrary direction. The effect of an external uniform magnetic field applied at an oblique angle ψ to the easy axis of magnetization of a single-domain ferromagnetic particle on the Néel [16] relaxation time for uniaxial anisotropy is investigated by numerically solving the Fokker–Planck equation for the smallest nonvanishing eigenvalue (λ_1) (as discussed in the latter part of Section I). λ_1 is the rate constant associated with the longest surviving decay mode of the magnetization, where its reciprocal (in the high-barrier limit) approximately describes the time of reversal of the magnetization over the potential barrier [whose potential is described by (Eq. 1.5)] and is determined by the crystalline anisotropy as illustrated by Fig. 5 and also the externally applied field. In the next few sections, both an exact numerical solution for λ_1 and an asymptotic estimate which is based on an extension of the Kramers [25] theory of the thermally activated escape of particles over potential barriers are presented. The formulas have been expressed in terms of spherical polars so that the theory may be applied to three-dimensional rotations. In the recent work of Coffey et al. [31], the numerical calculations presented proceeded from the assumption that the dimensionless damping parameter a [as given by Eq. (1.30)] was so large that the effect of the gyromagnetic term was negligible. However, in the following sections, we present cases where a is assigned a very large value, so that the gyromagnetic term is effectively switched off, but also cases where a is very small and so the gyromagnetic terms are boosted.

The omission of the gyromagnetic terms has the advantage that the set of differential recurrence relations for the aftereffect functions of the magnetization to which the Fokker–Planck equation may be reduced by expanding its solution as a series of spherical harmonics $X_{l,m}(\theta, \varphi)$ in the polar angles θ and φ (specifying the orientation of the magnetic moment) reduces to a 9-term rather than a 13-term recurrence relation between l and m (this is discussed in further detail in Section II.B). Furthermore, the recurrence relations for negative m may easily be expressed in terms of those for positive m by means of the symmetry properties of $X_{l,m}$ [because the expectation values of the spherical harmonics $x_{l,m}(t)$ are real when the gyromagnetic terms are ignored], thus greatly reducing the amount of computer time required for the solution.

If the gyromagnetic terms are included, the expectation values, $x_{l,m}(t)$ are complex. The recurrence relations for $x_{l,-m}(t)$ are then expressed in terms of that of its complex conjugate $x_{l,m}^*(t)$ as described by Eq. (2.29) (the star denotes the complex conjugate). Hence the recurrence relations may be solved only by expressing the solution of the Fokker–Planck equation in terms of surface spherical harmonics. These lead to recurrence relations for the real and imaginary parts of $x_{l,m}(t)$, which are described in more detail later. As far as the asymptotic estimate of λ_1 is concerned the omission of the gyromagnetic term indicated at first that an axially symmetric approximation [17,31,32] might be made in the Kramers [25] approach to the problem, thus reducing it to an effective one-variable problem as only the reaction rate coordinate θ (colatitude) will now be involved. This, however, leads to asymptotic estimates which deviate appreciably from the numerical solution so that a more rigorous treatment including both reaction coordinates θ, φ must be given (this is discussed fully in Section III).

Neglect of the gyromagnetic term, besides having the obvious disadvantage (when the treatment is extended to calculate the complex magnetic susceptibility) that ferromagnetic resonance [28] and other phenomena may not be included, implies that the calculation of λ_1 will be invalid for small values of the damping parameter a. This is important in view of the experimental results of Dormann et al. [33] on γ-Fe$_2$O$_3$ with weak interparticle interactions, which indicate that a may be as small as 0.05, hence the influence of a on λ_1 may be substantial [35].

In the following sections it is demonstrated that the gyromagnetic effects may be included in the calculation of λ_1 and hence the Néel relaxation time. Both an exact solution for λ_1 (based on the 13-term differential recurrence relations arising from the Fokker–Planck equation including the gyromagnetic term) and a rigorously derived estimate of λ_1 [based on a two-variable Kramers approach as used for a general potential by Brown [4] and Smith and de Rozario [34] (in a discussion of relaxation in cubic anisotropy potentials)], which allows one to include the influence of the gyromagnetic term in the prefactor of such an asymptotic estimate are presented. The two-variable [25,31] asymptotic estimate is essentially a modification of the Kramers [25,41] calculation of the escape rate for a particle diffusing in phase space. In addition, the integral relaxation time τ, which is the area under the curve of the decay of the magnetization [Eq. (1.71)] following a change in the applied field, will be calculated as far as terms linear in the perturbation. This means that the integral relaxation time may be identified with the correlation time of the autocorrelation function [Eq. (1.70)] of the magnetization [35].

We present expressions for the magnetic response functions and characteristic times for the cases where the Gibbs free energy per unit volume is given
by Eq. (1.5) by starting at the Fokker–Planck equation where the distribution
of orientations $W(\mathbf{r}, t)$ of the magnetization $\mathbf{M}$ of an assembly of noninteracting single-domain ferromagnetic particles on the unit sphere satisfies the
Fokker–Planck equation, namely

$$\dot{W} = L_{\mathrm{FP}} W \tag{2.1}$$

where $L_{\mathrm{FP}}W$ is as given by Eq. (1.51) and θ and φ (see Fig. 6) are the polar
angles specifying the orientation of $\mathbf{M}$ where

$$\mathbf{r} = \frac{\mathbf{M}}{M_s} \tag{2.2}$$

where M_s is the mean magnetization of a nonrelaxing particle. The diffusional
relaxation time is

$$\tau_{\mathrm{N}} = \frac{\beta}{2b} \tag{2.3}$$

Here $a = \eta \gamma M_s$ is the dimensionless damping parameter, η is the phenomenological damping constant from Gilbert's [21] equation [Eq. (1.21)], and γ is
the gyromagnetic ratio. The (Gibb's) free energy per unit volume for uniaxial
anisotropy is given by Eq. (1.3) and can be rewritten as

$$V(\mathbf{r}) = \beta^{-1}\sigma(1 - \langle \mathbf{r} \cdot \mathbf{e}_3 \rangle^2) - \beta^{-1}\xi\langle \mathbf{r} \cdot \mathbf{h} \rangle \tag{2.4}$$

where

$$\sigma = \frac{Kv}{kT} = \beta K \tag{2.5}$$

is the barrier height parameter (K is the anisotropy parameter)

$$\xi = \frac{HM_s v}{kT} \tag{2.6}$$

is the external field parameter, and $\mathbf{h}$ is the orientation of the external magnetic
field $\mathbf{H}$. The orthogonal coordinate system $\{\mathbf{e}_i\}_{i=1,2,3}$ (see Fig. 6) is chosen so
that $\mathbf{e}_3$ is parallel to the anisotropy axis and $\mathbf{h}$ is parallel to the plane containing
$\mathbf{e}_1$ and $\mathbf{e}_3$, hence

$$\mathbf{h} = \sin \psi \mathbf{e}_1 + \cos \psi \mathbf{e}_3 \tag{2.7}$$

The operator ∇^2 denotes the angular part of the Laplacian, that is

$$\nabla^2 = \frac{1}{\sin\theta}\frac{\partial}{\partial\theta}\left(\sin\theta\frac{\partial}{\partial\theta}\right) + \frac{1}{\sin^2\theta}\frac{\partial^2}{\partial\varphi^2} \tag{2.8}$$

where ∇ is the two-dimensional gradient operator on the surface of the unit sphere. Coffey et al. [31] assumed a to be so large that the influence of the gyromagnetic term on the Néel [16] relaxation time τ, given by

$$\tau = \frac{2\tau_N}{\lambda_1} = \frac{\beta/b}{\lambda_1} \tag{2.9}$$

is negligible and λ_1 is the smallest nonvanishing eigenvalue of $-(\beta/b)L_{FP}$.

In order to calculate λ_1 while including the effects of the gyromagnetic term, we first calculate the linear response of the system following a small decrease $\mathbf{H}_1$ in the field $\mathbf{H}$, where $\mathbf{H}_1$ is such that

$$\frac{vM_s\mathbf{H}_1}{kT} = \xi_1 \ll 1 \tag{2.10}$$

so that terms $O(\xi_1^2)$ may be ignored. Thus the initial condition for linear response may be imposed by replacing ξ with $\xi + \xi_1$ for $t \leq 0$ where $\xi_1 \ll \xi$. The system is assumed to be in thermodynamic equilibrium at $t = 0, \infty$, in which case both the initial and final distributions are of a Maxwell–Boltzmann type, and are given by Eq. (1.62), which may be expressed as

$$W(r, 0) = \frac{e^{-\beta V + \xi_1 \mathbf{r}\cdot\mathbf{h}}}{\int e^{-\beta V + \xi_1 \mathbf{r}\cdot\mathbf{h}}d\Omega} \tag{2.11}$$

$$W(r, \infty) = \frac{e^{-\beta V}}{\int e^{-\beta V}d\Omega} \tag{2.12}$$

The volume element is $d\Omega = \sin\theta\, d\theta\, d\varphi$ and the surface integrals are over the unit sphere. The relaxation behavior is described by the response function

$$f(t) = \lim_{\xi_1\to 0}\xi_1^{-1}(\langle\mathbf{r}\cdot\mathbf{h}\rangle - \langle\mathbf{r}\cdot\mathbf{h}\rangle_0) = \sum_i A_i e^{-b\beta^{-1}\lambda_i t}, \qquad t > 0 \tag{2.13}$$

where $\lambda_1, \lambda_2, \lambda_3, \ldots;\ \{0 < |\lambda_1| < |\lambda_2| < |\lambda_3| < \cdots\}$ are the eigenvalues and $A_1, A_2, A_3, \ldots$ the corresponding amplitudes of $-(\beta/b)L_{FP}$ [17,42] and

$$\langle\mathbf{r}\cdot\mathbf{h}\rangle = \int \mathbf{r}\cdot\mathbf{h}W(\theta, \varphi, t)d\Omega = \int(\cos\theta\cos\psi + \sin\theta\cos\varphi\sin\psi)Wd\Omega \tag{2.14}$$

represents the decay of the average cosine in the direction of the field. The subscript zero in Eq. (2.13) denotes that the statistical average is to be evaluatedin the absence of the perturbation ξ_1 [35].

The normalized complex susceptibility $\chi(\omega)$ may now be evaluated since according [42] to linear response theory $\chi(\omega)$ is given by

$$\frac{\chi(\omega)}{\chi'(0)} = 1 - i\omega \int_0^\infty e^{-ib\beta^{-1}\omega t} f(t)\mathrm{d}t = \frac{\sum_k \dfrac{A_k}{1 + i\omega/\lambda_k}}{\sum_k A_k} \tag{2.15}$$

Two global quantities associated with the decay of the magnetization in addition to the mean first passage time are the integral relaxation time (in this case the correlation time) and the effective eigenvalue (the initial slope of the normalized decay functions of the magnetization), which are given respectively by

$$T_c = b\beta^{-1} \int_0^\infty \frac{f(t)\mathrm{d}t}{f(0)} = \frac{\sum_k \lambda_k^{-1} A_k}{\sum_k A_k} \tag{2.16}$$

$$\lambda_{\text{eff}} = \frac{-\beta b^{-1} \dot{f}(0)}{f(0)} = \frac{\sum_i \lambda_i A_i}{\sum_i A_i} \tag{2.17}$$

Thus

$$T_c = \frac{\tilde{f}(0)}{f(0)} \tag{2.18}$$

where

$$\tilde{f}(0) = \lim_{s \to 0} \tilde{f}(s) \tag{2.19}$$

with the tilde denoting the Laplace transform of $f(t)$.

Both these quantities contain contributions from all the decay modes of the system. In order to calculate the aftereffect function in Eq. (2.13) and the smallest nonvanishing eigenvalue, it is first necessary (as in Coffey et al. [31]) to represent the Fokker–Planck equation in Eq. (2.1) as a differential recurrence relation that is then converted to a first-order matrix differential equation with constant coefficients. The smallest nonvanishing eigenvalue is then the smallest root of the characteristic equation of the system where a set of equations large enough to ensure convergence of the system is taken [35]. This may be accomplished by expanding the distribution function in spherical harmonics as shall be described in the following section.

B. Matrix Formulation of the Fokker–Planck Equation

The Fokker–Planck equation may now be reduced to a set of differential recurrence relations for the aftereffect functions of the magnetization by expanding the solution as a series of spherical harmonics $X_{l,m}(\theta, \varphi)$ in the polar angles θ and φ. The spherical harmonics in which we wish to express the solution to the Fokker–Planck equation are defined by (refer to Appendix B)

$$X_{l,m}(\theta, \varphi) \equiv P_l^m(\cos\theta)e^{im\varphi}, \qquad |m| \le l \qquad (2.20)$$

where P_l^m denote the associated Legendre functions. In addition, the normalized spherical harmonics $N_{l,m}X_{l,m}$ are required, where the normalization factor is given by

$$N_{l,m} = (-1)^m \sqrt{\frac{(2l+1)(l-m)!}{4\pi(l+m)!}} \qquad (2.21)$$

The normalized spherical harmonics form a complete biorthonormal basis for $L^2\{\Omega\}$, $\Omega = \{\mathbf{r}, \|\mathbf{r}\| = 1\}$ (the Hilbert space of square integrable functions, which are defined on the unit sphere). The spherical harmonics $X_{l,m}$ satisfy [43] the symmetry relation (the star denotes complex conjugate)

$$X_{l,-m} = \rho_{l,m}X_{l,m}^*, \qquad \rho_{l,m} = (-1)^m \frac{(l-m)!}{(l+m)!} \qquad (2.22)$$

Let us now suppose that the solution of Eq. (2.1) has a representation as a Fourier series

$$W(\mathbf{r}, t) = \sum_{|m| \le l} N_{l,m}^2 x_{l,m}^*(t)X_{l,m}(\mathbf{r}) \qquad (2.23)$$

Now Eq. (2.4) becomes

$$\beta V = \sigma(1 - \alpha_3^2) - \xi(\sin\psi\alpha_1 + \cos\psi\alpha_3) \qquad (2.24)$$

$$\beta\nabla^2 V = -2\sigma(1 - 3\alpha_3^2) + 2\xi(\sin\psi\alpha_1 + \cos\psi\alpha_3) \qquad (2.25)$$

where the $\alpha_i = \mathbf{r} \cdot \mathbf{e}_i$, that is

$$\alpha_1 = \sin\theta\cos\varphi, \qquad \alpha_2 = \sin\theta\sin\varphi, \qquad \alpha_3 = \cos\theta \qquad (2.26)$$

are the direction cosines of the magnetization. In addition, the Fokker–Planck operator, when it acts on a spherical harmonic (viz., $L_{FP}X_{l,m}$), has the following representation [Eq. (2.13) of Ref. 17 and Eq. (39) of Ref. 44):

$$
L_{FP}X_{l,m} = b\left[-l(l+1)\beta^{-1} + \frac{l(l+1)}{2}V + \frac{1}{2}\nabla^2 V\right]X_{l,m}
$$
$$
+ \frac{ib(l+m)(l-m+1)}{2a}\left(\frac{\partial V}{\partial\alpha_1} + i\frac{\partial V}{\partial\alpha_2}\right)X_{l,m-1}
$$
$$
- iba^{-1}m\frac{\partial V}{\partial\alpha_3}X_{l,m} + \frac{b}{2}\nabla^2 VX_{l,m} + \frac{ib}{2a}\left(\frac{\partial V}{\partial\alpha_1} - i\frac{\partial V}{\partial\alpha_2}\right)X_{l,m+1}
$$

$$(2.27)$$

Thus using the recurrence relations of the spherical harmonics as in Ref. 17 we then find that the expectation values $x_{l,m}$ of the spherical harmonics $X_{l,m}$ satisfy the differential recurrence relations [Eq. (3.28) and Eq. (3.29) of Ref. 17]

$$
\dot{x}_{l,m} = (X_{l,m}\dot{W}) = (X_{l,m}L_{FP}W) = b\beta^{-1}\sum_{p,q}e_{l,m,p,q}x_{p,q} \qquad (2.28)
$$

(where the braces denote the inner product). Then $x_{l,-m}$ may be written in terms of $x_{l,m}^*$ by means of the symmetry relation

$$
x_{l,-m} = \rho_{l,m}x_{l,m}^* \qquad (2.29)
$$

which follows immediately from Eq. (2.22), where the expressions for the matrix elements $e_{l,m,p,q}$ are as follows:

$$
e_{l,m,l-1,m-1} = \frac{\sigma h \sin\psi(l+1)(l+m)(l+m-1)}{(2l+1)}
$$
$$
e_{l,m,l,m-1} = \frac{i\sigma h \sin\psi(l+m)(l-m+1)}{a}
$$
$$
e_{l,m,l+1,m-1} = \frac{\sigma h \sin\psi l(l-m+2)(l-m+1)}{(2l+1)}
$$
$$
e_{l,m,l-2,m} = \frac{2\sigma(l+1)(l+m)(l+m-1)}{(2l-1)(2l+1)}
$$
$$
e_{l,m,l-1,m} = \frac{2\sigma(l+m)}{(2l+1)}[h\cos\psi(l+1) - ia^{-1}m]
$$
$$
e_{l,m,l+1,m} = -\frac{2\sigma(l-m+1)}{(2l+1)}[h\cos\psi l + ia^{-1}m]
$$
$$
e_{l,m,l+2,m} = -\frac{2\sigma(l-m+2)(l-m+1)}{(2l+1)(2l+3)}
$$

$$e_{l,m,l-1,m+1} = -\frac{\sigma h \sin \psi (l+1)}{(2l+1)}$$

$$e_{l,m,l,m+1} = \frac{i\sigma h \sin \psi}{a}$$

$$e_{l,m,l+1,m+1} = -\frac{\sigma h \sin \psi l}{(2l+1)}$$

$$e_{l,m,l,m} = -l(l+1) + \frac{2\sigma}{(2l-1)(2l+3)}[l(l+1) - 3m^2] - 2\sigma h \cos \psi ia^{-1}m$$

(2.30)

Equation (2.28) constitutes a set of differential recurrence relations for the expectation values $x_{l,m}$ of the spherical harmonics $X_{l,m}$. These allow us to form expressions for the differential recurrence relations for the aftereffect functions $f(t)$ in Eq. (2.13). In order to achieve this, we introduce the surface spherical harmonics [17,36] (the introduction of which avoids a complex coefficient matrix, hence greatly reducing the amount of CPU (central processing unit) time required to compute the lowest eigenvalue λ_1) namely

$$U_{l,0} = X_{l,0}, \qquad U_{l,m} = \frac{1}{\sqrt{2}}(X_{l,m} + X_{l,m}^*) = \sqrt{2}ReX_{l,m}$$

$$U_{l,-m} = \frac{1}{i\sqrt{2}}(X_{l,m} - X_{l,m}^*) = \sqrt{2}ImX_{l,m}, \qquad 0 < m \leq l \qquad (2.31)$$

and form differential recurrence relations in terms of these values. We remark that the normalized surface spherical harmonics $N_{l,|m|}X_{l,|m|}$ also form a complete biorthonormal basis for $L^2\{\Omega\}$. Thus we suppose that the solution of Eq. (2.1) has a representation given by

$$W(\mathbf{r}, t) = \sum_{|m| \leq l} N_{l,|m|}^2 u_{l,m}(t) U_{l,m}(\mathbf{r}) \qquad (2.32)$$

Then the expectation values $u_{l,m}$ of the surface harmonics $U_{l,m}$ satisfy another set of differential-recurrence relations, namely

$$\dot{u}_{l,m} = (\dot{W}, U_{l,m}) = (L_{FP}W, U_{l,m}) = b\beta^{-1} \sum_{p,q} a_{l,m,p,q} u_{p,q}. \qquad (2.33)$$

the new matrix elements of which are given by

$$a_{l,m,p,q} = \beta b^{-1} N_{p,|q|}^2 (U_{l,m}, L_{FP}U_{p,q}) \qquad (2.34)$$

These may easily be related to the matrix elements $e_{l,m,p,q}$ defined in Eq. (2.30) as follows. For $m, q > 0$ we form the matrices

$$A_{l,m,p,q} = \begin{pmatrix} a_{l,m,p,q} & a_{l,m,p,-q} \\ a_{l,-m,p,q} & a_{l,-m,p,-q} \end{pmatrix} \qquad A_{l,m,p,0} = \begin{pmatrix} a_{l,m,p,0} \\ a_{l,-m,p,0} \end{pmatrix}$$

$$A_{l,0,p,q} = (a_{l,0,p,q} \quad a_{l,0,p,-q}) \qquad A_{l,0,p,0} = a_{l,0,p,0}$$

$$R_{l,m} = \rho_{l,m} \begin{pmatrix} 1 & 0 \\ 0 & -1 \end{pmatrix} \qquad R_{l,0} = 1 \tag{2.35}$$

and construct the following matrices from the matrix elements $e_{l,m,p,q}$

$$E_{l,m,p,q} = \begin{pmatrix} \mathrm{Re}\, e_{l,m,p,q} & -\mathrm{Im}\, e_{l,m,p,q} \\ \mathrm{Im}\, e_{l,m,p,q} & \mathrm{Re}\, e_{l,m,p,q} \end{pmatrix} \qquad E_{l,m,p,0} = \sqrt{2} \begin{pmatrix} \mathrm{Re}\, e_{l,m,p,0} \\ \mathrm{Im}\, e_{l,m,p,0} \end{pmatrix}$$

$$E_{l,0,p,q} = \frac{1}{\sqrt{2}} (\mathrm{Re}\, e_{l,0,p,q} \quad -\mathrm{Im}\, e_{l,0,p,q}) \qquad E_{l,0,p,0} = \mathrm{Re}\, e_{l,0,p,0} \tag{2.36}$$

The various matrices above are related to one another by means of the equation

$$A_{l,m,p,q} = E_{l,m,p,q} + E_{l,m,p,-q} R_{p,q} \tag{2.37}$$

The matrix representation of the differential operator $(\beta/b)L_{FP}$ is now

$$A = \begin{bmatrix} A_{1,0,1,0} & A_{1,0,1,1} & A_{1,0,2,0} & A_{1,0,2,1} & A_{1,0,2,2} & A_{1,0,3,1} & \cdots \\ A_{1,1,1,0} & A_{1,1,1,1} & A_{1,1,2,0} & A_{1,1,2,1} & A_{1,1,2,2} & A_{1,1,3,0} & \cdots \\ A_{2,0,1,0} & A_{2,0,1,1} & A_{2,0,2,0} & A_{2,0,2,1} & A_{2,0,2,2} & A_{2,0,3,0} & \cdots \\ A_{2,1,1,0} & A_{2,1,1,1} & A_{2,1,2,0} & A_{2,1,2,1} & A_{2,1,2,2} & A_{2,1,3,0} & \cdots \\ A_{2,2,1,0} & A_{2,2,1,1} & A_{2,2,2,0} & A_{2,2,2,1} & A_{2,2,2,2} & A_{2,2,3,0} & \cdots \\ A_{3,0,1,0} & A_{3,0,1,1} & A_{3,0,2,0} & A_{3,0,2,1} & A_{3,0,2,2} & A_{3,0,3,0} & \cdots \\ \vdots & \vdots & \vdots & \vdots & \vdots & \vdots & \vdots \end{bmatrix} \tag{2.38}$$

the elements of which are computed by Eq. (2.37). This is now suitable for numerical computation of the response function [35,36] as is described in the following section.

C.   Computation of the Response Function and Initial Conditions

Having set up the Fokker–Planck equation coefficients in the matrix form of Eq. (2.38), equations for the calculation of the response function from Eqs. (2.33)–(2.37) can now be established by introducing the column vectors

$$U = (u_{1,0} \quad u_{1,1} \quad u_{1,-1} \quad u_{2,0} \quad u_{2,1} \quad \cdots)^T \tag{2.39}$$

and

$$B = \left(\frac{2\xi \cos \psi}{3} \quad \frac{2\sqrt{2}\xi \sin \psi}{3} \quad 0 \quad \frac{4\sigma}{5} \quad 0 \quad 0 \quad \cdots \right)^T \tag{2.40}$$

where U satisfies the set of simultaneous differential equations

$$\dot{U} = AU + B \tag{2.41}$$

and A is the Fokker–Planck matrix described by Eq. (2.38). On removal of the small perturbing field $\mathbf{H}_1$ (i.e. a small step decrease in the magnitude of the external uniform magnetic field), the final equilibrium condition is given by

$$\dot{U}(\infty) = 0 \tag{2.42}$$

and so the final equilibrium values vector, which is denoted by

$$U_\infty = U(\infty) \tag{2.43}$$

is found by solving [42] the set of simultaneous linear equations

$$AU_\infty = -B \tag{2.44}$$

The decay modes [31] of the magnetization, namely

$$C = \lim_{\xi_1 \to 0} (U - U_\infty) \tag{2.45}$$

[where ξ_1 is as defined by Eq. (2.10)] when expressed in matrix form satisfy the set of simultaneous differential equations

$$\dot{C} = AC \tag{2.46}$$

If we let $c_{l,m}$ denote the components of C, it then follows from Eqs. (2.11)–(2.13) that the initial values (just before the removal of the initial perturbing field) are given by

$$c_{l,m}(0) = \lim_{\xi_1 \to 0} \xi_1^{-1} \left[\frac{\int U_{l,m} e^{-\beta V + \xi_1 \mathbf{r} \cdot \mathbf{h}} \, d\Omega}{\int e^{-\beta V + \xi_1 \mathbf{r} \cdot \mathbf{h}} \, d\Omega} - \frac{\int U_{l,m} e^{-\beta V} \, d\Omega}{\int e^{-\beta V} \, d\Omega} \right]$$

$$= \frac{\int \mathbf{r} \cdot \mathbf{h} U_{l,m} e^{-\beta V} \, d\Omega}{\int e^{-\beta V} \, d\Omega} - \frac{\int \mathbf{r} \cdot \mathbf{h} e^{-\beta V} \, d\Omega}{\int e^{-\beta V}} \frac{\int U_{l,m} e^{-\beta V} \, d\Omega}{\int e^{-\beta V} \, d\Omega} \tag{2.47}$$

This equation may also be expressed in a more compact form [35]:

$$c_{l,m}(0) = \frac{\partial u_{l,m}}{\partial \xi}(\infty) = \langle \mathbf{r} \cdot \mathbf{h} U_{l,m} \rangle_0 - \langle \mathbf{r} \cdot \mathbf{h} \rangle_0 \langle U_{l,m} \rangle_0 = \sum_{p,q} b_{l,m,p,q} u_{p,q}(\infty) \tag{2.48}$$

In order to describe how the matrix elements $b_{l,m,p,q}$ arise, we firstly replace $x_{l,m}$ by $u_{l,m}$ and rename the matrix coefficients as $w_{l,m,p,q}$. We then compute $w_{l,m,p,q}$ by applying the spherical harmonics recurrence relations, and the expressions for each element are as follows [36,35]:

$$w_{l,m,l-1,m-1} = \frac{\sin \psi(l+m)(l+m-1)}{2(2l+1)}$$

$$w_{l,m,l+1,m-1} = -\frac{\sin \psi(l-m+2)(l-m+1)}{2(2l+1)}$$

$$w_{l,m,l-1,m} = \frac{\cos \psi(l+m)}{(2l+1)}$$

$$w_{l,m,l+1,m} = \frac{\cos \psi(l-m+1)}{(2l+1)}$$

$$w_{l,m,l-1,m+1} = -\frac{\sin \psi}{2(2l+1)}$$

$$w_{l,m,l+1,m+1} = \frac{\sin \psi}{2(2l+1)}$$

$$w_{l,m,l,m} = -\cos \psi u_{1,0}(\infty) - \sin \psi u_{1,1}(\infty) \tag{2.49}$$

The matrix elements $b_{l,m,p,q}$ are then determined from the elements $w_{l,m,p,q}$ using the transformation in Eq. 2.37, namely

$$B_{l,m,p,q} = W_{l,m,p,q} + W_{l,m,p,-q}R_{p,q} \tag{2.50}$$

The initial values vector C_0 is then given by

$$C_0 = C(0) = WU_\infty + \left(\frac{1}{3}\cos \psi \quad \frac{\sqrt{2}}{3}\sin \psi \quad 0 \quad \cdots \right)^T \tag{2.51}$$

where the matrix W is given by

$$W = \begin{bmatrix} W_{1,0,1,0} & W_{1,0,1,1} & W_{1,0,2,0} & W_{1,0,2,1} & W_{1,0,2,2} & W_{1,0,3,1} & \cdots \\ W_{1,1,1,0} & W_{1,1,1,1} & W_{1,1,2,0} & W_{1,1,2,1} & W_{1,1,2,2} & W_{1,1,3,0} & \cdots \\ W_{2,0,1,0} & W_{2,0,1,1} & W_{2,0,2,0} & W_{2,0,2,1} & W_{2,0,2,2} & W_{2,0,3,0} & \cdots \\ W_{2,1,1,0} & W_{2,1,1,1} & W_{2,1,2,0} & W_{2,1,2,1} & W_{2,1,2,2} & W_{2,1,3,0} & \cdots \\ W_{2,2,1,0} & W_{2,2,1,1} & W_{2,2,2,0} & W_{2,2,2,1} & W_{2,2,2,2} & W_{2,2,3,0} & \cdots \\ W_{3,0,1,0} & W_{3,0,1,1} & W_{3,0,2,0} & W_{3,0,2,1} & W_{3,0,2,2} & W_{3,0,3,0} & \cdots \\ \vdots & \vdots & \vdots & \vdots & \vdots & \vdots & \vdots \end{bmatrix} \tag{2.52}$$

The matrix W is analogous to the Fokker–Planck matrix A with $e_{l,m,p,q}$ in Eq. (2.37) replaced by $w_{l,m,p,q}$. The zero-frequency Laplace transforms vector is determined from Eq. (2.51) by solving the set of simultaneous equations

$$A\tilde{C}(0) = C(0) \tag{2.53}$$

The solution of Eq. (2.46) is given by

$$C(t) = Se^{b\beta^{-1}Lt}K \tag{2.54}$$

where L is the diagonal matrix whose components are the eigenvalues of the Fokker–Planck matrix A of Eq. (2.38) $(\lambda_1, \lambda_2, \lambda_3, \ldots)$ and

$$S = \begin{pmatrix} s_1^{1,0} & s_2^{1,0} & \cdots \\ s_1^{1,1} & s_2^{1,1} & \cdots \\ \vdots & \vdots & \vdots \end{pmatrix} \tag{2.55}$$

is the matrix whose column vectors are the eigenvectors of A. The vector

$$K = (k_1 \ k_2 \ldots k_N)^T \tag{2.56}$$

is found by solving the set of simultaneous linear equations

$$SK = C_0 \tag{2.57}$$

The response function in Eq. (2.13) may be determined from the decay modes because

$$f(t) = \cos \psi c_{1,0}(t) + \sin \psi c_{1,1}(t) = \sum_i A_i e^{-b\beta^{-1}\lambda_i t} \tag{2.58}$$

so that the amplitudes defined by Eq. (2.16) are given by

$$A_i = k_i \left[\cos \psi s_i^{1,0} + \frac{1}{\sqrt{2}} \sin \psi s_i^{1,1} \right] \tag{2.59}$$

and the expression for the correlation time in Eq. (2.16) becomes [35,36]

$$T_c = \frac{\cos \psi \tilde{c}_{1,0}(0) + \sin \psi \tilde{c}_{1,1}(0)}{\cos \psi c_{1,0}(0) + \sin \psi c_{1,1}(0)}. \tag{2.60}$$

Expressions for the response function $f(t)$ in Eq. (2.13), the complex susceptibility $\chi(\omega)$ in Eq. (2.15), the correlation time in Eq. (2.16), and the mean first passage time or Néel time as given by Eq. (2.9) have been formulated. These expressions may be used to describe the magnetization response following a small perturbation in the field $\mathbf{H}_1$. However, because of the order of the matrices required to obtain convergence, it is possible (at the moment) to compute exact values of these quantities for only a limited range of the parameters h, σ, a, ψ. The numerical aspects of the problem could be greatly simplified, however, if it were possible to use an approximate analytic expression for λ_1 in the high-barrier limit. Such a formula could provide a check on the validity of the numerical calculations as well as an analytic formula that may be compared with experimental results such as those of Dormann et al. [33], Wernsdorfer et al. [37] (the work of Wernsdorfer et al. is discussed further in Section III), and Kachkachi et al. [39]. Analytic expressions for λ_1 can also be used to compare with new Monte Carlo simulations and Langevin dynamics procedures [40] The application of such an asymptotic formula to the present nonaxially symmetric problem becomes much more difficult than in the case of axial symmetry, because the mean first-passage time now depends on two reaction coordinates rather than only one (viz., the latitude). However, the Kramers [25,41] treatment for the escape of particles (diffusing in phase space) over potential barriers is described by two variables: position and velocity. Kramers' treatment of such a system allows us to apply his methods to the current non-axially symmetric problem. This two-variable Kramers [25] procedure was first introduced into the theory of superparamagnetism by Smith and de Rozario [34] in 1976 in a discussion of relaxation in cubic anisotropy potentials. The procedure was also used by Eisenstein and Aharoni [48] in 1977, which was generalized by Brown [4] in 1979 to an arbitrary nonaxially symmetric potential and later discussed by Klik and Gunther [49] referring specifically to the Kramers [25] two-variable theory. In the following section it is the two-variable method of Kramers [25] that has been applied to the calculation of the asymptotic expressions for λ_1 in the high-barrier limit.

III. APPROXIMATION FORMULAS FOR NONAXIALLY SYMMETRIC PROBLEMS

A. Brown's High-Energy Approximation

It is imperative to have mathematically accurate asymptotic formulas for the reversal time for magnetocrystalline anisotropy potentials that are nonaxially symmetric in relation to the anisotropy axis in order to achieve a reliable comparison of the theory with experiment and computerized simulations [40]. An accurate formula for τ also allows us to deduce the values of other experimental parameters, for example, the blocking temperature [39].

So that the high-energy-barrier approximation of Brown [4] may be derived, it is first necessary to describe the behavior of the potential energy given by

Eq. (1.5) [also Eq. (2.4)], which may be expressed as

$$\beta V = \sigma \sin^2 \theta - \xi \cos \psi \cos \theta - \xi \sin \psi \sin \theta \cos \varphi \qquad (3.1)$$

since $\sigma = \beta K$. The behavior of the potential energy (as a function of the applied field **H**) as a calculation for rotation in a plane [rather than in three dimensions as given by Eq. (3.1)] is considered. Eq. (7.1) can be rewritten so that it satisfies the conditions $(0 \le \theta \le \pi, 0 \le \varphi \le 2\pi)$

$$U = \frac{\beta V}{\sigma} = \sin^2 \theta - 2h(\cos \psi \cos \theta + \sin \psi \sin \theta \cos \varphi) \qquad (3.2)$$

where

$$h = \frac{\xi}{2\sigma}$$

The stationary points occur for $\varphi = 0$ and $\varphi = \pi$. The stationary point for $\varphi = \pi$ corresponds to a maximum of Eq. (3.2) and so is of little or no interest. The stationary points for $\varphi = 0$, however, correspond to a saddle point of Eq. (3.2) at θ_m, and minima at θ_A and θ_B for h less than some critical value h_c. (The saddle point generally represents the equator, while θ_A and θ_B lie in the north and south polar regions, respectively). The two equilibrium directions of the magnetization and their associated polar angles θ_A and θ_B lie in the x–z plane $(\varphi = 0)$ (see Fig. 6) and are determined by the conditions for a minimum of U, namely

$$\frac{\partial U}{\partial \theta} = 0, \qquad \frac{\partial^2 U}{\partial \theta^2} = 0 \qquad (3.3)$$

The position of the saddle point follows from the conditions for a maximum of U, namely

$$\frac{\partial U}{\partial \theta} = 0, \qquad \frac{\partial^2 U}{\partial^2 \theta} < 0 \qquad (3.4)$$

and the critical value of the ratio of field to barrier height parameter (h_c) at which the potential loses its bistable character follows from the condition for a point of inflexion of U, namely

$$\frac{\partial U}{\partial \theta} = \frac{\partial^2 U}{\partial \theta^2} = 0 \qquad (3.5)$$

Equation (3.5) now yields

$$\tfrac{1}{2} \sin 2\theta = -h_c \sin(\theta - \psi) \qquad (3.6)$$

$$\cos 2\theta = -h_c \cos(\theta - \psi) \tag{3.7}$$

that is

$$\tan 2\theta = 2\tan(\theta - \psi) \tag{3.8}$$

or with

$$\tan \theta = t$$

$$t^3 = -\tan \psi \tag{3.9}$$

the only real root of which is

$$\tan \theta = -(\tan \psi)^{1/3} \tag{3.10}$$

so that with Eq. (3.6)

$$h_c = -\frac{\sin\theta\cos\theta}{\sin\theta\cos\psi - \cos\theta\sin\psi} = \frac{1}{\cos\psi[1 + (\sin\psi/\cos\psi)^{2/3}]^{3/2}} \tag{3.11}$$

and so

$$h_c = [(\sin\psi)^{2/3} + (\cos\psi)^{2/3}]^{-3/2} \tag{3.12}$$

which may also be expressed [31] in terms of $\tan\psi$ as

$$h_c = \sqrt{1 + \tan^2\psi}(1 + (\tan\psi)^{2/3})^{-3/2}, \quad 0 \le \psi \le \frac{\pi}{2} \tag{3.13}$$

If h satisfies the condition

$$h < h_c \tag{3.14}$$

where h_c is as given by Eq. (3.12) or (3.13), then the potential energy V remains bistable, having a maximum at V_0, separated by minima V_1, V_2. The smallest nonvanishing eigenvalue λ_1 of the Fokker–Planck equation, Eq. (2.1), is the rate of escape of magnetic moments over the potential barrier characterized by the maximum V_0. The condition in Eq. (3.14) was originally given by Stoner and Wohlfahrt [50] in their discussion of the construction of hysteresis loops from the potential given by Eq. (4.4) and by Pfeiffer [65], who studied the reversal of the magnetization (the present problem) in the discrete orientation [16] (Néel) approximation. In Ref. 31, we have already derived [by neglecting the gyromagnetic term in Eq. (1.55)] an approximate asymptotic formula for λ_1 in the limit of high potential barriers ($\sigma \gg 1$). This is given by Eq. (72) of Ref. 31, namely

$$\lambda_1 = \frac{1}{\sqrt{2\pi}} \sin\theta_0 \sqrt{-\beta V''(\theta_0)}$$

$$\times \left(\frac{\sqrt{\beta V''(\theta_1)}}{\sin\theta_1} e^{-\beta[V(\theta_0)-V(\theta_1)]} + \frac{\sqrt{\beta V''(\theta_2)}}{\sin\theta_2} e^{-\beta[V(\theta_0)-V(\theta_2)]} \right) \quad (3.15)$$

where an expression for τ is given by

$$\tau^{-1} = \frac{\lambda_1}{2\sqrt{2\pi}\tau_n} \approx \frac{1}{2\tau_n} \sin\theta_0 \sqrt{-\beta V''(\theta_0)}$$

$$\times \left(\frac{\sqrt{\beta V''(\theta_1)}}{\sin\theta_1} e^{-\beta[V(\theta_0)-V(\theta_1)]} + \frac{\sqrt{\beta V''(\theta_2)}}{\sin\theta_2} e^{-\beta[V(\theta_0)-V(\theta_2)]} \right) \quad (3.16)$$

Equation (3.15) for λ_1 is a generalization of Brown's original [20] (1963) calculation for an axially symmetric potential that has minima at $\theta = 0, \pi$ to an axially symmetric potential that has minima at $\theta = \theta_1, \theta_2$, where $0 \leq \theta \leq \pi$; thus it is the rotational analog of the translational single reaction coordinate problem considered by Kramers [25]. Such an axially symmetric calculation leads to a rough estimate of λ_1 in the high-barrier limit as described in Ref. 20; however, the analogy of such an axially symmetric approximation to a nonaxially symmetric problem can never be rigorously justified. The axially symmetric asymptote for λ_1 in equation (3.15) is by definition very restrictive. For example, for a uniform field, it may be used only if the field is parallel to the easy axis, and if there is uniaxial anisotropy. Moreover, it cannot be applied to higher-order anisotropies such as cubic, which are inherently nonaxially symmetric. Another restriction of Eq. (3.15) is that because it arises from a single variable ($x = \cos\theta$) Fokker–Planck equation, it is valid for all values of the damping factor a since a appears only in the diffusional time τ_n; thus there is no geometric coupling between the transverse and longitudinal modes of the magnetization. This is not generally true because two reaction coordinates — (x, ϕ) [51], ensuring coupling, and multiplicative noises — are involved [13,51,52], just as in the conventional Kramers theory [25] of escape of particles over potential barriers (for a mechanical system with a single reaction coordinate governed by the Klein–Kramers equation), in which the range of values of the damping factor a for which a particular escape rate formula is valid must be taken into account [53]. The axially symmetric formula [Eq. (3.15)], although superficially similar to the very high damping (Smoluchowski) limit of Kramers theory (derived from Kramers IHD formula), has a radically different origin from the high-damping Kramers formula as it arises from symmetry, and not from strong damping of the momentum as in the Kramers problem. In other words, the concept of a Smoluchowski equation is irrelevant in the magnetic problem, as that equation pertains to mechanical particles.

The first attempt to lift the restriction of axial symmetry was made by Smith and de Rozario [34], who derived an asymptotic formula for λ_1 for the particular case of cubic anisotropy, and later for a general nonaxially symmetric potential by Brown [4]. However, neither Smith and de Rozario [34], nor Brown [4] in their formulas, which are analogous to the IHD limit of Kramers [25] theory, addressed the problem of a range of values for a for which their results are valid, so that their papers contain no reference to a very low-damping formula analogous to that obtained by Kramers [25] for diffusion along the energy coordinate, in a single-degree-of-freedom mechanical system with additive noise governed by the Klein–Kramers equation in phase space.

A low-damping formula for the Kramers escape rate and so an asymptotic formula for λ_1 in the energy-controlled limit was first derived by Klik and Gunther [49,54], who bypassed the original Kramers low-damping approach entirely, by suitably adapting the uniform expansion of the first-passage time proposed by Matkowsy et al. [55] for the Klein–Kramers problem (in order to describe the crossover from the extremely weak-damping case, i.e., energy-controlled diffusion, to the moderate to high-damping case (details are given in Refs. 51 and 55)]. This calculation led Klik and Gunther to the concept of a range of values for a for which a particular asymptotic formula is valid in the magnetic problem. Moreover, they realized that the intermediate-to-high-damping (IHD) asymptotic formula in the magnetic problem is in essence, a particular example of the multireaction coordinate Kramers problem with additive white noise treated by Brinkman [56], Landauer and Swanson [57], and with the greatest degree of generality by Langer [58], reviewed in depth by Hängii et al. [69].

In the IHD approximation of Brown [4,17], we suppose that the potential V may be approximated close to the ith stationary point by a Taylor series truncated at the second-order terms so that

$$V(\mathbf{r}) = V_i + \tfrac{1}{2} c_1^{(i)} \left\langle \mathbf{re}_1^{(i)} \right\rangle^2 + \tfrac{1}{2} c_2^{(i)} \left\langle \mathbf{re}_2^{(i)} \right\rangle^2 \tag{3.17}$$

where the coordinate systems $\{\mathbf{e}_K^{(i)}\}_{k=1,2,3}$ are orientated so that $\mathbf{e}_3^{(i)}$ points in the direction of the stationary point (see Fig. 6). From this point forward, we use the parameters

$$u = h \cos \psi, \qquad v = h \sin \psi \tag{3.18}$$

and if we express the potential as in Eq. (3.1), namely

$$\beta V = \sigma \sin^2 \theta - \xi \cos \psi \cos \theta - \xi \sin \psi \sin \theta \cos \varphi \qquad (3.19)$$

then the condition for a stationary point is

$$\frac{\partial V}{\partial \theta} = 0 \qquad (3.20)$$

that is

$$\frac{\partial V}{\partial \theta} = \sin \theta \cos \theta + h \cos \psi \sin \theta - h \sin \psi \cos \theta \cos \varphi = 0 \qquad (3.21)$$

Using Eq. (3.18), Eq. (3.21) may also be written as

$$(x + u)\sqrt{1 - x^2} \pm vx = 0 \qquad (3.22)$$

where $x = \cos \theta$. This is Eq. (28) of Ref. 31, where the negative sign corresponds to the stationary points that occur for $\varphi = 0$ (and so these may be ignored) while the positive sign represents the local maximum that occurs for $\varphi = \pi$. Equation (3.17) yields [17]

$$V_i = \beta^{-1} \sigma \left(1 - x_i^2 - 2ux_i - 2v\sqrt{1 - x_i^2} \right) \qquad (3.23)$$

$$c_1^{(i)} = 2\beta^{-1} \sigma \left(x_i^2 + ux_i + v\sqrt{1 - x_i^2} \right) \qquad (3.24)$$

$$c_2^{(i)} = 2\beta^{-1} \sigma \left(-1 + 2x_i^2 + ux_i + v\sqrt{1 - x_i^2} \right) \qquad (3.25)$$

where $-1 \leq x_2 \leq x_0 \leq x_0' \leq x_1 \leq 1$ are the roots of the quartic equation

$$(x + u)^2 (1 - x^2) = v^2 x^2 \qquad (3.26)$$

which is obtained by squaring Eq. (3.22) [35,36,53].

Let us suppose that the ratios of barrier height to thermal energy become appreciable, specifically, $\beta(V_0 - V_i) \gg 1$, so that we may assume that the density of magnetic moment orientations W, if replaced by n_i (the number of particles in the ith orientation), rapidly achieves a state of quasiequilibrium [4]; thus n_i changes with time in accordance with the equation

$$\dot{n}_1 = -\dot{n}_2 = v_{2,1} n_2 - v_{1,2} n_1 \qquad (3.27)$$

and so

$$\lambda_1 \approx \beta b^{-1}(\nu_{1,2} + \nu_{2,1}) \tag{3.28}$$

where $\nu_{1,2}$, $\nu_{2,1}$ are Kramers [25] escape rates (transition probabilities for positive orientation 1 to negative orientation 2 and vice versa). Thus, after lengthy calculations (which are detailed in Ref. 17, Section V), we have essentially Brown's Eq. (84) of Ref. 4 or Eq. (5.60) of Ref. 17, namely

$$\lambda_1 \equiv \beta \left(\sqrt{c_1^{(1)} c_2^{(1)}} e^{-\beta(V_0 - V_1)} + \sqrt{c_1^{(2)} c_2^{(2)}} e^{-\beta(V_0 - V_2)} \right)$$
$$\times \frac{-c_1^{(0)} - c_2^{(0)} + \sqrt{(c_2^{(0)} - c_1^{(0)})^2 - 4a^{-2} c_1^{(0)} c_2^{(0)}}}{4\pi \sqrt{-c_1^{(0)} c_2^{(0)}}}. \tag{3.29}$$

where an expression for $\tau = 2\tau_n/\lambda_1$ may be given in this case as

$$\tau^{-1} \approx \frac{\lambda_1}{2\tau_{\hat{N}}} \equiv \frac{\gamma a}{M_s(1 + a^2)} \left(\sqrt{c_1^{(1)} c_2^{(1)}} e^{-\beta(V_0 - V_1)} + \sqrt{c_1^{(2)} c_2^{(2)}} e^{-\beta(V_0 - V_2)} \right)$$
$$\times \frac{-c_1^{(0)} - c_2^{(0)} + \sqrt{(c_2^{(0)} - c_1^{(0)})^2 - 4a^{-2} c_1^{(0)} c_2^{(0)}}}{4\pi \sqrt{-c_1^{(0)} c_2^{(0)}}} \tag{3.30}$$

This equation may be written more conveniently as [53]

$$\tau^{-1} \approx \frac{\lambda_1}{2\tau_{\hat{N}}} \approx \frac{\Omega_0}{2\pi\omega_0} \{\omega_1 \exp[-\beta(V_0 - V_1)] + \omega_2 \exp[-\beta(V_0 - V_2)]\} \tag{3.31}$$

where

$$\omega_1 = \frac{\gamma}{M_s} \sqrt{c_1^{(1)} c_2^{(1)}} \tag{3.32}$$

$$\omega_2 = \frac{\gamma}{M_s} \sqrt{c_1^{(2)} c_2^{(2)}} \tag{3.33}$$

$$\omega_0 = \frac{\gamma}{M_s} \sqrt{-c_1^{(0)} c_2^{(0)}} \tag{3.34}$$

and

$$\Omega_0 = \frac{a\gamma}{2(1 + a^2)M_s} \left[-c_1^{(0)} - c_2^{(0)} + \sqrt{(c_2^{(0)} - c_1^{(0)})^2 - 4a^{-2} c_1^{(0)} c_2^{(0)}} \right] \tag{3.35}$$

B. Analytic Solutions to Brown's High-Energy-Barrier Approximation

We may evaluate λ_1 analytically in three distinct cases. First $\psi = 0$, the axially symmetric case. Here the process depends on the single reaction coordinate [2] θ, and we have the axially symmetric results of Brown [20] and Aharoni [27] (an exact solution for the problem may be obtained from the theory of mean first passage times as described in Ref. 47)

$$\lambda_1 = \frac{2\sigma^{3/2}}{\sqrt{\pi}}(1 - h^2)[(1 + h)e^{-\sigma(1+h)^2} + (1 - h)e^{-\sigma(1-h)^2}] \tag{3.36}$$

and therefore

$$\tau = \frac{\tau_N \sqrt{\pi}}{(1 - h^2)}\sigma^{-3/2}\left[\frac{e^{\sigma(1+h)^2}}{(1 + h)} + \frac{e^{\sigma''(1-h)^2}}{(1 - h)}\right] \tag{3.37}$$

The derivation of this result is given in Section I. The other cases are $\psi = \pi/2$, $\pi/4$ which were also considered by Pfeiffer [65] in the discrete-orientation approximation. Here the quartic equation (3.26) may be easily factorized. For $\psi = \pi/2$, Eq. (3.26) becomes

$$x^2(1 - h^2 - x^2) = 0 \tag{3.38}$$

and so

$$x_2 = -\sqrt{1 - h^2}, \qquad x_0 = 0, \qquad x_1 = \sqrt{1 - h^2} \tag{3.39}$$

Equations (3.23)–(3.26) give

$$\beta(V_0 - V_i) = \sigma(1 - h)^2, \qquad i = 1, 2$$
$$c_1^{(1)} = c_1^{(2)} = 2K \qquad c_2^{(1)} = c_2^{(2)} = 2K(1 - h^2)$$
$$c_1^{(0)} = 2Kh \qquad c_2^{(0)} = -2K(1 - h) \tag{3.40}$$

and so

$$\lambda_1 = \frac{\sigma(1 - 2h + \sqrt{1 + 4a^{-2}h(1 - h)})\sqrt{(1 + h)}}{\pi\sqrt{h}}e^{-\sigma(1-h)^2} \tag{3.41}$$

where

$$\tau = \frac{2\tau_N\pi\sqrt{h}}{\sigma(1 - 2h + \sqrt{1 + 4a^{-2}h(1 - h)})\sqrt{(1 + h)}}e^{\sigma(1-h)^2} \tag{3.42}$$

Thus the $v_{i,j}$ of the two-level system described by Eq. (3.28) are degenerate in this case. We remark that the axially symmetric approximation Eq. (82) of Ref. 31 differs from Eq. (3.41) in the high-damping limit ($a \to \infty$), simply by the factor $\sqrt{h}$ so that both formulas become asymptotic to each other as $h \to 1$. If we consider the other analytically soluble case where $\psi = \pi/4$, then the stationary condition Eq. (3.22) becomes

$$x\sqrt{1-x^2} + \frac{h}{\sqrt{2}}(\sqrt{1-x^2} - x) = 0 \tag{3.43}$$

If we let

$$w = \sqrt{1-x^2} - x \tag{3.44}$$

then Eqs. (3.43) and (3.44) yield respectively the pair of simultaneous quadratic equations

$$2x^2 + 2wx + (w^2 - 1) = 0 \tag{3.45}$$

$$w^2 - h\sqrt{2}w - 1 = 0 \tag{3.46}$$

The roots of Eq. (3.46) are then

$$w_{\pm 1} = \frac{h \pm \sqrt{h^2 + 2}}{\sqrt{2}} \tag{3.47}$$

and so the simultaneous solutions of Eqs. (3.45) and (3.46) are the four roots of the pair of quadratic equations obtained on replacing w in Eq. (3.45) with $w_{\pm 1}$ in Eq. (3.47):

$$x_{\pm 1} = \frac{w_1}{2} \pm \frac{1}{\sqrt{2}}\sqrt{1 - \frac{w_1^2}{2}} = -\frac{h + \sqrt{h^2 + 2}}{2\sqrt{2}} \pm \frac{\sqrt{1 - h^2 + h\sqrt{h^2 + 2}}}{2} \tag{3.48}$$

$$x'_{\pm 1} = -\frac{w_{-1}}{2} \pm \frac{1}{\sqrt{2}}\sqrt{1 - \frac{w_{-1}^2}{2}} = \frac{-h - \sqrt{h^2 + 2}}{2\sqrt{2}} \pm \frac{\sqrt{1 - h^2 + h\sqrt{h^2 + 2}}}{2} \tag{3.49}$$

If we let

$$S_{\pm 1} = \sqrt{\frac{1 - h^2 \pm h\sqrt{h^2 + 2}}{2}} \tag{3.50}$$

then Eqs. (3.23)–(3.25) yield for the various constants in Eq. (3.29) [35]

$$\beta(V_0 - V_1) = \frac{\sigma}{2}\sqrt{h^2 + 2}(S_1 + S_{-1}) \qquad \beta(V_0 - V_2) = \sigma S_{-1}(\sqrt{h^2 + 2} - 3h)$$

$$c_1^{(1)} = \beta^{-1}\sigma[1 + (h + \sqrt{h^2 + 2})S_1] \qquad c_2^{(1)} = 2\beta^{-1}\sigma S_1\sqrt{h^2 + 2}$$

$$c_1^{(0)} = \beta^{-1}\sigma[1 + (h - \sqrt{h^2 + 2})S_{-1}] \qquad c_2^{(0)} = -2\beta^{-1}\sigma S_{-1}\sqrt{h^2 + 2}$$

$$c_1^{(2)} = \beta^{-1}\sigma[1 + (-h + \sqrt{h^2 + 2})S_{-1}] \qquad c_2^{(2)} = 2\beta^{-1}\sigma S_{-1}\sqrt{h^2 + 2}. \qquad (3.51)$$

which are explicit forms for the constants in Eq. (3.29). However, for all other values of ψ, it is not possible to apply the algebraic formula for the roots of the quartic equation due to the proximity of the roots at x_0 and x_0'. Instead, a numerical procedure is used to find the roots for $\psi \neq 0, \pi/4, \pi/2$.

C. Kramers Low-Damping and Transition-State Results

Although the work of Kramers [25] was based on a mechanical system, there exists a strong analogy between his various calculations and the superparamagnetic calculation [66]. The similarity between the two cases was highlighted by Klik and Gunther [49] in 1990 and exploited by them in order to find a solution for very small values of the dimensionless damping constant a of Eq. (1.30). Extensive work has been carried out on the problem by Brown [4] and Smith and de Rosario [34]. While appreciating the analogy of the nonaxially symmetric problem with the Kramers [25] calculation, they confined their calculations to the intermediate-to-high-damping (IHD) cases only. The obvious disadvantage of this approach is that it is not possible to provide an accurate range of values for the dimensionless parameter a for which any particular formula of Kramers [25] will be valid. The purpose of this section, then, is to assess the accuracy of the various nonaxially symmetric asymptotic formulas based on the Kramers theory [25] of the escape of particles over potential barriers, and to ascertain the range of friction in which a particular asymptotic formula is valid by comparison with exact numerical solution from the Fokker–Planck equation for the particular nonaxially symmetric problem of uniaxial anisotropy with a uniform field applied at an oblique angle to the anisotropy axis. This is one of the few nonaxially symmetric problems for which exact numerical solutions are available for relatively high potential barrier heights, where the asymptotic behavior of the solution is precisely known [53].

For Eq. (3.29) to be valid, a must be large [25,51] enough to ensure a Maxwell–Boltzmann distribution of orientations as one moves away from the saddle point. The derivation of Eq. (3.29) [and (Eq. 3.30)] also requires the solution of the linearized Fokker–Planck equation (FPE) in the vicinity of the minima that is the Maxwell–Boltzmann distribution. In writing Eqs. (3.29),

(3.30), and indeed Eq. (3.31), it is always supposed that the ratios of barrier height to thermal energy become appreciable [i.e., $\beta(V_0 - V_i) \gg 1$] so that one may assume that the density of magnetic moment orientations W rapidly achieves a state of quasiequilibrium, thus the FPE reduces to the master equation in 3.27. [53, 38]

Equation (3.29) [and (3.30)] (are) clearly of the same form as the IHD formula derived by Kramers [25] in the context of the mechanical problem obeying the Klein–Kramers equation; consequently, it is subject to the same limitations as that formula regarding the range of values of a for which it is applicable. As stated previously, a must be large ensure to ensure a Maxwell–Boltzmann distribution of orientations as one moves away from the saddle point or to rephrase, in one cycle of the motion of the orienting moments, the energy dissipated must be significantly greater than the thermal energy [53].

Equations essentially similar to Brown equations (3,23) to (3,29) were derived by Klik and Gunther [49,54] by supposing that the saddle point and minima of the potential lie on the equator. When the Fokker–Planck equation, or equivalently, the Langevin equation, is linearized at any point in the vicinity of the equator, the nonlinear system with multiplicative noise so linearized behaves as a two-reaction coordinate system with additive noise to which the formalism of Langer [58,69] may be directly applied, with the angular frequencies given by the Hessian matrix of the energy at the stationary points.

The transition-state-theory result may be written by simply taking the limit as $a \to 0$ in 3.29 [and Eq. (3.30)], since as $a \to 0$, then

$$\frac{-c_1^{(0)} c_2^{(0)} + \sqrt{(c_2^{(0)} - c_1^{(0)})^2 - 4a^{-2} c_1^{(0)} c_2^{(0)}}}{4\pi \sqrt{-c_1^{(0)} c_2^{(0)}}} \to \frac{1}{2\pi} \tag{3.52}$$

so that the saddle angular frequency is $\Omega_0 = \omega_0$, and so Eq. (3.31) becomes

$$\tau^{-1} \approx \frac{1}{2\pi} \{\omega_1 \exp[-\beta(V_0 - V_1)] + \omega_2 \exp[-\beta(V_0 - V_2)]\} \tag{3.53}$$

The equivalent expression for λ_1 is given by

$$\lambda_1 = \frac{\beta}{2\pi} \left(\sqrt{c_1^{(1)} c_2^{(1)}} e^{-\beta(V_0 - V_1)} + \sqrt{c_1^{(2)} c_2^{(2)}} e^{-\beta(V_0 - V_2)} \right) \tag{3.54}$$

In Eqs. (3.53) and (3.54) there is no longer any frictional dependence on the prefactor; thus Néel relaxation would occur in the absence of damping, which is impossible. We therefore require a formula that reduces τ^{-1} to zero in the low-damping limit. This is accomplished by using the uniform expansion of the first-passage time technique of Matkowsky et al. [55] [who essentially used it to treat the damping regime between the very low damping region where Eq. (3.53) applies, and (roughly), the aperiodic damping region beyond which Eq. (3.30) would be expected to apply] as adapted by Klik and Gunther [49,54] to the magnetic problem

$$\tau^{-1} \approx \frac{a}{2\pi}\{\omega_1\beta(V_0 - V_1)\exp[-\beta(V_0 - V_1)]$$

$$+ \omega_2\beta(V_0 - V_2)\exp[-\beta(V_0 - V_2)]\} \qquad (3.55)$$

which is valid if the energy loss per cycle is significantly less than the thermal energy [53]. The low-damping expression for λ_1 is then given by multiplying this equation by $2\tau_N$.

$$\lambda_1 = \frac{\beta(1 + a^2)}{2\pi}\left(\sqrt{c_1^{(1)}c_2^{(1)}}\beta(V_0 - V_1)e^{-\beta(V_0 - V_1)}\right.$$

$$\left. + \sqrt{c_1^{(2)}c_2^{(2)}}\beta(V_0 - V_2)e^{-\beta(V_0 - V_2)}\right) \qquad (3.56)$$

D. Validity of the Approximation Formulas of Brown and Kramers as a Function of the Damping Parameter

So that we may roughly establish the range of values of a in which (3.29) [also (3.31)] and Eq. (3.55) [also (3.56)] are valid, we apply the criterion of Kramers [25,69], namely, the crossover region in which neither the IHD nor the low-damping (LD) formulas hold, namely

$$\tau \approx \tau_{TS} \qquad (3.57)$$

where τ is as given by Eq. (3.55) and τ_{TS} is given by the $a = 0$ limit of (3.31), namely, Eq. (3.53). This immediately leads [51] to an estimate of the range of validity of Eq. (3.31); specifically, the friction parameter a must satisfy the relations

$$a\beta(V_0 - V_1) > 1 \qquad a\beta(V_0 - V_2) > 1 \qquad (3.58)$$

with

$$\beta(V_0 - V_1) > 1 \qquad \beta(V_0 - V_2) > 1 \qquad (3.59)$$

In the low-damping limit, on the other hand, these criteria become

$$a\beta(V_0 - V_1) < 1 \qquad a\beta(V_0 - V_2) < 1 \qquad (3.60)$$

and of course Eq. (3.59) still applies.

In addition to the situation described by Eqs. (3.58)–(3.60), it is also possible to have

$$a\beta(V_0 - V_1) \gg 1 \qquad (3.61)$$

with

$$\beta(V_0 - V_1) \gg 1 \qquad (3.62)$$

and

$$a\beta(V_0 - V_2) < 1 \qquad (3.63)$$

with

$$\beta(V_0 - V_2) > 1 \qquad (3.64)$$

corresponding to a deep lower minimum with barrier height given by Eq. (3.62) and a relatively shallow upper minimum, where the barrier height is given by Eq. (3.64). If this situation arises, then the IHD formula must be used to estimate the contribution to λ_1 for transitions from the lower minimum, while the LD formula must be used for transitions from the upper minimum [53].

This discussion serves to underline an important feature of the various asymptotic formulas for λ_1; in particular, it is possible to identify from them the separate contributions to λ_1 from transitions between the upper and lower minima and vice versa. This is not, in general, possible if one constructs the exact solution by numerically calculating the smallest nonvanishing eigenvalue of L_{FP}.

E.   Comparison of IHD and LD Formulas with the Exact Solution from the Fokker–Planck Equation

So that the validity criteria presented in the previous section may be investigated, the approximate eigenvalue in both low-damping and intermediate-to-high-damping situations [Eqs. (3.29) and (3.56), respectively] is compared with the eigenvalue yielded by the exact solution of the Fokker–Planck equation using the techniques described in detail in previous sections and in Refs. 35 and 36. The exact solutions apply for the particular case of a uniform field applied at an oblique angle to the polar axis where the potential is of the form of Eq. (3.1) for which accurate numerical solutions for λ_1 of the Fokker–Planck equation are available [35]. In this case ψ is the colatitude of the field that is assumed to be applied in the x–z plane.

The plane containing the relevant stationary points lies in the longitude $\phi = 0$, and at the minima, the truncated Taylor series for the potential has the form of an elliptic paraboloid, while at the saddle point, it describes a hyperbolic paraboloid [53]. The calculations for the low-damping approximation are somewhat easier since they require only the computation of the barrier height; however, exact calculations of λ_1 for very low damping (values of $a < 0.1$) are extremely difficult because of the order of the matrices required in order to obtain convergence. Some cases where $a < 0.1$ have been presented, however, in Ref. 59. For the moment, only those cases where $a >= 0.1$ are presented.

The results of the numerical calculations and comparison with the asymptotic formulas may be summarized as follows. In Fig. 9 and Table I,

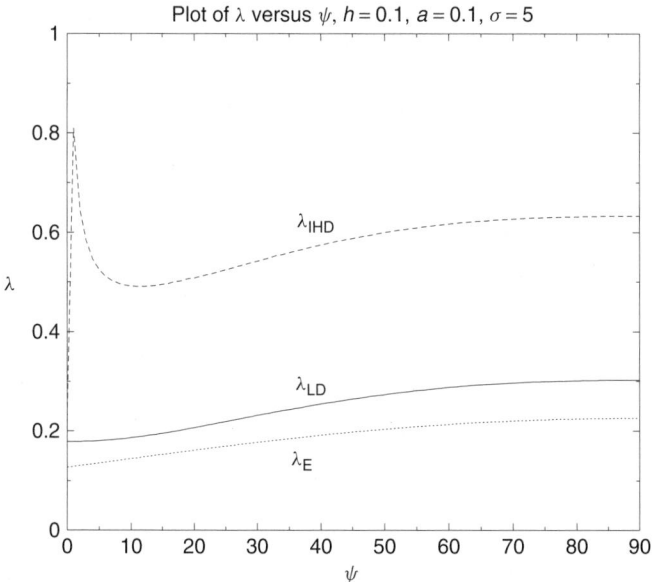

FIGURE 9. Smallest nonvanishing eigenvalue λ_1 as a function of relative orientation ψ in degrees for $h = 0.1$, $a = 0.1$, $\sigma = 5$. Here the low-damping asymptote (λ_{LD}) of Eq. (3.29) provides a much better approximation than does the IHD formula (λ_{IHD}) of Eq. (3.56), since $a\beta(V_0 - V_1) < 1$ and $a\beta(V_0 - V_2) < 1$ by inspection of Table I.

TABLE I
λ_E, λ_{LD}, λ_{IHD}, Barrier Heights, and Validity Condition for $h = 0.1$, $a = 0.1$, $\sigma = 5$

ψ	λ_E	λ_{LD}	λ_{IHD}	$\beta(V_0 - V_1)$	$\beta(V_0 - V_2)$	$a\beta(V_0 - V_2)$
0	0.1786	0.1273	0.2282	6.0500	4.0500	0.4050
15	0.1953	0.1530	0.4947	5.7581	3.8269	0.3827
30	0.2313	0.1773	0.5417	5.4170	3.6871	0.3687
45	0.2642	0.1979	0.5880	5.0503	3.6396	0.3640
60	0.2878	0.2132	0.6171	4.6834	3.6871	0.3687
75	0.2993	0.2226	0.6299	4.3421	3.8269	0.3827
90	0.3025	0.2257	0.6330	4.0500	4.0500	0.4050

results for relatively low $a = 0.1$ are presented. In this case the low-damping equation (3.56) may reasonably be expected to apply; this conjecture appears to be valid in this case since it is evident from Fig. 9 that the LD formula of Eq. (3.56) is a much better approximation to the exact value of λ_1 (denoted by λ_E) than the IHD approximation of Eq. (3.31). It is evident from Table I that the validity conditions for low damping are satisfied, and in particular, $a\beta(V_0 - V_2) < 1$. Figure 10 and Table II represent results for $a = 0.05$, and so we would again expect the LD formula of Eq. (3.56) to provide a good approximation to λ_E, even more so than the previous case since the value of the damping constant is smaller this time. The validity conditions for use of the LD formula have been satisfied, and again, in particular, $a\beta(V_0 - V_2) \ll 1$. It is clear from Fig. 10 that the LD formula of Eq. (3.56) provides an excellent approximation to the exact value of the lowest eigenvalue (λ_E) when compared with that of the IHD formula.

Figures 11 and 12 and Tables III and IV highlight those cases where either of the two approximation formulas provide a reasonable approximation to the exact value of λ_1. These cases (where $a = 0.1$ and 0.2, respectively), when taking the barrier-height parameter σ and the reduced field h into

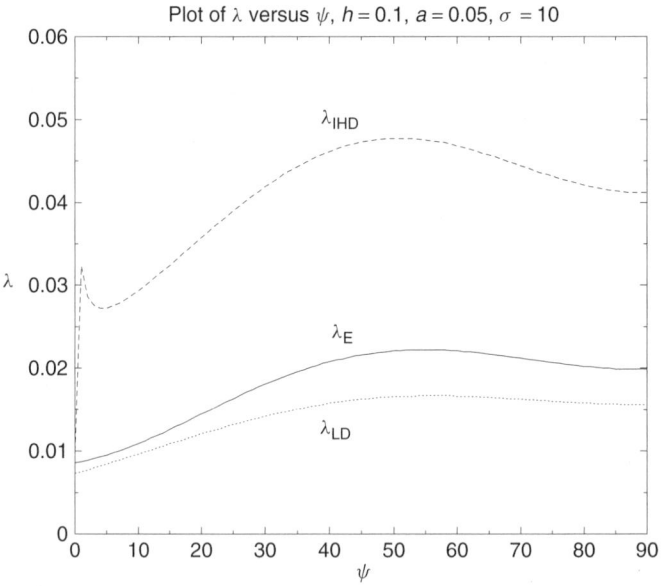

FIGURE 10. Smallest nonvanishing eigenvalue λ_1 as a function of relative orientation ψ in degrees for $h = 0.1$, $a = 0.1$, $\sigma = 10$. Again the low-damping asymptote (λ_{LD}) of Eq. (3.29) provides a much better approximation to λ_E than does the IHD formula (λ_{IHD}) of Eq. (3.56), since $a\beta(V_0 - V_1) < 1$ and $a\beta(V_0 - V_2) < 1$ by inspection of Table II.

TABLE II

λ_E, λ_{LD}, λ_{IHD}, Barrier Heights, and Validity Condition for $h = 0.1$, $a = 0.05$, $\sigma = 10$

ψ	λ_E	λ_{LD}	λ_{IHD}	$\beta(V_0 - V_1)$	$\beta(V_0 - V_2)$	$a\beta(V_0 - V_2)$
0	0.0086	0.0073	0.0099	12.1000	8.1000	0.4050
15	0.0126	0.0109	0.0324	11.5161	7.6537	0.3827
30	0.0181	0.0142	0.0419	10.8340	7.3743	0.3687
45	0.0216	0.0163	0.0473	10.1005	7.2792	0.3640
60	0.0221	0.0166	0.0468	9.3667	7.3743	0.3687
75	0.0207	0.0160	0.0432	8.6841	7.6537	0.3827
90	0.0199	0.0156	0.0412	8.1000	8.1000	0.4050

TABLE III

λ_E, λ_{LD}, λ_{IHD}, Barrier Heights, and Validity Condition for $h = 0.2$, $a = 0.2$, $\sigma = 10$

ψ	λ_E	λ_{LD}	λ_{IHD}	$\beta(V_0 - V_1)$	$\beta(V_0 - V_2)$	$a\beta(V_0 - V_2)$
0	0.0383	0.0282	0.0456	14.4000	6.4000	1.2800
15	0.0578	0.0586	0.0754	13.2447	5.5280	1.1056
30	0.0933	0.0922	0.1160	11.8830	4.9906	0.9981
45	0.1121	0.1106	0.1360	10.4083	4.8096	0.9619
60	0.0982	0.1031	0.1173	8.9295	4.9906	0.9981
75	0.0679	0.0809	0.0796	7.5595	5.5280	1.1056
90	0.0527	0.0690	0.0612	6.4000	6.4000	1.2800

TABLE IV

λ_E, λ_{LD}, λ_{IHD}, Barrier Heights, and Validity Condition for $h = 0.1$, $a = 0.2$, $\sigma = 10$

ψ	λ_E	λ_{LD}	λ_{IHD}	$\beta(V_0 - V_1)$	$\beta(V_0 - V_2)$	$a\beta(V_0 - V_2)$
0	0.0086	0.0076	0.0099	12.1000	8.1000	1.6200
15	0.0095	0.0113	0.0124	11.5161	7.6537	1.5307
30	0.0114	0.0147	0.0142	10.8340	7.3743	1.4749
45	0.0126	0.0169	0.0153	10.1005	7.2792	1.4558
60	0.0123	0.0173	0.0147	9.3667	7.3743	1.4749
75	0.0113	0.0166	0.0134	8.6841	7.6537	1.5307
90	0.0108	0.0162	0.0127	8.1000	8.1000	1.6200

account, represent the crossover region of between the IHD and LD formulas of Eqs. (3.29) and (3.56), namely, the region where $a\beta(V_0 - V_i) \approx 1$. This can be seen particularly in the last column of Tables III and IV. It is evident, then, also from Figs. 11 and 12 that either approximation formula is reasonable in this region. This crossover region is currently being investigated for

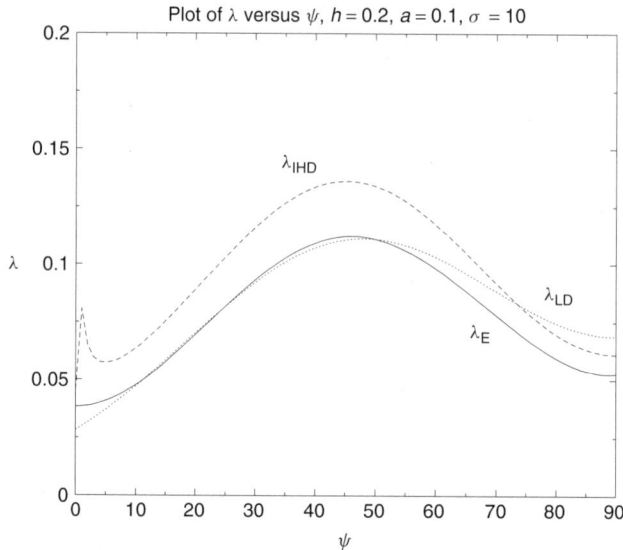

FIGURE 11. Smallest nonvanishing eigenvalue λ_1 as a function of relative orientation ψ in degrees for $h = 0.2$, $a = 0.1$, $\sigma = 10$. Here the low-damping formula of Eq. (3.56) is marginally better, since $a\beta(V_0 - V_2)$ is marginally less than unity as is apparent from Table III.

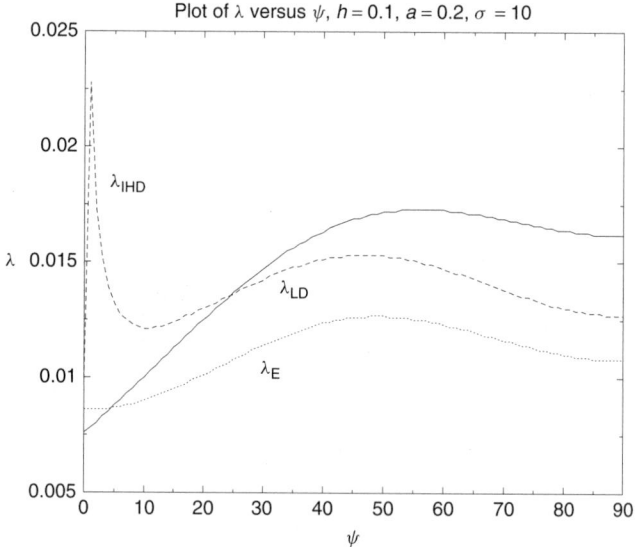

FIGURE 12. Smallest nonvanishing eigenvalue λ_1 as a function of relaxation orientation ψ in degrees for $h = 0.1$, $a = 0.2$, $\sigma = 10$. Here, both the IHD formula of Eq. (3.29) and the LD formula for Eq. (3.56) provide a reasonable approximation to λ_E, since on inspection of Table IV, $a\beta(V_0 - V_2)$ is in the range 1–2, and the IHD formula is marginally better.

the case of a transversely applied field [59], but for now, it is best to use either approximation formula for the region of validity in which it holds. It is therefore essential for each combination of the parameters involved, namely, σ, h, ψ, a, to verify the validity conditions given by Eqs. (3.58)–(3.64), and then use the appropriate asymptotic formula.

Figures 13 and 14 and Tables V and VI represent those cases where it is appropriate to use the IHD formula [Eq. (3.29)]. Here, a is assigned values of 1.0 and 3.0, respectively. One would expect, then, in these cases that the IHD formula would provide a better approximation to the exact value of λ_1, (λ_E than the LD formula of Eq. (3.56). The prediction is proved to be true then by examining Figs. 13 and 14, where it can be seen that the IHD asymptote of Eq. (3.29) provides an excellent approximation to the exact value of λ_1, whereas the LD asymptote of Eq. (3.56) provides a very poor approximation. It can be seen from Tables V and VI that the validity conditions for intermediate to high damping have been satisfied, and in particular, $a\beta(V_0 - V_2) > 1$.

Although it is apparent from Figs. 9–14 and Tables I–VI that both the asymptotes [Eqs. (3.29) and (3.56)] yield acceptable approximations to λ_1 in

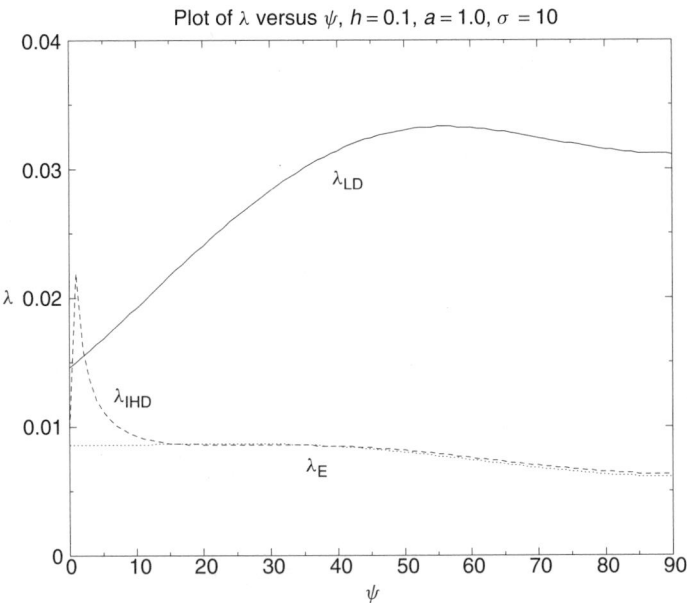

FIGURE 13. Smallest nonvanishing eigenvalue λ_1 as a function of relative orientation ψ in degrees for $h = 0.1$, $a = 1.0$, $\sigma = 10$. Here, the IHD asymptote, Eq. (3.29), provides a much better approximation to λ_E than does the LD formula of Eq. (3.56), since on inspection of Table V, $a\beta(V_0 - V_2) > 1$, with the exception of small values of ψ, which are discussed in the next section.

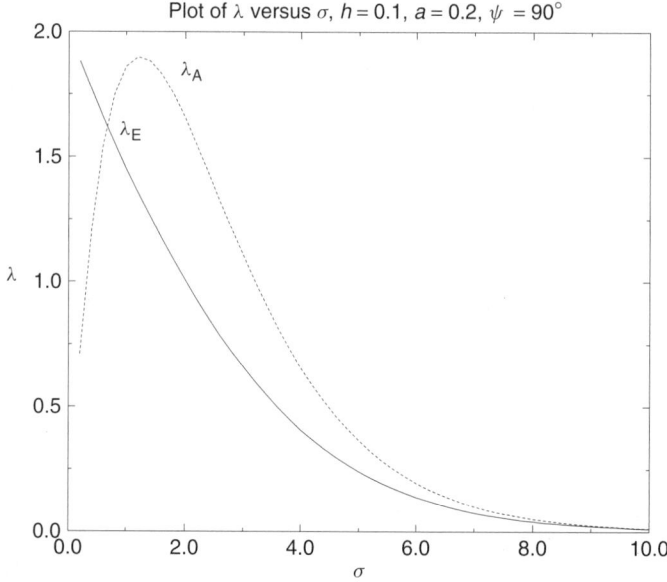

FIGURE 14. Comparison of exact $(\lambda_1 = \lambda_E)$ and asymptotic $(\lambda_1 = \lambda_A)$ expressions including the gyromagnetic term for $h = 0.1$.

TABLE V

λ_E, λ_{LD}, λ_{IHD}, Barrier Heights, and Validity Condition for $h = 0.1$, $a = 1.0$, $\sigma = 10$

ψ	λ_E	λ_{LD}	λ_{IHD}	$\beta(V_0 - V_1)$	$\beta(V_0 - V_2)$	$a\beta(V_0 - V_2)$
0	0.0086	0.0146	0.0099	12.1000	8.1000	8.1000
15	0.0087	0.0217	0.0087	11.5161	7.6537	7.6537
30	0.0087	0.0283	0.0086	10.8340	7.3743	7.3743
45	0.0082	0.0324	0.0084	10.1005	7.2792	7.2792
60	0.0074	0.0332	0.0076	9.3667	7.3743	7.3743
75	0.0065	0.0320	0.0067	8.6841	7.6537	7.6537
90	0.0061	0.0311	0.0063	8.1000	8.1000	8.1000

the range of values of a for which each one is applicable, a more refined estimate of λ_1 could be constructed by adapting the methods of Matkowsky et al. [55], or other techniques that attempt to provide asymptotic Kramers formulas that are valid for all ranges of friction for the Klein–Kramers problem (for a review, see Büttiker [60], Landauer [57], and Hängii et al. [69]). This is, however, likely to be more difficult to carry out than the Klein–Kramers problem [53].

TABLE VI

λ_E, λ_{LD}, λ_{IHD}, Barrier Heights, and Validity Condition for $h = 0.1$, $a = 3.0$, $\sigma = 10$

ψ	λ_E	λ_{LD}	λ_{IHD}	$\beta(V_0 - V_1)$	$\beta(V_0 - V_2)$	$a\beta(V_0 - V_2)$
0	0.0086	0.0728	0.0099	12.1000	8.1000	24.3000
15	0.0086	0.1083	0.0085	11.5161	7.6537	22.9611
30	0.0085	0.1416	0.0082	10.8340	7.3743	22.1228
45	0.0079	0.1621	0.0079	10.1005	7.2792	21.8376
60	0.0070	0.1659	0.0071	9.3667	7.3743	22.1228
75	0.0061	0.1598	0.0062	8.6841	7.6537	22.9611
90	0.0057	0.1557	0.0058	8.1000	8.1000	24.3000

F. Validity of Asymptotic Formulas as a Function of the Field Angle

A common feature to all results presented in Section III.E is the departure of the asymptotes calculated from the IHD formula [Eq. (3.29)], from the exact solution when the field angle is very small. In the case of a very small angle, we may assume that the problem becomes almost axially symmetric. The departure is particularly noticeable in Fig. 9. It is also apparent, however, from Fig. 13 that in the range of ψ between 0 to 10°, where one would expect the departure from axial symmetry to be small, and that the nonaxially symmetric asymptotes of equations (3.29) and (3.56) depart significantly from the exact solution λ_E. Such asymptotic behavior is in essence a consequence of the linearization of the Fokker–Planck equation about the minima and the saddle points because for very small angles, one is significantly outside the range of validity of the Taylor series expansion of the potential given by Eq. (3.23). Here it appears that Brown's axially symmetric solution, Eq. (3.36), provides a more accurate approximation to the exact solution λ_E for small angles, as is borne out by the results shown in Table VII.

An interesting feature of Figs. 9 and 13, and Tables I and V is that the asymptotic eigenvalue λ_{IHD} yielded by Eq. (3.29) for small angles requires different explanations in each case [53].

In Fig. 9, the gyroscopically modified well is shallow and the other is deep; hence we have rapid relaxation. In Fig. 13, both wells are comparatively deep, so that the gyroscopic term must artificially create a spurious resonance between the two deep wells. In both Figs. 9 and 13, we thus have artificial resonance for small ψ, for different reasons, however, essentially because the harmonic oscillator approximation for each cell is in general a poor one, for small ψ, particularly for Fig. 13 (two deep wells), but also for Fig. 9 (one deep well).

The axially symmetric formula of Eq. (3.36) gives a result of 0.0456 for λ_1, which is, in general, a better approximation to λ_E for $\psi \leq 10$ [53]. Although

TABLE VII

Values of λ_E and λ_{IHD} for $h = 0.2$,
$a = 10.0$, $\sigma = 10$

ψ	λ_E	λ_{IHD}
0	0.0383	0.0456
1	0.0384	0.0742
2	0.0384	0.0560
3	0.0385	0.0487
4	0.0386	0.0449
5	0.0388	0.0427
6	0.0389	0.0413
7	0.0392	0.0406
8	0.0394	0.0402
9	0.0398	0.0400
10	0.0401	0.0401
11	0.0404	0.0403
12	0.0408	0.0407
13	0.0412	0.0411
14	0.0416	0.0416
15	0.0420	0.0422
16	0.0424	0.0428
17	0.0429	0.0434
18	0.0433	0.0440
19	0.0438	0.0447
20	0.0442	0.0454

the exact solution λ_E is valid for all ψ, it is expensive in terms of CPU time and computer memory to compute λ_1 exactly for all ψ, h, σ, and a; however, it should be noted that it has been impossible [until the recent (at the time of writing) work of Garanin et al. [59]] to smoothly join the axially symmetric [Eq. (3.36)] and nonaxially symmetric [Eq. (3.29)] formulas since the latter arises from a two-variable Fokker–Planck equation, while the former arises from a single-space-variable Fokker–Planck equation that results from symmetry. However, the recent work of Garanin et al. [59] discusses only the transverse-field case, and further investigation into other angles is currently under way. The fact that the formulas have not yet been success-fully merged (except for the transverse-field case) is in marked contrast to the Klein–Kramers problem [25], where the one-variable equation arises from the strong damping of the momentum, and so the IHD asymptote derived from the Klein–Kramers equation goes over smoothly into the very high-damping asymptote. This mathematical constraint appears to be of little or no consequence; however, it is apparent by inspection of Table VII that in the region 0–4° of smallest departure from axial symmetry, the axially symmetric asymptote appears to give a reasonable approximation to λ_E.

G. Comparison of the Asymptotes with Experimental Observations

The derivation of reliable approximations for the relaxation time τ of the magnetic moment of fine particles for nonaxially symmetric potentials that have been presented throughout this section (III) have important technological benefits and can also be used to compare with experimental observations and computerised simulations [40]. An example of such experimental comparisons is that of the most common experiment used for characterizing fine-particle assemblies, namely, zero-field-cooled magnetization (M_{ZFC}) measurements. The temperature T_m of the maximum of M_{ZFC} is directly related to the blocking temperature T_B. T_B is obtained from a transcendental equation involving τ [61]. An analytic expression for T_B is useful for determining T_m, and its variation with experimental parameters and can be derived only if an analytic expression for τ is available. The expression for T_B may be extracted from the expressions for τ, and compared with the experimental observations. Details of these calculations are provided in Ref. 39.

As discussed earlier in this section, an expression for τ may be derived by suitably adapting the original Kramers [25,57] approach to the calculation of escape rates to nonaxially symmetric problems in superparamagnetism. This was achieved by Brown [4,20] essentially by calculating the flow of representative points across a saddle point of the potential. His general nonaxially symmetric IHD result [Eq. (3.29)] is in essence a particular practical example of the results of Langer [57,58,69] for the multireaction coordinate Kramers problem [53].

In using Brown's results, however, certain conditions must be fulfilled. First, for the shallower of the two minima

$$E = \beta(V_0 - V_2) \gg 1 \qquad (3.65)$$

Second, two expressions, Eqs. (3.29) and (3.56), may be derived according to the value of $E(a)$ with

$$E(a) = a\beta(V_0 - V_2) \qquad (3.66)$$

As previously explained, the first expression Eq. (3.29) corresponds to the high, or IHD damping regime, that is, for $E(a) \gg 1$, and the second, Eq. 3.56 to the LD limit, namely $E(a) \ll 1$. However, despite the reasonable results in Figs. 11 and 12, there is still no formula that is rigourously valid in the crossover region $E(a) \approx 1$ where the damping roughly lies between the very low-damping and aperiodic regimes. The reason for discussing this point in further detail is that $E(a) \approx 1$ corresponds to an actual experimental situation, and so it is not merely a theoretical question. In view of these considerations, it should be noted that experiments can investigate mainly the relaxation of the magnetic moment **m** of a particle when the relaxation time τ is of the order of magnitude of the measuring time τ_m. If $\tau \gg \tau_m$, **m** appears to be blocked and the measured properties correspond to the static properties of the particles. If $\tau \ll \tau_m$, on the other hand, the average of the properties over the measuring time is measured and so the properties do not depend on τ. Taking into account the values of the various parameters included in the expression for τ, the order of magnitude of E varies from 5 [in the case of Mössbauer spectroscopy where the measuring time is very small, $\tau_m \approx 10^{-8}$, to 30 (quasistatic measurements where the measuring time is very large, $\tau_m \approx 10^2$ s)]. This means that the high-barrier condition, Eq. (3.65), is always fulfilled so that the factor that determines the choice of formula is always the value of a [53].

Very few data have been published on a values for fine particles. For γFe_2O_3 particles in a polymer, a ranges between 0.05 and 1 depending on interparticle interaction strength [62]. For interacting $Fe_2\gamma Fe$ particles in an alumina matrix, $a \approx 1$. On the other hand, for bulk materials, $a \approx 0.01$ for Fe. From the recent experimental results of Coffey et al. [38] for BaFeCoTiO, it was found that $a \approx 0.03$, and for individual cobalt particles, $a \approx 0.5$ (for further details, see Ref. 38). Furthermore, lower values of a were observed for compounds such as yttrium garnet, as well as higher values depending on the compound. Nevertheless, in fine particles, a is a phenomenological constant in the Gilbert equation relative to the whole particle, including the defects inherent in the particle surface. Thus, one may expect that the smaller the particle is, the more pronounced will be the increase of a with respect to its bulk value. Therefore, one may reasonably expect that for large particles, or particles with few defects, a ranges between 0.05 and 5. As a consequence, one can see that the three cases are possible from an experimental point of view, namely, $E(a) \gg 1$ mainly for quasistatic measurements, $E(a) \ll 1$ mainly for short τ_m, which arise in Mössbauer spectroscopy, and $E(a) \approx 1$ for all experiments in which a is small. The fact that $E(a)$ is likely to be of experimental significance indicates that the present treatment should be extended (using one of the methods that have been devised [55,60,69] to yield a formula for the escape rate in the Klein–Kramers problem that is valid for all values of the friction), to yield a formula for the relaxation time that is valid for all values of a [53].

To conclude this section on the validity of the asymptotic formulas, it should be noted that the very low-damping formula of Kramers (Eq. 3.56) for magnetic relaxation has recently been rederived [63] in a very simple manner using the original Kramers energy-diffusion method. The undamped motion is considered as the rotation of a gyro in a uniform field (so that the harmonic oscillator equation applies) rather than the vibrational motion in a well, as in the original Kramers problem. It should also be noted that Eqs. (3.29) and (3.56) yield fairly good approximations to the angular variation of the prefactor of Co and BaFeCoTiO particles as well as providing a good description of the behaviour of the prefactor in the nonaxially symmetric problem of magnetic relaxation in a cubic anisotropy potential [38,53,64].

IV. NUMERICAL COMPUTATION OF THE FOKKER–PLANCK MATRIX AND THE LOWEST EIGENVALUE

A. Numerical Computation of λ_1

Exact values of λ_1 were computed by programming the matrix form of the Fokker–Planck equation [as described by Eq. (2.38)] in FORTRAN, and a complete listing of the code is provided in Ref. 36, (excluding the external subroutines, which have been called from the LAPACK and EISPACK libraries, which are available at the Internet address http://www.netlib.no/netlib/lapack). First, the Fokker–Planck matrix A of Eq. (2.38) is computed by observing that $0 \leq m \leq l$, $l \geq 1$; then for each l and m, the matrix elements given by Eq. (2.30) are computed and positioned accordingly into the Fokker–Planck matrix. The real and imaginary parts of all the eigenvalues are then computed and in particular, the lowest eigenvalue (λ_1). The set of equations in Eq. (2.30) were split into real and imaginary parts in a manner such that

$$e_{l,m,p,q} = f_{l,m,p,q} + i g_{l,m,p,q} \tag{4.1}$$

where

$$f_{l,m,l-2,m} = \frac{2\sigma(l+1)(l+m)(l+m-1)}{(2l-1)(2l+1)} \tag{4.2}$$

$$f_{l,m,l-1,m-1} = -l(l+1) + \frac{2\sigma}{(2l-1)(2l+3)}[l(l+1) - 3m^2] \tag{4.3}$$

$$f_{l,m,l+2,m} - \frac{2\sigma(l-m+2)(l-m+1)}{(2l+1)(2l+3)} \tag{4.4}$$

$$f_{l,m,l-1,m} = \frac{2\sigma(l+m)}{(2l+1)}[u(l+1)] \tag{4.5}$$

$$f_{l,m,l+1,m} = -\frac{2\sigma(l-m+1)}{(2l+1)}[ul] \tag{4.6}$$

$$f_{l,m,l-1,m-1} = \frac{\sigma v(l+1)(l+m)(l+m-1)}{(2l+1)} \tag{4.7}$$

$$f_{l,m,l+1,m-1} = \frac{\sigma v l(l-m+2)(l-m+1)}{(2l+1)} \tag{4.8}$$

$$f_{l,m,l-1,m+1} = -\frac{\sigma v(l+1)}{(2l+1)} \tag{4.9}$$

$$f_{l,m,l+1,m+1} = -\frac{\sigma v l}{(2l+1)} \tag{4.10}$$

$$g_{l,m,l-1,m} = -\frac{2\sigma(l+m)}{(2l+1)}ia^{-1}m \tag{4.11}$$

$$g_{l,m,l+1,m} = -\frac{2\sigma(l-m+1)}{(2l+1)}ia^{-1}m \tag{4.12}$$

$$g_{l,m,l,m} = -2\sigma u i a^{-1}m \tag{4.13}$$

$$g_{l,m,l,m-1} = \frac{i\sigma v(l+m)(l-m+1)}{a} \tag{4.14}$$

$$g_{l,m,l,m+1} = \frac{i\sigma v}{a} \tag{4.15}$$

where u and v are as given by Eq. (3.18).

We then wish to solve a matrix problem of the type

$$\dot{X} = AX \tag{4.16}$$

where X is the vector given by

$$\begin{bmatrix} f_{1,0} \\ f_{1,1} \\ g_{1,1} \\ f_{2,0} \\ f_{2,1} \\ g_{2,1} \\ f_{2,2} \\ \vdots \\ f_{n,n} \end{bmatrix} \tag{4.17}$$

and A is the Fokker–Planck matrix of Eq. (2.38). Note that in the vector represented by Eq. (4.17), the terms in $g_{l,0}$ are ignored, since if we examine Eqs. (4.11)–(4.13), it is clear that if $m = 0$, these terms then take on a value of 0. Having calculated all terms in Eqs. (4.2)–(4.15), we now compute the submatrices of Eqs. (2.35) and (2.36), which are related by Eq. (2.37). If we expand Eq. (2.37) in terms of Eq. (2.36), we obtain the submatrices $A_{l,m,p,q}$ as follows

$$A_{l,m,p,q} = \begin{pmatrix} f_{l,m,p,q} & -g_{l,m,p,q} \\ g_{l,m,p,q} & f_{l,m,p,q} \end{pmatrix} + \begin{pmatrix} f_{l,m,p,-q} & -g_{l,m,p,-q} \\ g_{l,m,p,-q} & f_{l,m,p,-q} \end{pmatrix} \rho_{p,q} \begin{pmatrix} 1 & 0 \\ 0 & -1 \end{pmatrix} \tag{4.18}$$

so that

$$A_{l,m,p,q} = \begin{pmatrix} f_{l,m,p,q} & -g_{l,m,p,q} \\ g_{l,m,p,q} & f_{l,m,p,q} \end{pmatrix} + \rho_{p,q} \begin{pmatrix} f_{l,m,p,-q} & g_{l,m,p,-q} \\ g_{l,m,p,-q} & -f_{l,m,p,-q} \end{pmatrix} \tag{4.19}$$

if $q = 1$, otherwise

$$A_{l,m,p,q} = \begin{pmatrix} f_{l,m,p,q} & -g_{l,m,p,q} \\ g_{l,m,p,q} & f_{l,m,p,q} \end{pmatrix} \tag{4.20}$$

also

$$A_{l,m,p,0} = \sqrt{2} \begin{pmatrix} f_{l,m,p,0} \\ g_{l,m,p,0} \end{pmatrix} \tag{4.21}$$

$$A_{l,0,p,q} = \left(\frac{1}{\sqrt{2}} \right) (f_{l,0,p,q} \quad - g_{l,0,p,q}) \tag{4.22}$$

$$A_{l,0,p,0} = f_{l,0,p,0} \tag{4.23}$$

If we examine carefully the matrix elements in Eqs. (4.2)–(4.15), it is clear that q takes only one negative value, which is $m - 1$. Obviously q is negative only when $m = 0$, and so for all such cases (i.e., $f_{l,m,l-1,m-1}$, $f_{l,m,l+1,m-1}$, $g_{l,m,l,m-1}$), then

$$\rho_{p,q} = (-1)^q \frac{(p-q)!}{(p+q)!} \tag{4.24}$$

and since $q = 1$ in each case, then

$$\rho_{p,q} = -\frac{(p-1)!}{(p+1)!} = -\frac{1}{p(p+1)} \tag{4.25}$$

We then use Eq. (4.25) where appropriate so that

$$A_{l,m,p,q} = \begin{pmatrix} f_{l,m,p,q} - \dfrac{f_{l,m,p,-q}}{p(p+1)} & -g_{l,m,p,q} - \dfrac{g_{l,m,p,-q}}{p(p+1)} \\ g_{l,m,p,q} - \dfrac{g_{l,m,p,-q}}{p(p+1)} & f_{l,m,p,q} + \dfrac{f_{l,m,p,-q}}{p(p+1)} \end{pmatrix} \tag{4.26}$$

Now if $p = l - 1$ and $q = m - 1$ as in Eq. (4.3), then

$$
A_{l,m,l-1,1} = \begin{pmatrix} f_{l,m,l-1,1} - \dfrac{f_{l,m,l-1,-1}}{l(l-1)} & -g_{l,m,l-1,1} - \dfrac{g_{l,m,l-1,1}}{l(l-1)} \\[2ex] g_{l,m,l-1,1} - \dfrac{g_{l,m,l-1,-1}}{l(l-1)} & f_{l,m,l-1,1} + \dfrac{f_{l,m,l-1,-1}}{l(l-1)} \end{pmatrix}
$$

$$
= \begin{pmatrix} f_{l,m,l-1,1} - \dfrac{f_{l,m,l-1,-1}}{l(l-1)} & 0 \\[2ex] 0 & 0 \end{pmatrix} \tag{4.27}
$$

since $g_{l,m,l-1,1}$ and $g_{l,m,l-1,-1}$ do not exist, and

$$
f_{l,m,l-1,1} + \frac{f_{l,m,l-1,-1}}{l(l-1)} = 0 \tag{4.28}
$$

then

$$
f_{l,m,l-1,1} - \frac{f_{l,m,l-1,-1}}{l(l-1)} = -\frac{\sigma v l}{(2l+1)} - \frac{\sigma v l(l+1)l(l-1)}{l(l-1)}
$$

$$
= -\frac{2\sigma v l(l+1)}{(2l+1)} = 2f_{l,m,l+1,1} \tag{4.29}
$$

Also, if $p = l - 1$ and $q = m - 1$ as in Eq. (4.7), then

$$
A_{l,m,l+1,1} = \begin{pmatrix} f_{l,m,l+1,1} - \dfrac{f_{l,m,l+1,-1}}{(l+1)(l+2)} & -g_{l,m,l+1,1} - \dfrac{g_{l,m,l+1,1}}{(l+1)(l+2)} \\[2ex] g_{l,m,l+1,1} - \dfrac{g_{l,m,l+1,-1}}{(l+1)(l+2)} & f_{l,m,l+1,1} + \dfrac{f_{l,m,l+1,-1}}{(l+1)(l+2)} \end{pmatrix}
$$

$$
= \begin{pmatrix} f_{l,m,l+1,1} - \dfrac{f_{l,m,l+1,-1}}{(l+1)(l+2)} & 0 \\[2ex] 0 & 0 \end{pmatrix} \tag{4.30}
$$

since $g_{l,m,l+1,1}$ and $g_{l,m,l+1,-1}$ do not exist, and

$$
f_{l,m,l+1,1} + \frac{f_{l,m,l+1,-1}}{(l+1)(l+2)} = 0 \tag{4.31}
$$

then

$$
f_{l,m,l+1,1} - \frac{f_{l,m,l+1,-1}}{(l+1)(l+2)} = -\frac{\sigma v l}{(2l+1)} - \frac{\sigma v l(l+2)(l+1)}{(l+2)(L+1)(2l+1)}
$$

$$
= -\frac{2\sigma v l}{(2l+1)} = 2f_{l,m,l-1,1} \tag{4.32}
$$

Also, if $p = l$ and $q = m - 1$ as in Eq. (4.14), then

$$
A_{l,m,l,1} = \begin{pmatrix} f_{l,m,l,1} - \dfrac{f_{l,m,l,-1}}{l(l+1)} & -g_{l,m,l,1} - \dfrac{g_{l,m,l,-1}}{l(l+1)} \\[3mm] g_{l,m,l,1} - \dfrac{g_{l,m,l,-1}}{l(l+1)} & f_{l,m,l,1} + \dfrac{f_{l,m,l,-1}}{l(l+1)} \end{pmatrix}
$$

$$
= \begin{pmatrix} 0 & 0 \\[3mm] 0 & -g_{l,m,l,1} - \dfrac{g_{l,m,l,-1}}{l(l+1)} \end{pmatrix} \tag{4.33}
$$

since $f_{l,m,l,1}$ and $f_{l,m,l,-1}$ do not exist, and

$$
g_{l,m,l,1} - \frac{g_{l,m,l,-1}}{l(l+1)} = 0 \tag{4.34}
$$

then

$$
-g_{l,m,l,1} - \frac{g_{l,m,l,-1}}{l(l+1)} = -\frac{i\sigma v}{a} - \frac{i\sigma v l(l+1)}{l(l+1)a} = -\frac{2i\sigma v}{a} = 2g_{l,m,l,1} \tag{4.35}
$$

Using Eqs. (4.19)–(4.23), the Fokker–Planck matrix is built up as follows:

$$
\begin{bmatrix}
f_{1,0,1,0} & \sqrt{2}f_{1,0,1,1} & -\sqrt{2}g_{1,0,1,1} & f_{1,0,2,0} & \sqrt{2}f_{1,0,2,1} & -\sqrt{2}g_{1,0,2,1} & \cdots \\
\sqrt{2}f_{1,1,1,0} & f_{1,1,1,1} & -g_{1,1,1,1,} & \frac{1}{\sqrt{2}}f_{1,1,2,0} & f_{1,1,2,1} & -g_{1,1,2,1} & \cdots \\
\sqrt{2}g_{1,1,1,0} & g_{1,1,1,1} & f_{1,1,1,1} & \frac{1}{\sqrt{2}}g_{1,1,2,0} & g_{1,1,2,1} & f_{1,1,2,1} & \cdots \\
f_{2,0,1,0} & 2f_{2,0,1,1} & -2g_{2,0,1,1} & f_{2,0,2,0} & 2f_{2,0,2,1} & -2g_{2,0,2,1} & \cdots \\
\sqrt{2}f_{2,1,1,0} & f_{2,1,1,1} & -g_{2,1,1,1} & \frac{1}{\sqrt{2}}f_{2,1,2,0} & f_{2,1,2,1} & -g_{2,1,2,1} & \cdots \\
\sqrt{2}g_{2,1,1,0} & g_{2,1,1,1} & f_{2,1,1,1} & \frac{1}{\sqrt{2}}g_{2,1,2,0} & g_{2,1,2,1} & f_{2,1,2,1} & \cdots \\
\sqrt{2}f_{2,2,1,0} & f_{2,2,1,1,} & -g_{2,2,1,1} & \frac{1}{\sqrt{2}}g_{2,2,2,0} & f_{2,2,2,1} & -g_{2,2,2,1} & \cdots \\
\sqrt{2}g_{2,2,1,0} & g_{2,2,1,1} & f_{2,2,1,1} & \frac{1}{\sqrt{2}}g_{2,2,2,0} & g_{2,2,2,1} & f_{2,2,2,1} & \cdots \\
\vdots & \vdots & \vdots & \vdots & \vdots & \vdots & \vdots
\end{bmatrix} \tag{4.36}
$$

The value of λ_1 is determined by diagonalizing the Fokker–Planck matrix using a LAPACK routine DGEEV, which calls the recommended sequence of subroutines from the eigensystem package (EISPACK) to find the eigenvalues, and optionally the eigenvectors for the real generalized problem $AX = (\lambda)BX$, where A is a real general matrix and B is a real general matrix (the identity matrix in this case). MATZ is set to zero if only the eigenvalues are required, ALFR and ALFI contain the real and imaginary parts of the eigenvalues respectively. The code allows us to calculate λ_1 for a range of h, σ, a, and ψ. To ensure convergence, each case is computed for successive values of l, where the order of the square matrix is given by $l*(l+2)$. The calculation is repeated until successive values of λ_1 lie within a tolerance of 10^{-4}. By using this procedure, we can ensure the accuracy of the results.

The barrier-height parameter σ tends to dominate the order of the matrix required in order to obtain convergence. For small values of $\sigma(\sigma < 5)$, the matrix size parameter l requires a value no larger than 12 (order of square Fokker–Planck matrix $= 168$), so we set l to $9,10,11,12, \ldots$ until convergence is obtained within a tolerance of 10^{-4}. Simple cases such as these can be computed on any machine architecture with ease requiring up to 1 min of CPU time. For higher values of $\sigma(5 < \sigma \leq 8)$, we require l to start at 13, and its value is then increased until convergence is obtained. For these cases, convergence (within a tolerance of 10^{-4}) is usually obtained when $l = 16$ for all values of the parameters h, σ, a, ψ, requiring no more than 30 min of CPU time.

For values of σ where $8 < \sigma \leq 10$, the matrix size parameter must be set to 17 initially, and can be increased up to $l = 25$ depending on the influence of the other parameters. Difficult cases such as these were computed using the CRAY Y-MP EL at QUB, and the CRAY-J90 at the Rutherford Appleton Laboratory using double-precision techniques (this is equivalent to quadruple precision on machine architectures such as the VAX 9000).

Tables VIII and IX provide a listing of the CPU times required to compute the lowest eigenvalue λ_1 for successive matrices when $\sigma = 5$, and $\sigma = 10$. The CPU times presented therefore represent the amount of computer time required to calculate λ_1 for two successive matrices. This step must be taken to ensure convergence of each result. The tolerances listed in the tables represent the tolerance within which successive results for λ_1 were obtained. All results presented in Tables VIII and IX were computed using the CRAY-J90 at Rutherford.

Examining Table VIII provides an indication of how the parameters a, h, and ψ influence the amount of CPU time required to compute λ_1. The values of l remain constant since it is generally the barrier-height parameter σ that

TABLE VIII
CPU Time Requirements for Parameters h, a, ψ When $\sigma = 5$

σ	h	a	ψ	Tolerance	l_{max}	CPU Time, s
5	0.9	10^6	$\pi/2$	10^{-5}	11,12	181.867
5	0.9	0.2	$\pi/2$	10^{-5}	11,12	122.106
5	0.2	10^6	$\pi/2$	10^{-5}	11,12	106.757
5	0.2	0.2	$\pi/2$	10^{-4}	11,12	132.992
5	0.45	10^6	$\pi/4$	10^{-4}	11,12	181.317
5	0.45	0.2	$\pi/4$	10^{-4}	11,12	216.024
5	0.2	10^6	$\pi/4$	10^{-4}	11,12	166.999
5	0.2	0.2	$\pi/2$	10^{-4}	11,12	215.222

TABLE IX
CPU Time Requirements for Parameters h, a, ψ When $\sigma = 10$

σ	h	a	ψ	Tolerance	l_{max}	CPU Time, s
10	0.9	10^6	$\pi/2$	10^{-7}	19,20	1997.207
10	0.9	0.2	$\pi/2$	10^{-5}	19,20	2105.301
10	0.2	10^6	$\pi/2$	10^{-6}	19,20	1784.916
10	0.2	0.2	$\pi/2$	10^{-5}	19,20	2120.387
10	0.45	10^6	$\pi/4$	10^{-7}	20,21	4007.247
10	0.45	0.2	$\pi/4$	10^{-5}	20,21	4651.132
10	0.2	10^6	$\pi/4$	10^{-7}	20,21	3838.569
10	0.2	0.2	$\pi/2$	10^{-5}	20,21	4806.466

determines the order of the matrices required to obtain convergence. When $a = 10^6$ (the gyromagnetic terms are effectively switched off), the amount of CPU time required to obtain convergence appears to decrease by approximately 20% when compared with the cases when the gyromagnetic terms are boosted by setting $a = 0.2$. When h is chosen close to its critical value h_c [see Eqs. (3.12) and (3.13)] ($h_c = 1.0$ for $\psi = \pi/2$, and $h_c = 0.5$ when $\psi = \pi/4$), and $a = 10^6$, then the CPU time required to obtain convergence increases by approximately 10% when compared with the cases where $h = 0.2$. When $\psi \neq 0$ or $\pi/2$ (i.e., $\psi = \pi/4$), then we would expect the calculation to become more difficult, and thus increasing the CPU time requirements. This trend is evident on inspection of Tables VIII and IX.

When the barrier-height parameter is increased from $\sigma = 5$ to $\sigma = 10$ (Table IX), the matrix size parameter l is increased to ensure convergence (it is slightly higher for the cases when $\psi = \pi/4$), and generally, CPU time requirements for $\sigma = 10$ are more than 10 times those of the cases where $\sigma = 5$. It has been observed in the calculation of λ_1, that the diagonal terms of the matrices involved increase as n^2, where n is the order of the square matrices. This would explain the increasing difficulty of the problem, and the increase in CPU time requirements as the dimensions of the matrices increase.

In the following section, the exact numerical results obtained using the CRAY Y-MP EL and the CRAY-J90 are compared with the IHD asymptotic estimate [Eq. (3.29)], which has been described in detail in Section III.

B. Comparison of the Numerical Computation of λ_1 with Asymptotic Estimates

In this section, the numerical calculations of λ_1 that provide a description of the Néel time from the formula

$$\tau = \frac{2\tau_N}{\lambda_1} \qquad (4.37)$$

will be compared with the asymptotic estimates for $\psi = \pi/2$, $\pi/4$ and other nonaxially symmetric cases, yielded by Eqs. (3.29), (3.36), and (3.42)–(3.51), which are the only cases where explicit solution of the quartic equation (3.22) is possible (except for $\psi = 0$, of course). Figures 14 and 15 illustrate the behavior of the lowest eigenvalue of the Fokker–Planck matrix λ_1 which is denoted by λ_E when compared with λ_A, which demonstrates the behavior yielded by the nonaxially symmetric formula Eq. (3.29) for $\psi = \pi/2$ [which leads to Eq. (3.41) for this particular angle], and the dimensionless damping parameter a takes on the value of 0.2. It is evident from Figs. 14 and 15 that the asymptotic formula, Eq. (3.29), provides a good approximation to the exact numerical result λ_E for larger barrier heights (i.e., larger values of σ).

This trend is confirmed by the results shown in Fig. 16, which are for infinite a values [$a = 10^6$ (the high-damping limit)]. It is apparent that the ratio of the estimated and exact values λ_A/λ_E is in the region of 1 for all h values of interest. This result should be compared with one yielded by the asymptotic formula using an axially symmetric approximation as given by

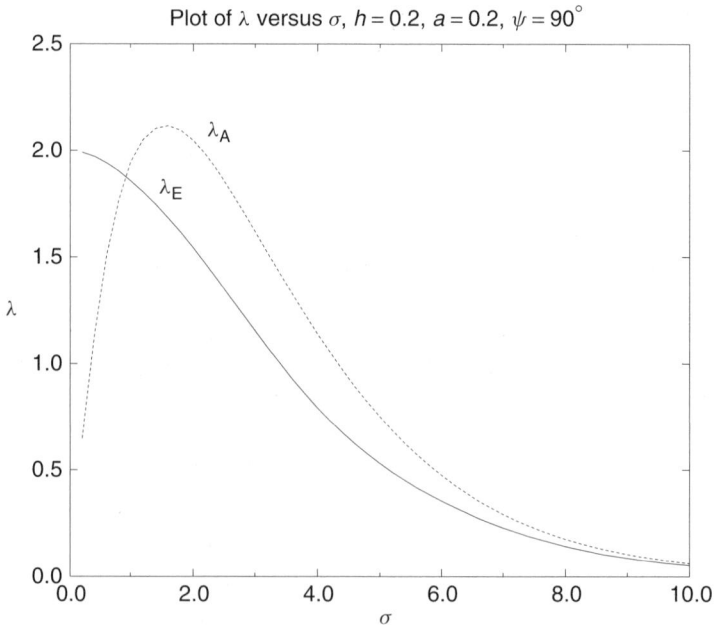

FIGURE 15. Comparison of exact ($\lambda_1 = \lambda_E$) and asymptotic ($\lambda_1 = \lambda_A$) expressions including the gyromagnetic term for $h = 0.2$. Analogous to Fig. 14, Eq. (3.41) yields a good approximation to the high-barrier solution in both cases.

Eq. (82) of Ref. 31, namely

$$\lambda_1 = \frac{\sigma(1-h)\sqrt{1+h}}{\pi h}e^{-\sigma(1-h)^2} \tag{4.38}$$

It is apparent that in the high-damping limit, the two formulas [Eqs. (4.38) and (3.41)] differ simply by a factor of $\sqrt{h}$ so that they become asymptotic to each other as $h \to 1$, the error then increasing as h decreases. The effect of the $\sqrt{h}$ correction is then to eliminate the constant error apparent in Fig. 13 of Ref. 46.

The results of Fig. 16 are for $a = 10^6$. In Fig. 17, we show λ_A/λ_E versus σ for finite $a = 0.2$. Again the constancy is reasonable, indicating that Eq. (3.29) provides a useful asymptotic estimate in this case (i.e., as borne out by Figs. 14 and 15).

In Fig. 18 we show the variation of the ratio of the asymptotic estimate of Eq. (3.41) for (a range of) finite and infinite $a(a = 10^6)[\lambda_A(a)/\lambda_A(\infty)]$ indicating that the general effect of the gyromagnetic term is to decrease the longitudinal relaxation time.

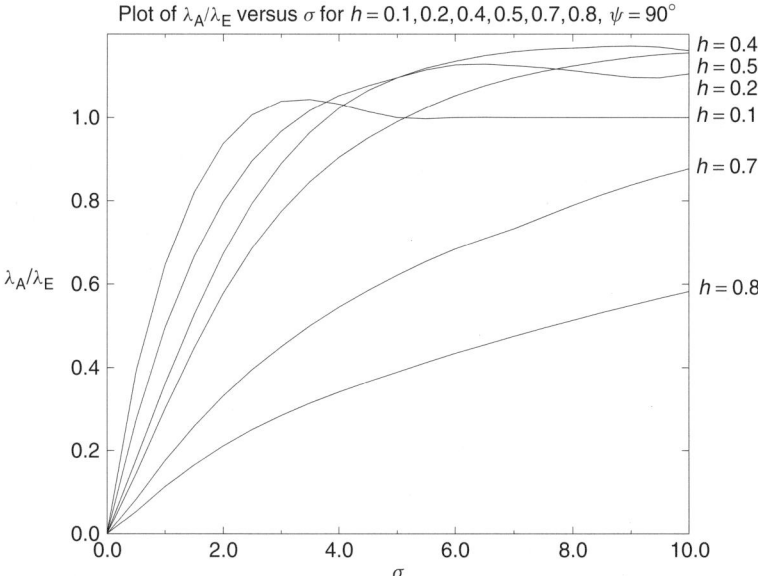

FIGURE 16. Comparison of the solution $\lambda_1(\pi/2)$ yielded by the asymptotic formula [Eq. (3.41)] in the high-damping limit $a \to \infty$ with the exact numerical solution. It is apparent that the systematic error in Fig. 13 of Ref. 46 is removed by the correct asymptotic formula.

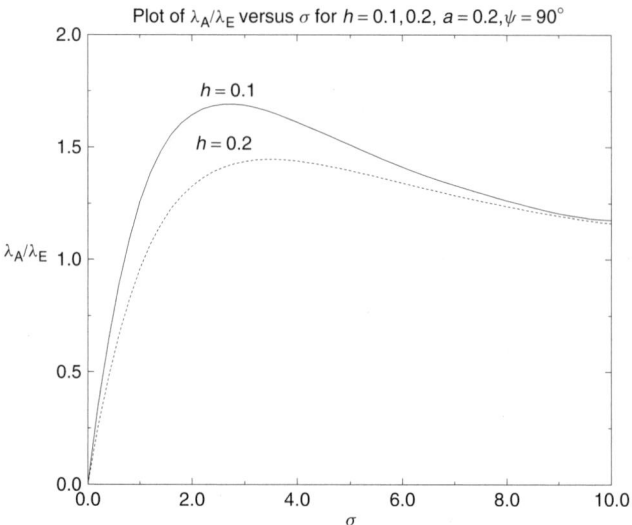

FIGURE 17. Ratio of the asymptotic [λ_A yielded by Eq. (3.41)] and exact expressions for λ_1 with a finite (= 0.2) providing reasonable constancy, and thus indicating that Eq. (3.41) provides a useful asymptotic estimate in these cases.

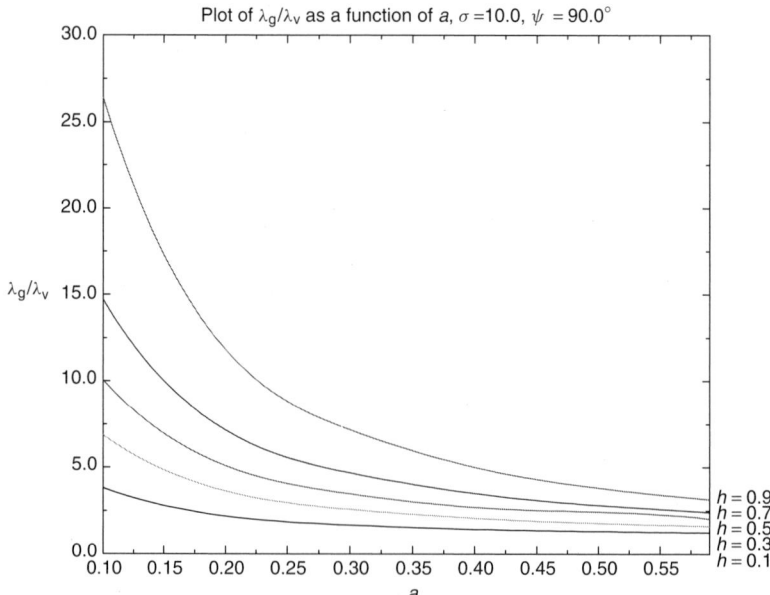

FIGURE 18. Ratio of [$\lambda_A(a) = \lambda_g$] and [$\lambda_A(\infty) = \lambda_v$] showing the variation of $\lambda_1(a)/\lambda_1(\infty)$ for various values of the parameter a. Clearly the effect of the gyromagnetic term is to decrease the longitudinal relaxation time.

This is an important observation and can be understood by means of the following discussion. Without an applied field, or with an applied field and $\psi = 0$, the gyromagnetic term in the Gilbert [21] equation [corresponding to the fourth term in Eq. (1.57)] has no effect on the calculation of the Néel [16] relaxation mode. This means that the longitudinal [16] and the transverse (precessional) modes are completely decoupled from each other. The effect of the dimensionless damping factor a on the longitudinal [16] relaxation mode manifests itself solely through the diffusional (τ_N) time, proportional to $(1 + a^2)/a$, which leads to the disappearance of the Néel [16] relaxation in the limiting cases ($a = \infty$ and zero). Now, with an applied field and ψ different from zero, then there is a strong (mode–mode) coupling between the two modes of motion. The mode–mode coupling manifests itself as a geometric dependence of the prefactor of λ_1 on a which for $\psi = \pi/2$ for example is

$$= \frac{\sigma\sqrt{1 + h}}{\pi\sqrt{h}}(1 - 2h + \sqrt{1 + 4a^{-2}h(1 - h)}) \qquad (4.39)$$

which shall be used for the purpose of discussion. It is evident that the precessional mode corresponds to very high frequencies [i.e., to very short times (much shorter than the Néel [16] time when σ is not too small; in fact, when Eq. (3.41) is valid] and that the coupling leads to a decrease of the longitudinal [16] time compared to the case where no coupling acts. Thus, the variation of the relaxation time of the Néel mode with a is governed by two factors: the diffusional time τ_N proportional to $(1 + a^2)/a$ and the mode–mode coupling effect (that is the decrease of $1/\lambda_1$ with a).

Figures 19–22 show the behavior of λ_E for $\psi = \pi/4$ compared with the asymptotic (λ_A) estimate of Eq. (3.28). It is apparent that the asymptotic estimate yielded by Eqs. (3.29) and (3.38)–(3.51) again provides a good approximation to the exact solution. We remark, however, that higher values of σ at given h have to be taken in this instance in order that the asymptotic formula should be valid as the effective barrier height is only half that for $\psi = 0$ or $\psi = \pi/2$. This makes numerical calculation more difficult to carry out as a larger matrix size must be taken in order to ensure convergence.

Inspection of Figs. 21 and 22 shows that the constancy of λ_A/λ_E is also remarkable for these cases for values of σ larger than 8 for the reduced field h given (noting that $h_c = 0.5$ for $\psi = \pi/4$). It is apparent from Figs. 19 and 20 that the systematic error induced by the axially symmetric approximation [Eqs. (101) and (102) and Fig. 14 of Ref. 46] is again eliminated by use of the correct asymptotic formula [Eqs. (3.29), (3.38)–(3.51)]. In Fig. 23 we show the variation of $\lambda_A(a)/\lambda_A(\infty)$ for $\psi = \pi/4$ with a, which again shows that the effect of the gyromagnetic term is to reduce the relaxation time.

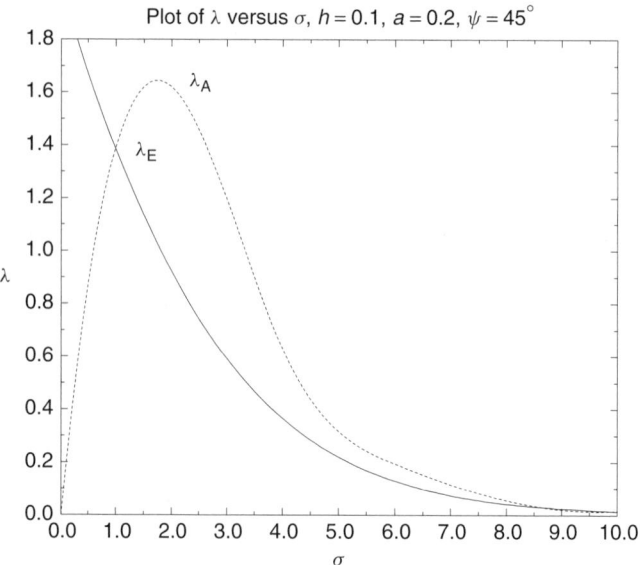

FIGURE 19. Comparison of exact $(\lambda_1 = \lambda_E)$ and asymptotic $(\lambda_1 = \lambda_A)$ expressions including the gyromagnetic term for $\psi = 45°$, $h = 0.1$.

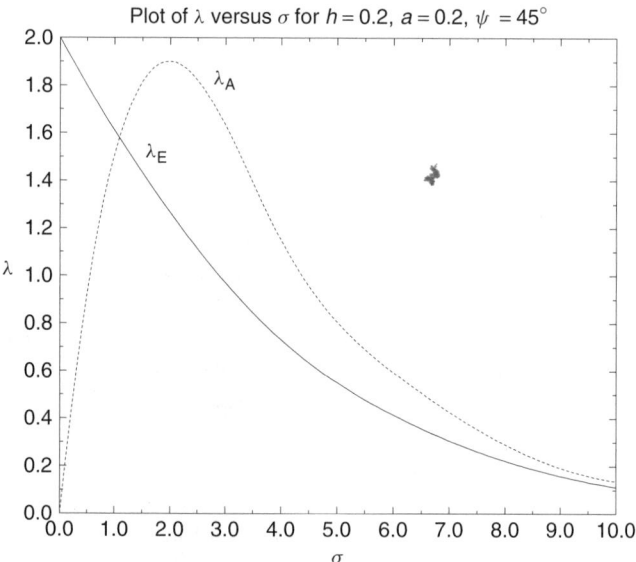

FIGURE 20. Comparison of exact $(\lambda_1 = \lambda_E)$ and asymptotic $(\lambda_1 = \lambda_A)$ expressions including the gyromagnetic term for $\psi = 45°$, $h = 0.2$. It is evident from Figs. 19 and 20 that Eq. (3.29) yields a good approximation to the exact solution.

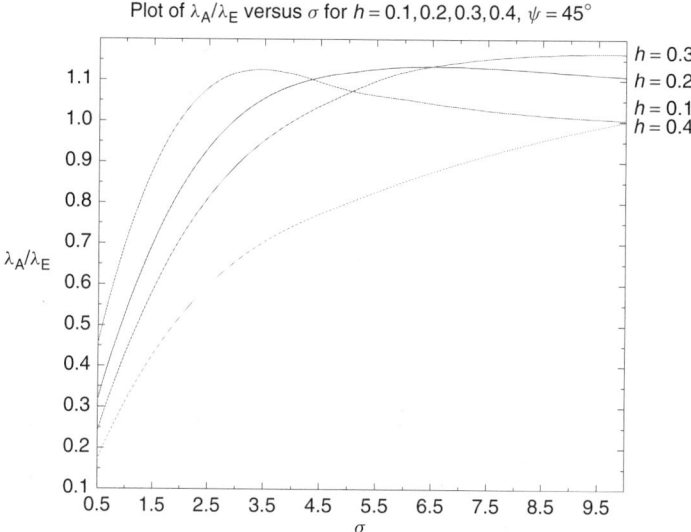

FIGURE 21. Comparison of the solution $[\lambda_1(\pi/4)]$ yielded by the asymptotic formula of Eq. (3.29) in the high-damping limit ($a \to \infty$) with the exact numerical solution; the constancy indicates that the asymptotic estimate again yields a good approximation to the exact solution.

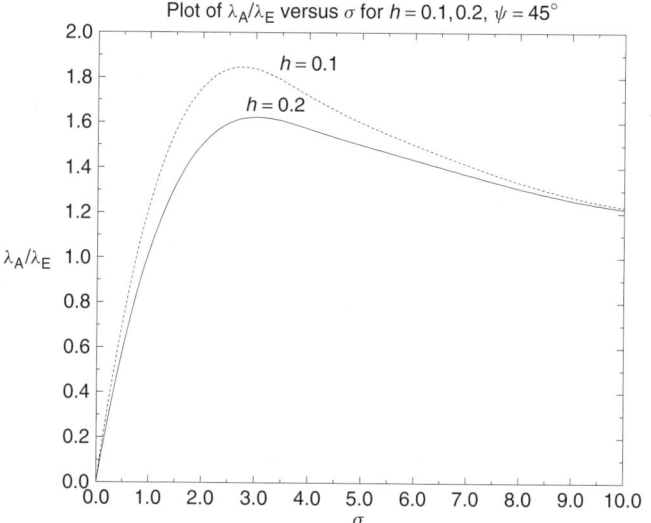

FIGURE 22. Ratio of the asymptotic expression for λ_1 yielded by Eq. (3.29) with the exact numerical solution as a function of σ for $\psi = 45°$. It is evident from the constancy of the plot that the asymptotic estimate of Eq. (3.29) provides a good approximation to the exact value of λ_1 for $\psi = 45°$ also.

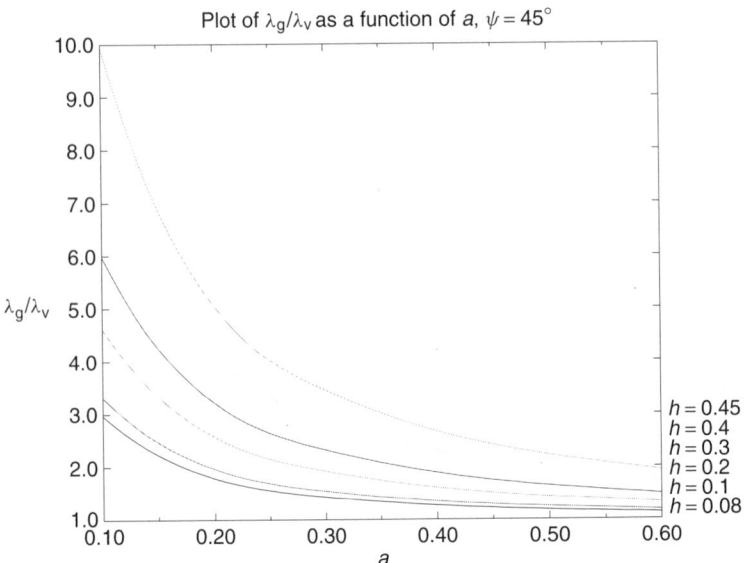

FIGURE 23. Ratio of $[\lambda_A(a) = \lambda_g]$ and $[\lambda_A(\infty) = \lambda_v]$ showing the variation of $\lambda_A(a)/\lambda_A(\infty)$ for various values of a, which indicates that just as for $\psi = 90°$, the effect of the gyromagnetic term is to decrease the relaxation time.

In situations where the value of ψ differs from 0, $\pi/4$, $\pi/2$, the quartic equation (3.26) is solved numerically and the roots ordered using $-1 \leq x_2 \leq x_0 \leq x_0' \leq x_1 \leq 1$. The root x_0' corresponding to the anisotropic maximum at $\varphi = \pi$ is not required. On proceeding in this way it is then possible to compare the variation of λ_1 with the angle ψ as predicted by the exact solution with that yielded by the asymptotic estimate Eq. (3.29), evidence of which is shown by Fig. 24. Figure 25, which is the reciprocal of Fig. 24, shows that in all cases the Néel [16] relaxation time has an absolute maximum at $\psi = \pi/2$ and a minimum near $\psi = \pi/4$. It is apparent that the asymptotic formula Eq. (3.29) again provides a reasonable estimate of λ_1. The deviation near $\psi = 0$ is consistent with the fact that the asymptotic estimate Eq. (3.29) is no longer valid in the region $\psi = 0$ (the validity of the asymptotic formula has been discussed in detail in Section III). The asymptote should be constructed in this case using the Brown–Aharoni formula, Eq. (3.36). This formula does not exhibit the characteristic $\sqrt{h}$ dependence of the nonaxially symmetric asymptote; nor does it contain the dimensionless damping factor a arising from the inclusion of the gyromagnetic term that occurs when one uses Eq. (3.29). The formula [Eq. (3.29)] is thus invalid for very small values of $h \ll 0.1$, and for $\psi = 0$ since there is no dependence on the gyromagnetic term a in this axially symmetric situation.

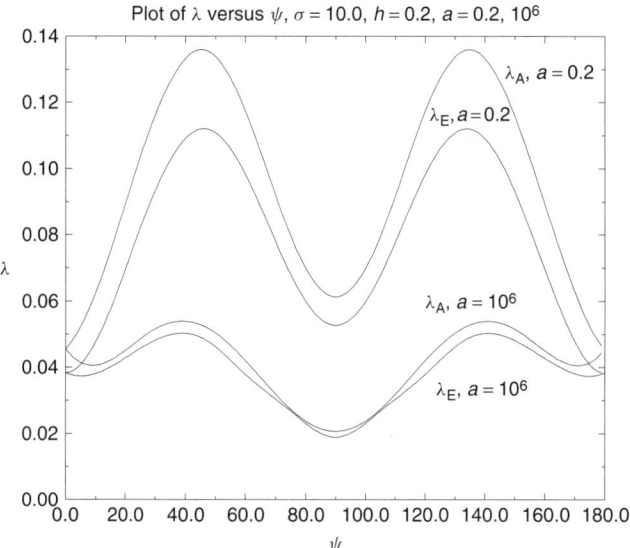

FIGURE 24.   Variation of the exact ($\lambda_1 = \lambda_E$) and asymptotic ($\lambda_1 = \lambda_A$) values of λ_1 with ψ. The asymptotic estimate given by Eq. (3.29) provides a good approximation to the exact solution by reasonably predicting the behavior of λ_1. The deviation near $\psi = 0°$ is due to the fact that the asymptotic estimate of Eq. (3.29) is no longer valid in the region $\psi = 0°$.

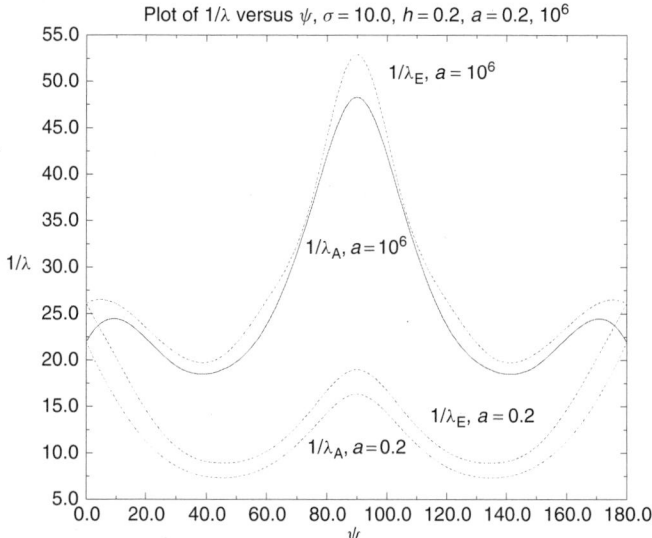

FIGURE 25.   Variation of the reciprocals of the exact and asymptotic values of λ_1 with ψ showing that the Néel relaxation time is an absolute maximum at $\psi = 90°$ and a minimum at $\psi = 45°$.

In view of the acceptable approximation provided by the asymptotic formula of Eq. (3.29), we have shown in Fig. 26 the variation of λ_A with ψ for $a = \infty$ and $a = 0.2$, confirming that the effect of the gyromagnetic term is to reduce the relaxation time. Finally, in Figs. 27 and 28 we show variation of $\lambda_A(a)/\lambda_A(\infty)$ with a as computed for $\psi = 30°$ and $\psi = 75°$ using the asymptotic formula, Eq. (3.29).

To conclude this section, the results provided by Figs. 14–28 confirm that the asymptotic formula following the Brown [4] approach [Eq. (3.29)] is capable of providing an accurate description of the Néel [16] relaxation time for the present nonaxially symmetric problem. Brown's [4] approach which appears to have been suggested by the earlier work of Smith and de Rozario [34] is in effect an adaptation to curvilinear coordinates of the intermediate to high damping asymptotic formula of Kramers [25] [Eq. (25) of his 1940 paper] for escape rates for a process governed by the Fokker–Planck equation in phase space (the Klein–Kramers [25] equation) to the present problem. In the limit of very small damping (i.e., a finite) Eq. (3.41) will reduce to a formula analogous to Eq. (26) of Kramers [25] (the transition-state value in his nomenclature). A discussion of the problem in the context of the Kramers [25] calculation has also been provided by Klik and Gunther [49] in their 1990 paper. In order to further assist the reader, a representative selection of λ_1 for various angles, and for $a = 10^6$ are presented in Tables X–XIII. The

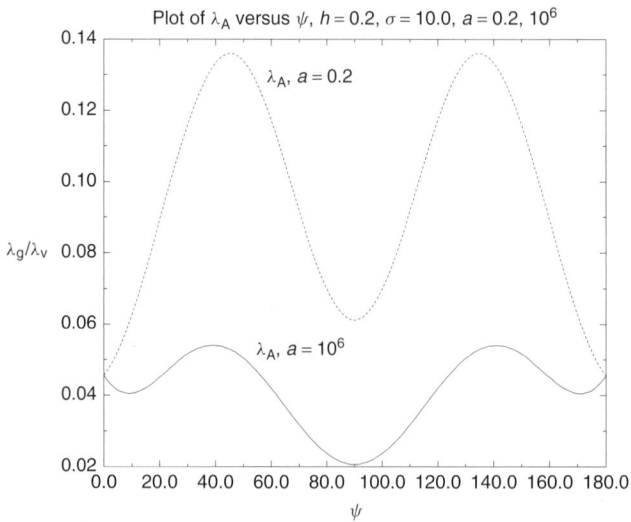

FIGURE 26. Variation of the asymptotic values of λ_1, when $\lambda_A(\infty) = \lambda_A, a = 10^6$, and $\lambda_A(0.2) = \lambda_A, a = 0.2$ with ψ. The plot confirms that the effect of the gyromagnetic term is to reduce the relaxation time.

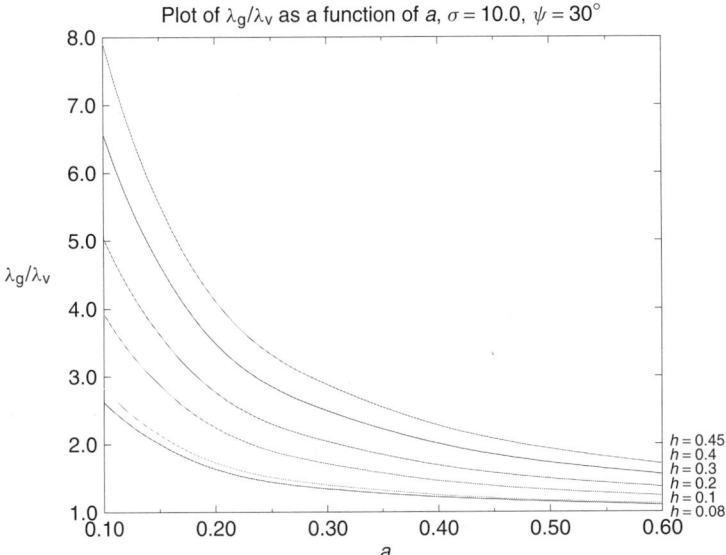

FIGURE 27. Ratio of $\lambda_A(a) = \lambda_g$ with $\lambda_A(\infty) = \lambda_v$ again showing the variation of $\lambda_A(a)/\lambda_A(\infty)$ for various values of a and $\psi = 30°$, confirming just as in Fig. 26, the effect of the gyromagnetic term is to reduce the relaxation time.

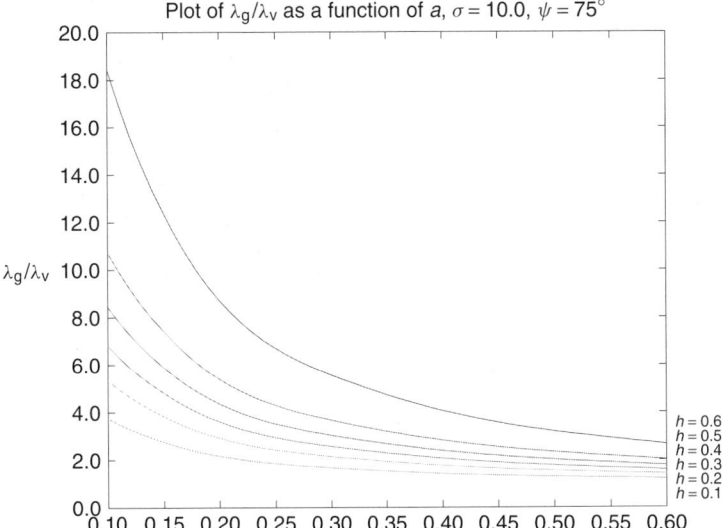

FIGURE 28. Ratio of $\lambda_A(a) = \lambda_g$, and $\lambda_A(\infty) = \lambda_v$ showing the variation of $\lambda_A(a)/\lambda_A(\infty)$ for various values of a and $\psi = 75°$. Again the effect of the gyromagnetic term is to decrease the relaxation time.

TABLE X

The Lowest Nonvanishing Eigenvalue λ_1 for Various Values of the Barrier Height (σ) and Field Parameters (h); $\psi = 0°$

h	$\sigma = 0.2$	$\sigma = 0.5$	$\sigma = 1.0$	$\sigma = 2.0$	$\sigma = 5.0$	$\sigma = 10.0$
0.01	1.844	1.627	1.306	0.8079	0.1354	0.002920
0.1	1.845	1.629	1.314	0.8322	0.1876	0.00860
0.2	1.846	1.634	1.336	0.9056	0.3134	0.03830
0.4	1.849	1.659	1.426	1.196	0.9109	0.3846
0.5	1.852	1.676	1.493	1.412	1.398	0.8829
0.7	1.860	1.723	1.671	1.974	2.782	2.927
0.8	1.864	1.753	1.783	2.317	3.670	4.524
1.0	1.876	1.823	2.049	3.121	5.824	8.777

TABLE XI

The Lowest Nonvanishing Eigenvalue λ_1 for Various Values of the Barrier Height (σ) and Field Parameters (h); $\psi = 30°$

h	$\sigma = 0.2$	$\sigma = 0.5$	$\sigma = 1.0$	$\sigma = 2.0$	$\sigma = 5.0$	$\sigma = 10.0$
0.01	1.844	1.627	1.306	0.8079	0.1358	0.002924
0.1	1.845	1.629	1.313	0.8297	0.1743	0.008450
0.2	1.845	1.634	1.334	0.8691	0.306	0.04819
0.4	1.850	1.656	1.416	1.165	1.022	0.8447
0.5	1.852	1.672	1.477	1.371	1.706	2.248
0.7	1.859	1.716	1.641	1.929	3.905	8.499
0.8	1.863	1.743	1.744	2.284	5.447	13.50
1.0	1.874	1.808	1.989	3.148	9.349	26.94

TABLE XII

The Lowest Nonvanishing Eigenvalue λ_1 for Various Values of the Barrier Height (σ) and Field Parameter (h); $\psi = 45°$

h	$\sigma = 0.2$	$\sigma = 0.5$	$\sigma = 1.0$	$\sigma = 2.0$	$\sigma = 5.0$	$\sigma = 10.0$
0.01	1.844	1.627	1.306	0.8079	0.1357	0.002918
0.1	1.845	1.628	1.312	0.8272	0.1694	0.007901
0.2	1.845	1.634	1.331	0.8861	0.2895	0.04788
0.4	1.849	1.654	1.405	1.127	0.9981	0.9523
0.5	1.851	1.669	1.461	1.313	1.712	2.594
0.7	1.858	1.708	1.609	1.825	4.108	10.05
0.8	1.862	1.733	1.701	2.153	5.850	16.22

TABLE XIII

The Lowest Nonvanishing Eigenvalue λ_1 for Various Values of the Barrier Height (σ) and Field Parameter (h); $\psi = 90°$

h	$\sigma = 0.2$	$\sigma = 0.5$	$\sigma = 1.0$	$\sigma = 2.0$	$\sigma = 5.0$	$\sigma = 10.0$
0.01	1.844	1.627	1.306	0.8073	0.1356	0.002906
0.1	1.845	1.628	1.311	0.8220	0.1577	0.005649
0.2	1.845	1.632	1.326	0.8649	0.2276	0.01889
0.4	1.848	1.649	1.383	1.034	0.5392	0.1676
0.5	1.850	1.661	1.426	1.158	0.7968	0.3918
0.7	1.856	1.693	1.540	1.478	1.5354	1.384
0.8	1.859	1.714	1.609	1.671	2.014	2.198
1.0	1.867	1.762	1.774	2.114	3.167	4.585

following section addresses the numerical calculation of the correlation time. Tables C.1–C.28 of Appendix C provide further results for the exact values for λ_1 for a range of σ and h. The values of a represented in these tables are $a = 0.2, 0.3, 0.4,$ and 1.0.

C. Numerical Calculation of the Correlation Time

The correlation time was computed by programming Eq. (2.60) in FORTRAN. The preceding calculations involved setting up the vector B of Eq. (2.40) (which is denoted by F in the subroutine WM) so that Eq. (2.44) may be used to evaluate the equilibrium values vector U_∞. A LAPACK linear equation solver DGESV was used to compute the solution to the real system of linear equations [Eq. (2.44)] so that

$$AU_\infty = -B \tag{4.40}$$

where A is the Fokker–Planck matrix of Eq. (2.38). LU decomposition with partial pivoting and row interchanges is used to factor the Fokker–Planck matrix A as

$$A = PLU \tag{4.41}$$

where A is a permutation matrix, L is a unit lower triangular, and U is an upper triangular. The factored form of A is then used to solve the system of equations $AU_\infty = -B$. The next step in the calculation of the correlation time is to compute the matrix of Eq. (2.52). The matrix is computed in the same manner as the Fokker–Planck matrix as described in detail in Section II where $e_{l,m,p,q}$ of Eq. (4.1) are replaced by $w_{l,m,p,q}$ of Eq. (2.49). So that the Fokker–Planck matrix A and the matrix used in the computation of the correlation time W are of the same order, all unused terms are replaced by zero, so that we have a set of equations for the matrix elements similar to those given by

Eqs. (4.2)–(4.15), namely

$$w_{l,m,l-2,m} = 0 \tag{4.42}$$

$$w_{l,m,l,m} = -\cos\psi u_{1,0}(\infty) - \sin\psi u_{1,1}(\infty) \tag{4.43}$$

$$w_{l,m,l+2,m} = 0 \tag{4.44}$$

$$w_{l,m,l-1,m} = \frac{\cos\psi(l+m)}{(2l+1)} \tag{4.45}$$

$$w_{l,m,l+1,m} = \frac{\cos\psi(l-m+1)}{(2l+1)} \tag{4.46}$$

$$w_{l,m,l-1,m-1} = \frac{\sin\psi(l+m)(l+m-1)}{2(2l+1)} \tag{4.47}$$

$$w_{l,m,l-1,m-1} = -\frac{\sin\psi(l-m+2)(l+m-1)}{2(2l+1)} \tag{4.48}$$

$$w_{l,m,l-1,m+1} = -\frac{\sin\psi}{2(2l+1)} \tag{4.49}$$

$$w_{l,m,l+1,m+1} = \frac{\sin\psi}{2(2l+1)} \tag{4.50}$$

and since there are no imaginary terms included in the matrix elements here, those terms in the corresponding array positions are set to zero. For each element in Eqs. (4.42)–(4.50), we compute the value of the indices p, q. This matrix may be computed using the same procedure used previously to compute the Fokker–Planck matrix of Eq. (2.38) and consequently, Eq. (4.36). Following Eqs. (4.18)–(4.39), the matrix W is obtained by replacing $A_{l,m,p,q}$ by $W_{l,m,p,q}$, $f_{l,m,p,q}$ by $w_{l,m,p,q}$ and $g_{l,m,p,q}$ by zero (since there are no imaginary terms involved here). $\rho_{p,q}$ is as before so that

$$W = \begin{bmatrix}
w_{1,0,1,0} & \sqrt{2}w_{1,0,1,1} & 0 & w_{1,0,2,0} & \sqrt{2}w_{1,0,2,1} & 0 & \cdots \\
\sqrt{2}w_{1,1,1,0} & w_{1,1,1,1}, & 0 & \frac{1}{\sqrt{2}}w_{1,1,2,0} & w_{1,1,2,1} & 0 & \cdots \\
0 & 0 & w_{1,1,1,1} & 0 & 0 & w_{1,1,2,1} & \cdots \\
w_{2,0,1,0} & 2w_{2,0,1,1} & 0 & w_{2,0,2,0} & 2w_{2,0,2,1} & 0 & \cdots \\
\sqrt{2}w_{2,1,1,0} & w_{2,1,1,1} & 0 & \frac{1}{\sqrt{2}}w_{2,1,2,0} & w_{2,1,2,1} & 0 & \cdots \\
0 & 0 & w_{2,1,1,1} & 0 & 0 & w_{2,1,2,1} & \cdots \\
\sqrt{2}w_{2,2,1,0} & w_{2,2,1,1}, & 0 & \frac{1}{\sqrt{2}}w_{2,2,2,0} & w_{2,2,2,1} & 0 & \cdots \\
0 & 0 & w_{2,2,1,1} & 0 & 0 & w_{2,2,2,1} & \cdots \\
\vdots & \vdots & \vdots & \vdots & \vdots & \vdots & \vdots
\end{bmatrix}$$

$$\tag{4.51}$$

After setting up the W matrix, C_0 of Eq. (2.51) is calculated. The zero-frequency Laplace transforms vector is then determined by solving Eq. (2.53) for $\tilde{C}(0)$. This is computed by again calling the LAPACK linear equation solver DGESV. The correlation time T_c is then computed by Eq. 2.60 (where $\tilde{c}_{1,0}(0)$ and $\tilde{c}_{1,1}(0)$ are the first and second elements of $\tilde{C}(0)$ respectively, and $c_{1,0}(0)$ and $c_{1,1}(0)$ are the first and second elements of $\tilde{C}(0)$ (after the call to subroutine DGESV)).

The results of numerical calculation of the correlation time (when $a = 10^6$) are shown in Figs. 29–32 and Tables XIV–XVII. Further results for the correlation time T_c (for a range of σ, a, h) are provided in Appendix D. An alternative procedure for calculating the correlation time T_c is to construct the matrix of Eq. (2.55) from the eigenvectors of the Fokker–Planck matrix A and then solve Eq. (2.57). If the amplitudes of Eq. (2.59) are calculated, then the correlation time is then as given by Eq. (2.60).

On examining Figs. 29–32, it is evident from each plot that the correlation time T_c exhibits significant departure from increasing exponential behavior for large values of the barrier-height parameter σ, where h is also sufficiently large, but complies with the condition $h < h_c$ [where h_c is as given by Eq. (3.12) or (3.13)] so that the bistable nature of the potential is preserved. Further evidence of this behavior is found by examining Tables XVIII–XXI and Figs. 33 and 34 where in all cases, the product of the correlation time

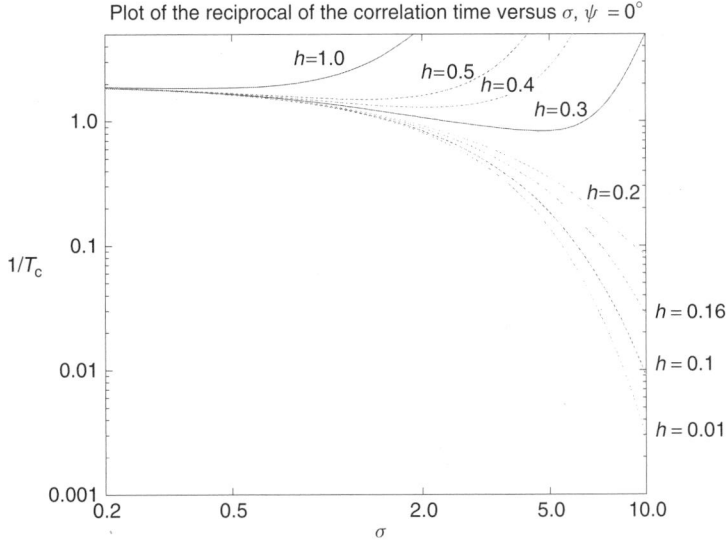

FIGURE 29. Variation of the reciprocal of the correlation time with the barrier-height parameter σ for $\psi = 0°$.

Plot of the reciprocal of the correlation time versus σ, $\psi = 45°$

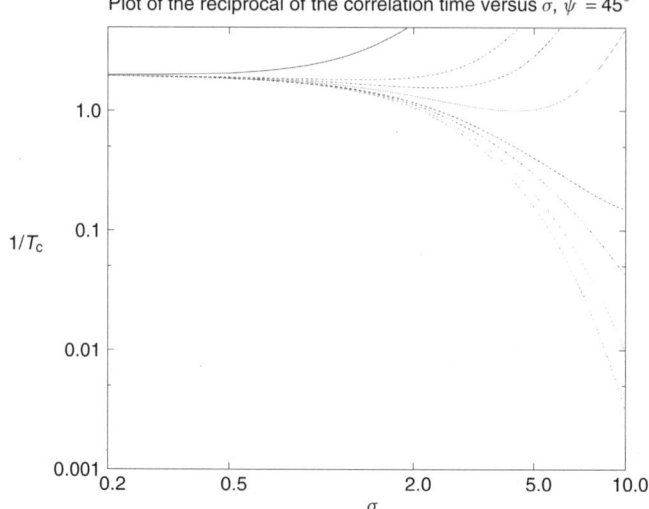

FIGURE 30. Variation of the reciprocal of the correlation time $1/T_c$ with the barrier-height parameter σ for $\psi = 45°$.

Plot of the reciprocal of the correlation time versus σ, $\psi = 60°$

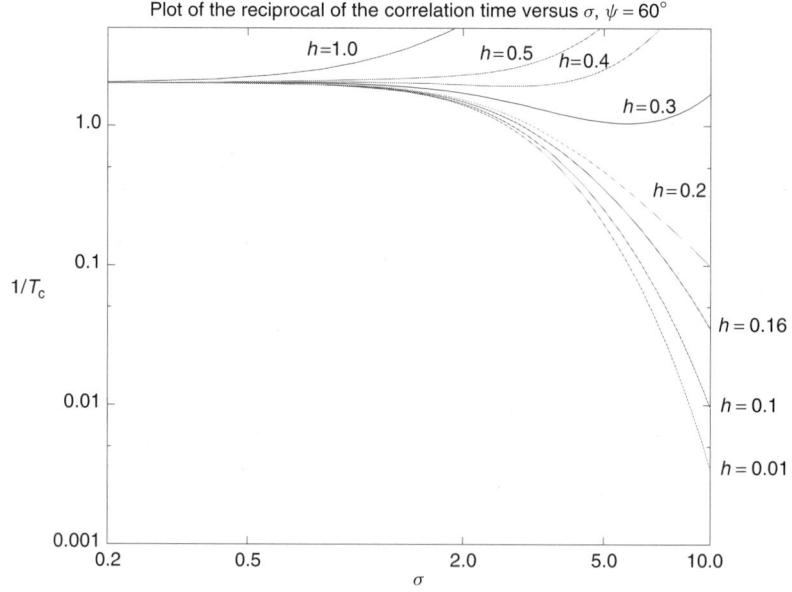

FIGURE 31. Variation of the reciprocal of the correlation time $1/T_c$ with the barrier-height parameter σ for $\psi = 60°$. It is evident from Figs. 29–31 that the correlation time T_c exhibits significant departure from increasing exponential behavior as σ increases and when h is large (but no larger than h_c, at which point the bistable structure of the potential is destroyed).

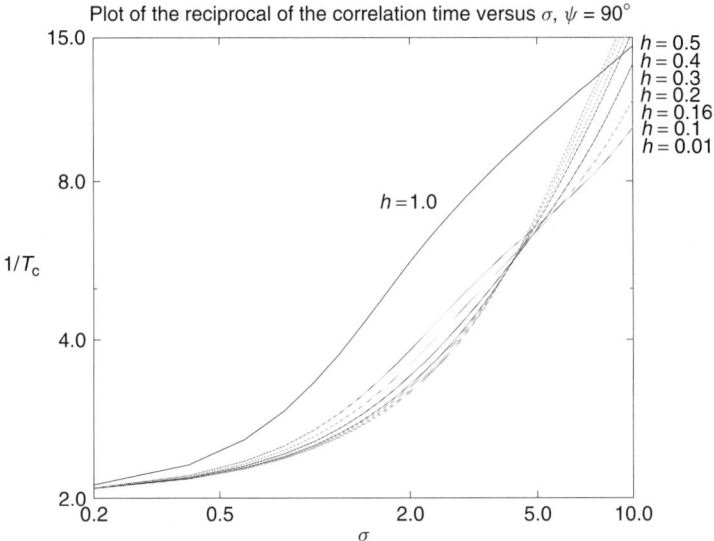

FIGURE 32. Variation of the reciprocal of the correlation time $1/T_c$ with the barrier-height parameter σ for the case when $\psi = 90°$. The correlation time exhibits the most significant departure from exponential behavior at $\psi = 90°$.

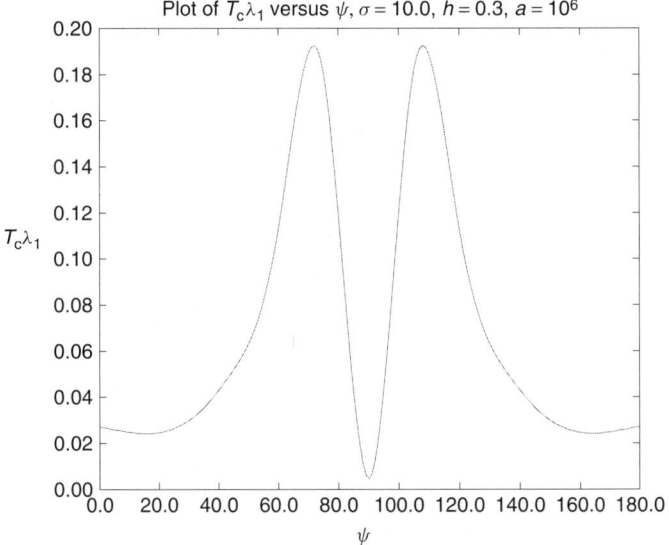

FIGURE 33. Variation of the product of the exact values of the lowest eigenvalue λ_1 and the correlation time T_c with ψ. The figure shows significant departure from the increasing exponential behavior at $\psi = 90°$.

TABLE XIV

Reciprocal of the Correlation Time $1/T_c$ for Various Values of the Barrier Height (σ) and Field (h) Parameters; $\psi = 0°$

h	$\sigma = 0.2$	$\sigma = 0.5$	$\sigma = 1.0$	$\sigma = 2.0$	$\sigma = 5.0$	$\sigma = 10.0$
0.01	1.844	1.628	1.309	0.8132	0.1375	0.002943
0.1	1.845	1.630	1.317	0.8404	0.1833	0.008923
0.2	1.846	1.637	1.343	0.9246	0.3496	0.08420
0.4	1.850	1.663	1.446	1.292	2.738	43.33
0.5	1.853	1.683	1.526	1.607	9.160	51.79
0.7	1.862	1.737	1.744	2.648	24.58	60.75
0.8	1.867	1.771	1.886	3.451	28.05	65.11
1.0	1.880	1.854	2.244	5.728	33.17	73.68

TABLE XV

Reciprocal of the Correlation Time $1/T_c$ for Various Values of the Barrier Height (σ) and Field (h) Parameters; $\psi = 30°$

h	$\sigma = 0.2$	$\sigma = 0.5$	$\sigma = 1.0$	$\sigma = 2.0$	$\sigma = 5.0$	$\sigma = 10.0$
0.01	1.896	1.726	1.432	0.8989	0.1444	0.002998
0.1	1.896	1.729	1.440	0.9269	0.1906	0.009579
0.2	1.897	1.735	1.466	1.014	0.3785	0.1647
0.4	1.901	1.762	1.571	1.396	3.102	27.49
0.5	1.904	1.782	1.652	1.720	8.464	30.96
0.7	1.913	1.836	1.872	2.756	18.31	35.75
0.8	1.918	1.870	2.015	3.518	20.54	38.16
1.0	1.931	1.952	2.371	5.560	23.88	43.09

TABLE XVI

Reciprocal of the Correlation Time $1/T_c$ for Various Values of the Barrier Height (σ) and Field (h) Parameters; $\psi = 45°$

h	$\sigma = 0.2$	$\sigma = 0.5$	$\sigma = 1.0$	$\sigma = 2.0$	$\sigma = 5.0$	$\sigma = 10.0$
0.01	1.952	1.850	1.614	1.053	0.1582	0.003109
0.1	1.952	1.852	1.623	1.083	0.2053	0.009641
0.2	1.953	1.859	1.649	1.175	0.4052	0.1510
0.4	1.957	1.885	1.756	1.576	2.892	22.26
0.5	1.960	1.906	1.837	1.909	6.923	26.33
0.7	1.969	1.959	2.059	2.919	14.77	29.85
0.8	1.974	1.993	2.201	3.621	16.75	36.62

TABLE XVII

Reciprocal of the Correlation Time $1/T_c$ for Various Values of the Barrier Height (σ) and Field (h) Parameters; $\psi = 90°$

h	$\sigma = 0.2$	$\sigma = 0.5$	$\sigma = 1.0$	$\sigma = 2.0$	$\sigma = 5.0$	$\sigma = 10.0$
0.01	2.083	2.220	2.484	3.172	6.949	17.31
0.1	2.084	2.222	2.492	3.198	6.860	16.81
0.2	2.084	2.229	2.517	3.277	6.657	15.37
0.4	2.089	2.255	2.617	3.588	6.434	11.38
0.5	2.092	2.274	2.692	3.818	6.575	10.10
0.7	2.100	2.327	2.892	4.415	7.448	9.975
0.8	2.106	2.340	3.018	4.776	8.162	10.938
1.0	2.118	2.439	3.318	5.608	10.08	14.441

TABLE XVIII

Product of the Lowest Eigenvalue λ_1 and the Correlation Time T_c for Various Values of the Barrier Height (σ) and Field (h) Parameters; $\psi = 0°$

h	$\sigma = 0.2$	$\sigma = 0.5$	$\sigma = 1.0$	$\sigma = 2.0$	$\sigma = 5.0$	$\sigma = 10.0$
0.01	1.0	0.9994	0.9977	0.9935	0.9847	0.9922
0.1	1.0	0.9994	0.9977	0.9902	0.9744	0.9638
0.2	1.0	0.9982	0.9948	0.9794	0.8965	0.4549
0.4	0.9994	0.9976	0.9862	0.9257	0.3326	0.008875
0.5	0.9995	0.9958	0.9784	0.8787	0.1526	0.01704
0.7	0.9989	0.9919	0.9581	0.7455	0.1132	0.0482
0.8	0.9984	0.9898	0.9454	0.6714	0.1308	0.06949
1.0	0.9979	0.9833	0.9131	0.5449	0.1756	0.1191

TABLE XVIX

Product of the Lowest Eigenvalue λ_1 and the Correlation Time T_c for Various Values of the Barrier Height (σ) and Field (h) Parameters; $\psi = 30°$

h	$\sigma = 0.2$	$\sigma = 0.5$	$\sigma = 1.0$	$\sigma = 2.0$	$\sigma = 5.0$	$\sigma = 10.0$
0.01	0.9730	0.9424	0.9123	0.8988	0.9401	0.9753
0.1	0.9729	0.9422	0.9116	0.8951	0.9145	0.8822
0.2	0.9729	0.9418	0.9095	0.8837	0.8087	0.2926
0.4	0.9726	0.9398	0.9008	0.8349	0.3293	0.03073
0.5	0.9723	0.9384	0.8943	0.7966	0.2015	0.07260
0.7	0.9717	0.9628	0.8765	0.6999	0.2133	0.2377
0.8	0.9713	0.9320	0.8656	0.6942	0.2652	0.3538
1.0	0.9703	0.9261	0.8389	0.5662	0.3916	0.6251

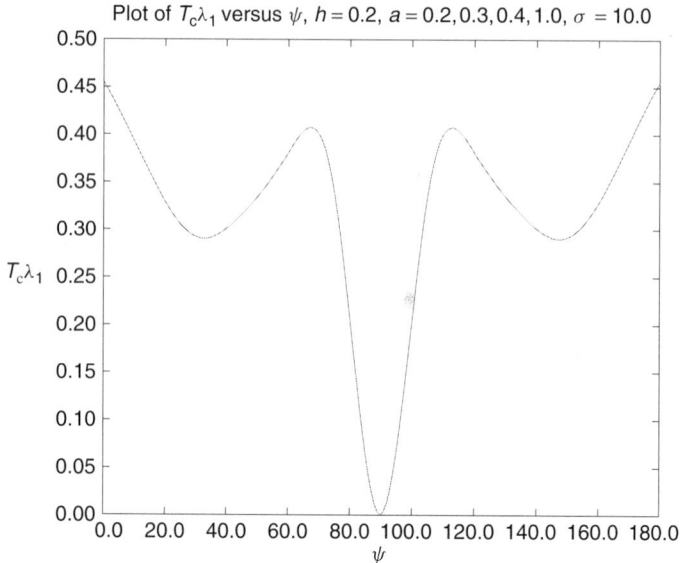

FIGURE 34. Variation on $T_c\lambda_1$ with ψ when the gyromagnetic term is included for various values of a. The figure shows that the gyromagnetic term has no effect on the behavior of $T_c\lambda_1$.

TABLE XX
Product of the Lowest Eigenvalue λ_1 and the Correlation Time T_c for Various Values of the Barrier Height (σ) and Field (h) Parameters; $\psi = 45°$

h	$\sigma = 0.2$	$\sigma = 0.5$	$\sigma = 1.0$	$\sigma = 2.0$	$\sigma = 5.0$	$\sigma = 10.0$
0.01	0.9450	0.8796	0.8094	0.7674	0.8542	0.9386
0.1	0.9449	0.8795	0.8086	0.7641	0.8251	0.8196
0.2	0.9448	0.8798	0.8071	0.7541	0.7146	0.3172
0.4	0.9444	0.8771	0.8002	0.7154	0.3451	0.04278
0.5	0.9441	0.8756	0.7950	0.6881	0.2473	0.09849
0.7	0.9434	0.8718	0.7812	0.6251	0.2782	0.3366
0.8	0.9429	0.8694	0.7727	0.5946	0.3493	0.4430

(T_c) and lowest eigenvalue λ_1 displays a marked departure from unity for sufficiently large h and σ.

Moreover, this effect has a strong dependence on the angle ψ, the most marked departure occurring at $\psi = \pi/2$. This is a vivid example of the effect of a strong uniform field that was first described in Ref. 46 and explained physically by Garanin [70] as being due to the depletion of the shallower of the two potential wells by the field so that the Néel [16] mode is completely swamped by the relatively fast modes in the deeper potential well.

TABLE XXI

Product of the Lowest Eigenvalue λ_1 and the Correlation Time T_c for Various Values of the Barrier Height (σ) and Field (h) Parameters; $\psi = 90°$

h	$\sigma = 0.2$	$\sigma = 0.5$	$\sigma = 1.0$	$\sigma = 2.0$	$\sigma = 5.0$	$\sigma = 10.0$
0.01	0.8854	0.7329	0.5258	0.2545	0.01952	0.0001679
0.1	0.8853	0.7328	0.5261	0.2571	0.02299	0.0003360
0.2	0.8852	0.7325	0.5266	0.2640	0.03418	0.001229
0.4	0.8848	0.7312	0.5285	0.2881	0.08380	0.01472
0.5	0.8844	0.7303	0.5298	0.3032	0.1212	0.03878
0.7	0.8835	0.7277	0.5322	0.3348	0.2062	0.1388
0.8	0.8829	0.7261	0.5333	0.3499	0.2468	0.2010
1.0	0.8815	0.7224	0.5346	0.3770	0.3141	0.3174

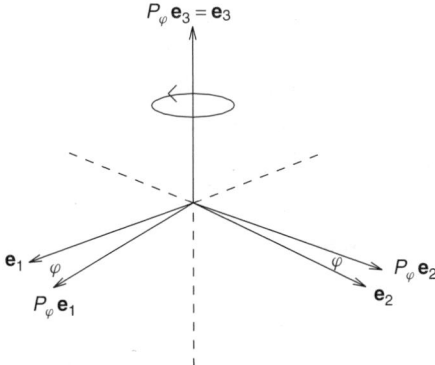

FIGURE 35.   Orthogonal transformation for rotation through an angle φ about the $\mathbf{e}_3$ axis.

The behavior of the correlation time T_c for small values of h is similar to that of $1/\lambda_1$, except for values of ψ in the range of 60–90°, where a strong decrease is observed with T_c becoming very small. This is also true of the product $T_c\lambda_1$ (Figs. 34 and 35 and may be explained as follows. For $\psi = \pi/2$, the two minima of the energy are symmetric with respect to the direction of the applied field, and so their populations are equal in thermodynamic equilibrium. Starting from such a situation, on slightly decreasing the field, there is a quasi-instantaneous small change of the angle of the minima and thus the magnetic moment direction (so that only rapidly damped fast oscillations occur). Thus, the area under the magnetization decay curve (i.e., T_c) is very small. On the other hand, when ψ decreases from 90°, the two minima become asymmetric with respect to the applied field direction, and the decay of the magnetization appears along this direction. Here T_c increases, and takes on values of the order of $1/\lambda_1$ at $\psi \approx 75°$.

In fact, for $\psi \leq 70°$, the product $T_c\lambda_1$ has only a weak dependence on the value of the dimensionless damping factor a; a slight increase is observed for small values of σ. As far as the dependence on the other parameters is concerned, $T_c\lambda_1$ decreases from unity when ψ increases, with an enhanced decrease if σ is small. On the other hand, $T_c\lambda_1$ slightly increases with σ, and finally is almost independent of h for small values of σ, while it decreases for high values of σ.

It should be emphasized that values of $T_c\lambda_1$ sensibly smaller than unity indicate that the higher (faster) relaxation modes (corresponding to $\lambda_2, \lambda_3, \ldots$) are of importance. Nevertheless, for those experiments that measure the decay of the magnetization over a long period of time, such as field-cooled and zero-field-cooled magnetizations, and thermoremanant magnetization experiments, λ_1 is the correct parameter to calculate since the magnetization is governed by

$$\sum_k A_k e^{-\lambda_k t/2\tau_N} \tag{4.52}$$

and as $\lambda_1 < \lambda_k (k \geq 2)$, the term in $e^{-\lambda_1 t/2\tau_N}$, (i.e., the longest-lived relaxation mode) dominates. However, in those experiments that measure the magnetization behavior over a short time interval (such as susceptibility measurements at high frequency), T_c must be used. Consequently, if $T_c\lambda_1$ is markedly different from unity, then measurements at short and long times do not examine the same phenomena, and therefore the results are not directly comparable.

All the diagrams and results in this section were obtained by using the FORTRAN code for which a full listing is provided in Ref. 36, and a copy may be obtained from the corresponding author. As for the calculation of λ_1, the code allows the calculation of T_c for a range of h, σ, a and ψ. To ensure convergence, each case is computed for successive values of l (where the order of the square matrices A and W is given by $l * (l + 2)$). The calculations are repeated until successive values of T_c lie within a tolerance of 10^{-4}.

All results were obtained using the CRAY-J90 at the Rutherford Appleton Laboratory, requiring no more than two minutes of CPU time for even the most difficult cases. For example, when $\psi = \pi/4$, $h = 0.4$, $a = 0.2$, and $\sigma = 10$, the amount of CPU time required to complete the calculation for successive matrix sizes given by $l = 25$ and $l = 26$, within a tolerance of 10^{-7}, was 78.01 s.

Having computed the lowest eigenvalue λ_1 and the correlation time T_c for a range of h, σ, a, and ψ, and having compared the Néel time with the asymptotic estimates of Section III, we can now present some conclusions, and provide a discussion of possible further research.

V. CONCLUSIONS

A. Possible Further Research

The results presented in previous chapters indicate that significant advancements in the field of ferromagnetics have been obtained. From the 1960s to the 1990s, researchers have attempted to fit an asymptotic formula to the exact value of the relaxation time τ given by $2\tau_N/\lambda_1$. This requires the computation of exact values of λ_1 (the lowest eigenvalue of the associated Fokker–Planck matrix) for a field applied at an oblique angle, for a range of barrier heights (and therefore a range of values of σ) and also including the effect of the dimensionless damping constant a. However, a lack of computing facilities in those years prevented researchers from obtaining exact values of λ_1 for cases of interest.

Present computer technology however has allowed for a comprehensible wide range of exact values for λ_1 and the correlation time T_c using the matrix methods discussed in detail in Sections II–IV to be obtained. Computation of the lowest eigenvalue was expensive in terms of CPU time, although a new procedure for computing these values has recently been presented by Kalmykov et al. [74]. Another method that may be used (to compute T_c) is the matrix continued-fraction method. The procedure requires the use of matrices that are of a much lower order than the matrices required in the methods discussed earlier in chapters Sections II–IV. The new method allows for the computation of T_c for higher barrier heights (and thus higher values of σ); however, the procedure has yet to be successfully adapted for the extraction of the lowest eigenvalue λ_1. Present techniques accommodate the computation of λ_1 for $\sigma \leq 10$ within a reasonable amount of time and without increasing precision levels to an unrealistic limit. However, successful extraction of λ_1 from the continued-fraction method will provide exact values of λ_1 for much higher barrier heights very quickly and accurately in future calculations, although the method is not yet complete.

The problems discussed in this chapter involve only those having uniaxially anisotropic potentials. However, another area that has great influence in the field of ferromagnetics is that of cubic anisotropy. A great deal of work has already been carried out on this topic by Eisenstein and Aharoni [48,76], and their publications provide interesting reading on the subject. The difference between a uniaxial crystal and a crystal possessing cubic anisotropy is that a uniaxial crystal allows for only two orientations (this has been discussed in detail in Section I), whereas for more than two orientations, the case of greatest interest is when the free energy per unit volume V arises from cubic anisotropy. A detailed discussion is provided in Section C of Ref. 17. The problems arising from cubic anisotropy, however, include the fact that the matrix required to compute the lowest eigenvalue λ_1 arises from a 27-term recurrence relation rather than a 13-term one as is the case for uniaxial anisotropy. This, of course, entails the calculation of a much larger matrix, and hence CPU time requirements become unrealistic if we wish to obtain results for a range of parameters of any interest. Successful adaptation of the continued-fraction method, however, could prove to be beneficial in such computations.

Successful derivation of the asymptotes in Section III for both low-damping (LD) and intermediate-to-high-damping (IHD) situations and comparison of the results with exact data for λ_1 has allowed for experimentalists to successfully compare the asymptotes with their experimental data [38,39,40]. At the end of Section III, the work of Wernsdorfer et al. [37] was mentioned; however, successful comparison of the asymptotes with experimental data will lead to further collaborations with experimentalists, and the comparison of various other experimental parameters, such as the blocking temperature T_B. Successful extraction of the blocking temperature from the asymptotes has led to a publication with Kachkachi et al. [39] on the temperature at the peak of the zero-field-cooled magnetization. However, the asymptotes may be used in another area of research, namely, the comparison with computerized simulations, especially Monte Carlo simulations and Langevin dynamics. Some successful work has just been carried out by Nowek et al. [40].

APPENDIX A. ORTHOGONAL TRANSFORMATIONS

If we define a triad of unit vectors in R^3 as

$$E = (\mathbf{e}_1, \mathbf{e}_2, \mathbf{e}_3) \tag{A.1}$$

where the dot products are expressed as

$$\mathbf{e}_i \mathbf{e}_j = \delta_{i,j}, \quad 1 \leq i, \quad j \leq 3 \tag{A.2}$$

then E is an *orthonormal basis* for R^3. Any vector $\mathbf{x} \in R^3$ can thus be uniquely represented in terms of E by the equation

$$\mathbf{x} = \sum x_i \mathbf{e}_i \tag{A.3}$$

The scalars x_i are called the *coordinates of* $\mathbf{x}$ *with respect to* E and are arranged as a coordinate vector

$$X = [x_i]_{1 \leq i \leq 3} = \begin{bmatrix} x_1 \\ x_2 \\ x_3 \end{bmatrix} \tag{A.4}$$

For the purpose of introducing the concept of an orthogonal transformation, it is convenient to express Eq. (A.3) in terms of Eqs. (A.1) and (A.4) by writing

$$\mathbf{x} = EX \tag{A.5}$$

When $\|\mathbf{x}\| = 1$, then x_i are called the *direction cosines of* $\mathbf{x}$ *with respect to E.* Choosing another vector $\mathbf{y} \in R^3$ having coordinate vector $Y = [y_i]_{1 \leq i \leq 3}$ with respect to E, then the dot product can be expressed as

$$\mathbf{x}\mathbf{y} = X^T Y \tag{A.6}$$

In particular, we have

$$x_i = \mathbf{x}\mathbf{e}_i \tag{A.7}$$

We then define a linear transformation

$$T : R^3 \to R^3, \quad T\mathbf{e}_i = \sum A_{j,i}\mathbf{e}_j \tag{A.8}$$

where the scalars $A_{i,j}$ are called the *matrix elements* of T. They are arranged as a matrix

$$A = [A_{i,j}]_{1 \leq i, j \leq 3} \tag{A.9}$$

which is called the *representing matrix* or *matrix representation* of T. The transformation equations in Eq. (A.8) can be expressed in terms of Eqs. (A.1) and (A.9) as

$$TE = EA^T \tag{A.10}$$

T is called an *orthogonal transformation* if

$$T\mathbf{x}T\mathbf{y} = \mathbf{x}\mathbf{y} \tag{A.11}$$

for all $\mathbf{x}, \mathbf{y} \in R^3$, which, with the aid of Eqs. (A.6) and (A.9), becomes

$$(AX)^T AY = X^T Y \tag{A.12}$$

Hence the representing matrix is orthogonal:

$$A^{-1} = A^T \tag{A.13}$$

The triad of vectors

$$TE = (T\mathbf{e}_1, T\mathbf{e}_2, T\mathbf{e}_3) \tag{A.14}$$

is also an orthonormal basis for R^3 since

$$Te_iTe_j = e_ie_j = \delta_{i,j} \tag{A.15}$$

Suppose that T_1 and T_2 are orthogonal transformations, $T_i : R^3 \to R^3$; then by Eq. (A.11), the composition T_2T_1 is also orthogonal, since

$$T_2T_1\mathbf{x}T_2T_1\mathbf{y} = T_1\mathbf{x}T_1\mathbf{y} = \mathbf{xy} \tag{A.16}$$

Let A_1 denote the representing matrix for T_1 with respect to E, namely

$$T_1E = EQ_1^T \tag{A.17}$$

and A_2 denote the representing matrix for T_2 with respect to T_1E, specifically

$$T_2(T_1E) = (T_1E)A_2^T \tag{A.18}$$

then the representing matrix for T_2T_1 with respect to E, is given by the matrix product A_2A_1 since

$$T_2T_1E = (T_1E)A_2^T = (EA_1^T)A_2^T = E(A_1^TA_2^T) = E(A_2A_1)^T \tag{A.19}$$

We observe that

$$(A_2A_1)^{-1} = A_1^{-1}A_2^{-1} = A_1^TA_2^T = (A_2A_1)^T \tag{A.20}$$

hence the representing matrix A_2A_1 satisfies orthogonally as required. The orthogonal transformation T gives rise to an *orthogonal coordinate transformation*

$$\mathbf{e}_i' = T\mathbf{e}_i = \sum A_{j,i}\mathbf{e}_j \tag{A.21}$$

which can be expressed in matrix form

$$E' = EA^T \tag{A.22}$$

Application of the orthogonal coordinate transformation to Eq. (A.2) gives the following expression for a vector expressed in terms of the new coordinate system

$$\mathbf{x} = EX = EA^{-1}AX = EA^TAX = E'X' \tag{A.23}$$

where

$$X' = AX = [x_i']_{l \leq i \leq 3} \tag{A.24}$$

is the coordinate vector of $\mathbf{x}$ with respect to E'.

Consider the orthogonal transformations P_φ and Q_θ defined on $\{\mathbf{r} \in R^3, \|\mathbf{r}\| = 1\}$, where P_φ is defined by the finite rotation through an azimuthal

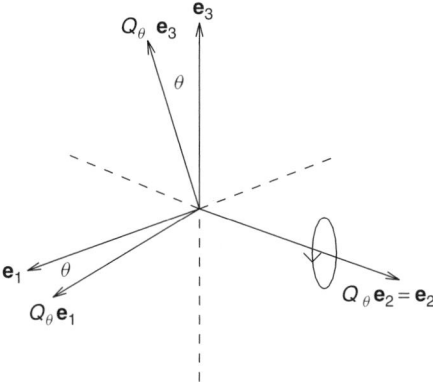

FIGURE 36. Orthogonal transformation for rotation through an angle θ about the $\mathbf{e}_2$ axis.

angle φ about the $\mathbf{e}_3$ axis in the $\mathbf{e}_1\mathbf{e}_2$-plane (see Fig. 36), and Q_θ is defined by the finite rotation through a polar angle θ about the $\mathbf{e}_2$ axis in the $\mathbf{e}_1\mathbf{e}_3$-plane (see Fig. 37). The angle of rotation in both cases is positive for counterclockwise rotation. The transformation equations are

$$P_\varphi \mathbf{e}_1 = \cos\varphi \, \mathbf{e}_1 + \sin\varphi \, \mathbf{e}_2$$
$$P_\varphi \mathbf{e}_2 = -\sin\varphi \, \mathbf{e}_1 + \cos\varphi \, \mathbf{e}_2$$
$$P_\varphi \mathbf{e}_3 = \mathbf{e}_3 \qquad (A.25)$$
$$Q_\theta \mathbf{e}_1 = \cos\theta \, \mathbf{e}_1 - \sin\theta \, \mathbf{e}_3$$
$$Q_\theta \mathbf{e}_2 = \mathbf{e}_2$$
$$Q_\theta \mathbf{e}_3 + = \sin\theta \, \mathbf{e}_1 + \cos\theta \, \mathbf{e}_3 \qquad (A.26)$$

Hence the representing matrices of P_φ and Q_θ are, respectively

$$A_\varphi = \begin{bmatrix} \cos\varphi & \sin\varphi & 0 \\ -\sin\varphi & \cos\varphi & 0 \\ 0 & 0 & 1 \end{bmatrix} \qquad (A.27)$$

$$B_\theta = \begin{bmatrix} \cos\theta & 0 & -\sin\theta \\ 0 & 1 & 0 \\ \sin\theta & 0 & \cos\theta \end{bmatrix} \qquad (A.28)$$

We observe that

$$C_\varphi^{-1} = C_\varphi^T = C_{-\varphi} \qquad \det C_\varphi = 1 \qquad (A.29)$$

where C is A or B. The *orthogonal transformation to spherical polar coordinates* is defined by the composition

$$T_{\theta,\varphi} = Q_\theta P_\varphi \tag{A.30}$$

From Eq. (A.19), the representing matrix for $T_{\theta,\varphi}$ is thus

$$S_{\theta,\varphi} = B_\theta A_\varphi = \begin{bmatrix} \cos\theta\cos\varphi & \cos\theta\sin\varphi & -\sin\theta \\ -\sin\varphi & \cos\varphi & 0 \\ \sin\theta\cos\varphi & \sin\theta\sin\varphi & \cos\theta \end{bmatrix} \tag{A.31}$$

The transformation equations are

$$T_{\theta,\varphi}E = E' = ES_{\theta,\varphi}^T \tag{A.32}$$

The new triad of orthogonal unit vectors is denoted

$$E' = (\mathbf{e}_1', \mathbf{e}_2', \mathbf{e}_3') = (\mathbf{e}_\theta, \mathbf{e}_\varphi, \mathbf{e}_r) \tag{A.33}$$

In general, any orthogonal transformation may be specified by three parameters (φ, θ, ψ) and written as a composition of the form

$$T_{\psi,\theta,\varphi} = P_\psi Q_\theta P_\varphi \tag{A.34}$$

The angles (φ, θ, ψ) are called the *Eulerian angles* of S. The presenting matrix for $T_{\psi,\theta,\varphi}$ is

$$S_{\psi,\theta,\varphi} = A_\psi B_\theta A_\varphi$$

$$= \begin{bmatrix} \cos\psi\cos\theta\cos\varphi - \sin\psi\sin\varphi & \cos\psi\cos\theta\sin\varphi + \sin\psi\cos\varphi & -\cos\psi\sin\theta \\ -\sin\psi\cos\theta\cos\varphi - \cos\psi\sin\varphi & -\sin\psi\cos\theta\sin\varphi + \cos\psi\cos\varphi & -\sin\psi\sin\theta \\ \sin\theta\cos\varphi & \sin\theta\sin\varphi & \cos\theta \end{bmatrix} \tag{A.35}$$

The transformation equations are

$$T_{\psi,\theta,\varphi}E = \overline{E} = ES_{\psi,\theta,\varphi}^T. \tag{A.36}$$

The new triad of orthogonal unit vectors $\overline{E} = (\overline{\mathbf{e}}_1, \overline{\mathbf{e}}_2, \overline{\mathbf{e}}_3)$ results from the three successive rotations

Rotation φ about the $\mathbf{e}_3$ axis giving $E' = (\mathbf{e}_1', \mathbf{e}_2', \mathbf{e}_3')$

Rotation θ about the $\mathbf{e}_2'$ axis giving $E'' = (\mathbf{e}_1'', \mathbf{e}_2'', \mathbf{e}_3'')$

Rotation ψ about the $\mathbf{e}_3''$ axis giving $\overline{E} = (\overline{\mathbf{e}}_1, \overline{\mathbf{e}}_2, \overline{\mathbf{e}}_3)$

APPENDIX B. ASSOCIATED LEGENDRE FUNCTIONS AND SPHERICAL HARMONICS

The associated Legendre functions (of the first kind) are defined by

$$P_l^m = \frac{1}{2^l l!}(1 - x^2)^{m/2}\frac{d^{l+m}}{dx^{l+m}}(x^2 - 1)^l \tag{B.1}$$

They satisfy the symmetry relation

$$P_l^{-m} = (-1)^m \frac{(l-m)!}{(l+m)!}P_l^m \tag{B.2}$$

the recurrence relations

$$xP_l^m = \frac{(l+m)}{(2l+1)}P_{l-1}^m + \frac{(l-m+1)}{(2l+1)}P_{l+1}^m \tag{B.3}$$

$$\sqrt{1-x^2}P_l^m = \frac{(l+m)(l+m-1)}{(2l+1)}P_{l-1}^{m-1} - \frac{(l-m+1)(l-m+2)}{(2l+1)}P_{l+1}^{m-1}$$

$$= -\frac{1}{(2l+1)}P_{l-1}^{m+1} + \frac{1}{(2l+1)}P_{l+1}^{m+1} \tag{B.4}$$

$$(l+m)P_{l-1}^m - (l-m)xP_l^m = \sqrt{1-x^2}P_l^{m+1} \tag{B.5}$$

$$(l-m+1)P_{l+1}^m - (l-m+1)xP_l^m = -\sqrt{1-x^2}P_l^{m+1} \tag{B.6}$$

$$P_{l+1}^m - xP_l^m = -(l-m+1)\sqrt{1-x^2}P_l^{m-1} \tag{B.7}$$

$$P_{l+1}^m - xP_l^m = (l+m)\sqrt{1-x^2}P_l^{m-1} \tag{B.8}$$

$$\sqrt{1-x^2}P_l^{m+1} = 2mxP_l^m - (l+m)(l-m+1)\sqrt{1-x^2}P_l^{m-1} \tag{B.9}$$

and the differential recurrence relations

$$(1-x^2)\frac{d}{dx}P_l^m = -\sin\theta\frac{d}{d\theta}P_l^m = (l+m)P_{l-1}^m - lxP_l^m$$

$$= (l+1)xP_l^m - (l-m+1)P_{l+1}^m$$

$$= (2l+1)^{-1}[(l+1)(l+m)P_{l-1}^m - l(l-m+1)P_{l+1}^m] \tag{B.10}$$

Expressions for the associated Legendre functions are obtained from the definition of Eq. (B.1) or from the recurrence relations in Eqs. (B.3)–(B.9). In particular

$$P_0^0 = 1 \quad P_1^0 = x \quad P_1^1 = \sqrt{1-x^2}$$

$$P_2^0 = \tfrac{1}{2}(3x^2 - 1) \quad P_2^1 = 3x\sqrt{1 - x^2} \quad P_2^2 = 3(1 - x^2)$$

$$P_3^0 = \tfrac{1}{2}(5x^3 - 3x) \quad P_3^1 = \tfrac{1}{2}\sqrt{1 - x^2}(15x^2 - 3)$$

$$P_3^2 = 15x(1 - x^2) \quad P_3^3 = 15(1 - x^2)^{3/2}$$

$$P_4^0 = \tfrac{1}{8}(35x^4 - 30x^2 + 3) \quad P_4^1 = \tfrac{1}{2}\sqrt{1 - x^2}(35x^3 - 15x)$$

$$P_4^2 = \tfrac{1}{2}(1 - x^2)(105x^2 - 15) \quad P_4^3 = 105x(1 - x^2)^{3/2}$$

$$P_4^4 = 105(1 - x^2)^2 \tag{B.11}$$

The eigenfunctions of ∇^2 are the spherical harmonics

$$X_{l,m}(\theta, \varphi) = e^{im\varphi} P_l^m(\cos\theta) \tag{B.12}$$

in that

$$\nabla^2 X_{l,m} = l(l + 1)X_{l,m} \tag{B.13}$$

From Eq. (B.2), the spherical harmonics satisfy the symmetry relation

$$X_{l,-m} = (-1)^m \frac{(l - m)!}{(l + m)!} X_{l,m}^* \tag{B.14}$$

The spherical harmonics can be defined in terms of the direction cosines by

$$X_{l,m} = \frac{1}{2^l l!}(\alpha_1 + i\alpha_2)^m \frac{d^{l+m}}{d\alpha_3^{l+m}}(\alpha_3^2 - 1)^l \tag{B.15}$$

The following recurrence relations are obtained from Eqs. (B.3) and (B.4):

$$\alpha_1 X_{l,m} = \frac{(l + m)(l - m + 1)}{2(2l + 1)} X_{l-1,m-1} - \frac{(l - m + 2)(l - m + 1)}{2(2l + 1)} X_{l+1,m-1}$$

$$- \frac{1}{2(2l + 1)} X_{l-1,m+1} + \frac{1}{2(2l + 1)} X_{l+1,m+1} \tag{B.16}$$

$$i\alpha_2 X_{l,m} = -\frac{(l + m)(l + m - 1)}{2(2l + 1)} X_{l-1,m-1} + \frac{(l - m + 2)(l - m + 1)}{2(2l + 1)} X_{l+1,m-1}$$

$$- \frac{1}{2(2l + 1)} X_{l-1,m+1} + \frac{1}{2(2l + 1)} X_{l+1,m+1} \tag{B.17}$$

$$\alpha_3 X_{l,m} = \frac{(l+m)}{(2l+1)} X_{l-1,m} + \frac{(l-m+1)}{(2l+1)} X_{l+1,m} \tag{B.18}$$

Equations (B.16) and (B.17) may be expressed in a more compact form as

$$(\alpha_1 + i\alpha_2)X_{l,m} = -\frac{1}{(2l+1)} X_{l-1,m+1} + \frac{1}{(2l+1)} X_{l+1,m+1} \tag{B.19}$$

$$(\alpha_1 - i\alpha_2)X_{l,m} = \frac{(l+m)(l+m-1)}{(2l+1)} X_{l-1,m-1}$$

$$- \frac{(l-m+2)(l-m+1)}{(2l+1)} X_{l+1,m-1} \tag{B.20}$$

The following recurrence relation is then obtained by successive application of Eq. (B.18):

$$\alpha_3^2 X_{l,m} = \frac{(l+m)(l+m-1)}{(2l-1)(2l+1)} X_{l-2,m} + \frac{2(l+m)(l-m)+2l-1}{(2l-1)(2l+3)} X_{l,m}$$

$$+ \frac{(l-m+2)(l-m+1)}{(2l+1)(2l+3)} X_{l+2,m} \tag{B.21}$$

The *product formula for spherical harmonics* is

$$X_{s,t} X_{l,m} = \sum_{r=-s}^{s} \begin{bmatrix} s & l & l+r \\ t & m & -m-t \end{bmatrix} X_{l+r,m+t} \tag{B.22}$$

where the product coefficients are expressed in terms of the *Wigner 3j symbols* by

$$\begin{bmatrix} s & l & l+r \\ t & m & -m-t \end{bmatrix} = (-1)^m \frac{2l+2r+1}{4\pi} \sqrt{\frac{(l+r-m-t)!(l+m)!(s+t)!}{(l+r+m+t)!(l-m)!(s-t)!}}$$

$$\cdot \begin{bmatrix} s & l & l+r \\ 0 & 0 & 0 \end{bmatrix} \begin{bmatrix} s & l & l+r \\ t & m & -m-t \end{bmatrix}. \tag{B.23}$$

The *vector-addition* rules imply that unless

$$r \equiv s(\mathrm{mod}2) \tag{B.24}$$

the leading Wigner 3j symbol in Eq. (B.23) vanishes, so that the summation in Eq. (B.22) is taken over values of r for which Eq. (B.24) is satisfied.

E. E. C. KENNEDY

APPENDIX C. FURTHER RESULTS FOR λ_1

TABLE C.1
λ_1 for Various σ and h, $\psi = 0°$, $a = 0.2$

	$\sigma = 0$	$\sigma = 2$	$\sigma = 4$	$\sigma = 6$	$\sigma = 8$	$\sigma = 10$
$h = 0.0$	2.0000	0.80768	0.25967	0.067226	0.014724	0.0028821
$h = 0.05$	2.0000	0.81382	0.27124	0.075648	0.018469	0.0041333
$h = 0.1$	2.0000	0.83221	0.30601	0.10164	0.030735	0.0086005
$h = 0.15$	2.0000	0.86282	0.36416	0.14725	0.054588	0.018579
$h = 0.2$	2.0000	0.90560	0.44596	0.21560	0.095013	0.038339

TABLE C.2
λ_1 for Various σ and h, $\psi = 15°$, $a = 0.2$

	$\sigma = 0$	$\sigma = 2$	$\sigma = 4$	$\sigma = 6$	$\sigma = 8$	$\sigma = 10$
$h = 0.0$	2.0000	0.80768	0.25967	0.067226	0.014724	0.0028821
$h = 0.05$	2.0000	0.81690	0.27868	0.076506	0.018757	0.0042187
$h = 0.1$	2.0000	0.84482	0.31542	0.10674	0.033072	0.0095379
$h = 0.15$	2.0000	0.89230	0.39019	0.16523	0.064950	0.023670
$h = 0.2$	2.0000	0.96066	0.50442	0.26466	0.12851	0.057770

TABLE C.3
λ_1 for Various σ and h, $\psi = 30°$, $a = 0.2$

	$\sigma = 0$	$\sigma = 2$	$\sigma = 4$	$\sigma = 6$	$\sigma = 8$	$\sigma = 10$
$h = 0.0$	2.0000	0.80768	0.25967	0.067226	0.014724	0.0028821
$h = 0.05$	2.0000	0.82527	0.27868	0.07871	0.019462	0.0041609
$h = 0.1$	2.0000	0.87860	0.33889	0.11834	0.037979	0.0011405
$h = 0.15$	2.0000	0.96922	0.44978	0.20176	0.084666	0.033145
$h = 0.2$	2.0000	1.0994	0.62697	0.35627	0.18890	0.093277

TABLE C.4
λ_1 for Various σ and h, $\psi = 45°$, $a = 0.2$

	$\sigma = 0$	$\sigma = 2$	$\sigma = 4$	$\sigma = 6$	$\sigma = 8$	$\sigma = 10$
$h = 0.0$	2.0000	0.80768	0.25967	0.067226	0.014724	0.0028821
$h = 0.05$	2.0000	0.83661	0.28575	0.081406	0.020238	0.0046051
$h = 0.1$	2.0000	0.92313	0.36619	0.12938	0.041699	0.012564
$h = 0.15$	2.0000	1.0665	0.50902	0.22888	0.096370	0.038131
$h = 0.2$	2.0000	1.2661	0.73106	0.41384	0.221852	0.112013

TABLE C.5
λ_1 for Various σ and h, $\psi = 60°$, $a = 0.2$

	$\sigma = 0$	$\sigma = 2$	$\sigma = 4$	$\sigma = 6$	$\sigma = 8$	$\sigma = 10$
$h = 0.0$	2.0000	0.80768	0.25967	0.067226	0.014724	0.0028821
$h = 0.05$	2.0000	0.84784	0.29247	0.08376	0.020811	0.0047065
$h = 0.1$	2.0000	0.96593	0.38864	0.13567	0.042463	0.012328
$h = 0.15$	2.0000	1.1561	0.54814	0.23527	0.093566	0.035084
$h = 0.2$	2.0000	1.4122	0.78236	0.41117	0.205811	0.098243

TABLE C.6
λ_1 for Various σ and h, $\psi = 75°$, $a = 0.2$

	$\sigma = 0$	$\sigma = 2$	$\sigma = 4$	$\sigma = 6$	$\sigma = 8$	$\sigma = 10$
$h = 0.0$	2.0000	0.80768	0.25967	0.067226	0.014724	0.0028821
$h = 0.05$	2.0000	0.85598	0.29718	0.085266	0.021107	0.0047256
$h = 0.1$	2.0000	0.99622	0.40234	0.13761	0.041326	0.011338
$h = 0.15$	2.0000	1.2174	0.56620	0.22843	0.082967	0.027984
$h = 0.2$	2.0000	1.5093	0.79285	0.37456	0.16468	0.067879

TABLE C.7
λ_1 for Various σ and h, $\psi = 90°$, $a = 0.2$

	$\sigma = 0$	$\sigma = 2$	$\sigma = 4$	$\sigma = 6$	$\sigma = 8$	$\sigma = 10$
$h = 0.0$	2.0000	0.80768	0.25967	0.067226	0.014724	0.0028821
$h = 0.05$	2.0000	0.85895	0.29885	0.085776	0.021190	0.0047229
$h = 0.1$	2.0000	1.0071	0.40682	0.13779	0.040568	0.010810
$h = 0.15$	2.0000	1.2390	0.57081	0.22362	0.077324	0.024386
$h = 0.2$	2.0000	1.5432	0.79155	0.35406	0.142272	0.052690

TABLE C.8
λ_1 for Various σ and h, $\psi = 0°$, $a = 0.3$

	$\sigma = 0$	$\sigma = 2$	$\sigma = 4$	$\sigma = 6$	$\sigma = 8$	$\sigma = 10$
$h = 0.0$	2.0000	0.80768	0.25967	0.067226	0.014724	0.0028821
$h = 0.05$	2.0000	0.81382	0.27124	0.07565	0.018469	0.0041333
$h = 0.1$	2.0000	0.83221	0.30601	0.10164	0.030735	0.0086005
$h = 0.15$	2.0000	0.86282	0.36416	0.14725	0.054588	0.018579
$h = 0.2$	2.0000	0.90560	0.44598	0.21560	0.095013	0.038339

TABLE C.9
λ_1 for Various σ and h, $\psi = 15°$, $a = 0.3$

	$\sigma = 0$	$\sigma = 2$	$\sigma = 4$	$\sigma = 6$	$\sigma = 8$	$\sigma = 10$
$h = 0.0$	2.0000	0.80768	0.25967	0.067226	0.014724	0.0028821
$h = 0.05$	2.0000	0.81574	0.27234	0.07605	0.018577	0.0041666
$h = 0.1$	2.0000	0.84013	0.31129	0.10435	0.031994	0.0091271
$h = 0.15$	2.0000	0.88141	0.37937	0.15778	0.060877	0.021789
$h = 0.2$	2.0000	0.94051	0.48145	0.24612	0.116504	0.051307

TABLE C.10
λ_1 for Various σ and h, $\psi = 30°$, $a = 0.3$

	$\sigma = 0$	$\sigma = 2$	$\sigma = 4$	$\sigma = 6$	$\sigma = 8$	$\sigma = 10$
$h = 0.0$	2.0000	0.80768	0.25967	0.067226	0.014724	0.0028821
$h = 0.05$	2.0000	0.82099	0.27528	0.07706	0.018865	0.0042363
$h = 0.1$	2.0000	0.86152	0.32464	0.11054	0.034612	0.010165
$h = 0.15$	2.0000	0.93095	0.41537	0.17988	0.073282	0.028003
$h = 0.2$	2.0000	1.03182	0.56021	0.30682	0.158191	0.076404

TABLE C.11
λ_1 for Various σ and h, $\psi = 45°$, $a = 0.3$

	$\sigma = 0$	$\sigma = 2$	$\sigma = 4$	$\sigma = 6$	$\sigma = 8$	$\sigma = 10$
$h = 0.0$	2.0000	0.80768	0.25967	0.067226	0.014724	0.0028821
$h = 0.05$	2.0000	0.82814	0.27914	0.07828	0.019135	0.0042825
$h = 0.1$	2.0000	0.89018	0.34047	0.11623	0.036348	0.010684
$h = 0.15$	2.0000	0.99580	0.45250	0.19600	0.080125	0.030975
$h = 0.2$	2.0000	1.14788	0.63065	0.34473	0.17983	0.088838

TABLE C.12
λ_1 for Various σ and h, $\psi = 60°$, $a = 0.3$

	$\sigma = 0$	$\sigma = 2$	$\sigma = 4$	$\sigma = 6$	$\sigma = 8$	$\sigma = 10$
$h = 0.0$	2.0000	0.80768	0.25967	0.067226	0.014724	0.0028821
$h = 0.05$	2.0000	0.83523	0.28283	0.079306	0.019282	0.0042734
$h = 0.1$	2.0000	0.91821	0.35364	0.11893	0.036082	0.010221
$h = 0.15$	2.0000	1.0576	0.47716	0.19753	0.076284	0.027944
$h = 0.2$	2.0000	1.2553	0.66451	0.33739	0.16420	0.076586

TABLE C.13
λ_1 for Various σ and h, $\psi = 75°$, $a = 0.3$

	$\sigma = 0$	$\sigma = 2$	$\sigma = 4$	$\sigma = 6$	$\sigma = 8$	$\sigma = 10$
$h = 0.0$	2.0000	0.80768	0.25967	0.067226	0.014724	0.0028821
$h = 0.05$	2.0000	0.84043	0.28543	0.07993	0.019314	0.0042320
$h = 0.1$	2.0000	0.93834	0.36169	0.11911	0.034673	0.0092849
$h = 0.15$	2.0000	1.10117	0.48799	0.19004	0.066995	0.022067
$h = 0.2$	2.0000	1.32945	0.66882	0.33739	0.12972	0.052426

TABLE C.14
λ_1 for Various σ and h, $\psi = 90°$, $a = 0.3$

	$\sigma = 0$	$\sigma = 2$	$\sigma = 4$	$\sigma = 6$	$\sigma = 8$	$\sigma = 10$
$h = 0.0$	2.0000	0.80768	0.25967	0.067226	0.014724	0.0028821
$h = 0.05$	2.0000	0.84229	0.28635	0.08014	0.019309	0.0042097
$h = 0.1$	2.0000	0.94563	0.36432	0.11881	0.033907	0.0088189
$h = 0.15$	2.0000	1.11675	0.49053	0.18555	0.062254	0.019169
$h = 0.2$	2.0000	1.35574	0.66638	0.28792	0.11230	0.040572

TABLE C.15
λ_1 for Various σ and h, $\psi = 0°$, $a = 0.4$

	$\sigma = 0$	$\sigma = 2$	$\sigma = 4$	$\sigma = 6$	$\sigma = 8$	$\sigma = 10$
$h = 0.0$	2.0000	0.80768	0.25967	0.067226	0.014724	0.0028821
$h = 0.05$	2.0000	0.81382	0.27124	0.075648	0.018469	0.0041333
$h = 0.1$	2.0000	0.83221	0.30601	0.10164	0.030735	0.0086005
$h = 0.15$	2.0000	0.86282	0.36416	0.14725	0.054588	0.018579
$h = 0.2$	2.0000	0.90560	0.44597	0.21560	0.095013	0.038339

TABLE C.16
λ_1 for Various σ and h, $\psi = 15°$, $a = 0.4$

	$\sigma = 0$	$\sigma = 2$	$\sigma = 4$	$\sigma = 6$	$\sigma = 8$	$\sigma = 10$
$h = 0.0$	2.0000	0.80768	0.25967	0.067226	0.014724	0.0028821
$h = 0.05$	2.0000	0.81518	0.27189	0.075831	0.018509	0.0041429
$h = 0.1$	2.0000	0.83780	0.30929	0.10321	0.031487	0.0089377
$h = 0.15$	2.0000	0.87592	0.37408	0.15415	0.058913	0.020893
$h = 0.2$	2.0000	0.93007	0.47005	0.23688	0.11075	0.048113

E. E. C. KENNEDY

TABLE C.17
λ_1 for Various σ and h, $\psi = 30°$, $a = 0.4$

	$\sigma = 0$	$\sigma = 2$	$\sigma = 4$	$\sigma = 6$	$\sigma = 8$	$\sigma = 10$
$h = 0.0$	2.0000	0.80768	0.25967	0.067226	0.014724	0.0028821
$h = 0.05$	2.0000	0.81890	0.27364	0.076279	0.018585	0.0041533
$h = 0.1$	2.0000	0.85297	0.31757	0.106667	0.032953	0.0095605
$h = 0.15$	2.0000	0.91112	0.39753	0.16844	0.067343	0.025352
$h = 0.2$	2.0000	0.99516	0.52405	0.27961	0.14139	0.067330

TABLE C.18
λ_1 for Various σ and h, $\psi = 45°$, $a = 0.4$

	$\sigma = 0$	$\sigma = 2$	$\sigma = 4$	$\sigma = 6$	$\sigma = 8$	$\sigma = 10$
$h = 0.0$	2.0000	0.80768	0.25967	0.067226	0.014724	0.0028821
$h = 0.05$	2.0000	0.82396	0.27592	0.076769	0.018608	0.0041306
$h = 0.1$	2.0000	0.87343	0.32727	0.10942	0.033590	0.0097276
$h = 0.15$	2.0000	0.95794	0.42149	0.17777	0.071211	0.027139
$h = 0.2$	2.0000	1.0802	0.57197	0.30447	0.15398	0.076108

TABLE C.19
λ_1 for Various σ and h, $\psi = 60°$, $a = 0.4$

	$\sigma = 0$	$\sigma = 2$	$\sigma = 4$	$\sigma = 6$	$\sigma = 8$	$\sigma = 10$
$h = 0.0$	2.0000	0.80768	0.25967	0.067226	0.014724	0.0028821
$h = 0.05$	2.0000	0.82900	0.27809	0.077123	0.018541	0.0040658
$h = 0.1$	2.0000	0.89361	0.33516	0.10997	0.032693	0.0091223
$h = 0.15$	2.0000	1.00342	0.43659	0.17584	0.066593	0.024073
$h = 0.2$	2.0000	1.16152	0.59267	0.29341	0.14030	0.064661

TABLE C.20
λ_1 for Various σ and h, $\psi = 75°$, $a = 0.4$

	$\sigma = 0$	$\sigma = 2$	$\sigma = 4$	$\sigma = 6$	$\sigma = 8$	$\sigma = 10$
$h = 0.0$	2.0000	0.80768	0.25967	0.067226	0.014724	0.0028821
$h = 0.05$	2.0000	0.83268	0.27961	0.077295	0.018434	0.0039912
$h = 0.1$	2.0000	0.90820	0.33983	0.10901	0.031084	0.0082090
$h = 0.15$	2.0000	1.03593	0.44231	0.16763	0.057968	0.018849
$h = 0.2$	2.0000	1.21913	0.59189	0.23650	0.11001	0.043926

TABLE C.21

λ_1 for Various σ and h, $\psi = 90°$, $a = 0.4$

	$\sigma = 0$	$\sigma = 2$	$\sigma = 4$	$\sigma = 6$	$\sigma = 8$	$\sigma = 10$
$h = 0.0$	2.0000	0.80768	0.25967	0.067226	0.014724	0.0028821
$h = 0.05$	2.0000	0.83402	0.28015	0.077339	0.018383	0.0039585
$h = 0.1$	2.0000	0.91350	0.34131	0.10838	0.030296	0.0077644
$h = 0.15$	2.0000	1.04768	0.44331	0.16321	0.053720	0.016331
$h = 0.2$	2.0000	1.23989	0.58841	0.24804	0.095010	0.033911

TABLE C.22

λ_1 for Various σ and h, $\psi = 0°$, $a = 1.0$

	$\sigma = 0$	$\sigma = 2$	$\sigma = 4$	$\sigma = 6$	$\sigma = 8$	$\sigma = 10$
$h = 0.0$	2.0000	0.80768	0.25967	0.067226	0.014724	0.0028821
$h = 0.05$	2.0000	0.81382	0.27124	0.075649	0.018469	0.0041333
$h = 0.1$	2.0000	0.83221	0.30601	0.10164	0.030735	0.0086005
$h = 0.15$	2.0000	0.86282	0.36416	0.14725	0.054588	0.018579
$h = 0.2$	2.0000	0.90560	0.44596	0.21560	0.095013	0.038339

TABLE C.23

λ_1 for Various σ and h, $\psi = 15°$, $a = 1.0$

	$\sigma = 0$	$\sigma = 2$	$\sigma = 4$	$\sigma = 6$	$\sigma = 8$	$\sigma = 10$
$h = 0.0$	2.0000	0.80768	0.25967	0.067226	0.014724	0.0028821
$h = 0.05$	2.0000	0.81411	0.27117	0.075501	0.018395	0.0040199
$h = 0.1$	2.0000	0.83341	0.30602	0.10146	0.030743	0.0086682
$h = 0.15$	2.0000	0.86563	0.36524	0.14847	0.055968	0.019596
$h = 0.2$	2.0000	0.91084	0.45048	0.22205	0.10165	0.043337

TABLE C.24

λ_1 for Various σ and h, $\psi = 30°$, $a = 1.0$

	$\sigma = 0$	$\sigma = 2$	$\sigma = 4$	$\sigma = 6$	$\sigma = 8$	$\sigma = 10$
$h = 0.0$	2.0000	0.80768	0.25967	0.067226	0.014724	0.0028821
$h = 0.05$	2.0000	0.81491	0.27096	0.075074	0.018175	0.0040369
$h = 0.1$	2.0000	0.83669	0.30577	0.100570	0.030448	0.0086794
$h = 0.15$	2.0000	0.87324	0.36683	0.14968	0.057954	0.021285
$h = 0.2$	2.0000	0.92493	0.45858	0.23272	0.11340	0.052647

TABLE C.25
λ_1 for Various σ and h, $\psi = 45°$, $a = 1.0$

	$\sigma = 0$	$\sigma = 2$	$\sigma = 4$	$\sigma = 6$	$\sigma = 8$	$\sigma = 10$
$h = 0.0$	2.0000	0.80768	0.25967	0.067226	0.014724	0.0028821
$h = 0.05$	2.0000	0.81601	0.27064	0.074432	0.017831	0.0039151
$h = 0.1$	2.0000	0.84113	0.30481	0.098394	0.029297	0.0082858
$h = 0.15$	2.0000	0.88345	0.36586	0.14636	0.056445	0.020991
$h = 0.2$	2.0000	0.94362	0.45972	0.23097	0.14267	0.054671

TABLE C.26
λ_1 for Various σ and h, $\psi = 60°$, $a = 1.0$

	$\sigma = 0$	$\sigma = 2$	$\sigma = 4$	$\sigma = 6$	$\sigma = 8$	$\sigma = 10$
$h = 0.0$	2.0000	0.80768	0.25967	0.067226	0.014724	0.0028821
$h = 0.05$	2.0000	0.81710	0.27028	0.073723	0.017437	0.0037683
$h = 0.1$	2.0000	0.84552	0.30314	0.095124	0.027286	0.0074250
$h = 0.15$	2.0000	0.89346	0.36130	0.13732	0.050089	0.017727
$h = 0.2$	2.0000	0.96173	0.44967	0.21057	0.097676	0.044334

TABLE C.27
λ_1 for Various σ and h, $\psi = 75°$, $a = 1.0$

	$\sigma = 0$	$\sigma = 2$	$\sigma = 4$	$\sigma = 6$	$\sigma = 8$	$\sigma = 10$
$h = 0.0$	2.0000	0.80768	0.25967	0.067226	0.014724	0.0028821
$h = 0.05$	2.0000	0.81790	0.26998	0.073161	0.017116	0.0036449
$h = 0.1$	2.0000	0.84871	0.30146	0.092034	0.025273	0.0065041
$h = 0.15$	2.0000	0.90065	0.35570	0.12710	0.042398	0.013528
$h = 0.2$	2.0000	0.97464	0.44534	0.18389	0.074648	0.029411

TABLE C.28
λ_1 for Various σ and h, $\psi = 90°$, $a = 1.0$

	$\sigma = 0$	$\sigma = 2$	$\sigma = 4$	$\sigma = 6$	$\sigma = 8$	$\sigma = 10$
$h = 0.0$	2.0000	0.80768	0.25967	0.067226	0.014724	0.0028821
$h = 0.05$	2.0000	0.81819	0.26987	0.072947	0.016992	0.0035965
$h = 0.1$	2.0000	0.84988	0.30075	0.090756	0.024422	0.0061068
$h = 0.15$	2.0000	0.90326	0.35316	0.12259	0.038948	0.011627
$h = 0.2$	2.0000	0.97930	0.42864	0.17166	0.063969	0.022540

APPENDIX D. FURTHER RESULTS FOR T_C

TABLE D.1
T_c for Various σ and h, $\psi = 0°$, $a = 0.2$

	$\sigma = 0$	$\sigma = 2$	$\sigma = 4$	$\sigma = 6$	$\sigma = 8$	$\sigma = 10$
$h = 0.0$	0.50000	1.2302	3.8030	14.715	67.446	345.508
$h = 0.05$	0.50000	1.2199	3.6320	13.032	53.563	239.92
$h = 0.1$	0.50000	1.1899	3.1923	9.5663	31.556	112.07
$h = 0.15$	0.50000	1.1427	2.6353	6.3567	16.491	44.587
$h = 0.2$	0.50000	1.0816	2.0820	3.9541	7.4175	11.8767

TABLE D.2
T_c for Various σ and h, $\psi = 15°$, $a = 0.2$

	$\sigma = 0$	$\sigma = 2$	$\sigma = 4$	$\sigma = 6$	$\sigma = 8$	$\sigma = 10$
$h = 0.0$	0.50000	1.1947	3.7470	14.590	67.061	344.02
$h = 0.05$	0.50000	1.1808	3.5506	12.761	52.322	233.284
$h = 0.1$	0.50000	1.1408	3.0479	8.9826	28.814	93.394
$h = 0.15$	0.50000	1.0787	2.4188	5.5382	13.242	31.670
$h = 0.2$	0.50000	1.0006	1.8117	3.1041	4.9385	6.2174

TABLE D.3
T_c for Various σ and h, $\psi = 30°$, $a = 0.2$

	$\sigma = 0$	$\sigma = 2$	$\sigma = 4$	$\sigma = 6$	$\sigma = 8$	$\sigma = 10$
$h = 0.0$	0.50000	1.0819	3.5561	14.154	65.694	338.72
$h = 0.05$	0.50000	1.0601	3.3013	11.992	49.117	217.60
$h = 0.1$	0.50000	0.99955	2.6832	7.7507	23.916	77.359
$h = 0.15$	0.50000	0.91235	1.9785	4.2572	9.2396	19.428
$h = 0.2$	0.50000	0.81305	1.3712	2.1110	2.8843	3.1319

TABLE D.4
T_c for Various σ and h, $\psi = 45°$, $a = 0.2$

	$\sigma = 0$	$\sigma = 2$	$\sigma = 4$	$\sigma = 6$	$\sigma = 8$	$\sigma = 10$
$h = 0.0$	0.50000	0.87625	3.1477	13.151	62.450	325.93
$h = 0.05$	0.50000	0.84918	2.8467	10.729	44.592	198.88
$h = 0.1$	0.50000	0.77844	2.1879	6.4816	20.170	65.237
$h = 0.15$	0.50000	0.68673	1.5314	3.3762	7.3836	15.759
$h = 0.2$	0.50000	0.59432	1.0236	1.6205	2.3060	2.8270

TABLE D.5

T_c for Various σ and h, $\psi = 60°$, $a = 0.2$

	$\sigma = 0$	$\sigma = 2$	$\sigma = 4$	$\sigma = 6$	$\sigma = 8$	$\sigma = 10$
$h = 0.0$	0.50000	0.57160	2.3431	10.8449	54.392	292.74
$h = 0.05$	0.50000	0.54934	2.0691	8.5729	37.589	173.69
$h = 0.1$	0.50000	0.49468	1.5299	5.0432	17.043	59.268
$h = 0.15$	0.50000	0.43046	1.0508	2.6664	6.6493	16.395
$h = 0.2$	0.50000	0.37205	0.70364	1.3379	2.3608	3.8765

TABLE D.6

T_c for Various σ and h, $\psi = 75°$, $a = 0.2$

	$\sigma = 0$	$\sigma = 2$	$\sigma = 4$	$\sigma = 6$	$\sigma = 8$	$\sigma = 10$
$h = 0.0$	0.50000	0.23582	0.98611	5.5409	31.902	188.10
$h = 0.05$	0.50000	0.22808	0.85882	4.3047	21.781	111.52
$h = 0.1$	0.50000	0.21022	0.62663	2.5489	10.418	42.676
$h = 0.15$	0.50000	0.19131	0.43502	1.4217	4.6393	14.901
$h = 0.2$	0.50000	0.17609	0.30088	0.77863	1.9997	4.9749

TABLE D.7

T_c for Various σ and h, $\psi = 90°$, $a = 0.2$

	$\sigma = 0$	$\sigma = 2$	$\sigma = 4$	$\sigma = 6$	$\sigma = 8$	$\sigma = 10$
$h = 0.0$	0.50000	0.073990	0.021173	0.0075381	0.0036071	0.0023493
$h = 0.05$	0.50000	0.075280	0.021624	0.0077585	0.0037242	0.0024034
$h = 0.1$	0.50000	0.078860	0.022706	0.0082957	0.0040438	0.0025647
$h = 0.15$	0.50000	0.084021	0.023938	0.0089665	0.0045273	0.0028422
$h = 0.2$	0.50000	0.089959	0.025044	0.0096837	0.0051499	0.0032501

TABLE D.8

T_c for Various σ and h, $\psi = 0°$, $a = 0.3$

	$\sigma = 0$	$\sigma = 2$	$\sigma = 4$	$\sigma = 6$	$\sigma = 8$	$\sigma = 10$
$h = 0.0$	0.50000	1.2302	3.8031	14.715	67.446	345.51
$h = 0.05$	0.50000	1.2199	3.6320	13.032	53.563	239.92
$h = 0.1$	0.50000	1.1899	3.1923	9.5663	31.556	112.07
$h = 0.15$	0.50000	1.1427	2.6353	6.3567	16.491	44.587
$h = 0.2$	0.50000	1.0816	2.0820	3.9540	7.4175	11.877

TABLE D.9

T_c for Various σ and h, $\psi = 15°$, $a = 0.3$

	$\sigma = 0$	$\sigma = 2$	$\sigma = 4$	$\sigma = 6$	$\sigma = 8$	$\sigma = 10$
$h = 0.0$	0.50000	1.1958	3.7472	14.590	67.061	344.02
$h = 0.05$	0.50000	1.1836	3.5627	12.838	52.799	236.20
$h = 0.1$	0.50000	1.1480	3.0875	9.1859	29.782	102.82
$h = 0.15$	0.50000	1.0925	2.4857	5.7962	14.123	34.398
$h = 0.2$	0.50000	1.0218	1.8943	3.3328	5.4357	6.9975

TABLE D.10

T_c for Various σ and h, $\psi = 30°$, $a = 0.3$

	$\sigma = 0$	$\sigma = 2$	$\sigma = 4$	$\sigma = 6$	$\sigma = 8$	$\sigma = 10$
$h = 0.0$	0.50000	1.0867	3.5569	14.154	65.694	338.72
$h = 0.05$	0.50000	1.0702	3.3420	12.247	50.669	226.83
$h = 0.1$	0.50000	1.0231	2.7989	8.2926	26.237	86.792
$h = 0.15$	0.50000	0.95269	2.1384	4.7685	10.668	22.988
$h = 0.2$	0.50000	0.86827	1.5297	2.4457	3.4397	3.8211

TABLE D.11

T_c for Various σ and h, $\psi = 45°$, $a = 0.3$

	$\sigma = 0$	$\sigma = 2$	$\sigma = 4$	$\sigma = 6$	$\sigma = 8$	$\sigma = 10$
$h = 0.0$	0.50000	0.88769	3.1500	13.151	62.450	325.93
$h = 0.05$	0.50000	0.86868	2.9149	11.155	47.159	213.85
$h = 0.1$	0.50000	0.81660	2.3516	7.2091	23.132	76.709
$h = 0.15$	0.50000	0.74355	1.7207	3.9377	8.8748	19.394
$h = 0.2$	0.50000	0.66304	1.1860	1.9430	2.8421	3.5627

TABLE D.12

T_c for Various σ and h, $\psi = 60°$, $a = 0.3$

	$\sigma = 0$	$\sigma = 2$	$\sigma = 4$	$\sigma = 6$	$\sigma = 8$	$\sigma = 10$
$h = 0.0$	0.50000	0.59290	2.3482	10.847	54.393	292.74
$h = 0.05$	0.50000	0.57778	2.1429	9.0522	40.564	191.29
$h = 0.1$	0.50000	0.53793	1.6826	5.7491	20.050	71.474
$h = 0.15$	0.50000	0.48560	1.2091	3.1738	8.1515	20.579
$h = 0.2$	0.50000	0.43222	0.83219	1.6308	2.9587	4.9707

TABLE D.13
T_c for Various σ and h, $\psi = 75°$, $a = 0.3$

	$\sigma = 0$	$\sigma = 2$	$\sigma = 4$	$\sigma = 6$	$\sigma = 8$	$\sigma = 10$
$h = 0.0$	0.50000	0.26798	0.99597	5.5448	31.904	7188.10
$h = 0.05$	0.50000	0.26275	0.90252	4.5934	23.803	124.580
$h = 0.1$	0.50000	0.24945	0.70352	2.9453	12.414	52.110
$h = 0.15$	0.50000	0.23308	0.51087	1.7105	5.7440	18.894
$h = 0.2$	0.50000	0.21763	0.36324	0.95831	2.5284	6.4049

TABLE D.14
T_c for Various σ and h, $\psi = 90°$, $a = 0.3$

	$\sigma = 0$	$\sigma = 2$	$\sigma = 4$	$\sigma = 6$	$\sigma = 8$	$\sigma = 10$
$h = 0.0$	0.50000	0.11139	0.034413	0.013668	0.0072347	0.0049468
$h = 0.05$	0.50000	0.11216	0.034687	0.013888	0.0073806	0.0050215
$h = 0.1$	0.50000	0.11434	0.035402	0.14486	0.0078057	0.0052513
$h = 0.15$	0.50000	0.11759	0.036338	0.15346	0.0084847	0.0056530
$h = 0.2$	0.50000	0.12143	0.037321	0.016358	0.0093842	0.0062476

TABLE D.15
T_c for Various σ and h, $\psi = 0°$, $a = 0.4$

	$\sigma = 0$	$\sigma = 2$	$\sigma = 4$	$\sigma = 6$	$\sigma = 8$	$\sigma = 10$
$h = 0.0$	0.50000	1.2302	3.8031	14.715	67.446	345.51
$h = 0.05$	0.50000	1.2199	3.6320	13.032	53.563	239.92
$h = 0.1$	0.50000	1.1899	3.1923	9.5663	31.556	112.07
$h = 0.15$	0.50000	1.1427	2.6353	6.3567	16.491	44.587
$h = 0.2$	0.50000	1.0816	2.0820	3.9541	7.4175	11.877

TABLE D.16
T_c for Various σ and h, $\psi = 15°$, $a = 0.4$

	$\sigma = 0$	$\sigma = 2$	$\sigma = 4$	$\sigma = 6$	$\sigma = 8$	$\sigma = 10$
$h = 0.0$	0.50000	1.1969	3.7474	14.591	67.061	344.024
$h = 0.05$	0.50000	1.1854	3.5686	12.874	53.023	237.55
$h = 0.1$	0.50000	1.1521	3.1071	9.2865	30.260	105.01
$h = 0.15$	0.50000	1.0999	2.5199	5.9309	14.592	35.871
$h = 0.2$	0.50000	1.0334	1.9386	3.4601	5.7228	7.4605

TABLE D.17

T_c for Various σ and h, $\psi = 30°$, $a = 0.4$

	$\sigma = 0$	$\sigma = 2$	$\sigma = 4$	$\sigma = 6$	$\sigma = 8$	$\sigma = 10$
$h = 0.0$	0.50000	1.0912	3.5579	14.154	65.694	338.72
$h = 0.05$	0.50000	1.0772	3.3625	12.373	51.432	231.36
$h = 0.1$	0.50000	1.0373	2.8605	8.5913	27.555	92.275
$h = 0.15$	0.50000	0.97642	2.2321	5.0886	11.604	25.389
$h = 0.2$	0.50000	0.90171	1.6324	2.6796	3.8454	4.3347

TABLE D.18

T_c for Various σ and h, $\psi = 45°$, $a = 0.4$

	$\sigma = 0$	$\sigma = 2$	$\sigma = 4$	$\sigma = 6$	$\sigma = 8$	$\sigma = 10$
$h = 0.0$	0.50000	0.89834	3.1525	13.152	62.451	325.93
$h = 0.05$	0.50000	0.88338	2.9505	11.374	48.492	221.71
$h = 0.1$	0.50000	0.84153	2.4463	7.6548	25.027	84.248
$h = 0.15$	0.50000	0.78058	1.8461	4.3385	9.9813	22.131
$h = 0.2$	0.50000	0.71004	1.3066	2.1972	3.2745	4.1576

TABLE D.19

T_c for Various σ and h, $\psi = 60°$, $a = 0.4$

	$\sigma = 0$	$\sigma = 2$	$\sigma = 4$	$\sigma = 6$	$\sigma = 8$	$\sigma = 10$
$h = 0.0$	0.50000	0.61272	2.3536	10.848	54.394	292.74
$h = 0.05$	0.50000	0.60120	2.1838	9.3085	42.184	201.04
$h = 0.1$	0.50000	0.56981	1.7780	6.2156	22.124	80.081
$h = 0.15$	0.50000	0.52614	1.3234	3.5632	9.3345	23.885
$h = 0.2$	0.50000	0.47840	0.93550	1.8747	3.4612	5.8862

TABLE D.20

T_c for Various σ and h, $\psi = 75°$, $a = 0.4$

	$\sigma = 0$	$\sigma = 2$	$\sigma = 4$	$\sigma = 6$	$\sigma = 8$	$\sigma = 10$
$h = 0.0$	0.50000	0.29790	1.0064	5.5492	31.907	188.10
$h = 0.05$	0.50000	0.29387	0.93074	4.7534	24.938	132.07
$h = 0.1$	0.50000	0.28315	0.75648	3.2197	13.846	58.996
$h = 0.15$	0.50000	0.26881	0.57035	1.9409	6.6378	22.118
$h = 0.2$	0.50000	0.25385	0.41680	1.1124	2.9817	7.6436

TABLE D.21

T_c for Various σ and h, $\psi = 90°$, $a = 0.4$

	$\sigma = 0$	$\sigma = 2$	$\sigma = 4$	$\sigma = 6$	$\sigma = 8$	$\sigma = 10$
$h = 0.0$	0.50000	0.14618	0.048426	0.020747	0.011655	0.0081784
$h = 0.05$	0.50000	0.14659	0.048622	0.020997	0.011839	0.0082781
$h = 0.1$	0.50000	0.14775	0.049160	0.021708	0.012387	0.0085861
$h = 0.15$	0.50000	0.14947	0.049929	0.022782	0.013278	0.0091267
$h = 0.2$	0.50000	0.15146	0.050818	0.024099	0.014473	0.0099286

TABLE D.22

T_c for Various σ and h, $\psi = 0°$, $a = 1.0$

	$\sigma = 0$	$\sigma = 2$	$\sigma = 4$	$\sigma = 6$	$\sigma = 8$	$\sigma = 10$
$h = 0.0$	0.50000	1.2302	3.8031	14.715	67.446	345.51
$h = 0.05$	0.50000	1.2199	3.6320	13.032	53.563	239.92
$h = 0.1$	0.50000	1.1899	3.1923	9.5663	31.556	112.07
$h = 0.15$	0.50000	1.1427	2.6353	6.3567	16.492	44.587
$h = 0.2$	0.50000	1.0816	2.0820	3.9541	7.4175	11.877

TABLE D.23

T_c for Various σ and h, $\psi = 15°$, $a = 1.0$

	$\sigma = 0$	$\sigma = 2$	$\sigma = 4$	$\sigma = 6$	$\sigma = 8$	$\sigma = 10$
$h = 0.0$	0.50000	1.2003	3.7484	14.591	67.061	344.02
$h = 0.05$	0.50000	1.1902	3.5790	12.930	53.351	239.45
$h = 0.1$	0.50000	1.1604	3.1405	9.4462	30.992	108.26
$h = 0.15$	0.50000	1.1136	2.5798	6.1558	15.357	38.245
$h = 0.2$	0.50000	1.0532	2.0195	3.6878	6.2330	8.2840

TABLE D.24

T_c for Various σ and h, $\psi = 30°$, $a = 1.0$

	$\sigma = 0$	$\sigma = 2$	$\sigma = 4$	$\sigma = 6$	$\sigma = 8$	$\sigma = 10$
$h = 0.0$	0.50000	1.1056	3.5624	14.156	65.695	338.72
$h = 0.05$	0.50000	1.0959	3.3996	12.572	52.590	238.04
$h = 0.1$	0.50000	1.0676	2.9725	9.1101	29.818	101.64
$h = 0.15$	0.50000	1.0230	2.4172	5.7213	13.479	30.237
$h = 0.2$	0.50000	0.96582	1.8584	3.2126	4.7919	5.5469

TABLE D.25
T_c for Various σ and h, $\psi = 45°$, $a = 1.0$

	$\sigma = 0$	$\sigma = 2$	$\sigma = 4$	$\sigma = 6$	$\sigma = 8$	$\sigma = 10$
$h = 0.0$	0.50000	0.93284	3.1646	13.156	62.453	325.93
$h = 0.05$	0.50000	0.92443	3.0188	11.733	50.606	233.91
$h = 0.1$	0.50000	0.90003	2.6337	8.5107	28.689	98.902
$h = 0.15$	0.50000	0.86196	2.1291	5.2640	12.587	28.610
$h = 0.2$	0.50000	0.81349	1.6220	2.8915	4.4684	5.7914

TABLE D.26
T_c for Various σ and h, $\psi = 60°$, $a = 1.0$

	$\sigma = 0$	$\sigma = 2$	$\sigma = 4$	$\sigma = 6$	$\sigma = 8$	$\sigma = 10$
$h = 0.0$	0.50000	0.67691	2.3806	10.859	54.399	292.75
$h = 0.05$	0.50000	0.67118	2.2720	9.7461	44.858	216.92
$h = 0.1$	0.50000	0.65463	1.9864	7.1894	26.505	98.383
$h = 0.15$	0.50000	0.62899	1.6144	4.5642	12.407	32.434
$h = 0.2$	0.50000	0.59663	1.2431	2.6145	4.9725	8.5877

TABLE D.27
T_c for Various σ and h, $\psi = 75°$, $a = 1.0$

	$\sigma = 0$	$\sigma = 2$	$\sigma = 4$	$\sigma = 6$	$\sigma = 8$	$\sigma = 10$
$h = 0.0$	0.50000	0.39482	1.0584	5.5751	31.921	188.11
$h = 0.05$	0.50000	0.39258	1.0139	5.0457	26.871	144.65
$h = 0.1$	0.50000	0.38609	0.89776	3.8337	17.038	74.389
$h = 0.15$	0.50000	0.37600	0.72824	2.5777	9.0831	30.822
$h = 0.2$	0.50000	0.36322	0.60008	1.6117	4.4032	11.421

TABLE D.28
T_c for Various σ and h, $\psi = 90°$, $a = 1.0$

	$\sigma = 0$	$\sigma = 2$	$\sigma = 4$	$\sigma = 6$	$\sigma = 8$	$\sigma = 10$
$h = 0.0$	0.50000	0.25887	0.11829	0.062228	0.039540	0.029092
$h = 0.05$	0.50000	0.25842	0.11837	0.062716	0.039947	0.029336
$h = 0.1$	0.50000	0.25707	0.11858	0.064127	0.041166	0.030088
$h = 0.15$	0.50000	0.25485	0.11886	0.066309	0.043170	0.031405
$h = 0.2$	0.50000	0.25180	0.11911	0.069031	0.045893	0.033353

APPENDIX E. USEFUL DEFINITIONS

- *Anisotropic.* Not isotropic, that is, not possessing a property that varies with direction.

- *Axially Symmetric.* The external magnetic field is termed axially symmetric when it is applied parallel to the easy axis of magnetization.

- *Curie Point.* For ferromagnetic solids, there is a change from ferromagnetic behavior to paramagnetic behavior above a particular temperature, and the paramagnetic material then obeys the Curie–Weiss law above this temperature (the law shows that the susceptibility is proportional to the excess of temperature over a fixed temperature θ, where θ is known as the *Weiss constant*, and is a temperature characteristic of the material). This is known as the *Curie point* for the material.

- *Degenerate.* The molecular system has two or more distinct states of the same energy.

- *Dipole Moment.* A dipole describes a system of two equal and opposite charges placed at a very short distance apart. The product of either of the charges and the distance between them is known as the *electric dipole moment.*

- *Discrete Orientation.* Noncontinuous orientation.

- *Gibbs Free Energy.* In a reversible change occurring at constant temperature and pressure, the change in the Gibbs function of a system is equal to the work done on it.

- *Gyromagnetic.* The relationship between the magnetization of a body and its rotation.

- *Gyroscopic.* If a couple is applied to the frame of a gyroscope, the resulting motion of precession tends to align the gyro with the axis of the couple. The rate of turning or precession is proportional to the moment of the applied couple, and inversely proportional to the angular momentum of the gyro.

- *Hysterisis.* A delay in the change of an observed effect in response to a change in the mechanism producing the effect. This is a phenomenon shown by ferromagnetic substances whereby the magnetic flux through the medium depends not only on the existing magnetizing field but also on the previous state or states of the substance. The existence of permanent magnets is due to hysterisis.

- *Magnetic Dipole Moment.* The torque experienced when the magnet is set with its axis at right angles to a magnetic field of unit size. The torque can thus be expressed as the vector product of magnetic moment and magnetic field strength.

- *Magnetic Flux.* The product of a particular area under consideration and the component normal to the area of the average magnetic flux density over it. For an element of area $d\mathbf{A}$ the flux $d\Phi$ is the scalar product $\mathbf{B} \cdot d\mathbf{A}$, where $\mathbf{B}$ is the magnetic flux density.

- *Maxwell–Boltzmann Distribution.* Maxwell's law of the distribution of velocities, based on classical statistics, states that for a gas in equilibrium the number of molecules whose total velocity lies in the range $c \rightarrow (c + dc)$ is given by the expression

$$dN_c = 4\pi N \left\{ \frac{hm}{\pi} \right\}^{3/2} (e^{-hmc^2})c^2 dc$$

  where N is the total number of molecules, m is the mass of a molecule, h is a constant equal to $1/(2kT)$, and T is the thermodynamic temperature. This relation is called the *Maxwell–Boltzmann distribution.*

- *Nonaxially Symmetric.* If the external magnetic field is applied at an oblique angle to the easy axis of magnetization, then it is said to be nonaxially symmetric.

- *Paleomagnetism.* The study of the residual magnetization of certain rocks in order to determine the direction of polarization of the earth's magnetic field at the time of the rock's formation.

- *Paramagnetism.* A paramagnetic substance is regarded as an assembly of magnetic dipoles that have random orientation. In the presence of a field, the magnetization is determined by competition between the effect of the field, in tending to align the magnetic poles, and the random thermal agitation.

- *Precession.* If a body is spinning about an axis of symmetry Oc (where O is a fixed point) and C is rotating around an axis Oz fixed outside the body, the body is said to be precessing around Oz, where Oz is the precession axis.

- *Remanance.* The residual magnetic flux density in a substance when the magnetizing field strength is returned to zero.

- *Stochastic Process.* A process resulting from random behavior of its generators.

- *Thermal Equilibrium.* The condition of a system in which the net rate of exchange of heat between the components is zero [77].

- *Unixial Anistropy.* The anisotropy is a function of only one paramater, namely, the angle between the easy axis and the direction of the magnetization [2].

APPENDIX F. LIST OF SYMBOLS

$a_{l,m}$	Fourier coefficients
A	The Fokker–Planck matrix
$A_{l,m,p,q}$	Submatrices of the Fokker–Planck matrix A
A_0	Normalization constant
a, b	Brown's constants ($a = \eta\gamma M_S$)
c_k^i	Coefficients in Taylor series of the free energy near the ith stationary point
$c_{l,m}$	Decay modes
$C(t)$	Autocorrelation function
C_0	Initial values vector
e	Charge of electron
$e_{l,m,p,q}$	Differential recurrence coefficients
$f_{l,m,p,q}$, $g_{l,m,p,q}$	Real and imaginary parts of $e_{l,m,p,q}$
$f(t)$	Response function
G	Young's modulus
$\mathbf{h}$	External magnetic field orientation
$h = \frac{HM_s}{2K}$	Ratio of external field to anisotropy parameter
h_c	Critical value of h above which the bistable structure of the potential disappears
$\mathbf{h}(t)$	Random field term
$\mathbf{H}$	External magnetic field
$\mathbf{H}_1$	Perturbing field
H_{eff}	Effective magnetic field
$\mathbf{J}(\mathbf{r}, t)$	Current density of points moving around the surface of a unit sphere
k	Boltzmann's constant
kT	Thermal energy
K	Anisotropy energy
L_{FP}	Fokker–Planck operator
m	Mass of electron
M_0	Magnetization of the suspension when in thermodynamic equilibrium
M_s	Spontaneous magnetization of the ferromagnetic material
n_i	Number of particles in ith orientation
$N_{l,m}$	Normalization constants
p_{eff}	Effective eigenvalue

p_n, F_n, A_n	Corresponding eigenvalues, eigenfunctions, and amplitudes of the Fokker–Planck equation
pref	Prefactor of the approximation formulas c, δ — constants relating to the prefactor
$\mathbf{r}$	Magnetization orientation vector
S	Matrix of the eigenvectors of the Fokker–Planck matrix A
T	Absolute temperature
T_c	Correlation time
T_N	Néel relaxation time
$u = h \cos \psi$	
U	Stochastic variable
$U_{l,m}$	Surface spherical harmonics
V_i	Values of V at stationary points
$V(\mathbf{r})$	Gibbs free energy per unit volume
$w_{l,m,p,q}$	Matrix coefficients for W matrix
W	Matrix used in computing the correlation time T_c
$W(\mathbf{r},t)$	Distribution of magnetization orientations
W_0	Equilibrium solution to the Fokker–Planck equation
$x = \cos \theta$	
$X_{l,m}$	Spherical harmonics
$\langle X_{l,m} \rangle$	Expectation values of spherical harmonics
$Y_{l,m}$	Normalized spherical harmonics
α_i	Direction cosines of the magnetization
$\beta = v/kT$	v = magnetic particle volume
γ	Gyromagnetic ratio
$\epsilon = h_c - h$	
η	Phenomenological damping constant
θ, φ	Angular spherical polar coordinates
λ_1	Lowest eigenvlaue of the associated Fokker–Planck matrix
λ_n	Eigenvalues of corresponding Sturm–Liouville equation
λ_s	Empirical magnetization constant
μ	Magnetic moment (magnetization × volume)
$\nu = h \sin \psi$	
$\nu_{i,j}$	Transition probability
ξ	External field parameter
$\xi_1 = \beta H_1 / M_s$	
$\sigma = \beta K$	Barrier-height parameter
τ	Relaxation time $(1/p_1)$
τ_N	Néel relaxation time

τ^{-1}	Switching rate
$\tau_\parallel$	Longitudinal relaxation time
$\tau_\perp$	Transverse relaxation time
$\tau = 2\tau_N / \lambda_1$	
ψ	Angle between the easy axis and the externally applied magnetic field $\mathbf{h}$
$\chi(\omega)$	Normalized complex susceptibility
ω	Angular momentum

ACKNOWLEDGMENTS

The author would like to thank Professors D. S. F. Crothers and W. T. Coffey for their assistance. E. E. C. Kennedy would like to thank EPSRC for financial assistance.

REFERENCES

1. R. A. Serway, *Physics for Scientists and Engineers with Modern Physics*, CBS Publishing, 1986.

2. A. Aharoni, *Introduction to the Theory of Ferromagnetism*, Oxford Univ. Press, 1996.

3. A. Menon, "Towards 100Gb/in^2 Recording Systems: The Evolution of High Density Magnetic Storage," Ewing Lecture (1996).

4. W. F. Brown, *IEEE Trans. Mag.* **15**, 1196 (1979).

5. P. J. Weiss, *Phys.* **6**, 661 (1907).

6. J. Frenkel and J. Dorfman, *Nature* **126**, 274 (1930).

7. L. Landau and E. Lifshitz, *Physik. Zes. Sowjetunion* **8**, 153 (1935).

8. W. C. Elmore, *Phys. Rev.* **54**, 1092 (1938).

9. C. Kittel, *Phys. Rev.* **70**, 965 (1946).

10. C. P. Bean and J. D. Livingston, *J. Appl. Phys. Suppl.* **30**, 120s (1959).

11. C. G. Montgomery, *Phys. Rev.* **38**, 1782 (1931).

12. C. G. Montgomery *Phys. Rev.* **39**, 163 (1932).

13. P. Langevin, *Comptes Rend.* **146**, 530 (1908).

14. W. T. Coffey, D. S. F. Crothers, J. L. Dormann, L. J. Geoghegan, Yu. P. Kalmykov, J. T. Waldron, and A. W. Wickstead, *J. Magn. Mater.*, **145**, L263 (1995).

15. P. Debye, *Polar Molecules*, Chemical Catalog Co. Inc., 1929.

16. L. Néel, *Ann. Geophys.* **5**, 99, (1949).

17. L. J. Geoghegan, W. T. Coffey, and B. Mulligan, *Adv. Chem. Phys.* **100**, 475 (1997).

18. P. M. S. Blackett, *Lectures on Rock Magnetism*, Weizmann Science Press, Jerusalem, 1956.

19. L. Weil, *J. Chem. Phys.* **51**, 715 (1954).

20. W. F. Brown, *Phys. Rev.* **130**, 1677 (1963).

21. T. L. Gilbert, *Phys. Rev.* **100**, 1243 (1955).

22. W. T. Coffey, P. J. Cregg, and Yu. P. Kalmykov, *Adv. Chem. Phys.* **83**, 263 (1993).

23. M. C. Wang and G. E. Uhlenbeck, *Rev. Mod. Phys.* **17**, 323 (1945).

24. A. Einstein, *Investigations on the Theory of Brownian Movement*, Meuthen, London, 1926 (reprinted by Dover, New York, 1956).

25. H. A. Kramers, *Physica* **7**, 284 (1940).

26. A. Aharoni, *Phys. Rev.* **135**, A447 (1964).

27. A. Aharoni, *Phys. Rev.* **177**, 793 (1969).

28. Yu. L. Raîkher and M. I. Shliomis, in *Relaxation Phenomena in Condensed Matter*, Vol. **87**, W. T. Coffey, ed., Wiley Scientific, New York, 1994.

29. C. N. Scully, *Semiclassical Phase Integral Solutions of a Fokker-Planck Equation for a Ferromagnetic Particle*, Ph.D. thesis, The Queen's Univ. of Belfast, 1993.

30. M. I. Shliomis, *Sov. Phys. Usp.* **17**, 153, (1974).

31. W. T. Coffey, D. S. F. Crothers, J. L. Dormann, L. J. Geoghegan, Yu. P. Kalmykov, J. T. Waldron, and A. W. Wickstead, *Phys. Rev. B* **52**, 15951 (1995).

32. L. J. Geoghegan, Ph.D. thesis, University of Dublin, 1995.

33. J. L. Dormann, F. D' Orazio, F. Lucari, E. Tronc, P. Prene, J. P. Jolivet, D. Fiorani, R. Ch erkaoui, and M. Nogués, *Phys. Rev. B* **53**, 14297 (1996).

34. D. A. Smith and F. A. de Rozario, *J. Magn. Magn. Mater.* **3**, 219 (1976).

35. W. T. Coffey, D. S. F. Crothers, J. L. Dormann, L. J. Geoghegan, and E. C. Kennedy, *Phys. Rev B* **58**(6), 3249–3266 (1998).

36. E. E. C. Kennedy, *Relaxation Times for Single-Domain Ferromagnetic Particles*, Ph.D. thesis, The Queen's University of Belfast, 1997.

37. W. Wernsdorfer, E. Bonet Orozco, K. Hasselbach, A. Benoit, B. Barbara, N. Demoncy, A. Loiseau, D. Boivin, H. Pascard, and D. Mailly, *Phys. Rev. Lett.* **78**, 1791 (1997).

38. W. T. Coffey, D. S. F. Crothers, J. L. Dormann, Yu. P. Kalmykov, E. C. Kennedy, and W. Wernsdorfer, *Phys. Rev. Lett.* **80**(25), 5655–5658 (1998).

39. H. Kachkachi, W. T. Coffey, D. S. F. Crothers, A. Ezzir, E. C. Kennedy, and M. Noguès, *J. Phys. C.* (in press).

40. U. Nowek, R. Chantrell, and E. E. C. Kennedy, *Phys. Rev. Lett.* (in press).

41. S. Chandrasekhar, *Rev. Mod. Phys.* **15**, 1 (1943).

42. W. T. Coffey, Yu. P. Kalmykov, and J. T. Waldron, *The Langevin Equation*, World Scientific, Singapore, 1996.

43. A. R. Edmonds, *Angular Momentum in Quantum Mechanics*, Princeton Univ. Press, Princeton, NJ, 1957.

44. W. T. Coffey and L. J. Geoghegan, *J. Mol. Liq.* **69**, 53 (1996).

45. W. T. Coffey, Yu. P. Kalmykov, and E. S. Massawe, in *Modern Non-Linear Optics*, Vol. **85**, Part 2, Wiley-Interscience, New York, 1993 (reprinted 1996), p. 667.

46. W. T. Coffey, D. S. F. Crothers, Yu. P. Kalmykov, and J. T. Waldron, *Phys. Rev. B* **51**, 15947, (1995).

47. W. T. Coffey, *J. Mol. Struct.* (in press); *Adv. Chem. Phys.* (in press).

48. I. Eisenstein and A. Aharoni, *Phys. Rev. B* **16**, 1278, 1285 (1977).

49. I. Klik and L. Gunther, *J. Stat. Phys.* **60**, 473 (1990).

50. E. C. Stoner and E. P. Wohlfahrt, *Phil. Trans. Roy. Soc. Lond. A* **240**, 599 (1948).

51. W. T. Coffey, *Adv. Chem. Phys.* **103**, 259 (1998).

52. H. Risken, *The Fokker-Planck Equation*, 2nd ed., Springer-Verlag, 1989.

53. W. T. Coffey, D. S. F. Crothers, J. L. Dormann, L. J. Geoghegan, E. C. Kennedy, and W. Wernsdorfer, *J. Phys. Condens. Matter* **10**, 9093–9109 (1998).

54. I. Klik and L. Gunther, *J. Appl. Phys.* **67**, 4505 (1990).

55. B. J. Matkowsky, Z. Schuss, and C. Tier, *J. Stat. Phys.* **35**, 443 (1984).

56. H. C. Brinkman, *Physica* **22**, 29,149 (1956).

57. R. Landauer and J. A. Swanson, *Phys. Rev.* **121**, 1668 (1961).

58. J. S. Langer, *Ann. Phys. (Leipzig)* **54**, 258 (1969).

59. D. A. Garanin, E. C. Kennedy, W. T. Coffey, *Phys. Rev. E.* (in press).

60. M. Büttiker, in *Noise in Non-Linear Systems*, Vol. 1, F. Moss and P. V. E. McKlintock, eds., Cambridge Univ. Press London, 1989.

61. J. L. Dormann, D. Fiorani, and E. Tronc, *Adv. Chem. Phys.* **98**, 283 1997.

62. J. L. Dormann, F. D'Orazio, F. Lucari, E. Tronc, P. Prené, J. P. Jolivet, D. Fiorani, R. Cherkaoui, and M. Nogués, *Phys. Rev. B*, **53**, 14297 1996.

63. W. T. Coffey, *J. Mol. Struct.* (in press).

64. Yu. P. Kalmykov, S. V. Titov, and W. T. Coffey, *Phys. Rev. B* **58**, 3267 (1998).

65. H. Pfeiffer, *Phys. Stat. Sol.* **118**, 295 (1990); **122**, 377 (1990).

66. W. T. Coffey, *Adv. Chem. Phys.* **103**, 259–333, 1998.

67. H. Risken, *The Fokker-Planck Equation*, 2nd ed., Springer, Berlin, 1989.

68. V. I. Mel'nikov and S. V. Meshkov, *J. Chem. Phys.* **85**, 1018 (1996).

69. P. Hängii, P. Talkner, and M. Borovec, *Rev. Mod. Phys.* **62**, 251 (1990).

70. D. A. Garanin, *Phys. Rev. E* **54**, 3250 (1996).

71. J. Kurkijarvi, *Phys. Rev. B* **6**, 832 (1971).

72. L. Gunther and B. Barbara, *Phys. Rev. B* **49**, 3926 (1994).

73. A. Garg, *Phys. Rev. B* **51**, 15592 (1995).

74. Yu. P. Kalmykov, S. A. Titov, and P. M. Déjardin, private communication.

75. J. L. Dormann, D. Fiorani, and E. Tronc, *Adv. Chem. Phys.* **98**, 283 (1997).

76. A. Aharoni and J. P. Jakubovics, *IEEE Trans. Magn.* **24**, 11892 (1988).

77. V. Illingworth, *Dictionary of Physics*, Penguin, 1991.

ONE-DIMENSIONAL ISING MODEL FOR SPIN SYSTEMS OF FINITE SIZE

ANDRZEJ R. ALTENBERGER AND JOHN S. DAHLER

Departments of Chemistry and of Chemical Engineering and Materials Science, University of Minnesota, Minneapolis, Minnesota

CONTENTS

I. INTRODUCTION

A system with exchange interactions that favor the parallel alignment of spins can exhibit a spontaneous transition from a paramagnetic state, characterized by randomly oriented spins, to a ferromagnetic state of virtually complete alignment. The paramagnetic states of such systems possess only residual magnetization, with average squared magnetizations (ASM) of the order $\overline{M^2}(T \to \infty) = N\mu^2$. Here N denotes the number of spins and μ is the magnitude of the moment associated with an individual spin. In contrast to this, the ASM of the ferromagnetic state is of the order $\overline{M^2}(T \to 0) = N^2\mu^2$. The transition between these two occurs in the vicinity of the Curie temperature, at which temperature the value of the ASM exhibits an almost steplike discontinuity.

The one-dimensional Ising spin–lattice system is the simplest microscopic model for a system of this type. The model was originated in 1920 by

Advances in Chemical Physics, Volume 112, Edited by I. Prigogine and Stuart A. Rice
ISBN 0-471-38002-4.   © 2000 John Wiley & Sons, Inc.

Lenz [1], who proposed it as an elementary representation for a ferromagnetic system. It subsequently was solved in 1925 by his doctoral student, Ising [2], who produced exact analytic expressions for the partition function and for the magnetization $\overline{M}(T, B)$ in the presence of an external magnetic induction B. Ising found that this magnetization vanished identically in the absence of the field $[\overline{M}(T, 0) = 0$ for $T > 0]$ and that the entropy of the spin system was a continuous and differentiable function of temperature. From these observations, he concluded that the one-dimensional model did not predict spontaneous magnetization, contrary to Lenz expectations and to the predictions of the Weiss mean-field theory [3]. (For a more complete historical commentary, see Ref. 4.) The Weiss theory (proposed in 1907) was at this time a widely recognized microscopic theory of the magnetic phase transition. It provided a nonlinear equation for the magnetization and predicted that this magnetization would be nonvanishing for temperatures below a certain Curie temperature, $T_c = zJ/k_B$, which was proportional to the lattice coordination number z and the pair-exchange energy J but was independent of the dimensionality of the system.

From a present-day (at the time of writing) perspective it appears that Ising's conclusions were a bit too hasty. The magnetization, which is the basic quantity treated by the Weiss molecular field theory, is in general a vector valued property of a spin system. As such, it vanishes in the absence of external magnetic fields because of the symmetry of the spin system Hamiltonian operator. By itself, this fact does not preclude the occurrence of spontaneous magnetization. It simply means that in the absence of an external field there is no preferred spatial orientation for the vector magnetization. To proceed further under these circumstances, it is natural to adopt the average *squared* magnetization, in place of the average magnetization itself, as the appropriate order parameter of the system. As we shall see, this leads directly to the conclusion that the one-dimensional Ising model exhibits a well-defined para $\leftrightarrow$ ferro magnetic transition for all systems of finite size. The Curie temperature is found to depend on the number of spins. Its value falls very slowly as the size of the spin system increases and approaches absolute zero in the limit $N \to \infty$.

Contemporary theories of the phase transitions of many-body systems attach great importance to the occurrence of singularities exhibited by the thermodynamic potentials and/or their derivatives in the vicinity of the transition temperature. For example, one finds in a monograph by Yeomans [5] the following definition: "A phase transition is signalled by a singularity in a thermodynamic potential such as the free energy. If there is a finite discontinuity in one or more of the first derivatives of the appropriate thermodynamic potential, the transition is termed first order . . . . If the first derivatives are continuous but second derivatives are discontinuous or infinite the transition will be described as higher order, continuous or critical." This definition obviously was inspired by the earlier classification of phase transitions due to Ehrenfest [6], who drew on the experimental information available to him at the time. As the years passed, phase transformations were observed that could not be categorized unequivocally within the Ehrenfest scheme. To remedy this situation, more elaborate classifications were proposed by Tisza [7] and Münster [8]. However, in all these schemes the temperature of a phase transformation was identified as one at which some relevant thermodynamic potential exhibited either a discontinuity or a singularity. This implies that the system consists of an infinite number of particles or, more precisely, possesses an infinite number of mechanical degrees of freedom. The partition functions of finite systems are continuous and differentiable functions of the temperature and of parameters on which the system Hamiltonian is dependent [9,10]. Consequently, the thermodynamic response functions of finite systems will be free from discontinuities and singularities, although the ranges of parameter values over which these functions exhibit large variations can and often will diminish as the size of the system increases. In fact, the "singularities" associated with infinitely large systems spring from simple origins — the size of the "divergence" that a property exhibits near a transition temperature is a monotone increasing function of the number of participating degrees of freedom and so tends to infinity as the latter becomes unbounded [10]. Since phase transformations certainly do occur in systems of finite size, the basic objectives of theories for these systems should be to determine the number of transitions that a particular system can undergo and to identify the corresponding set of transition temperatures. Systems of infinite extent may be of mathematical interest but are rarely encountered, whereas the systems studied in laboratory experiments and computer simulations necessarily consist of finite numbers of particles.

The purpose of the present chapter is to show (by very elementary calculations) that even a system as simple as the one-dimensional Ising model exhibits a more complex phase behavior than generally is suspected. Some related issues already have been treated by Ciach [11], who focused on the density profile and pair correlation function of an open-chain subject to boundary conditions. Our own objectives are to identify the number of transformations that occur in a one-dimensional Ising spin system and to calculate the corresponding set of characteristic transition temperatures.

The following section is devoted to a presentation of the basic tools and criteria we use to accomplish these goals. In Section III these general procedures are applied to the one-dimensional Ising model. Although this model has limited relevance to real magnetic systems, it is one of the few microscopic models for which exact analytic formulas can be obtained for all the relevant thermodynamic properties [12]. Thus, it has significant instructional value and, additionally, may produce conclusions that have applicability to Ising spin systems of higher dimensionality.

II. STATISTICAL THERMODYNAMICS OF LATTICE–SPIN SYSTEMS

The exchange energy characteristic of a ferromagnetic system favors parallel alignment of the spins. At temperatures such that the value of $k_B T$ is greatly in excess of the pair-exchange energy J, the spin orientations will be very weakly correlated and the mean-squared magnetization will exhibit only a small residual value characteristic of a paramagnetic system. As the temperature diminishes, there will be a corresponding increase in the probability of forming clusters or microdomains of identically oriented spins. Under these conditions the system will consist of fewer independent elements, but most of these will possess magnetic moments significantly larger than those of the individual spins. However, since it is expected that the relatively large magnetic moments of these aggregates will be randomly oriented, the system will continue to exhibit paramagnetic behavior. As the temperature drops still further, the clusters become ever larger and fewer in number until, at last, all the spins are locked into a single ferromagnetic phase with an enormous magnetic moment of order N.

This phenomenological description suggests that the successive changes of state that occur during the cooling of a many-spin system are somewhat akin to those that accompany the gas–liquid–solid transitions characteristic of a system of simple molecules. At high temperatures the spins form a para-magnetic "spin gas" phase. The several-spin microdomains that form as the temperature falls are analogous to a dispersed, spatially uniform collection of liquid drops. This change of state is accompanied by the evolution of a significant amount of energy, identifiable with the heat of condensation of the high-temperature spin gas. Finally, when the temperature has become sufficiently low the magnetic moments of the relatively few remaining clusters (microdomains) will assume a state of parallel alignment, forming a ferromagnetic "spin solid." This final transformation is accompanied by a small but finite transfer of energy, the magnitude of which is negligibly small for a one-dimensional system in the limit of an infinitely large number of spins.

The mathematical implementation of these observations requires a more detailed and specific definition of the spin system. Accordingly, we assume that the spins themselves are located at the sites $i = 1, 2, \ldots, N$ of a regular lattice. Associated with each is a dimensionless spin variable and a magnetic moment μ. A microscopic state of the system is characterized by the collection of spin variables, $\mathbf{s}^N = \{\mathbf{s}_1, \mathbf{s}_2, \ldots, \mathbf{s}_N\}$, each of which can assume only a discrete set of eigenvalues.

The total magnetic moment and Hamiltonian of the system are

$$\mathbf{M}_N = \sum_{i=1}^{N} \mu \mathbf{s}_i \tag{2.1}$$

and

$$H_N = H_N^0 - \mathbf{B} \cdot \mathbf{M}_N \tag{2.2}$$

respectively. Here $\mathbf{B}$ denotes the magnetic induction and $H_N^0(\mathbf{s}^N)$ is the Hamiltonian in the absence of this external field. As such, H_N^0 will exhibit the symmetry of the lattice and incorporate an appropriate set of exchange energies.

We associate with each configuration $\mathbf{s}^N$ the corresponding probability density

$$P_N(\mathbf{s}^N, \mathbf{B}; T) = Q_N^{-1}(T, \mathbf{B}) \exp \frac{-H_N(\mathbf{s}^N, \mathbf{B})}{k_B T} \tag{2.3}$$

Here

$$Q_N(T, \mathbf{B}) = \sum_{\{\mathbf{s}^N\}} \exp \frac{-H_N(\mathbf{s}^N, \mathbf{B})}{k_B T} \tag{2.4}$$

denotes the canonical partition function, T the absolute temperature, and k_B the Boltzmann constant. The ensemble average of a function $A(\mathbf{s}^N)$ is then

$$\overline{A}(T, \mathbf{B}) = \sum_{\{\mathbf{s}^N\}} P_N(\mathbf{s}^N, \mathbf{B}; T) A(\mathbf{s}^N) \tag{2.5}$$

In terms of the partition function the free energy F_N, the thermodynamic internal energy $U_N = \overline{H}_N$, and the entropy S_N are given by the formulas

$$F_N(T, \mathbf{B}) = -k_B T \ln Q_N$$

$$U_N(T, \mathbf{B}) = k_B T^2 \frac{\partial}{\partial T} \ln Q_N \tag{2.6}$$

and

$$S_N(T, B) = T^{-1}(U_N - F_N) = -\frac{\partial F_N}{\partial T}$$

respectively. These and other thermodynamic properties of the spin system depend on the temperature T and the externally imposed magnetic induction $\mathbf{B}$.

Because of the symmetry of the Hamiltonian H_N^0, the ensemble-averaged longitudinal magnetization

$$\overline{M}_N(T, B) = -\frac{\partial F_N}{\partial B} \tag{2.7}$$

vanishes identically in the absence of external fields. However, this lack of spatial anisotropy does not mean that the system is incapable of exhibiting spontaneous magnetization. It simply is necessary to examine $\overline{M_N^2}(T, B = 0)$, the averaged square of the microscopic magnetic moment. This average squared magnetization (ASM) is finite-valued in both the presence and absence of external fields and so provides a natural measure of the zero-field spontaneous magnetization. Closely related to the ASM is the isothermal magnetic susceptibility

$$\chi_N(T, B) = \frac{\partial \overline{M}_N}{\partial B} = -\frac{\partial^2 F_N}{\partial B^2} = \frac{1}{k_B T}[\overline{M_N^2}(T, B) - \overline{M}_N^2(T, B)] \tag{2.8}$$

which, in the absence of external fields, reduces to

$$\chi_N(T, 0) = \frac{1}{k_B T}\overline{M_N^2}(T, 0) \tag{2.9}$$

Associated with $\chi_N(T, 0)$ is the order parameter

$$x_N^2(T) = \frac{\overline{M_N^2}(T)}{\mu^2 N^2} \le 1 \tag{2.10}$$

This finite-valued, smoothly varying steplike function of T monitors the magnetic behavior of the system under the field-free conditions for which $\overline{M}_N$ does not qualify as a thermodynamic property.

As previously mentioned, theories of phase transitions for systems with infinitely many degrees of freedom invariably rely on divergences and discontinuities of the basic thermodynamic response functions as indicators of phase transformation temperatures. In particular, it usually is assumed that the specific heat and magnetic susceptibility both diverge at the Curie temperature characteristic of the para $\leftrightarrow$ ferro magnetic transition. This assumption obviously is inapplicable to systems of finite size, for which transitions from one state to another occur over *bands* of temperature and involve *finite* changes of the thermodynamic response functions. To deal with situations of this type, we shall identify the characteristic temperature associated with a phase transition as one belonging to an interval over which the system interacts with its environment in a particularly strong way, either by exchanging uncommonly large amounts of energy (in the forms of heat or work) or by exhibiting a dramatic, near-steplike variation of an order parameter. Large amounts of absorbed or evolved energy generally are required for the activation or freezing of the microscopic degrees of freedom that usually accompany dissolution or formation of a new phase. However, the energetic requirements for an order–disorder transition are expected to be small since a transition of this type usually will have been preceded (as the temperature falls) by the formation of bound microdomains. Nevertheless, stabilization of the ferromagnetic phase certainly requires the extraction of a finite amount of energy.

Henceforth we focus our attention on phase transitions occurring in an initially unpolarized (i.e., zero-field) spin system. The characteristic temperatures of this system are to be identified by using probes that either supply or withdraw energy from the system. One such probe could transmit energy to the system by a heat–transfer process that does not involve the imposition of a magnetic field. This will result in a rise of temperature. Alternatively, the temperature can be maintained at a fixed value as one perturbs the system with a small externally applied magnetic field. The corresponding characteristic transition temperatures can be identified by examining the following expression for the incremental internal energy:

$$\Delta U_N(T; \delta T, \delta B) \equiv U_N(T + \delta T, \delta B) - U_N(T, 0)$$

$$= C_N(T)\delta T + \frac{1}{2}\left[\frac{d}{dT}C_N(T)\right]\delta T^2$$

$$+ \frac{1}{2}T^2\frac{d}{dT}\left[\frac{\overline{M_N^2}}{k_B T^2}\right]\delta B^2 + \cdots \tag{2.11}$$

Analogous to this is the formula (MacLaurin series)

$$\Delta S_N(T; \delta T, \delta B) \equiv S_N(T + \delta T, \delta B) - S_N(T, 0)$$

$$= \frac{C_N(T)}{T}\delta T + \frac{1}{2}\left[\frac{d}{dT}\frac{C_N(T)}{T}\right]\delta T^2$$

$$+ \frac{1}{2}\frac{d}{dT}\left[\frac{\overline{M_N^2}}{k_B T}\right]\delta B^2 + \cdots \tag{2.12}$$

for the incremental entropy.

To proceed further, we assume that the values of the increments δT and δB are fixed and concentrate on finding the reference state temperatures at which the coefficients appearing in these formulas exhibit especially large values. In the case of a pure heat transfer ($\delta B = 0$), the largest value of the incremental energy is that for which the heat capacity $C_N(T) = dU_N/dT$ is a maximum. The existence of such a maximum is evidence of the occurrence of a microscopic reorganization of the system, namely, the dissociation or formation of clusters of parallel-oriented spins. The rise and subsequent fall of the heat capacity indicates the range of temperatures over which significant amounts of energy are required to produce this transformation. Thus, the characteristic temperature T_0 of a condensation-type, first-order transition is obtained by solving the equation

$$\frac{dC_N(T)}{dT} = 0 \tag{2.13}$$

and the associated transition band (T_{01}, T_{02}) is bounded by the roots of the equation

$$\frac{d^2 C_N(T)}{dT^2} = 0 \tag{2.14}$$

Finally, the heat associated with formation of the corresponding microdomains is given by the formula

$$Q_{\text{CON}} = \int_{T_{01}}^{T_{02}} dT\, C_N(T) \tag{2.15}$$

The degree of spin ordering is characterized by the value of the order parameter $x_N^2(T) = \overline{M_N^2}(T)/\mu^2 N^2$, and so T_{m}, the temperature of the spontaneous (zero-field) para$\leftrightarrow$ferro magnetic transition, is that at which

$$\frac{d^2}{dT^2}\overline{M^2}(T) = 0 \tag{2.16}$$

The corresponding transition band $(T_{\text{m1}}, T_{\text{m2}})$ is determined by the roots of the equation

$$\frac{d^3}{dT^3}\overline{M^2}(T) = 0 \tag{2.17}$$

and the heat associated with this second-order transition is given by the formula

$$Q_{\text{OR--DIS}} = \int_{T_{\text{m1}}}^{T_{\text{m2}}} dT\, C_N(T) \tag{2.18}$$

Alternatively, the system can be maintained at a fixed temperature and probed by exposing it to small variations of the external magnetic induction.

The spin system then will be least stable (with respect to variations of the induction) at that temperature for which the absorption or emission of magnetic field energy is the greatest. This temperature, T_B, corresponds to an extremum of the *magnetic field absorption coefficient*

$$\kappa_N(T) = \frac{1}{2}T^2\frac{d}{dT}\left[\frac{x_N^2(T)}{k_B T^2}\right] \tag{2.19}$$

or, equivalently, to the temperature that satisfies the equation

$$\frac{d^2}{dT^2}\overline{M_N^2}(T,0) - \frac{2}{T}\frac{d}{dT}\overline{M_N^2} + \frac{2}{T^2}\overline{M_N^2} = 0 \tag{2.20}$$

It is important to recognize that by varying the external magnetic field one only obtains information about the para $\leftrightarrow$ ferro magnetic transition. Although this allows us to locate the Curie temperature T_m associated with the order-disorder transition, it apparently is unrelated to the characteristic temperatures defined by Eqs. (2.13) and (2.14); these are specific to a quasi-first-order trans-formation of a different kind. It is demonstrated in the following section that the transformation characterized by the temperatures (T_0, T_{01} and T_{02}) asso-ciated with the latter of these transitions corresponds to a "condensation" of individual spins into randomly oriented paramagnetic clusters, a process that is separate from that of spontaneous magnetization.

III. ONE-DIMENSIONAL ISING MODEL

The partition function of a many-spin system can be calculated exactly for the one-dimensional Ising model for which $H_N^0 = -J\Sigma'_{ij}s_is_j$ consists of a sum over spins assigned to nearest-neighbor lattice sites and $s_i = \pm 1, \forall i$. Different results are obtained for open-ended chains and for "Ising rings" satisfying the periodic condition $s_{N+1} = s_1$ [12]. We first consider an N-spin ring and then proceed to an open-ended chain.

A. The One-Dimensional Ising Ring

The partition function for this system can be written in the form [12]

$$Q_N(K,v) = \sum_{\{s_i=\pm 1\}}\prod_{i=1}^{N}\exp\left[Ks_is_{i+1} + \frac{1}{2}v(s_i + s_{i+1})\right]$$
$$= \lambda_+^N(K,v) + \lambda_-^N(K,v) \tag{3.1}$$

with $K = J/k_BT$ and $v = \mu B/k_BT$. The positive-valued parameter J denotes the nearest-neighbor, spin-coupling (exchange) energy, and the functions $\lambda_\pm$

are defined as follows

$$\lambda_{\pm} = \sigma_1 \pm \sigma_2 \tag{3.2}$$

with

$$\sigma_1 = \exp(K)\cosh(v)$$

and

$$\sigma_2 = [\exp(2K)\sinh^2(v) + \exp(-2K)]^{1/2} \tag{3.3}$$

Exact formulas also can be constructed for the various thermodynamic quantities considered in the previous section.

In the absence of external magnetic fields, all of these formulas become significantly simpler. Thus, the expression (3.1) for the partition function reduces to

$$Q_N(K, 0) = 2^N [\cosh^N K + \sinh^N K] \tag{3.4}$$

and the average squared magnetization (ASM) and heat capacity per spin are given by the formulas

$$\overline{M_N^2}(K, 0) = \mu^2 N \exp(2K)\frac{1 - \tanh^N K}{1 + \tanh^N K} \tag{3.5}$$

and

$$\frac{C_N}{k_B} = -\frac{2K^2}{\sinh^2(2K)}\frac{1}{(1 + \tanh^N K)^2}[1 - 2(N - 1)\tanh^N K + \tanh^{2N} K$$
$$+ \cosh(2K)\{\tanh^{2N} K - 1\}] \tag{3.6}$$

respectively. These last two quantities provide a complete description of all the thermodynamic and magnetic properties of a field-free N-spin Ising ring.

The curves of Fig. 1 illustrate how several properties vary with temperature for a ring consisting of $N = 10^5$ spins. One of these curves (labeled "4") is the internal energy per spin, which rises from a low-temperature limit characteristic of the ferromagnetic state (in which the spins are fully aligned) and tends monotonically to the value zero, associated with the paramagnetic state consisting of randomly oriented spins, namely, $-JN \leq U_N \leq 0$. The order parameter (labeled "1"), which is proportional to the ASM, varies from the low-temperature value of 1 corresponding to the completely ordered state, to the high-temperature value of N^{-1} descriptive of the residual magnetization of the paramagnetic phase. It clearly can be seen that there is a finite range of temperatures ($0 \leq k_B T/J \leq 0.2$ for $N = 10^5$) over which there exists a well-defined ferromagnetic state. The inflection point of curve 1 is that at which the

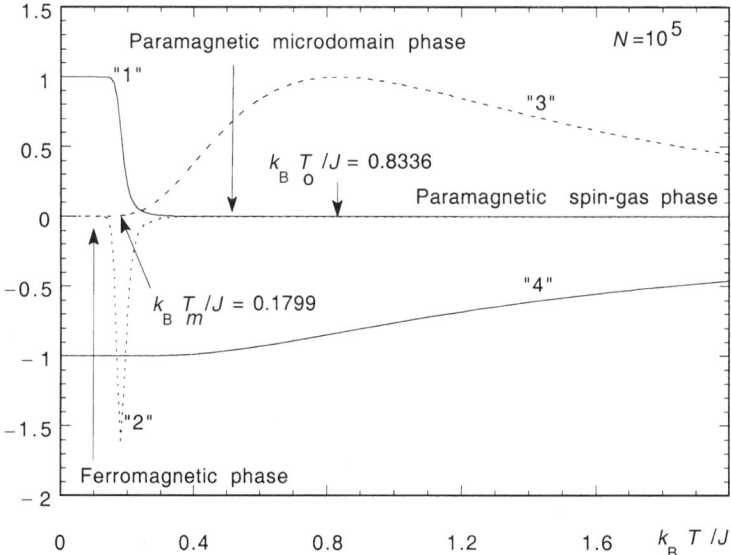

FIGURE 1. The temperature dependences of several thermodynamic properties for a one-dimensional Ising ring with $N = 10^5$ spins. The curve 1 represents the order parameter $x_N^2(T)$ defined by Eq. (2.10); curve 2 is $(dx_N^2/dT)/19$. Curves 3 and 4, respectively, show the heat capacity (divided by 2.3) and internal energy per spin in units of k_B. The scaling factors of 19 and 2.3 have been introduced to facilitate the graphical presentation.

function dx_N^2/dT of curve 2 exhibits an extremum. This occurs at a temperature $T_m \simeq 0.18 \ J/k_B$ characteristic of the order–disorder transition specific to an Ising ring of $N = 10^5$ spins. This temperature is significantly lower than $T_0 = 0.8336 \ J/k_B$, the temperature at which the heat capacity (curve labeled "3") reaches its maximum value.

The numerical value of the order–disorder transition temperature does, of course, depend on the size of the Ising ring. This is illustrated by Fig. 2, where order parameters are presented for rings of various lengths. As the ring size increases, the order–disorder transition temperature steadily falls, tending very slowly (see Table I and Fig. 3) to the limiting value of zero as N becomes extremely large. Thus, over the interval from $N = 5$ to $N = 10^6$ the value of $k_B T_m/J$ falls only from 1.156 to 0.1494. It also is apparent from Fig. 2 that as the ring length increases, the para $\leftrightarrow$ ferro magnetic transition becomes more sharply defined, with the value of the ASM dropping from its highly ordered to disordered limits over steadily diminishing intervals of temperature. In contrast to the steady decline of T_m with increasing N, the characteristic temperature of the transition responsible for the maximum of the heat capacity reaches its "asymptotic" value of $k_B T_0/J = 0.8336$ for chains of length $N = 10^2$, (see Table I).

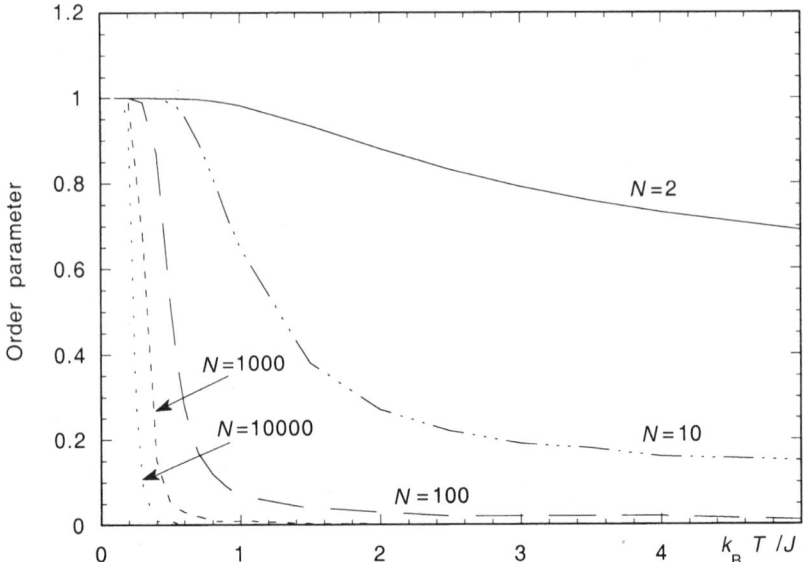

FIGURE 2. The temperature dependence of the order parameter for one-dimensional Ising rings of various sizes.

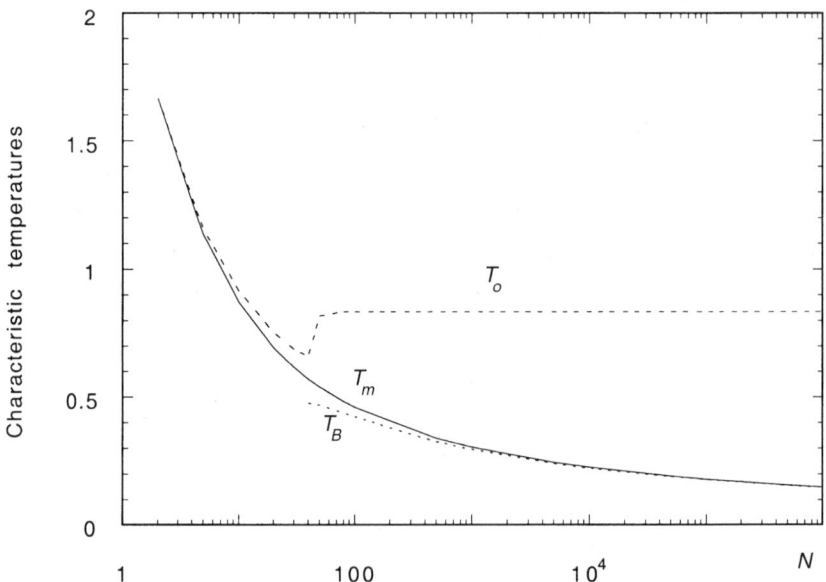

FIGURE 3. The variations with ring size of the three characteristic temperatures: T_0 (condensation), T_m (Curie), and T_B (magnetic absorptivity).

TABLE I

Characteristic Temperatures T_0, T_m, and T_B for One-Dimensional Ising Rings of Various Sizes[a]

N	$k_B T_0/J$	$k_B T_m/J$	$k_B T_B/J$	Q_{CON}/NJ	Q_{OR-DIS}/NJ
2	1.6671	1.6671	No absorption	0.3189	0.3189
5	1.1551	1.1358	No absorption	0.3379	0.3288
10	0.9147	0.8714	No absorption	0.2530	0.2317
20	0.7531	0.6932	No absorption	0.1672	0.1427
30	0.6873	0.6160	No absorption	0.1249	0.1035
40	0.6594	0.5700	0.4744	0.09759	0.08146
50	0.8176	0.5385	0.4682	0.2200	0.06724
80	0.8334	0.4817	0.4371	0.2563	0.04437
100	0.8336	0.4584	0.4213	0.2687	0.03625
500	0.8336	0.3386	0.3255	0.3139	9.871×10^{-3}
1,000	0.8336	0.3039	0.2947	0.3188	4.094×10^{-3}
5,000	0.8336	0.2452	0.2405	0.3190	8.514×10^{-4}
10,000	0.8336	0.2262	0.2226	0.3190	4.316×10^{-4}
50,000	0.8336	0.1917	0.1896	0.3190	8.839×10^{-5}
100,000	0.8336	0.1799	0.1781	0.3190	4.457×10^{-5}
500,000	0.8336	0.1573	0.1561	0.3190	9.039×10^{-6}
1,000,000	0.8336	0.1492	0.1482	0.3190	4.535×10^{-6}

[a] Also included are values of Q_{CON}/NJ and Q_{OR-DIS}/NJ, the heats per spin associated with the condensation and order–disorder transitions.

For Ising rings with $N \leq 30$ the condensation temperature T_0 and the order–disorder characteristic temperature T_m very nearly equal one another and the magnetic absorptivity $\kappa_N(T)$ fails to exhibit a local maximum; that is T_B is nonexistent. Once the chain length has risen beyond the transition value $N \simeq 30$, the values of T_0 and T_m progressively diverge from one another, while those of T_m and T_B tend toward a common limit. These behaviors are revealed by the curves of Fig. 3 and the entries in Table I.

Figure 4 shows how the temperature dependence of the specific heat varies with the size of the Ising ring. The shapes of these curves (of c_N/k_B versus $k_B T/J$) change rapidly as N increases from 2 to 10. They then much more gradually approach a limit that is fully established for $N \simeq 100$. This transition from small-system to large-system behavior also is visible in Fig. 3.

A more detailed picture is provided by Fig. 5 where the heat capacity per spin and its derivative $d(c_N/k_B)/dT$ are plotted versus temperature for an Ising ring of $N = 5000$. Also included in this figure is a plot of the order parameter $x_N^2(T)$. The first two extrema (a maximum followed by a minimum) on the curve of $d(c_N/k_B)/dT$ apparently are related to the order–disorder transition (see Table II) since their locations so nearly coincide with the temperatures (T_{m1}, T_{m2}) defining the boundaries of the para ↔ ferro magnetic transition

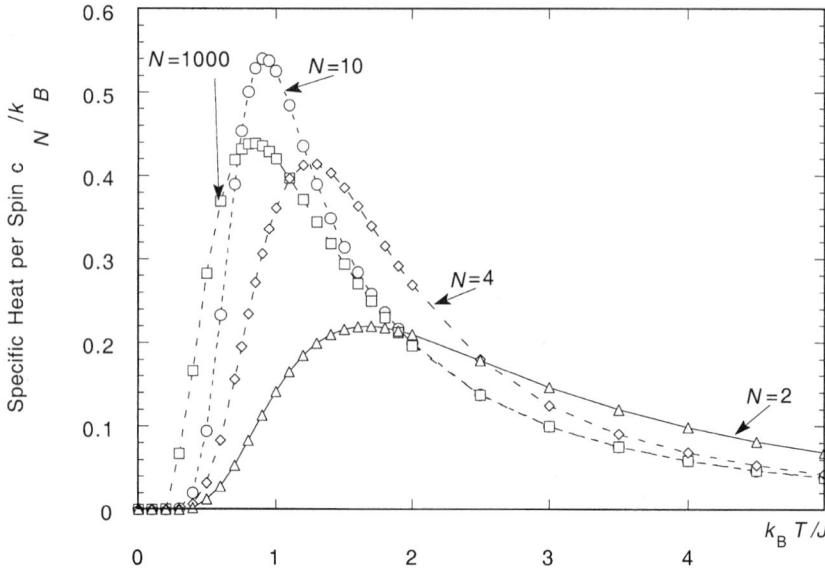

FIGURE 4. Temperature dependence of the heat capacity per spin (in units of k_B) for Ising rings of various sizes.

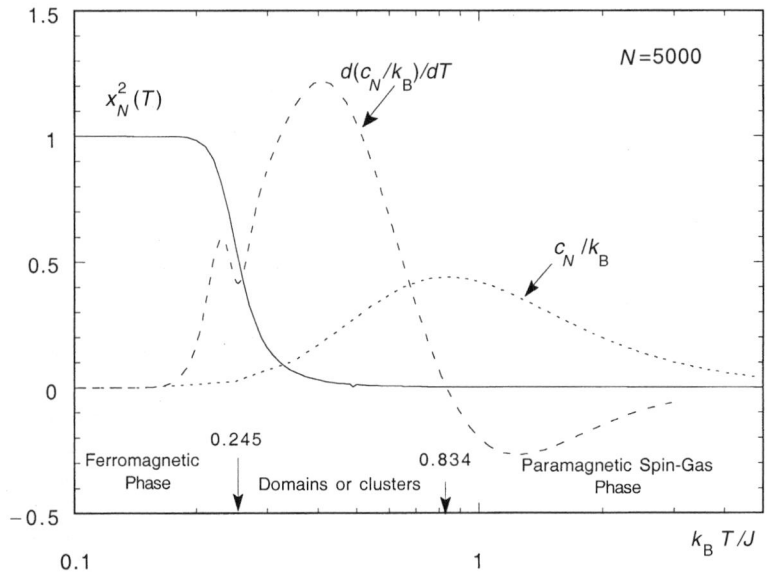

FIGURE 5. Properties of a one-dimensional Ising ring with $N = 5000$ spins. The solid curve represents the order parameter $x_N^2(T)$; the dotted curve represents $c_N(T)/k_B$, the heat capacity per spin in units of k_B. Finally, the dashed curve indicates the derivative $d(c_N/k_B)/dT$. Also indicated are the locations of the (reduced) transition temperature $k_B T_m/J$ and $k_B T_0/J$.

TABLE II
Temperatures Defining the Transition Bands for One-dimensional Ising Rings[a]

| | Extrema of dc_N/dT | | | | Extrema of $d^2\overline{M^2}/dT^2$ | |
| | $k_B T/J$ | | | | $k_B T_{m1}/J$ | $k_B T_{m2}/J$ |
N	Max	Min	Max	Min	Min	Max
2	0.8158	2.4873	none	none	0.8158	2.4873
5	0.7159	1.5553	none	none	0.7127	1.5293
10	0.6261	1.1564	none	none	0.6175	1.1060
20	0.5427	0.8985	none	none	0.5313	0.8428
30	0.4998	0.7854	0.9055	1.2323	0.4878	0.7344
40	0.4721	0.7131	0.7760	1.2432	0.4601	0.6716
50	0.4522	0.6596	0.7132	1.2436	0.4404	0.6292
80	0.4144	0.5637	0.6232	1.2437	0.4030	0.5543
100	0.3982	0.5271	0.5899	1.2437	0.3872	0.5243
500	0.3081	0.3625	0.4342	1.2437	0.2997	0.3872
1,000	0.2800	0.3205	0.4086	1.2437	0.2725	0.3329
5,000	0.2304	0.2532	0.4079	1.2437	0.2247	0.2640
10,000	0.2139	0.2324	0.4079	1.2437	0.2088	0.2423
50,000	0.1833	0.1952	0.4079	1.2437	0.1792	0.2033
100,000	0.1726	0.1826	0.4079	1.2437	0.1689	0.1901
500,000	0.1519	0.1589	0.4079	1.2437	0.1489	0.1651
1,000,000	0.1444	0.1506	0.4079	1.2437	0.1416	0.1526

[a]Columns 1–4 locate extrema of dc_N/dT; columns 5 and 6, extrema of $d^2\overline{M_N^2}/dT^2$.

zone. As the value of N rises beyond 20, a second pair of extrema appears. These become progressively more pronounced as N increases still further.

The response of the system to an external magnetic field probe is governed by the absorptivity $\kappa_N(T)$, defined by Eq. (2.19). The temperature characterizing the maximum of this response function is T_B, the solution of Eq. (2.20). How this characteristic temperature varies with N is shown in Fig. 3 and by the entries appearing in the last column of Table I. The curves of Fig. 6 illustrate the temperature dependence of $\kappa_N(T)$ for rings of various sizes. Only for rings that exceed a critical size of about 40 does the magnetic field absorption coefficient exhibit local extrema. It can be seen from Fig. 3 and Table I that for rings larger than this critical size, the temperature of maximum absorption very nearly coincides with the Curie temperature T_m, which is associated with the onset of spontaneous magnetization.

Shown in Fig. 7 (and tabulated in Table I) are the size dependences of the heats per spin, Q_{CON}/NJ and Q_{OR-DIS}/NJ [see Eqs. (2.15) and (2.18)], associated with the phase transitions of one-dimensional Ising rings. These two heats are virtually identical for rings that exhibit the previously mentioned small-system behavior. However, as N increases beyond 100, Q_{CON}/NJ

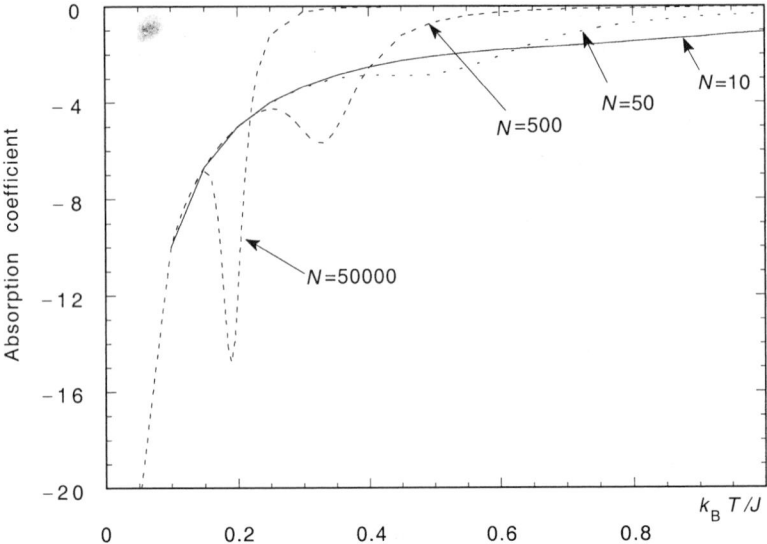

FIGURE 6. The variation with temperature of the magnetic field absorption coefficient $\kappa_N(T)$ for one-dimensional Ising rings of various sizes.

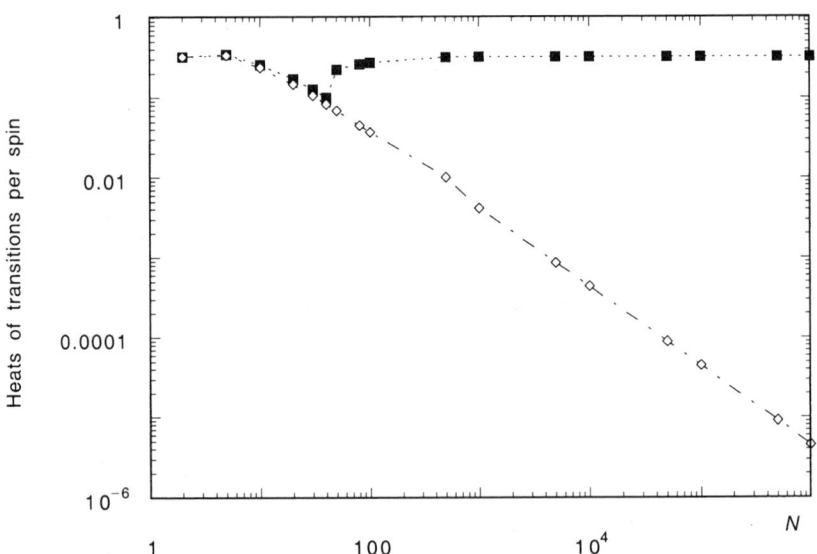

FIGURE 7. Variations with ring size of the heats Q_{CON}/NJ and Q_{OR-DIS}/NJ characteristic of the condensation and order–disorder transitions. The former are labeled by solid squares; the latter, by diamonds. Note that the asymptotic N^{-1} dependence of the heat of the order–disorder transition per spin shown in this figure indicates that the total heat of transition becomes independent of ring size for $N > 100$.

quickly stabilizes at a value of about 0.319 whereas the value of Q_{OR-DIS}/NJ diminishes at a rate that is very nearly proportional to N^{-1}. The direct proportionality between the total heat of transition Q_{CON} and the number of spins is a clear signature of a first-order, condensationlike transition. On the other hand, the fact that Q_{OR-DIS} remains constant, independent of the size of the system (once the "macroscopic" size of order $N = 100$ has been exceeded), is a typical feature of a second-order transition. It can be interpreted as meaning that at the Curie temperature T_m, the system consists of only two domains with antiparallel composite magnetic moments. The para $\leftrightarrow$ ferro magnetic transition then corresponds to a final alignment of these two domains to form a single-domain ferromagnetic phase. This assertion is confirmed by the fact that the computational data displayed in Fig. 7 are fitted (for $N > 100$) by the formula $Q_{OR-DIS}/NJ = 3.926\ N^{-0.9893}$. Thus, the value of Q_{OR-DIS} is $2(2J)$; $2J$ equals the spin-flip energy associated with a single-domain boundary, and the Ising ring has two such boundaries.

B.   The One-Dimensional Ising Chain

The partition function for a one-dimensional, open-ended Ising chain is given by the formula [12]

$$Q_N(K, v) = \lambda_+^{N-1}\{\cosh v + [\sinh^2 v + \exp(-2K)][\sinh^2 v + \exp(-4K)]^{-1/2}\}$$
$$+ \lambda_-^{N-1}\{\cosh v - [\sinh^2 v + \exp(-2K)][\sinh^2 v + \exp(-4K)]^{-1/2}\} \tag{3.7}$$

with K, v and $\lambda_\pm$ defined as they were in Section III.B. In the case of $B = 0$, this formula (3.7) reduces to the remarkably simple expression

$$Q_N(K, 0) = 2^N \cosh^{N-1} K \tag{3.8}$$

Corresponding to this are the formulas

$$\overline{M_N^2}(K, 0) = \mu^2 N \left[\exp(2K) + \frac{1}{2N}\{\exp(4K) - 1\}\{\tanh^N K - 1\}\right] \tag{3.9}$$

and

$$\frac{c_N}{k_B} = \left(1 - \frac{1}{N}\right)\frac{K^2}{\cosh^2 K} \tag{3.10}$$

for the ASM and heat capacity per spin of a field-free Ising chain.

The phase transition behavior of the open chain is simpler than that of the ring. The chain-length dependences of the characteristic temperatures T_0; T_m, and T_B are summarized in Table III. Most remarkable, perhaps, is the fact that $k_B T_0/J$ has the same numerical value, 0.8336, for all N! Except for $N = 2$, the values of T_0 and T_m are significantly different. The magnetic field absorption coefficient $\kappa_N(T)$ does not exhibit a local maximum until the chain length reaches a value of several hundred. As N increases beyond this threshold, the values of T_m and T_B draw progressively closer to one another, and, for sufficiently long chains, these values differ very little from those specific to Ising rings. Both approach zero for chains (or rings) of infinite length.

IV. SUMMARY AND DISCUSSION

By studying otherwise similar one-dimensional Ising spin systems with different numbers of degrees of freedom, we have been able to chart the dependences on size of the characteristic transition temperatures and to observe how these and other properties of the spin system approach the hypothetical thermodynamic limit of $N \to \infty$. It is well known [9] that the phase transitions occurring in systems of finite size never are as sharply delineated as those associated with this thermodynamic limit. And indeed, for finite systems it would seem more appropriate to adopt the definition of a phase transition proposed by Waldram [13], according to whom "A phase transition is a process in which a thermodynamic system changes, *over a negligible range of temperature, pressure or some other intensive variable*, from one state into another which has different properties." Our insertion of italics into this statement is intended to draw attention to the small but presumably finite range of intensive variable

TABLE III
Characteristic Temperatures of One-dimensional
Ising Chains (Open-Ended)

N	$k_B T_0/J$	$k_B T_m/J$	$k_B T_B/J$
2	0.8336	0.8336	—
5	0.8336	0.7160	—
10	0.8336	0.6257	—
20	0.8336	0.5443	—
30	0.8336	0.5024	—
40	0.8336	0.4753	—
50	0.8336	0.4557	—
80	0.8336	0.4183	—
100	0.8336	0.4023	—
500	0.8336	0.3119	0.2654
1,000	0.8336	0.2834	0.2536
5,000	0.8336	0.2331	0.2190
10,000	0.8336	0.2163	0.2055
50,000	0.8336	0.1851	0.1788
100,000	0.8336	0.1742	0.1691
500,000	0.8336	0.1532	0.1498
1,000,000	0.8336	0.1456	0.1428

values that Waldram associates with the occurrence of phase transitions. We have introduced prescriptions [cf. Eqs. (2.14) and (2.17)] for identifying these phase-transition bands and then used them to determine the corresponding "latent heats of transformation" defined by Eqs. (2.15) and (2.18).

Since the relevant thermodynamic properties of finite systems are not expected to display singularities or discontinuities, one must use alternative mathematical criteria to signal the occurrence of phase transformations and to locate the intensive variable (temperature) values associated with these transitions. The procedure we have adopted is based on the observation that all condensation phenomena (a category that includes the formation of spin microdomains) involve uncommonly large exchanges of energy between the system and its environment. This exchange is needed in order to stabilize the condensed phase (here a paramagnetic collection of microdomains) as it is formed by cooling the high-temperature spin-gas phase consisting of weakly correlated, randomly oriented individual spins. In contrast to this, the spontaneous zero-field order–disorder transition that converts the paramagnetic microdomain phase into a single-domain ferromagnetic phase, involves a relatively small amount of energy but is accompanied by an enormous increase of the average squared magnetization. The conditions required for creation of the ferromagnetic phase also can be probed by perturbing the microdomain phase with a small externally imposed magnetic field. The temperature at which a strong absorption of magnetic energy occurs coincides with the Curie temperature characteristic of the order–disorder transition.

It is important to note that our method does not imply or use the notion that phase transformations somehow are related to extrema of the free energy or of other thermodynamic potentials. Furthermore, we do not relate the occurrence of phase transitions to the limits of thermodynamic or mechanical stability of the system. In these respects our approach differs from the more general theories of phase transitions proposed by Tisza [7] and Münster [8]. What we do instead is identify the temperatures characteristic of phase transitions as those at which the spin system strongly interacts with external probes, either by exchanging large amounts of energy or by displaying uncommonly large variations of its magnetic properties.

A new and interesting observation resulting from this investigation is that even a system so simple as a finite-sized, one-dimensional Ising model can exhibit relatively complex phase behavior. As this system is cooled from its high-temperature paramagnetic phase, we first observe a quasi-first-order transition that produces a paramagnetic dispersion of microdomains, each consisting of a number of spins in parallel alignment. Further cooling leads to the occurrence of a second-order type of transition that results in the formation of an ordered ferromagnetic phase. This sequence of events resembles the familiar pattern of gas–liquid–solid transitions characteristic of systems composed of simple, structureless particles. As the size of the spin system increases, the temperature range over which the ferromagnetic phase exists gradually shrinks, tending to zero as $N \to \infty$. In this limit there is a ferromagnetic phase only at the absolute zero of temperature, but at higher temperatures there still can be two paramagnetic phases: (1) the microdomain phase, which is dominant for $T < T_0 = 0.8336 \, J/k_B$, and (2) the completely disordered spin-gas phase for $T > T_0$. It is natural to wonder whether similar, three-phase behaviors occur for Ising models of higher dimension. In this context, attention once again should be directed to the fact that the finite, one-dimensional Ising systems studied here exhibit *two* very distinct transition temperatures, one characteristic of the quasi-first-order "condensation" associated with the maximum of the heat capacity and another that is characteristic of the much less energetic, quasi-second-order para $\leftrightarrow$ ferro magnetic transition. All treatments of two- and three-dimensional Ising models of which we are aware assume there to be a single "Curie point" and so do not distinguish between these two transitions.

ACKNOWLEDGMENT

This research has been supported by a grant from the Theoretical and Computational Chemistry Program of the National Science Foundation.

REFERENCES

1. W. Lenz, *Physik Z.* **21**, 613–615 (1920).

2. E. Ising, *Z. Physik* **31**, 253–258 (1925).

3. P. E. Weiss, *J. Phys.* **6**, 661–690 (1907).

4. S. G. Brush, *Rev. Mod. Phys.* **39**, 883–893 (1967).

5. J. M. Yeomans, *Statistical Mechanics of Phase Transitions*, Clarendon Press, Oxford, 1992, Chapter 2, pp. 21–22.

6. P. E. Ehrenfest, *Proc. Acad. Sci. Amsterdam* **36**, 153–157 (1933).

7. L. Tisza, in *Phase Transformations in Solids*, (R. Smoluchowski, J. E. Mayer, and W. A. Weyl, eds., Wiley, New York, 1951, pp. 1–37 and *Generalized Thermodynamics*, MIT Press, Cambridge, MA, 1966 pp. 102–193.

8. A. Münster, *Statistical Thermodynamics*, Vol. 1, (Springer, Berlin, 1969, Chapter 4, pp. 256–268.

9. T. L. Hill, *Thermodynamics of Small Systems*, Benjamin, 1963, Chapter 5, pp. 114–132.

10. A. E. Ferdinand and M. E. Fisher, *Phys. Rev.* **185**, 832–846 (1969).

11. A. Ciach, *J. Stat. Phys.* **40**, 593–606 (1985).

12. B. M. McCoy and T. T. Wu, *The Two-Dimensional Ising Model*, Harvard Univ. Press, Cambridge, MA, 1973, Chapter 1, pp. 33–40.

13. J. R. Waldram, *The Theory of Thermodynamics*, Cambridge Univ. Press, Cambridge, UK, 1985, Chapter 18, p. 246.

QUANTUM ELECTRODYNAMICS OF RESONANCE ENERGY TRANSFER

GEDIMINAS JUZELIŪNAS

Institute of Theoretical Physics and Astronomy, Vilnius, Lithuania

DAVID L. ANDREWS

School of Chemical Sciences, University of East Anglia, Norwich, England

CONTENTS

Advances in Chemical Physics, Volume 112, Edited by I. Prigogine and Stuart A. Rice
ISBN 0-471-38002-4.   © 2000 John Wiley & Sons, Inc.

I. INTRODUCTION

More than a half century has elapsed (at the time of writing) since the first pioneering attempts by Fermi [1], Heitler and Ma [2], and Hamilton [3] (following a related study by Kikuchi [4]) to address the methods of quantum electrodynamics (QED) to resonance energy transfer (ET), a process in which the excitation energy of an atom, molecule, or other species D (to be referred to as the *donor*) is transferred to another entity A (the acceptor) separated from it by some distance R. The early QED theories of this process [1–3] concentrated on the radiative ET principally manifest at far-zone distances $R \gg \lambda$, where $\lambda = \lambda/2\pi$ and λ is the wavelength associated with the photon conveying the energy from D to A. In these early works it was shown that the transfer probability properly behaves as R^{-2}. Furthermore there was a relativistic time lag of R/c associated with propagation of the photon to the acceptor; in other words, the process satisfies the basic requirements of causality. This much is in agreement with what is expected from classical electrodynamics. Radiative transfer of energy can take place over various distances, including those of cosmic scale (in the case of energy coming from distant stars). At the other extreme, however, radiative energy transport can also occur between donor and acceptor sites within a single object (as long as $R \gg \lambda$); in particular, it is a mechanism leading to the re-absorption of emitted photons in optically thick samples [5,6].

Almost at the same time as the first work on the QED of radiative ET [1–3], another radiationless mechanism was suggested by Perrin [7], Förster [8], Dexter [9] and Galanin [10] to deal with near-zone ($R \ll \lambda$) energy transport. This mechanism treats the process as one induced by the instantaneous (Coulomb) interaction between the transfer species D and A. As a result, Förster arrived at an R^{-6} dependence for the transfer rate in the case where the process is governed by dipole-dipole coupling between the donor and acceptor [8]. Radiationless ET takes place in a wide variety of condensed systems, such as concentrated solutions [6,11] and, most importantly, in photosynthetic light-harvesting antennas [12–14]. Through its contextual significance in advancing both the fundamental understanding of biological photosynthesis, and informing strategies for synthetic mimicry of the latter, the process has attracted a greatly increasing interest in recent years.

In the 1960s Avery [15] and Gomberoff and Power [16] attempted to use the QED approach to generalize the Förster theory of short-range (radiation-less) ET [8,10,11] to arbitrary transfer distances. Such a unified approach to radiationless and radiative ET, in which the process is considered to be mediated by intermolecular propagation of a virtual photon, has received a considerable boost in the 1980s [17–22] and 1990s [22–32]. Through these studies it has been demonstrated [15,16,18,19,21,27,31] that the unified mechanism reproduces the Förster R^{-6} dependence of the transfer rates in the near zone — whereas in the far zone, resonance ET equates to emission of a photon by a donor molecule and subsequent recapture of the photon by an acceptor. It has also been shown [33] how the unified theory may be extended to include the effects of very short-range nondipolar coupling, associated with localized molecular-orbital interactions through the involvement of charge-transfer configurations. This approach offers a seamless extension of the Förster theory to regions of strong orbital overlap, where excitation transfers as in the Dexter exchange interaction [9]. In other developments [25,28,30,34–38], the time evolution of the transfer dynamics has been explicitly considered within the framework of the unified theory, addressing in detail the transfer dynamics beyond the rate regime. Among other issues, an intricate problem of causality in ET (initially raised by Fermi [1]) has been reexamined [25,39–52]. Another related problem concerns collective emission (super- and subradiance) by two atoms interacting via the resonance dipole–dipole coupling [34,35,53–58] a phenomenon finding an interest in laser cooling [59,60]. It is also worth mentioning the spontaneous emission by a molecule near a metal surface [61–65]. Such a process might be considered in certain respects as the resonance ET between the molecule and its image, although a more rigorous analysis of the phenomenon is to be based on the coupling of the excited molecule to surface plasmons [64,65]. It is noteworthy that the resonance ET belongs to a wider class of bimolecular photophysical processes [66,67] mediated by a virtual photon. Other bimolecular processes, such as the (third-order) single-photon cooperative absorption [68–70] or emission [71], involve creation or annihilation of real photons as well.

Another important raft of issues relates to the incorporation within the unified theory of the effects of the surrounding medium. Although a handful of sporadic attempts to accommodate medium effects in excitation transfer appeared previously [15,22,24], it was the case until quite recently that most QED theories totally ignored the influence of such effects. The 1989 treatment by Craig and Thirunamachandran [22], incorporating effects of a third molecule in the resonance ET between a selected pair of molecules, led to a new discussion of the way to include dielectric characteristics. It was suggested from macroscopic arguments that the vacuum dielectric permittivity ε_0 entering the rate of excitation transfer in vacuo should be replaced by its medium counterpart ε to represent the screening. Nevertheless, in using this prescriptive approach, other important medium effects, such as local fields, energy losses due to the absorbing medium, and influences on the character of the transfer rates in passing from the near zone to the far zone, were not considered.

More recently a QED theory has been developed [27,28,31] that systematically deals with these issues. A fully second-quantized formalism has been adopted to treat the effects of the molecular medium at the microscopic level.

In contrast to conventional QED theories for the resonance ET between a pair of species in vacuum [15–26], in which the process is considered to be mediated by the intermolecular propagation of virtual photons, the new theory [27,28,31] has been formulated by invoking the concept of bath polaritons ("medium-dressed photons") mediating the ET. To this end, the approach has some analogies with related work [72] on intermolecular forces and superradiance, in which a lossless medium was modeled by an ensemble of two-level species. The theory developed in Refs. 27 and, 28 (as well as subsequent work [31,73]) accommodates an arbitrary number of excited states of electronic, vibrational, and other origin for each molecule of the medium. This includes, in particular, a case of special interest, where the excited-state spectrum of the molecules is sufficiently dense and smooth for it to be treated as a quasicontinuum. In such absorbing regions of the spectrum, the exponential (Beer law) decay factor emerges in the microscopically derived rates for ET between a selected pair of species (pair-transfer rates). This feature also solves the problem of potentially infinite rates of the resonance ET in the ensemble, due to the far-zone inverse-square law, featured in the pair-transfer rates [27,28,31] (see Section V for more detail). Furthermore, the pair rates contain other refractive modifications, such as the local field factors.

Other recent studies have analyzed the resonance coupling between molecules situated in photonic bandgap crystals [71,74–78], in plasma media [79], between metal or dielectric plates [80,81], in dielectric microparticles (Mie particles) [82–86], and in other cavities [87,88]. Such systems can exhibit an optical response that fails to support the propagation of certain optical frequencies.* The spontaneous emission is inhibited within the bandgaps of the photon spectra [71,89–93] and no radiative limit exists for the resonance ET. Nevertheless, it has been demonstrated that ET can take place quite efficiently between species separated by distances up to those comparable with the wavelength of light or even somewhat larger [71]. This might serve as an alternative relaxation channel for excited atoms or molecules, the transition frequencies of which are tuned to the photonic bandgap.

It is the purpose of this chapter to review the QED approach to energy transfer taking place both in the free space (electromagnetic vacuum) and in media. The plan in the following sections is to work as follows:

- In Section II we define the basic concepts of quantum electrodynamics of resonance energy transfer, subsequently providing the derivation of the transition matrix element for the ET between a pair of chromophores

---

* On the other hand, the photon spectra might also exhibit singularities (such as Mie resonances [81–86]) at some other frequencies.

separated at an arbitrary distance R. At this stage the medium surrounding the transfer species, is not explicitly specified.

- The second step (Section III) is to calculate the tensor for the resonance dipole–dipole coupling between a pair of molecules (or atoms) in the electromagnetic vacuum, in a photonic bandgap crystal, as well as in a homogeneous dielectric medium. The latter dielectric medium is considered to be comprised of discrete atoms or molecules, so that the analysis is free of phenomenological assumptions.

- The subsequent analysis concentrates on the ET in the dielectric medium. In Section IV the energy-level structure is included for each donor and acceptor. Accordingly, the transfer rate is represented in terms of the overlap integral between the donor fluorescence and the acceptor absorption spectra (Section IV.A), establishing a firm connection with both the Förster R^{-6} radiationless result and the radiative limit. The transfer rate contains the required refractive modifications, including, inter alia, the local field factors. The consideration of the range dependence of the fluorescence depolarisation (Section IV.B) illustrates the approach.

- The next element (Section V) is to analyze the total rate of decay of an initially excited molecule, due to the ET to the surrounding medium. The contribution associated with the ET to the far-zone species is then identified as the rate of spontaneous emission in the absorbing medium. The emission rate contains the local field factors that support the previous phenomenological results obtained using other methods.

- Section VI analyses in detail the transfer dynamics for a pair of species in the dielectric medium. The section not only extends consideration beyond the rate description, but also reexamines conditions for that regime itself. That leads to incorporation of the shifts in energy for both the ground and excited states of the transfer species, due to the interaction of these species with the molecules belonging to the medium and also with each other. Attention is then focused on situations that do not fit into the rate regime, and where different dynamical aspects are apparent. The problem of causality in the ET is also here discussed.

- Finally, the concluding section, Section VII, summarizes the material presented in this chapter.

II. GENERAL FORMULATION

A. Hamiltonian

The framework within which the unified theory naturally emerges is nonrelativistic (molecular) QED [94–96], best implemented in its multipolar (Power–Zienau–Woolley) formulation [96–98]. In such a multipolar formulation, the Coulomb interaction between electrically neutral molecules (or atoms) completely cancels, and it is the coupling of the molecules with the transverse photons that is responsible not only for molecular absorption and emission but also for the intermolecular coupling via the exchange of virtual photons. Consequently, retardation is naturally accommodated, reflecting the finite speed of signal propagation. It is such retardation features, for example, that are responsible for modifying at mesoscopic distances the inverse sixth-power distance dependence of the London potential (the attractive part of the 6–12 Lennard-Jones potential) to the correct and experimentally verified form given by the Casimir–Polder formula — in which the asymptotic behavior at large distances proves to be of inverse seventh power form [96,99].

In the multipolar formulation of QED, the Hamiltonian for the system can generally be written as

$$H = H_{rad} + \sum_X H_{mol}(X) + \sum_X H_{int}(X) \qquad (2.1)$$

where H_{rad} is the Hamiltonian for the radiation field [explicitly presented in Eq. (2.4) and $H_{mol}(X)$ is the Hamiltonian for the molecule X, with summations taken over all the molecules of the system. Note that the term *molecule* is used here generically to encompass other electrically neutral species as well, such as atoms. The coupling between the molecular subsystem and the quantized field is represented by a set of terms $H_{int}(X)$ describing interaction of the field with the individual molecules. For present purposes it is sufficient to express the interaction terms in the electric dipole approximation, though the formalism we employ is perfectly amenable to the incorporation of higher mutipole terms [30]. Thus we write

$$H_{int}(X) = -\varepsilon_0^{-1}\boldsymbol{\mu}(X) \cdot \mathbf{d}^{\perp}(\mathbf{R}_X) \qquad (2.2)$$

where $\boldsymbol{\mu}(X)$ is the electric dipole operator of the molecule X positioned at $\mathbf{R}_X$ and $\mathbf{d}^{\perp}(\mathbf{R}_X)$ is the electric displacement field operator at the molecular site. The latter displacement operator and the radiation Hamiltonian may be cast as [96]:

$$\mathbf{d}^{\perp}(\mathbf{R}) = i \sum_{\mathbf{k},\lambda} \left(\frac{\hbar c k \varepsilon_0}{2V}\right)^{1/2} \mathbf{e}^{(\lambda)}(\mathbf{k})\{a^{(\lambda)}(\mathbf{k})e^{i\mathbf{k}.\mathbf{R}} - a^{(\lambda)+}(\mathbf{k})e^{-i\mathbf{k}.\mathbf{R}}\} \qquad (2.3)$$

and

$$H_{rad} = \sum_{\mathbf{k},\lambda} \left[a^{(\lambda)+}(\mathbf{k})a^{(\lambda)}(\mathbf{k})\hbar c k + \frac{1}{2}\right] \qquad (2.4)$$

where in each expression a sum is taken over radiation modes character-ized by wavevector $\mathbf{k}$ and polarization vector $e^{(\lambda)}(\mathbf{k})$ (with $\lambda = 1, 2$); $a^{(\lambda)^+}(\mathbf{k})$ and $a^{(\lambda)}(\mathbf{k})$ are the corresponding operators for creation and annihilation of a photon, and V is an arbitrarily large quantization volume.

B. Resonance Energy Transfer

We shall deal with the resonance ET between a selected pair of species (to be referred to as donor and acceptor) labeled D and A. For this purpose, the full system will be divided into two parts, one subsystem consisting of the transfer species D and A, and the other referred to as the *radiative* (*polariton*) *bath*. The latter bath comprises the quantized electromagnetic field and the remaining molecules that constitute the surrounding medium. Note that the molecules of the medium may, but do not necessarily, differ in type from the donor and the acceptor. With regard to the chosen partitioning of the system, the full Hamiltonian (2.1) splits into the zero-order Hamiltonian H^0 and the interaction term V, as

$$H = H^0 + V \tag{2.5}$$

with

$$V = H_{\text{int}}(D) + H_{\text{int}}(A) \tag{2.6}$$

and

$$H^0 = H_{\text{bath}} + H_{\text{mol}}(D) + H_{\text{mol}}(A) \tag{2.7}$$

and where

$$H_{\text{bath}} = H_{\text{rad}} + \sum_{X \neq D,A} [H_{\text{mol}}(X) + H_{\text{int}}(X)] \tag{2.7'}$$

is the "bath" Hamiltonian containing the radiation Hamiltonian H_{rad} and contributions from all molecules other than the donor and the acceptor.

To represent the ET from the donor to the acceptor, the initial and final statevectors are chosen to be the following eigenvectors of the zero-order Hamiltonian H^0:

$$|I\rangle = |D^*\rangle|A\rangle|0\rangle, \qquad |F\rangle = |D\rangle|A^*\rangle|0\rangle \tag{2.8}$$

where the corresponding eigenenergies are

$$E_I = e_{D^*} + e_A + e_{\text{vac}}, \qquad E_F = e_D + e_{A^*} + e_{\text{vac}} \tag{2.9}$$

and $|0\rangle$ represents the ground state of the polariton bath. Note that the subse-quent calculation of the transition matrix element (Section II.C) is not affected if the quantity $|0\rangle$ is replaced by any other radiative statevector characterized by a definite number of medium-dressed photons.* The modifications can emerge only in the fourth order of perturbation [101], which is beyond of the scope of the present analysis. In is intructive that the same radiative part $|0\rangle$ enters the initial and final statevectors [Eq. (2.8)], representing the ET. The change in the radiative state would lead to the higher-order bimolecular processes, such as the single-photon cooperative absorption [66–70] or emission [71].

Here e_{vac} is the zero-point energy of the bath; $|D^*\rangle$ and $|D\rangle$ label the initial and final states of the donor, $|A\rangle$ and $|A^*\rangle$ are the corresponding statevectors of the acceptor (where the asterisk refers to a molecule in an electronically excited state), and e_{D^*} and e_A (e_D and e_{A^*}) are the appropriate energies of the donor and acceptor in their initial (final) states. For generality the statevectors of donor and acceptor are considered to implicitly contain vibrational contributions that are normally separable from the electronic parts on the basis of the Born–Oppenheimer principle, both for the ground and excited electronic molecular states. The vibronic sublevels are explicitly included in the theory in Section IV.A.

It is noteworthy that conventional (semiclassical) theories of radiationless ET [6,8–11] do not invoke any photon states, and the ET appears as a first-order process induced by an instantaneous Coulomb interaction. Such an approach is justified in the near zone, when the distance R of donor–acceptor separation is much less than the reduced wavelength of light $\lambda = \lambda/2\pi$, where λ is the wavelength of light corresponding to the transfer energy. In the QED formalism employed here, the quantized electromagnetic field is treated on an equal footing to the molecular subsystem, both subsystems comprising a united dynamical system described by the full Hamiltonian H.

C. Transfer Rates

Let us suppose that the process can be described in terms of transfer rates, a condition satisfied by weakly coupled (i.e., most) ET systems. The detailed criteria for this assumption and extensions of the analysis beyond the rate regime are considered in Section VI. The rate of ET, associated with the initial and final states (2.8), can generally be written using the Fermi golden rule (see Section VI for more detail and Refs. 1 and 102 for examples):

$$W_{FI} = \frac{2\pi}{\hbar} |\langle F|T|I\rangle|^2 \delta(E_I - E_F) \qquad (2.10)$$

---

* RDDI is not altered in the squeezed radiative vacuum as well [100].

where T is the transition operator:

$$T = T^{(1)} + T^{(2)} + \cdots = V + V \frac{1}{E_I - H^0 + is'} V + \cdots, \qquad s' \to +0 \quad (2.11)$$

where the higher-order terms can for our purposes be neglected. The first-order term $T^{(1)} \equiv V$ does not contribute to the transfer rate in the multipolar formulation of QED employed; such a term describes processes involving the emission or absorption of a medium-dressed photon accompanied by transitions of individual molecules (see Fig. 1). It is the second-order contribution $T^{(2)}$ that is the leading term responsible for the resonance ET in question, a process involving changes in quantum states of both species D and A without alteration of the state of the radiative bath.

Using Eqs. (2.4)–(2.9), the second-order transition matrix element reads

$$\langle F|T^{(2)}|I \rangle = \frac{1}{\hbar} \sum_{\sigma} \sum_{q=1}^{2} \frac{\langle F|V|M_q(\sigma)\rangle \langle M_q(\sigma)|V|I \rangle}{(-1)^{q+1}\omega_{D^*} - \Pi_\sigma + is}, \qquad s \to +0 \quad (2.12)$$

where

$$\hbar\omega_{D^*} \equiv e_{D^*} - e_D = e_{A^*} - e_A > 0 \qquad (2.13)$$

is the excitation energy of the donor (as gained by the acceptor) and

$$\hbar\Pi_\sigma = e_\sigma - e_{\text{vac}} \qquad (2.14)$$

is the difference in energies between the excited and ground states of the bath. Here also $|M_q(\sigma)\rangle$ denote intermediate states in which both the donor and the

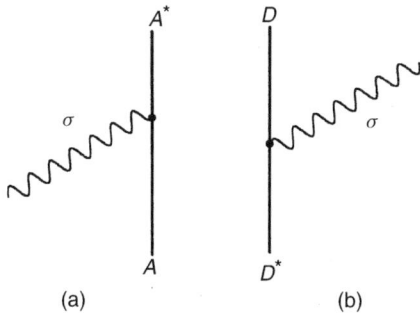

FIGURE 1. Time-ordered diagrams for (a) photoabsorption and (b) photoemission of a medium-dressed photon labeled by index σ.

acceptor are either in the ground ($q = 1$) or in the excited ($q = 2$) electronic states:

$$|M_1(\sigma)\rangle = |D\rangle|A\rangle|\sigma\rangle \qquad (2.15)$$

and

$$|M_1(\sigma)\rangle = |D^*\rangle|A^*\rangle|\sigma\rangle \qquad (2.16)$$

where $|\sigma\rangle$ labels the excited states of the radiative bath that are accessible though a single action of the interaction operator V on the ground-state vector $|0\rangle$. Accordingly, the ET is generally regarded as being mediated by such elementary excitations (the bath polaritons) representing photons "dressed" by the material medium.

The two types of intermediate state $|M_q(\sigma)\rangle$ ($q = 1,2$) correspond to the two possible sequences of transitions undergone by the donor and acceptor. In the first case ($q = 1$), the transition $D^* \rightarrow D$ precedes the transition $A \rightarrow A^*$, as depicted in Fig. 2a, whereas in the second case ($q = 2$) one has the opposite ordering (Fig. 2b). The latter sequence describes an apparently anomalous situation where the upward transition of the acceptor A is accompanied by the creation of a virtual polariton σ, and the subsequent annihilation of the polariton induces the downward transition by the donor $D^* \rightarrow D$. Nevertheless, both types of transition must be included in the theory according to the normal rules of time-dependent perturbation theory. The contribution associated with Fig. 2b plays an important role in the near zone, where the uncertainty in energy of a mediating polariton greatly exceeds the excitation energy transferred.

Substituting Eqs. (2.2), (2.8), (2.9) and (2.14)–(2.16) into Eq. (2.12), one arrives at the following expression for the transition matrix element:

$$\langle F|T^{(2)}|I\rangle = \mu_{A_l}^{\text{full}}\theta_{lj}(\omega_{D^*}, \mathbf{R})\mu_{D_j}^{\text{full}} \qquad (2.17)$$

where implied summation is assumed over the repeated Cartesian indices l and j, and where

$$\theta_{lj}(\omega, \mathbf{R}) = \frac{1}{\hbar\varepsilon_0^2}\sum_\sigma \left[\frac{\langle 0|d_l^\perp(\mathbf{R}_A)|\sigma\rangle\langle\sigma|d_j^\perp(\mathbf{R}_D)|0\rangle}{\omega - \Pi_\sigma + is} - \frac{\langle 0|d_j^\perp(\mathbf{R}_D)|\sigma\rangle\langle\sigma|d_l^\perp(\mathbf{R}_A)|0\rangle}{\omega + \Pi_\sigma - is}\right]$$
$$(\mathbf{R} = \mathbf{R}_A - \mathbf{R}_D) \qquad (2.18)$$

is the tensor for the resonance dipole–dipole interaction (RDDI) between a pair of molecules within the medium. The retarded tensor $\theta_{lj}(\omega, \mathbf{R})$ is influenced by the material medium through the yet unspecified intermediate polariton

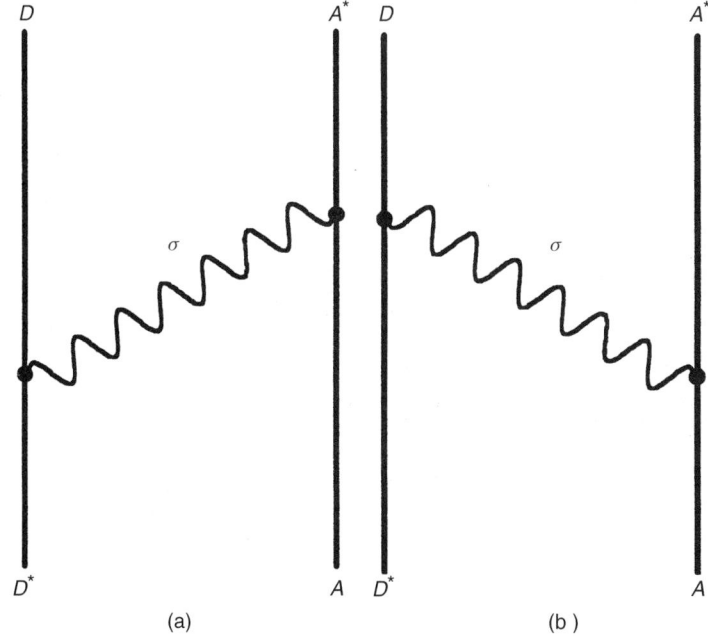

FIGURE 2. The two time-ordered diagrams for resonance ET, time progressing upward. In both cases the transfer of energy from the initially excited donor D^* to the acceptor A is mediated by a photon "dressed" by the medium (virtual polariton), as labeled by index σ.

states σ featured in Eq. (2.18). Here also

$$\mu_D^{\text{full}} = \langle D | \boldsymbol{\mu}(D) | D^* \rangle, \qquad \mu_A^{\text{full}} = \langle A^* | \boldsymbol{\mu}(A) | A \rangle \qquad (2.19)$$

are the transition dipole moments of the donor and acceptor, respectively. The superscript "full" indicates that the molecular state vectors entering the transition dipoles (2.19) contain both electronic and vibrational contributions, as made explicit later.

III.   ANALYSIS OF THE RDDI TENSOR $\theta_{lj}(w, R)$

A.   Vacuum Case

We commence the analysis of the RDDI tensor with the simplest case, where the transfer species are situated in the electromagnetic vacuum. Calculations of a similar kind can be traced back to the papers by Stephen [35], Avery [15], Gomberoff and Power [16], and others [17–21]. In such a vacuum case, the bath Hamiltonian reduces to that of the "free" radiation field, $H_{\text{bath}} = H_{\text{rad}}$, in which the corresponding eigenstates and eigenenergies take the form

$$|\sigma\rangle = |\mathbf{k}, \lambda\rangle = a^{(\lambda)^+}(\mathbf{k})|0\rangle \tag{3.1}$$

and

$$\Pi_\sigma = \omega_k = ck \tag{3.2}$$

where the state vector $|0\rangle$ now describes the electromagnetic vacuum. Here the intermediate (virtual) photons are considered to be the plane waves characterized by the wavevector $\mathbf{k}$ and polarization λ. One can alternatively represent the intermediate photons in terms of spherical harmonics [38]; however, this does not alter the results.

Calling on Eq. (2.3) for the displacement operator and performing summation over the photon polarizations ($\lambda = 1, 2$), the tensor (2.18) reduces to

$$\theta_{lj}(\omega, \mathbf{R}) \equiv \theta_{lj}^{vac}(\omega, \mathbf{R}) = \sum_{\mathbf{k}} \frac{(\delta_{lj} - \hat{k}_l\hat{k}_j)\omega_k}{2\varepsilon_0 V} \left[\frac{e^{i\mathbf{k}\cdot\mathbf{R}}}{\omega - \omega_k + is} - \frac{e^{-i\mathbf{k}\cdot\mathbf{R}}}{\omega + \omega_k + is} \right] \tag{3.3}$$

Here $\hat{\mathbf{k}} = \mathbf{k}/k$ labels a unit vector along the wavevector $\mathbf{k}$, where the sign of the imaginary infinitesimal is reversed in the second (nonresonant) term of Eq. (3.3). Replacing the summation over $\mathbf{k}$ by an integral and performing the angular integration, Eq. (3.3) takes the form

$$\theta_{lj}^{vac}(\omega, \mathbf{R}) = \frac{1}{16\pi^2\varepsilon_0} \int_0^\infty k^2 \, dk \left[\frac{\omega_k}{\omega - \omega_k + is} - \frac{\omega_k}{\omega + \omega_k + is} \right] \tau_{lj}(k, R) \tag{3.4}$$

with

$$\tau_{lj}(k, R) = \left[(\delta_{ij} - \hat{R}_i\hat{R}_j)\frac{\sin(kR)}{kR} + (\delta_{ij} - 3\hat{R}_i\hat{R}_j)\left(\frac{\cos(kR)}{k^2R^2} - \frac{\sin(kR)}{k^3R^3} \right) \right] \tag{3.5}$$

Since $\omega_{-k} = -\omega_k$ and $\tau_{lj}(-k, R) = \tau_{lj}(k, R)$, the integration contour can be expanded to negative values of k in Eq. (3.4), subsequently closing it up by a large semicircle on the upper or lower complex half-plane [depending on the sign of the exponents comprising the geometric functions in Eq. (3.5)]. Calculating the residues at $k = \pm(\omega + is)/c$, one arrives at the final result for the retarded RDDI tensor in vacuum for $R \neq 0$:

$$\theta_{ij}^{vac}(\omega, \mathbf{R}) = \frac{\omega^3 e^{iKR}}{4\pi\varepsilon_0 c^3} \left[(\delta_{ij} - 3\hat{R}_i\hat{R}_j)\left(\frac{c^3}{\omega^3R^3} - \frac{ic^2}{\omega^2R^2} \right) - (\delta_{ij} - \hat{R}_i\hat{R}_j)\frac{c}{\omega R} \right] \tag{3.6}$$

where use has been made of the dispersion relationship (3.2).

The above tensor contains an R^{-3} term characteristic of the near-zone ($\omega R/c \ll 1$), a radiative R^{-1} term operating in the far zone ($\omega R/c \gg 1$), as well as an R^{-2} contribution manifesting at critical retardation distances: $\omega R/c \cong 1$. In calculating the RDDI tensor, an important role is played by the imaginary infinitesimal "is" featured in Eq. (2.12) for the transition matrix element, and in the subsequent equations. It is noteworthy that such an infinitesimal emerges intrinsically from the time-dependent analysis of the problem (to be considered explicitly in Section VI). It is the presence of the imaginary infinitesimal that ensures the correct bypassing of a pole in Eqs. (3.4). Thus one automatically avoids analytic problems [19] associated with the choice of the integration contour.

Note, too, that in some studies, such as Ref. 96, the RDDI tensor (and hence the transition matrix element) has been calculated by taking the principal part of the integral: this corresponds to the real part of the RDDI tensor given by Eq. (3.6). In the near zone, the real part of the transition matrix element describes the shift in excitation energy for a pair of species coupled via the resonant dipole–dipole interaction, although such an interpretation can be inappropriate in the far zone [35].

B. Photonic Bandgap Structures

Analysis of the RDDI tensor for photonic bandgap crystals is generally a tough problem. In such structures the photons are characterized by a much more complex dispersion relationship ω_k compared to that in vacuum, and the mode functions are now spatially modulated [103]. The spatial modulation renders the transfer energy sensitive to the specific sites of the transfer species within an elementary cell of the photonic bandgap crystal (the superlattice) [75].* In order to establish some characteristic features of the RDDI coupling in photonic bandgap crystals, we neglect the spatial modulation [71,76], assuming the photon modes to represent transverse plane waves. Under this condition, one can repeat the preceding analysis of the vacuum case up to Eqs. (3.4) and (3.5), in which ω_k is now understood as a dispersion relationship for a photon in a bandgap structure.

Various forms have been used to represent the dispersion of ω_k. Kweon and Lawandy [76] adopted the vacuum dispersion $\omega_k = ck$ everywhere except a range of frequency magnitudes $\omega_0 \pm \delta\omega$ where no modes exist, and the center of the bandgap ω_0 was chosen to coincide with the transfer frequency ω_{D^*}. On the other hand, John and Wang [71] employed another form of photon

---

*The site-sensitivity features also for the RDDI in Mie particles [82–86], in waveguides [76], and for a system restricted by the two mirrors [80].

dispersion:

$$\omega_k = \frac{c}{4na}\arccos\left[\frac{4n\cos(kL) + (1-n)^2}{(1+n)^2}\right] \tag{3.7}$$

which follows from the isotropic model for electromagnetic waves in a three-dimensional periodic dielectric. Here n is the refractive index of the scatterer, a is its radius, and $2a + b = L$ is the constant of the superlattice; furthermore, Eq. (3.7) corresponds to the special case where $2na = b$. The function ω_k has branchcut singularities at $k = m\pi/L$, the bandgaps appearing for odd values of m, as illustrated in Fig. 3.

Calculations [71,76] have shown that the suppression of the RDDI can be appreciable only in the far zone: $\omega R/c \gg 1$, whereas the transition matrix element appears to be close to its vacuum values in the near zone. In fact, the precision with which the law of energy conservation has to apply to the virtual photons is determined by the time–energy uncertainty principle: $\delta E \delta t \geq \hbar$, where δt is the time necessary for a photon to cover the distance between the donor and the acceptor. In the near zone ($\omega R/c \ll 1$), the uncertainty in energies of virtual photons is large: $\delta E \gg \hbar\omega$, so that the presence of the bandgap makes little difference to the RDDI tensor.

In the far zone ($\omega R/c \gg 1$) one has the opposite effect: $\delta E \ll \lambda cK$, so that the virtual photon acquires a somewhat real character. As a result, the absence

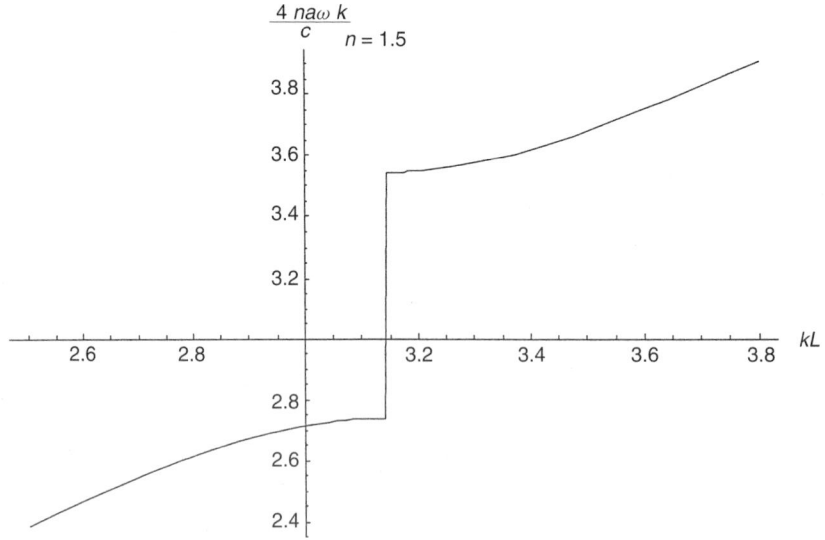

FIGURE 3. Dispersion relationship for a photon in an isotropic bandgap structure, calculated using Eq. (3.7).

of resonant photons can considerably diminish the transition matrix element for the excitation ET in the far zone.

To illustrate this fact, we shall present the analytic expressions for the RDDI tensor [and hence the transition matrix element given by Eq. (2.17)] by using the *effective-mass* approximation for the photon dispersion [71], this is appropriate for transition frequencies very close to the photonic bandedge and for far-zone distances $KR \gg 1$. Expanding Eq. (3.7) around the upper bandedge ω_c one has [71]:

$$\omega_k \approx \omega_c + A(k - k_c)^2 \tag{3.8}$$

where $k_c = \pi/L$,

$$\omega_c = \frac{c}{4na}\left[2\pi - \cos^{-1}\left(\frac{1 - 6n + n^2}{1 + 2n + n^2}\right)\right] \tag{3.9}$$

and

$$A = \frac{-cL^2}{2a(1 + n)^2}\frac{1}{\sin(4na\omega_c/c)} \tag{3.10}$$

Substituting the dispersion (3.8) into Eq. (3.4), the RDDI tensor takes the following form for $KR \gg 1$

$$\theta_{lj}(\omega, \mathbf{R}) = -(\delta_{lj} - \hat{R}_l\hat{R}_j)\frac{i\omega_0 k_c \sin(k_c R)}{16\pi\varepsilon_0\sqrt{A(\omega_c - \omega)}}\frac{e^{-R/\xi}}{R} \tag{3.11}$$

where

$$\xi = \sqrt{A/(\omega_c - \omega)} \tag{3.12}$$

can be interpreted [71,104] as the length of photon localization around embedded atoms or molecules. Near the bandedge, the localization length considerably exceeds the reduced wavelength: $\xi \gg \lambda = c/\omega$. Consequently, efficient ET can take place over far-zone distances up to the localization length, the process being ensured by the overlap of spatially extended clouds of virtual photons surrounding both donor and acceptor.

It is noteworthy that the straightforward application of the effective-mass approximation to the near-zone distances leads to a drastic modification of the RDDI [75]. Such an approach has been criticized by John and Wang [71], who pointed out that the effective-mass approximation is not applicable for the distance scales $KR \ll 1$. In fact, here the uncertainty in energies of virtual photons is large, so the relevant intermediate photons are no longer restricted to the bandedge. Note also that Refs. 71 and 76 (as well as other studies [77–78,105] on the resonance ET in the photonic bandgap crystals) are based on the isotropic model for the photon dispersion. Such a model overestimates the number of photon states near the bandedge.

Finally, we briefly mention recent analysis of the temporal behavior of excitation exchange between a pair of atoms in high-Q cavities [77,87] and in photonic bandgap crystals [77]. Such a process might be completely different from the one occurring in vacuum or in dielectric media (away from the exciton resonances). In fact, even in the single-center case, the rapidly varying density of radiative modes leads to nonexponential spontaneous atomic decay [92,93,106–108] in cavities or photonic crystals. For instance, for atomic frequencies close to the bandedge, the splitting of the atomic level (vacuum Rabi splitting) [71,104] leads to oscillatory spontaneous decay [92]. In the case of a two-center system, the kinetics can be even more complicated [77,87], the vacuum Rabi splitting now competing with the RDDI.

In what follows we shall concentrate on the quantum electrodynamics of energy transfer in dielectric media.

C. RDDI in Dielectric Media

We consider next, without making any phenomenological assumptions, the form of the RDDI in a discrete molecular medium. For this purpose, let us return to the general relationship (2.18) for the tensor $\theta_{lj}(\omega, \mathbf{R})$ in which the influences of the material medium arise through the intermediate polariton states labeled by the index σ. The analysis of the intermediate states can be bypassed in calculating the RDDI tensor by invoking the Green function formalism [27]. As an alternative, one can make use of the explicit mode expansion of the operators for the local displacement field entering Eq. (2.18). Such a mode expansion has been recently derived by Juzeliūnas, giving [109–111]

$$\mathbf{d}^{\perp}(\mathbf{R}_X) = i \sum_{\mathbf{k}} \sum_{m=1}^{M+1} \sum_{\lambda=1}^{2} \left(\frac{\varepsilon_0 \hbar \omega_k^{(m)} v_g^{(m)}}{2cV_0 n(\omega_k^{(m)})} \right)^{1/2} \left[\frac{[n(\omega_k^{(m)})]^2 + 2}{3} \right]$$
$$\times \mathbf{e}^{(\lambda)}(\mathbf{k})(e^{i\mathbf{k}\cdot\mathbf{R}_X} P_{\mathbf{k},m,\lambda} - e^{-i\mathbf{k}\cdot\mathbf{R}_X} P_{\mathbf{k},m,\lambda}^{+}) \qquad (3.13)$$

with $(\mathbf{k}, m, \lambda) \equiv \sigma$ labeling the normal (polariton) modes. Here $P_{\mathbf{k},m,\lambda}^{+}(P_{\mathbf{k},m,\lambda})$ is the Bose operator for creation (annihilation) of a polariton characterized by a wavevector $\mathbf{k}$, a polarization index λ, with an extra index m specifying a dispersion branch, and where

$$\Pi_{\sigma} \equiv \omega_k^{(m)} = \frac{ck}{n(\omega_k^{(m)})} \qquad (3.14)$$

is the polariton frequency in the mth dispersion branch. Here, too, $v_g^{(m)} = d\omega_k^{(m)}/dk$ is the branch-dependent group velocity and $n(\omega)$ is the refractive index:

$$[n(\omega)]^2 = 1 + \frac{\alpha(\omega)\rho/\varepsilon_0}{1 - \alpha(\omega)\rho/3\varepsilon_0} \qquad (3.15)$$

where ρ is the molecular density and $\alpha(\omega)$ is the molecular polarizability, given by

$$\alpha(\omega) = \frac{1}{\hbar} \sum_{\gamma=1}^{M} \left(\frac{\mu_\gamma^2}{\Omega_\gamma - \omega} + \frac{\mu_\gamma^2}{\Omega_\gamma + \omega} \right) \qquad (3.16)$$

The local operator (3.13) has been obtained through the diagonalization of a microscopic Hamiltonian for a discrete molecular medium coupled to a quantized radiation field. The molecules of the medium have been assumed to be all of the same type, regularly placed to form a simple cubic lattice, and characterized by the same isotropic polarizabilities. Such a model is believed to describe fairly well the optical properties of some other isotropic and homogeneous media that are not necessarily ordered, such as a common situation where the nonisotropic species are randomly oriented in their sites.

Equations (3.14)–(3.16) yield $M + 1$ dispersion branches, where M is the number of excitation frequencies Ω_γ accommodated by each molecule forming the medium, and μ_γ is the corresponding transition dipole moment. For instance, the single-frequency (Hopfield) model provides two polariton branches. Examples of dispersion curves with various values of M are presented in Fig. 4. As the number of molecular frequencies increases, the dispersion branches may start to form dense sets, as illustrated in Fig. 4c. Hence, the formalism may also adequately accommodate the formation of quasicontinua of vibrational, rotational, and other sublevels in the molecular spectra. This makes the approach applicable to both lossless and absorbing media. It is noteworthy that a similar multifrequency model of the medium has been considered very recently by Drummond and Hillery [113]. Yet, their theory has been restricted to the mode expansion of the macroscopic field operators. On the other hand, Juzeliūnas treated the operators for local [109–111], microscopic [111], and macroscopic [110] fields. It is the operator for the local displacement field that describes various molecule–radiation processes [109] in condensed media including, inter alia, the resonance ET.

The expansion (3.13) extends to the modes characterized by relatively large wavelength:

$$k \ll \frac{2\pi}{a} \qquad (3.17)$$

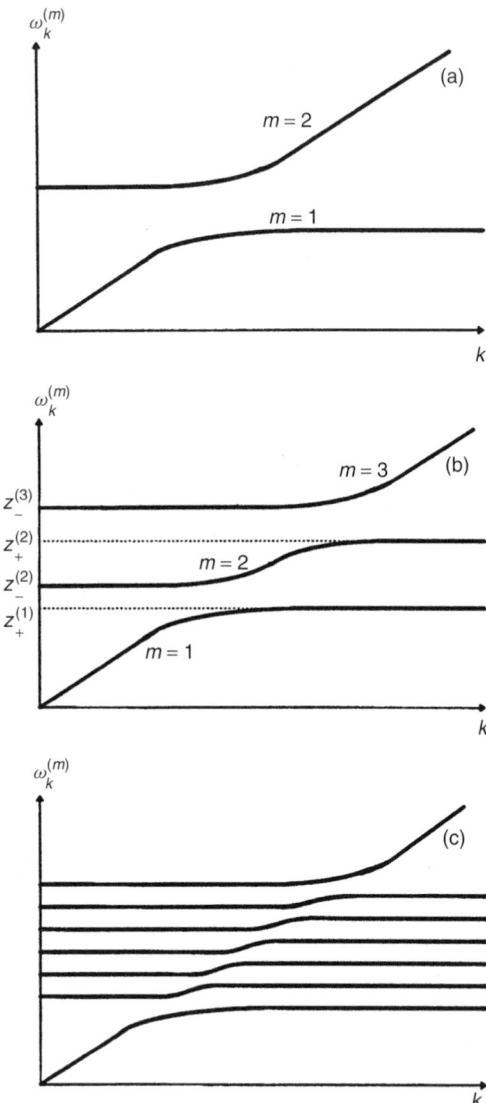

FIGURE 4. Schematic plot of the dispersion curves $\omega_k^{(m)}$ $(m = 1, \ldots, M + 1)$ for (a) $M = 1$ (Hopfield model) and (b) $M = 2$; the third diagram (c) illustrates a situation in which a dense set of dispersion curves is featured. Note that all the diagrams represent the long-wavelength region of the spectrum ($k \ll 2\pi/a$) of interest. For greater values of k, the effects of spatial dispersion are to be considered. The quantities $z = z_+^{(m)}$ and $z = z_-^{(m)}$ label the upper and lower boundaries of polariton bands, for which $n(z_+^{(m)}) = \infty$ and $n(z_-^{(m)}) = 0$, respectively. Diagrams (a)–(c) have been reproduced from a paper by G. Juzeliūnas (Ref. 110) with the kind permission of the American Physical Society. Copyright 1996 by the American Physical Society.

with the lattice constant $a = \rho^{-1/3}$ representing a characteristic distance between the molecules constituting the medium. Under this condition, the expanded displacement operator (3.13) is described exclusively in terms of the refractive index. The excluded short-wavelength modes play little role in the formation of the RDDI between donor and acceptor, as long as their separation distance R is large compared to a.

It is noteworthy that the mode expansion (3.13) holds for the local displacement field operator calculated both at a lattice site [109,110] and also at an intersitial location. This has been demonstrated in Appendix A using the mode expansions of microscopic field-operators derived in Ref. [111].* It is instructive that the (virtual cavity) local field factors explicitly feature in the microscopically derived mode expansion (3.11) of the local operator, leading to the appropriate local field corrections in the subsequent equations (3.18), (4.9), (4.10), (5.1), and (5.14).

Substituting Eq. (3.13) for $\mathbf{d}^{\perp}(\mathbf{R}_X)$ (with $X = D, A$) into Eq. (2.18), and performing the appropriate summation over the intermediate states, the RDDI tensor takes the form (see Appendix B):

$$\theta_{lj}(\omega, \mathbf{R}) = \frac{1}{n^2} \left(\frac{n^2 + 2}{3} \right)^2 \theta_{lj}^{vac}(n\omega, \mathbf{R}) \qquad (3.18)$$

where θ_{lj}^{vac} is the vacuum RDDI tensor given by Eq. (3.6). The relationship (3.18) is equivalent to the result by Juzeliūnas and Andrews [27] obtained using the Green function technique. The tensor (3.18) accommodates a screening contribution n^{-2} and a local field (Lorenz) factor $(n^2 + 2)/3$. In addition, the argument of the vacuum tensor $\theta_{lj}^{vac}(n\omega, \mathbf{R})$ is now scaled by the complex refractive at the frequency ω:

$$n \equiv n(\omega + is) = n' + in'' \qquad (3.19)$$

Note that in the present formalism, the initial refractive index $n(\omega_k^{(m)})$ entering the mode expansion (3.13) is a real branch-dependent quantity. In fact the poles of $n(\omega)$, given by (3.15), emerge at frequencies other than the eigenfrequencies $\omega_k^{(m)}$, so $n(\omega_k^{(m)})$ is a well-defined quantity without adding any imaginary part to the argument $\omega_k^{(m)}$. On the other hand, the refractive index $n \equiv n(\omega + is)$, featured in the RDDI tensor, would be ill-defined within the absorbing regions if a strict mathematical limit $s \to +0$ were taken. That

---

* A somewhat similar conclusion has been reached by De Vries and Lagendijk in their theory of resonant scattering of classical light by impurity atoms inside dielectric cubic lattices [112]. It has been demonstrated [112] that the local field factors associated with a virtual cavity apply for molecules in pure systems and for intersital impurities.

is why the quantity s is to be regarded as a finite (although small) parameter reflecting the natural widths of each molecular line; the replacement of an infinitesimal s by its finite counterpart can be justified on rigorous dynamical grounds, as discussed in Section VI. In the absorbing areas of the spectrum, the width s is considered to be larger than the characteristic distance between the densely spaced (quasicontinuum) molecular sublevels of vibrational or other origin. This makes the refractive index a smooth quantity containing a finite imaginary part in the absorbing areas.

IV. ENERGY TRANSFER IN A DIELECTRIC MEDIUM

A. Inclusion of the Vibrational Structure for the Transfer Species

Up to now the vibrational structure has been kept implicit for the molecular state vectors of donor and acceptor. To explicitly display the vibrational sublevels, we make use of the Born–Oppenheimer approximation [114], according to which the molecular state vectors are separated into electronic and vibrational parts, as

$$|D^*\rangle = |D^*_{\mathrm{el}}\rangle|\varphi^{(n)}_{D^*}\rangle, \qquad |D\rangle = |D_{\mathrm{el}}\rangle|\varphi^{(r)}_D\rangle \tag{4.1}$$

$$|D\rangle = |A_{\mathrm{el}}\rangle|\varphi^{(m)}_A\rangle, \qquad |A^*\rangle = |A^*_{\mathrm{el}}\rangle|\varphi^{(p)}_{A^*}\rangle \tag{4.2}$$

where the subscript "el" refers to the electronic part of the state vectors and the indices n, r, m, and p specify the vibrational, rotational, and other sublevels of the transfer species D and A. The transition dipole moments (2.19) then split into electronic and vibrational contributions, as:

$$\boldsymbol{\mu}^{full}_D = \langle D_{\mathrm{el}}|\boldsymbol{\mu}(D)|D^*_{\mathrm{el}}\rangle\langle\varphi^{(r)}_D|\varphi^{(n)}_{D^*}\rangle \equiv \boldsymbol{\mu}_D\langle\varphi^{(r)}_D|\varphi^{(n)}_{D^*}\rangle \tag{4.3}$$

$$\boldsymbol{\mu}^{full}_A = \langle A^*_{\mathrm{el}}|\boldsymbol{\mu}(A)|A_{\mathrm{el}}\rangle\langle\varphi^{(p)}_{A^*}|\varphi^{(m)}_A\rangle \equiv \boldsymbol{\mu}_A\langle\varphi^{(p)}_{A^*}|\varphi^{(m)}_A\rangle \tag{4.4}$$

where, according to the Condon principle, the dipole operators of the donor and acceptor, $\boldsymbol{\mu}(D)$ and $\boldsymbol{\mu}(A)$, respectively, are assumed not to depend on the vibrational degrees of freedom, where $\boldsymbol{\mu}_D$ and $\boldsymbol{\mu}_A$ are the appropriate electronic parts of the transition dipoles.

Substituting the transition dipoles (4.3) and (4.4) into Eq. (2.17), the full rate of donor–acceptor transfer reads — after performing the necessary averaging over the initial molecular states and summing over the final molecular states in Eq. (2.10)

$$W_{DA} = \frac{2\pi}{\hbar}\sum_{n,m,r,p}\rho^{(n)}_{D^*}\rho^{(m)}_A|\langle\varphi^{(r)}_D|\varphi^{(n)}_{D^*}\rangle\langle\varphi^{(p)}_{A^*}|\varphi^{(m)}_A\rangle|^2|\mu_{A_l}\mu_{D_j}|^2$$

$$\times |\theta_{lj}(\omega_{D^*_{nr}},\mathbf{R})|^2\delta(\hbar\omega_{D^*_{nr}} - \hbar\omega_{A^*_{pm}}) \tag{4.5}$$

Here $\rho_{D^*}^{(n)}$ and $\rho_A^{(m)}$ are the population distribution functions of the initial vibrational states of the donor and the acceptor, respectively; the vibrational indices are also included in the excitation energies of donor and acceptor $(\hbar\omega_{D_{nr}^*} \equiv \hbar\omega_{D^*} = e_{D_n^*} - e_{D_r}$ and $\hbar\omega_{A_{pm}^*} \equiv \hbar\omega_{A^*} = e_{A_p^*} - e_{A_m})$ that feature in the energy-conserving delta function.

Using Eqs. (3.18) and (3.6) for the RDDI tensor in the dielectric medium, the pair rate (4.5) can be expressed in terms of the overlap integral between the donor and acceptor spectra:

$$W_{DA} = \frac{9}{8\pi c^2 \tau_D} \int_0^\infty F_D(\omega)\sigma_A(\omega)\omega^2 g(\omega, \mathbf{R}) e^{-2n''\omega R/c} d\omega \qquad (4.6)$$

with

$$g(\omega, \mathbf{R}) = |n|^2 \left| \eta_3 \left[\left(\frac{c}{n\omega R}\right)^3 - i \left(\frac{c}{n\omega R}\right)^2 \right] - \eta_1 \frac{c}{n\omega R} \right|^2$$

$$= \frac{1}{|n|^4} \left\{ \eta_3^2 \frac{c^6}{\omega^6 R^6} + 2\eta_3^2 n'' \frac{c^5}{\omega^5 R^5} + \left[\eta_3^2 |n|^2 - 2\eta_1\eta_3(n'^2 - n''^2) \right] \right.$$

$$\left. \times \frac{c^4}{\omega^4 R^4} + 2\eta_1\eta_3 n'' |n|^2 \frac{c^3}{\omega^3 R^3} + \eta_1^2 |n|^4 \frac{c^2}{\omega^2 R^2} \right\} \qquad (4.7)$$

and where

$$\eta_q = (\hat{\boldsymbol{\mu}}_A . \hat{\boldsymbol{\mu}}_D) - q(\hat{\mathbf{R}} . \hat{\boldsymbol{\mu}}_A)(\hat{\mathbf{R}} . \hat{\boldsymbol{\mu}}_D), \qquad q = 1, 3 \qquad (4.8)$$

are the orientational factors (the carets referring to unit vectors), $n \equiv n(\omega + is) = n' + in''$ is the complex refractive index* at frequency ω, and

$$\sigma_A(\omega) = \frac{\pi\omega\mu_A^2}{3\varepsilon_0 c} \frac{1}{n'} \left| \frac{n^2+2}{3} \right|^2 \sum_{m,p} \rho_A^{(m)} |\langle \varphi_{A^*}^{(p)} | \varphi_A^{(m)} \rangle|^2 \delta(\hbar\omega_{A_{mp}^*} - \hbar\omega) \qquad (4.9)$$

and

$$F_D(\omega) = \frac{\omega^3 \tau_D \mu_D^2}{3\varepsilon_0 \pi c^3} n' \left| \frac{n^2+2}{3} \right|^2 \sum_{n,r} \rho_{D^*}^{(n)} |\langle \varphi_D^{(r)} | \varphi_{D^*}^{(n)} \rangle|^2 \delta(\hbar\omega_{D_{nr}^*} - \hbar\omega) \qquad (4.10)$$

can be identified (see Section V.B), respectively, as the absorption cross section of the acceptor and the emission spectrum of donor. The latter $F_D(\omega)$

---

* See the discussion following Eq. (3.19) regarding the introduction of the complex refractive index in the present formalism.

is normalized to unity in the Following sense:

$$\int F_D(\omega)d\omega = 1 \qquad (4.11)$$

It is noteworthy that both $\sigma_A(\omega)$ and $F_D(\omega)$ contain refractive corrections due to the medium, including the local field factors. In this way, dielectric influences of the material medium are introduced into the pair rate (4.6) through the refractive modifications of the spectral functions $F_D(\omega)$ and $\sigma_A(\omega)$, as well as through the factors $g(\omega, \mathbf{R})$ and $e^{-2n''\omega R/c}$. The latter exponential factor represents the Beer law losses in the absorbing medium. This factor will be demonstrated to play a vital role at large separations between the transfer species, providing a physically sensible total rate of ET to all the surrounding acceptors (see Section V.B).

The pair rate (4.6) applies to arbitrary transfer distances, covering both near and far zone, as well as intermediate distances. It can be presented meaningfully as the following sum of three terms:

$$W_{DA} = W_{DA}^{\text{Först}} + W_{DA}^I + W_{DA}^{\text{farzone}} \qquad (4.12)$$

The first term namely

$$W_{DA}^{\text{Först}} = \frac{9c^4\eta_3^2}{8\pi\tau_D R^6} \int_0^\infty F_D(\omega)\sigma_A(\omega)|n|^{-4}\omega^{-4}d\omega \qquad (4.13)$$

represents the familiar Förster rate of ET characterized by an R^{-6} distance dependence and featuring the orientational factor η_3^2; the latter factor is commonly labaled by κ^2 in the theory of radiationless ET (see, e.g., Ref. 115). The rate $W_{DA}^{\text{Först}}$ is the dominant contribution in the near zone ($|n|\bar{\omega}R/c \ll 1$). Here the exponential factor $e^{-2n''\omega R/c}$ is close to unity and has therefore been disregarded in Eq. (4.13), where $\bar{\omega}$ is an averaged transfer frequency. Consequently, the spectral integral (4.13) is weighted by the factor $|n|^{-4}$ in the near zone, in agreement with the standard theory of radiationless ET [6,11].

The third term in (4.12), namely

$$W_{DA}^{\text{farzone}} = \frac{9\eta_1^2}{8\pi\tau_D R^2} \int_0^\infty F_D(\omega)\sigma_A(\omega)e^{-2n''\omega R/c}d\omega \qquad (4.14)$$

dominating in the far zone ($|n|\bar{\omega}R/c \gg 1$), is characterized by the an R^{-2} behavior corrected by the exponential decay factor. The rate (4.14) has been represented through the overlap integral between the donor emission and acceptor absorption spectra, weighted by the exponential Beer law factor. The result (4.14) can be identified as the rate of radiative (far-zone) ET involving spontaneous emission by a donor, propagation of the emitted photon through the absorbing medium, followed by its absorption at the acceptor. The factors $F_D(\omega)$, $e^{-2n''\omega R/c}$ and $\sigma_A(\omega)$ characterize the corresponding processes. It is noteworthy that the local field factors are contained in the spectral functions $F_D(\omega)$ and $\sigma_A(\omega)$. Note also that the near and far zone rates differ not only in the distance dependence but also in their orientational factors. This leads to a completely different transfer-induced fluorescence depolarization in these two cases, an issue to be discussed in detail in the following section.

Finally, the middle term of the pair rate (4.12), W_{DA}^{I}, due to the remaining terms in the function (4.7), becomes important at intermediate distances where $|n|\overline{\omega}R/c \cong 1$. In general this intermediate contribution contains not only the usual R^{-4} term [15–23] but also additional terms in odd powers of R (i.e., R^{-3} and R^{-5}). However, in the case of weakly absorbing medium ($n'' \ll n'$), one can disregard the latter odd rank terms to arrive at the following approximate expression:

$$W_{DA}^{I} = \frac{9c^2(\eta_3^2 - 2\eta_1\eta_3)}{8\pi\tau_D R^4} \int_0^\infty F_D(\omega)\sigma_A(\omega)|n|^{-2}\omega^{-2}e^{-2n''\omega R/c}\,d\omega \qquad (4.15)$$

For such a weakly absorbing medium, the function $g(\omega, \mathbf{R})$ entering the pair rate (4.6) can be written approximately as

$$g(\omega, \mathbf{R}) = |n|^2 \left[\eta_3^2 \left(\frac{c}{|n|\omega R} \right)^6 + (\eta_3^2 - 2\eta_1\eta_3) \left(\frac{c}{|n|\omega R} \right)^4 + \eta_1^2 \left(\frac{c}{|n|\omega R} \right)^2 \right]$$
$$(4.16)$$

Note that in most situations the media are indeed weakly absorbing. The condition $n'' \ll n'$ can fail only near sharp resonances of the medium. Yet, in such a case the whole idea of description of the ET in terms of transfer rates is to be questioned.

We conclude this section with a brief remark concerning nonrigid systems having fast rotational motion of the donor and the acceptor. In such a situation the factor $g(\omega, \mathbf{R})$ given above should be replaced by its orientational average:

$$g_{av}(\omega, \mathbf{R}) = \frac{2}{9}|n|^2 \left[3 \left(\frac{c}{|n|\omega R} \right)^6 + \left(\frac{c}{|n|\omega R} \right)^4 + \left(\frac{c}{|n|\omega R} \right)^2 \right] \qquad (4.17)$$

B. Range Dependence of the Fluorescence Depolarization Due to ET

In this section we analyze the range dependence of the fluorescence depolarization, applying the unified approach to the radiationless and radiative ET.* For

---

*This issue has been first addressed in Ref. 23 neglecting the effects of the surrounding medium. Here we include the medium effects as well.

the near-zone (nonradiative) ET, the transfer rate depends only weakly on the average mutual orientation of the donors and acceptors [see Eq. (4.19) for the average of the appropriate orientational factor]. This is the reason for the well-known and considerable ($\frac{1}{25}$) reduction of fluorescence anisotropy following a single act of ET in an isotropic or randomly oriented system [6,11]. By contrast in the radiative mechanism, the ET between molecules with parallel transition dipoles is greatly preferred — as one can see from Eq. (4.18), in which the angle-dependent term is weighted by a factor of 7. This leads to a substantially smaller loss of polarization for an ensemble of randomly oriented transfer species; specifically, the residual anisotropy following a single act of photon reabsorption is 7 times greater than in the case of the short-range (nonradiative) ET.

Here we present a general result for the residual fluorescence anisotropy that connects and accommodates the two limiting cases, also providing the behavior at the intermediate distances where neither radiative nor radiation-less mechanism dominates. Note that the rotational depolarization is assumed to be negligible. In order to obtain a formula exhibiting the effects of the relative donor-acceptor orientation in an ensemble, it is necessary to average the pair rate (4.6) over the orientation of the radius vector $\mathbf{R}$ keeping a fixed mutual orientation between donor and acceptor. We then arrive at the following rotational averages of the orientational factors featured in the function $g(\omega, \mathbf{R})$ entering the rate (4.6):

$$\langle \eta_1^2 \rangle = \tfrac{1}{15}(7\cos^2\theta + 1) \tag{4.18}$$

$$\langle \eta_3^2 \rangle = \tfrac{1}{5}(\cos^2\theta + 3) \tag{4.19}$$

and

$$\langle \eta_1\eta_3 \rangle = \tfrac{1}{3}\langle \eta_3^2 \rangle \tag{4.20}$$

where $\cos\theta = \hat{\boldsymbol{\mu}}_A \cdot \hat{\boldsymbol{\mu}}_D$.

We shall concentrate on the situation where a medium is sufficiently weakly absorbing that one can make use of Eq. (4.16) for the function $g(\omega, \mathbf{R})$. Substituting the averages (4.18)–(4.20) into Eqs. (4.6) and (4.16), we then obtain the following orientationally averaged rate, for the case of a fixed angle θ between the transition dipoles of donor and acceptor:

$$\langle W(R) \rangle \propto (3y_6 + y_4 + 7y_2)\cos^2\theta + (9y_6 + 3y_4 + y_2) \tag{4.21}$$

where the range characteristics y_n are spectral overlap moments of the form

$$y_n = \frac{c^{n-2}}{R^n} \int_0^\infty F_D(\omega)\sigma_A(\omega)|n\omega|^{(2-n)}e^{-2n''\omega R/c}d\omega, \qquad n = 2, 4, 6 \tag{4.22}$$

Therefore, the properly normalized function for the orientational distribution of excited acceptors is given by

$$f(\theta, R) = \frac{3}{10} \frac{(3 + \overline{k^2}R^2 + 7\overline{k^4}R^4)\cos^2\theta + (9 + 3\overline{k^2}R^2 + \overline{k^4}R^4)}{3 + \overline{k^2}R^2 + \overline{k^4}R^4} \tag{4.23}$$

where the bar over k^2 and k^4 signifies the spectral averages:

$$\overline{k^n} = \frac{\int_0^\infty |n\omega/c|^n F_D(\omega)\sigma_A(\omega)|n\omega|^{-4} e^{-2n''\omega R/c} d\omega}{\int_0^\infty F_D(\omega)\sigma_A(\omega)|n\omega|^{-4} e^{-2n''\omega R/c} d\omega} \tag{4.24}$$

Of special interest is the fluorescence anisotropy commonly defined by

$$r = \frac{I_\parallel - I_\perp}{I_\parallel + 2I_\parallel} \tag{4.25}$$

where $I_\parallel$ and $I_\perp$ are components of the fluorescence intensity polarized, respectively, parallel and perpendicular to the polarization of the excitation light. In the case where fluorescence occurs directly from the molecule that absorbs the incident light (the donor), the anisotropy is designated r_0; where fluorescence occurs following a single-step transfer of energy to another molecule (the acceptor), the anisotropy is designated r_1. The value of r_0 reaches its theoretical maximum of 0.4 if intramolecular relaxation of the donor produces no change of electronic state [6,11].

Here it is the result for r_1 that is of principal interest; the fluorescence anisotropy following a chain of ET events can be directly calculated from this result. In terms of r_0, the acceptor anisotropy r_1 can be expressed as (see, e.g., Ref. 116)

$$r_1 = \langle\langle P_2(\cos\theta)\rangle\rangle r_0 \tag{4.26}$$

where $P_2(\cos\theta) = \frac{1}{2}(3\cos^2\theta - 1)$ is the second-order Legendre polynomial, and the double angular brackets denote the orientational average over the mutual orientations of donor and acceptor:

$$\langle\langle P_2(\cos\theta)\rangle\rangle = \frac{1}{2}\int_0^\pi P_2(\cos\theta)f(\theta)\sin\theta d\theta \tag{4.27}$$

Substituting the distribution function (4.23) into Eq. (4.27), we obtain the following most general result for the fluorescence anisotropy:

$$r_1(R) = \frac{r_0}{25} \frac{7\overline{k^4}R^4 + \overline{k^2}R^2 + 3}{\overline{k^4}R^4 + \overline{k^2}R^2 + 3} \tag{4.28}$$

This equation is valid for arbitrary separations R. As shown in Eqs. (4.34) and (4.35), the familiar near- and far-zone results are the asymptotes of this formula.

In the general case, the residual anisotropy depends not only on the transfer distance but also on the shapes of the spectral lines through the averages $\overline{k^2}$ and $\overline{k^4}$ featured in Eq. (4.28). However, as the widths of the absorption and emission lines are considerably less than the photon frequency, Eq. (4.28) can be rewritten without a significant loss of generality as

$$r_1(R) = \frac{r_0}{25} \frac{7\overline{k}^4 R^4 + \overline{k}^2 R^2 + 3}{\overline{k}^4 R^4 + \overline{k}^2 R^2 + 3} \tag{4.29}$$

where the average wavenumber $\overline{k}$ is related approximately to the averaged transfer frequency $\overline{\omega}$ as

$$\overline{k} = \frac{|n|\overline{\omega}}{c} \tag{4.30}$$

In the case where the absorption and emission lines are of Gaussian shape

$$F_D(\omega) \propto \omega^3 \exp\frac{(\omega - \overline{\omega}_{D^*})^2}{2\sigma^2}, \qquad \sigma_A(\omega) \propto \omega\exp\frac{(\omega - \overline{\omega}_{A^*})^2}{2\sigma^2} \tag{4.31}$$

we have, for $\overline{\omega}_{D^*} = \overline{\omega}_{A^*}$

$$\overline{k^2} = \overline{k}^2 \left[\frac{1 + (\sigma/\overline{\omega})^2}{2} \right] \tag{4.32}$$

$$\overline{k^4} = \overline{k}^4 \left[1 + 3\left(\frac{\sigma}{\overline{\omega}}\right)^2 + 3\frac{(\sigma/\overline{\omega})^4}{4} \right]. \tag{4.33}$$

This means that if, for instance, the ratio $\sigma/\overline{\omega}$ equals 0.1, the error made using the relationship (4.29) instead of the exact result (4.28) is less than a few percent. Furthermore, both formulas provide the corrects asymptotes at small and large distances:

1. $\overline{k}R = |n|\overline{\omega}R/c \ll 1$. Here Eqs. (4.28) and (4.29) reproduce the usual Galanin result:

$$r_1 \equiv r_1^{\text{nonrad}} = \frac{r_0}{25} \tag{4.34}$$

2. $\overline{k}R = |n|\overline{\omega}R/c \gg 1$. Here we arrive at the result for the fluorescence depolarization associated with radiative ET from the donor to an acceptor in the far zone:

$$r_1 \equiv r_1^{\text{rad}} = \frac{7r_0}{25} \tag{4.35}$$

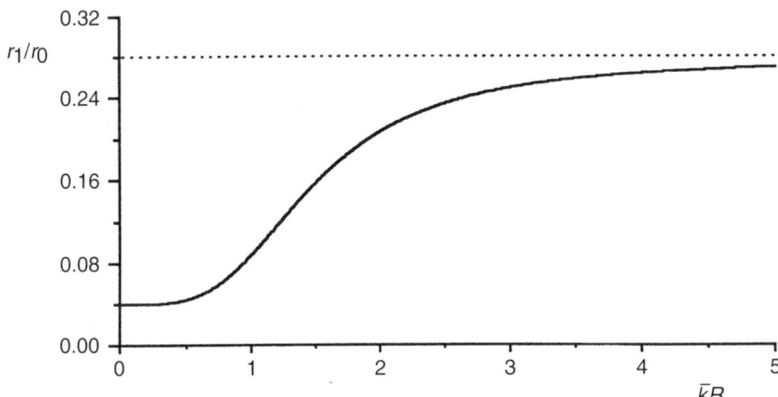

FIGURE 5.   The relative anisotropy r_1/r_0 versus $\bar{k}R$ calculated according to Eq. (4.29).

The distance dependence of the relative anisotropy r_1/r_0 calculated according to Eq. (4.29) is presented in Fig. 5. One can see the anisotropy rise to significant values at distances much less than those normally associated with radiative ET. For instance, with a donor–acceptor separation $R = 1.5/\bar{k} = 0.75\lambda/\pi$, the relative anisotropy r_1/r_0 attains the value of 3/25, considerably higher than the result for radiationless transfer, as follows from Eq. (4.29). In the vacuum limit $|n| \to 1$, the results for the fluorescence anisotropy reduce to those presented in Ref. 23. In the general case ($|n| \neq 1$), the wavevector $\bar{k} = |n|\bar{\omega}/c$ is scaled by $|n|$, adjusting the characteristic scale of distances. As such, the position at which the onset of significant retardation modifies the fluorescence anisotropy can be still closer than the vacuum formula would suggest.

In connection with multiphoton fluorescence ET [117], any microscopically disordered system exhibits the same sevenfold increase in fluorescence anisotropy as the donor–acceptor distance increases from the near-zone to the far-zone range. However, the detailed dependence on the relative orientations of the participating donor transition moments adds considerable complexity to the results even in the two-photon case [117].

V.   SPONTANEOUS EMISSION AS FAR-ZONE ENERGY TRANSFER

A.   Background to the Problem

Spontaneous emission is a phenomenon that has played an important role in establishing several key concepts of modern quantum theory [99]. Over recent years there has been a great deal of interest in modified spontaneous emission by atoms (molecules) in various environments, such as in photonic bandgap crystals [89–92], near metal surfaces [61–65], and near dielectric interfaces [118–121]. The characteristics of spontaneous emission are also modified in homogeneous dielectrics [72,109,110,122–128]. The rate for the process, in the case of photon emission into transparent areas of a dielectric, reads [72,109,110,125,126]*

$$W(D \leftarrow D^*) = \frac{\omega_{D^*}^3 \mu_D^2}{3\pi\varepsilon_0 \hbar c^3} n \left(\frac{n^2 + 2}{3}\right)^2 \qquad (5.1)$$

where the transition frequency ω_{D^*} is also modified by the surrounding medium [109,126].

The emission rate (5.1) can be obtained using various methods. For instance, the explicit mode expansion (3.13) for the local operator $\mathbf{d}^\perp(\mathbf{R}_D)$ offers a straightforward way to derive the emission rate (5.1) [109,110] through application of the Fermi golden rule (3.1), in which the spontaneous emission appears as a first-order process described by the transition operator: $T^{(1)} = -\varepsilon_0^{-1}\boldsymbol{\mu}.\mathbf{d}^\perp(\mathbf{R}_D)$. The result (5.1) is relevant to the case where only long-wavelength polariton modes ($k \ll a^{-1}$) are involved in the emission process, where a is a characteristic distance between the species constituting the medium. In fact, only these modes are accommodated in the mode expansion (3.13); see the discussion following Eq. (3.17). Consequently, Eq. (5.1) does not generally hold where the emission takes place into absorbing areas of the dielectric, in which case an important role is played by excitonlike modes of the medium with larger values of k, as discussed earlier [31,109,110].

Spontaneous emission in lossy dielectrics may be considered in a number of ways. The phenomenon might be dealt with by analyzing the macroscopic Maxwell equations for a classical dipole in an absorbing medium [129]. A quantum-mechanical analysis closer in tune with the modern theoretical formalism has also been carried out [123,124]. Specifically, in a paper by Barnett et al. [124], the decay rate has been calculated for an excited atom embedded in an absorbing dielectric. Applying the Green function technique, the rate was separated into "transverse" and "longitudinal" components: these correspond, respectively, to the rate of emission of a transverse photon and the

---

*Equation (5.1) applies to situations where the emitting molecule does not significantly disturb the electromagnetic modes of the medium. This is known as the *virtual-cavity case* (Ref. 11, Section I.4). An alternative is the *real-cavity model*, which has been applied to spontaneous emission by Glauber and Lewenstein [122]. The latter model, introducing another local field factor into the emission rate, is relevant to situations where the emitting dipoles are large species characterised by low polarisability, such as in recent experiments by Rikken and Kessener [127] and by Schuurnams et al. [128].

rate of nonradiative decay via longitudinal coupling of the atom to the dielectric [124]. As the absorbing medium was described macroscopically, effects due to discreteness of the medium (such as the local field effects) were not intrinsically reflected; local field corrections were introduced phenomenologically at later stages [123,124]. Yet, a comprehensively microscopic approach is to be preferred.

In this section, following Refs. 31 and 73, we present a microscopic analysis of the phenomenon based on the unified approach to radiationless and radiative ET reviewed in the previous section.

B. Decay of an Excited Molecule in the Absorbing Medium

Consider the decay of an excited-state donor D induced by the transfer of energy to the surrounding species X. By summing up all appropriate pair rates, the full decay rate is

$$\Gamma_D = \sum_{X \neq D} W_{DX} \tag{5.2}$$

where W_{DX} is the rate of excitation transfer between a pair of molecules D and X, as explicitly presented in the previous section (with $X \equiv A$). Using the partition (4.12) for the pair rates, the decay rate (5.2) can be cast as

$$\Gamma_D = \tilde{\Gamma}_D^{\text{Först}} + \Gamma_D^{\text{far zone}} \tag{5.3}$$

with

$$\tilde{\Gamma}_D^{\text{Först}} = \sum_{X \neq D} (W_{DX}^{\text{Först}} + W_{DX}^{I}) \tag{5.4}$$

and

$$\Gamma_D^{\text{far zone}} = \sum_{X \neq D} W_{DX}^{\text{far zone}} \tag{5.5}$$

The former decay rate, $\tilde{\Gamma}_D^{\text{Först}}$, contains contributions due to the near-zone (Förster) pair rates modified by the intermediate terms W_{DX}^{I}; the tilde over Γ reflects such a modification. The constituent pair rates have been explicitly presented by Eqs. (4.13) and (4.15), giving the explicit refractive dependence of $\tilde{\Gamma}_D^{\text{Först}}$.

A cautionary note should be made concerning such a straightforward derivation of $\tilde{\Gamma}_D^{\text{Först}}$ in the case where most of the surrounding species X are efficient energy acceptors. In this situation, the major contribution to $\tilde{\Gamma}_D^{\text{Först}}$ comes from ET to the closest species X (up to a few configurational spheres around D). At such small distances, description of the pair rates in terms of the refractive index is questionable. Yet, one can make use of a certain "effective" index n_{eff} in the pair rates (4.13) and (4.15) constituting the decay rate $\tilde{\Gamma}_D^{\text{Först}}$. The quantity, $n_{\text{eff}} \equiv n_{\text{eff}}(R)$, approaches the true refractive index n as the transfer distance R extends into the range where it considerably exceeds the characteristic distance between the species in the medium [11]. On the other hand, the far zone decay rate $\Gamma_D^{\text{far zone}}$ is built up of a macroscopically large number of pair rates, making its description through the refractive index quite legitimate (see discussion below). The same applies to the near-zone rate (5.4) in the case where the medium constituting the transparent background species X_1 is diluted with a small fraction of the absorbing acceptor molecules X_2. In such a case, the space between the donor and any acceptor X_2 is filled with a large number of background molecules, the latter species X_1 providing the major contribution to the refractive index n. The acceptors X_2 contribute much more weakly to n, yet ensuring the existence of some imaginary part $n'' \neq 0$ that plays an important role in the far zone through the exponential factor featured in corresponding pair rates (4.14).

In the remaining part of this section we concentrate on the decay rate $\Gamma_D^{\text{far zone}}$ associated with far-zone (radiative) ET. The far-zone transfer may be viewed [15,18,21,31,73] as spontaneous emission of a photon followed by its subsequent recapture by a distant acceptor. Adopting such a concept, we shall regard the contribution $\Gamma_D^{\text{far zone}}$ as the rate of spontaneous emission in the absorbing medium. The approach is in a certain sense related to absorber theory [130–132] in which spontaneous emission is seen to be the result of direct interaction between the emitting atom and "the Universe", the latter acting as a perfect absorber at all emitted frequencies. In our situation, the surrounding medium does indeed act as a perfect absorber even at extremely low concentrations of the absorbing species (or, alternatively, for an almost transparent condensed medium), so long as the system dimensions are large enough to ensure eventual recapture of the emitted photon. For such a weakly absorbing medium, the rate $\Gamma_D^{\text{far zone}}$ will be demonstrated to reproduce exactly the familiar rate [72,109,110,125,126] for spontaneous emission in a transparent dielectric.

To obtain the proper decay rate $\Gamma_D^{\text{far zone}}$ using Eq. (5.5), the pair-transfer rates $W_{DX}^{\text{far zone}}$ should not only reflect effects due to retardation but also incorporate influences of the surrounding medium, as in the previous section. Following Refs. 31 and 73, we shall demonstrate that $\Gamma_D^{\text{far zone}}$ does indeed represent the rate of spontaneous emission in the absorbing medium. For this purpose, we substitute Eq. (4.14) for the pair-rate $W_{DA}^{\text{far zone}}$ (with $A \equiv X$) into Eq. (5.5) to yield

$$\Gamma^{\text{far zone}} = \frac{9}{8\pi\tau_D} \int_0^\infty F_D(\omega) \sum_X R^{-2} \eta_1^2 \sigma_X(\omega) e^{-2n''\omega R/c} d\omega \qquad (5.6)$$

with $\mathbf{R} \equiv \mathbf{R}_{XD}$. The decay rate $\Gamma_D^{\text{far zone}}$ is made up of a large number of the pair rates operating predominantly in the far zone. Hence the summation over the molecules X can be changed to an integral over the radius vector $\mathbf{R}$, giving

$$\Gamma^{\text{far zone}} = \frac{1}{\tau_D} \int_0^\infty d\omega F_D(\omega)\sigma(\omega) \int_0^\infty d\mathbf{R} e^{-2n''\omega R/c} \tag{5.7}$$

$$= \frac{c}{2\tau_D} \int_0^\infty d\omega F_D(\omega)\sigma(\omega)/\omega n'' \tag{5.8}$$

where [in Eq. (5.7)] the appropriate orientational averaging has been carried out ($[\eta_1^2 \to \langle \eta_1^2 \rangle = \frac{2}{9}]$) and the cross section of the molecular absorption $\sigma_X(\omega)$ has been replaced by its ensemble average as given by

$$\sigma(\omega) \equiv \overline{\sigma}_X(\omega) = \frac{\omega}{c} \frac{1}{n'} \left| \frac{n^2 + 2}{3} \right|^2 \frac{\overline{\alpha}_X''}{\varepsilon_0} \tag{5.9}$$

Here, the quantity $\overline{\alpha}_X$ is understood as an averaged polarizability for all the species X constituting the medium:

$$\overline{\alpha}_X = N^{-1} \sum_X \alpha_X \tag{5.10}$$

where N is the total number of molecules in the system. The relationship (5.9) has been written exploiting Eq. (4.9) for $\sigma_A(\omega)$ (with $A \equiv X$). Here, constraints due to the energy-conserving delta functions have been expressed in terms of the imaginary part of the averaged molecular polarizability $\overline{\alpha}_X'' \equiv \alpha''$, as given by Eq. (5.10) with

$$\alpha_X \equiv \alpha_X(\omega + is) = \frac{1}{3\hbar} \sum_{m,p} \rho_X^{(m)} \left[\frac{\mu_X^2 |\langle \varphi_{X^*}^{(p)} | \varphi_X^{(m)} \rangle|^2}{\omega_{X_{pm}^*} - \omega - is} + \frac{\mu_X^2 |\langle \varphi_{X^*}^{(p)} | \varphi_X^{(m)} \rangle|^2}{\omega_{X_{pm}^*} + \omega + is} \right] \tag{5.11}$$

and where $\omega_{X_{pm}^*} = \omega_{X_p^*} - \omega_{X_m}$ is the excitation frequency of the molecule X and $|\langle \varphi_{X^*}^{(p)} | \varphi_X^{(m)} \rangle|^2$ are the Condon factors defined in Section IV.A (with $A \equiv X$). The imaginary part α'' is in turn related to the complex refractive index $n \equiv n(\omega)$, as

$$n'' = \frac{1}{2n'} \left| \frac{n^2 + 2}{3} \right|^2 \frac{\alpha'' \rho}{\varepsilon_0} \tag{5.12}$$

so that Eq. (5.9) ultimately reduces to

$$\sigma(\omega) \equiv \overline{\sigma}_X(\omega) = \frac{n'' 2\omega}{c\rho} \tag{5.13}$$

This equation can be identified as the usual relationship between the imaginary part of the complex refractive index and the absorption cross section. In this way, the quantity $\sigma(\omega)$ defined by Eq. (4.9) is indeed seen to represent the cross section of molecular absorption in the lossy medium. Substituting Eq. (5.13) into (5.8), one finds, using Eq. (4.10) for $F_D(\omega)$ and the normalization condition (4.11)

$$\Gamma^{\text{far zone}} = \tau_D^{-1} = \sum_{n,r} (\tau_D^{rn})^{-1} \rho_{D*}^{(n)} \tag{5.14}$$

with

$$(\tau_D^{rn})^{-1} = \frac{\omega_{D*_{nr}}^3 \mu_D^2}{3\hbar\varepsilon_0\pi c^3} n' \left| \frac{n^2 + 2}{3} \right|^2 |\langle \varphi_D^{(r)} | \varphi_{D*}^{(n)} \rangle|^2 \tag{5.15}$$

The rate $\Gamma^{\text{far zone}} = \tau_D^{-1}$ as given by Eqs. (5.14) and (5.15) represents the full rate of spontaneous emission by the donor, and so involves summation over the final levels and averaging over the initial levels of the donor, labeled respectively r and n, where $\rho_{D*}^{(n)}$ is the population distribution of the vibrational levels of donor in the initially excited electronic state. The constituent terms $(\tau_D^{rn})^{-1}$ represent the partial rates of spontaneous emission associated with downward transitions of the donor between the specific levels n and r, where in each of these terms the refractive index is to be calculated at the appropriate emission frequency $\omega_{D*} \equiv \omega_{D*_{nr}} = (e_{D_n^*} - e_{D_r})/\hbar$.

The emission rates (5.15) manifestly accommodate contributions due to the absorbing dielectric medium, including the local field factors. In addition, the emission frequency ω_{D*} is to be renormalized to include the effects of the medium, as follows from the strict analysis of ET dynamics presented in the next section. It is noteworthy that our analysis is based on a microscopic QED theory [28,31], the relationship (5.15) supporting previous phenomenological methods [123,124] used to introduce local field corrections to the rates of spontaneous emission in an absorbing medium. In the limit where $\alpha'' \to 0$, $n'' \to 0$, the far-zone rate [Eqs. (5.14) and (5.15)] reduces smoothly to the usual result for spontaneous emission in a transparent medium given by Eq. (5.1) (in which Condon factors are not included); in such a situation there is a vanishing contribution due to the decay in the near and intermediate zones: $\tilde{\Gamma}_D^{\text{Först}} \to 0$.

In this way, the present analysis reproduces in full the rate of spontaneous emission in transparent dielectrics $n'' = 0$, including, inter alia, the case of free space: $n'' = 0$, $n' = 1$. Here, free space is be to viewed as a limit in which the density of the absorbing species goes to zero, while the size of the system goes to infinity, so that the emitted photon is eventually recaptured somewhere in the system. It is notable that in order to arrive at a sensible result for

$\Gamma_D^{\text{far zone}} = \tau_D^{-1}$, such as that given by Eqs. (5.14) and (5.15), the influences of the absorbing medium are necessarily to be reflected in the pair-transfer rates constituting the far-zone decay rate (5.5). In fact, it is the exponential factor $\exp(-2n''\omega R/c)$, representing absorption losses at the intervening medium, that helps avoid the potentially infinite decay rate $\Gamma_D^{\text{far zone}}$ due to the R^{-2} factor featured in the constituent pair rates (4.14). Note also that the same result, (5.14) and (5.15), can be reproduced for the rate of spontaneous emission τ_D^{-1}, in terms of another microscopic method involving calculation of the quantum flow from the emitting molecule D in the absorbing medium [73].

It is noteworthy that the separation of the full decay rate into far- and near-zone components corresponds to a division of the rate into its "transverse" and "longitudinal" parts, to use the terms adopted by Barnett et al. [124]. Our description in terms of far- and near-zone rates goes along with the multipolar formulation of QED employed here; in such a formulation, [96], the coupling between the molecules is mediated exclusively via transverse photons, as there is no instantaneous (longitudinal) contribution to the intermolecular coupling. Finally, the analysis presented yields the exponential decay with time of the excited-state population for a selected molecule D, as long as the backtransfer of the excitation energy from the surrounding species X is negligible. The near-zone contribution to the decay rate, $\tilde{\Gamma}_D^{\text{Först}}$ [Eq. (5.4)], may depend markedly on a specific distribution of acceptors around the donor in the case of a somewhat inhomogeneous medium. At short times, this leads to the well-known [11] nonexponential time decay of the excited-state population averaged over the ensemble. On the other hand, the total rate of radiative ET $\Gamma_D^{\text{far zone}}$ depends much more weakly on the specific site of the emitter D, yielding an exponential contribution to the time decay of the excited-state population in the ensemble. Exponential decay dominates the kinetics at sufficiently large times.

VI. DYNAMICS OF ET BETWEEN A PAIR OF MOLECULES IN A DIELECTRIC MEDIUM

Up to now the transfer dynamics has been described in terms of well-defined rates for intermolecular ET. It is the purpose of the current section to pursue in more detail temporal aspects of the ET between a pair of species in a dielectric medium. The consideration is based on the QED study [28], a distinctive aspect of which is a combined analysis of rate- and nonrate regimes in the context of examining the influence of the dielectric medium on a microscopic basis. The theory is built on the foundation established in the previous sections. Again, the approach exploits the concept of ET mediated by bath polaritons. The theory also makes use of the microscopically derived tensor (3.18) for the retarded and medium-dressed dipole–dipole coupling, now with regard to the dynamical behavior. The present section not only extends consideration beyond the rate description, but also reexamines conditions for that regime itself. That leads to incorporation of the shifts in energy for both the ground and excited states of the transfer species, due to the interaction of these species with the molecules belonging to the medium and also with each other. That is a feature not reflected in direct application of the ordinary Fermi golden rule.

The section is organised as follows. In the next subsection the Heitler–Ma method [2,54,95,133,134] for describing the quantum time evolution is first outlined, and subsequently reformulated to suit our current purposes. Consequently, the basic equations for time evolution acquire a form more symmetric with respect to the initial and final states. Section VI.B presents the general analysis on the transfer dynamics between a pair of molecules in the molecular medium. At the end of the Section VI.B, we discuss the problem of causality in ET, an issue that has received a great deal of interest in the literature [39–52]. Section VI.C deals with the specific situations including both the rate regime and beyond. Note that the non-rate regime features in situations lacking an intrinsic density of molecular states for the participating species.

A. Time Evolution of a Quantum System

Consider the quantum dynamics of a system with a time-independent Hamiltonian separable as the sum of a zeroth-order Hamiltonian H^0 and an interaction term V [such as that defined by Eq. (2.5)], where the eigenvectors of H^0 include both the initial state $|I\rangle$ and the final state $|F\rangle$ for the process. For reasons which will become apparent later, we shall commence work in the Schrödinger representation rather than the more common interaction representation. The state vector of the system then evolves at positive times from the state $|I\rangle$ at $t = +0$ as

$$S(t)|I\rangle = \Theta(t)e^{-iHt/t}|I\rangle \qquad (6.1)$$

$$= -\frac{1}{2\pi i}\int_{-\infty}^{+\infty} d\varepsilon e^{-i\varepsilon t/\hbar}(\varepsilon - H + is')^{-1}|I\rangle, \qquad s' \to +0 \quad (6.2)$$

where $\Theta(t)$ is the unit step (Heaviside) function. Strictly speaking, the quantity $s' \equiv \hbar s$ is a positive infinitesimal. Yet, for finite times it may be considered a finite quantity obeying the following: $st \ll 1$, that is, s should be kept much less than the inverse lifetime for the excited states. Under this condition, introduction of a finite s does not influence the quantum dynamics governed by Eq. (6.2). Retention of a finite value for s plays an important role in smoothing of spectral lines. This makes the refractive index given by Eq. (3.19) [together with Eqs. (3.15) and (3.16)] a complex quantity in absorbing areas of the spectrum (characterized by densely spaced molecular sublevels of vibrational or other origin for each electronic transition).

The Heitler–Ma method [2,54,95,133,134] may now be employed, giving

$$\langle F|(\varepsilon - H + is')^{-1}|I\rangle = \frac{U_{FI}(\varepsilon)}{(\varepsilon - E_F + is')[\varepsilon - E_I + \frac{1}{2}i\hbar\Gamma_1(\varepsilon) + is']} \quad (6.3)$$

Here $U_{FI}(\varepsilon) \equiv \langle F|U(\varepsilon)|I\rangle$ and $\Gamma_I(\varepsilon) \equiv \langle I|\Gamma(\varepsilon)|I\rangle$ are, respectively, the matrix elements of the off-diagonal transition operator $U(\varepsilon)$ and the diagonal damping operator $\Gamma(\varepsilon)$, both determined by the following recurrence relation:

$$U(\varepsilon) - \tfrac{i}{2}\hbar\Gamma(\varepsilon) = V + V(\varepsilon - H^0 - V + is')^{-1}U(\varepsilon) \quad (6.4)$$

For the present purposes it is more convenient to represent this relation in a nonrecursive format as

$$[U(\varepsilon) - \tfrac{i}{2}\hbar\Gamma(\varepsilon)]|I\rangle = \lfloor V + VP_I(\varepsilon - H^0 - P_IVP_I + is')^{-1}P_IV\rfloor I\rangle \quad (6.5)$$

where the projection (idempotent) operator

$$P_I = 1 - |I\rangle\langle I| \quad (6.6)$$

identifies the exclusion of contributions by the initial state in the perturbation expansion of Eq. (6.5). Recasting the transition matrix element in a form where the perturbational contribution by the final state also no longer explicitly features, one arrives at (see Appendix B of Ref. 134)

$$U_{FI}(\varepsilon) = \frac{\varepsilon - E_F + is'}{\varepsilon - E_F + \frac{i}{2}\hbar\Gamma'_F + is'}U'_{FI} \quad (6.7)$$

where the newly defined quantities on the right, U'_{FI} and Γ'_F, both have implicit ε dependence and are given by

$$U'_{FI} \equiv \langle F|\lfloor V + VP_IP_F(\varepsilon - H^0 - P_IP_FVP_IP_F + is')^{-1}P_IP_FV\rfloor I\rangle \quad (6.8)$$

$$-\tfrac{i}{2}\hbar\Gamma'_F \equiv \langle F|\lfloor V + VP_IP_F(\varepsilon - H^0 - P_IP_FVP_IP_F + is')^{-1}P_IP_FV\rfloor I\rangle \quad (6.9)$$

with

$$P_F = 1 - |F\rangle\langle F| \quad (6.10)$$

Finally, calling on Eqs. (6.2), (6.3), and (6.7), one finds the following probability amplitude for the transition $|I\rangle \to |F\rangle$:

$$\langle F|S(t)|I\rangle = -\frac{1}{2\pi i}\int_{-\infty}^{+\infty} d\varepsilon \frac{U'_{FI}e^{-i\varepsilon t/\hbar}}{(\varepsilon - E_F + \frac{i}{2}\hbar\Gamma'_F + is')(\varepsilon - E_I + \frac{i}{2}\hbar\Gamma_I + is')} \quad (6.11)$$

which is an exact result. Here the presence of both Γ_I and Γ'_F in the energy denominators explicitly accommodates the damping corrections and energy renormalization of the initial and final states. Consequently, the transfer amplitude as presented above has a form obviously more symmetric with respect to the initial and final states than would result from direct substitution of Eq. (6.3) into (6.2). Still, there is some asymmetry with respect to these states, reflected by the prime on Γ'_F. The retention of such a asymmetry will be of vital importance in the case of sharp energy levels for donor and acceptor, that is, where the participating transfer species lack an intrinsic density of molecular states; this aspect is to be considered in the Section VI.C.2.

B. Dynamics of Energy Transfer

The dynamical system of interest has been defined by the Hamiltonian (2.5)–(2.7′). For the representation of resonance ET, the initial and final state vectors and their energies are considered to have the form of Eqs. (2.8) and (2.9). Because of the two-center character of the interaction operator (2.6), it is convenient to carry out the corresponding partitioning in Eqs. (6.5) and (6.9), writing

$$-\tfrac{i}{2}\hbar\Gamma_I = \Delta e_{D^*} + \Delta e_A - \tfrac{1}{2}\hbar\gamma_{D^*} - \tfrac{i}{2}\hbar\Gamma_{D^*A} \tag{6.12}$$

$$-\tfrac{i}{2}\hbar\Gamma'_F = \Delta e_D + \Delta e_{A^*} - \tfrac{1}{2}\hbar\gamma_{A^*} - \tfrac{i}{2}\hbar\Gamma'_{DA^*} \tag{6.13}$$

Here one-center contributions, denoted by a single index D (or A), are due to the terms containing only one operator V_D (or V_A) in the perturbation expansions of Eqs. (6.5) and (6.9). Such contributions have already been separated into real energy shifts and imaginary damping terms in the preceding equations. For instance, Δe_{D^*} and γ_{D^*} represent, respectively, the bath-induced level shift (energy renormalization) and the damping factor for the excited molecular state $|D^*\rangle$ (as there are no imaginary (damping) contributions for the ground molecular states $|D\rangle$ and $|A\rangle$). Each such energy renormalization (Δe_{D^*}, Δe_A, Δe_D, and Δe_{A^*}) embodies not only the radiative (Lamb) shift [94–96,99] but also the contribution due to the dispersion interaction between the donor D (or acceptor A) and the molecular medium. Note that the dispersion energy appears now in the second order of perturbation, rather than the usual fourth order [94–96,135], since the coupling of the radiation field with the medium has already been included in the zero-order Hamiltonian H^0 given by Eqs. (2.7) and (2.7′). Here we do not consider the explicit structure of these energy shifts, which are to be treated as the parameters of the theory. The remaining (complex) quantities Γ_{D^*A} and Γ'_{DA^*} are two-center contributions resulting from cross-terms (containing both V_D and V_A) that emerge in the perturbation expansions of Eqs. (6.5) and (6.9).

By making use of Eqs. (6.12) and (6.13), the probability amplitude (6.11) for the ET takes the form

$$\langle F|\tilde{S}(t)|I\rangle = -\frac{1}{2\pi i\hbar}\int_{-\infty}^{+\infty}d\omega$$

$$\frac{U'_{FI}(\omega)e^{-i(\omega-\omega_{A*})t}}{(\omega-\omega_{A*}+(i/2)\gamma_{A*}+(i/2)\Gamma_{DA*}+is)(\omega-\omega_{D*}+(i/2)\gamma_{D*}+(i/2)\Gamma_{D*A}+is)} \tag{6.14}$$

where

$$\omega = \frac{\varepsilon-e_D-e_A-\Delta e_D-\Delta e_A-e_{\text{vac}}}{\hbar} \tag{6.15}$$

is a new variable, and

$$\omega_{D*} = \frac{e_{D*}+\Delta e_{D*}-e_D-\Delta e_D}{\hbar} \tag{6.16}$$

$$\omega_{A*} = \frac{e_{A*}+\Delta e_{A*}-e_A-\Delta e_A}{\hbar} \tag{6.17}$$

are the excitation frequencies of the donor and the acceptor. The frequencies ω_{D*} and ω_{A*} incorporate level shifts for both the ground and excited molecular states. Finally, in Eq. (6.14) transformation has been carried out to a modified interaction representation, as

$$\langle F|\tilde{S}(t)|I\rangle = \langle F|S(t)|I\rangle \exp\frac{-i(E_F+\Delta E_F)t}{\hbar} \tag{6.18}$$

where the term "modified" refers to change of the final-state energy $E_F = e_D+e_{A*}$ by the amount $\Delta E_F = \Delta e_D+\Delta e_{A*}$.

Now we turn our attention to the transition matrix element $U'_{FI}(\omega)$ featured in Eq. (6.14). For the present purposes it is sufficient to represent it through an effective second-order contribution as

$$U'_{FI}(\omega) \approx U^{(2)}_{FI}(\omega) = \mu^{\text{full}}_{A_l}\theta'_{lj}(\omega,\mathbf{R})\mu^{\text{full}}_{D_j} \tag{6.19}$$

with

$$\theta_{lj}(\omega,\mathbf{R})=\frac{1}{\hbar\varepsilon_0^2}\sum_\sigma\left[\frac{\langle 0|d_l^\perp(\mathbf{R}_A)|\sigma\rangle\langle\sigma|d_j^\perp(\mathbf{R}_D)|0\rangle}{\omega-\Pi_\sigma+is}+\frac{\langle 0|d_j^\perp(\mathbf{R}_D)|\sigma\rangle\langle\sigma|d_l^\perp(\mathbf{R}_A)|0\rangle}{\omega-\omega_{D*}-\omega_{A*}-\Pi_\sigma+is}\right] \tag{6.20}$$

($\mathbf{R}=\mathbf{R}_A-\mathbf{R}_D$), where implied summation over the repeated Cartesian indices (l and j) is assumed, and μ^{full}_D and μ^{full}_A are the transition dipoles given by Eq. (2.19). As in the previous sections, here $\hbar\Pi_\sigma = e_\sigma - e_{\text{vac}}$ is the excitation energy of the bath, the index σ denoting excited (single polariton) states of the bath accessible from the ground state $|0\rangle$ by single action of the local displacement operator $\mathbf{d}^{\perp}(\mathbf{R}_X)(X = D, A)$. In what follows we replace the energy denominator $(\omega - \omega_{D^*} - \omega_{A^*} - \Pi_\sigma + is)$ by $(-\omega - \Pi_\sigma + is)$ in the nonresonant term of Eq. (6.20). The approximation holds for the range of frequencies $|\omega - \omega_{D^*}|, |\omega - \omega_{A^*}| \ll \omega_{D^*}$. Such a frequency range yields the major contribution to the integral (6.14) for times greater than the inverse molecular transition frequency $\omega_{D^*}^{-1}$, generally on the femtosecond timescale. This leads the causal result, as discussed later. Adopting the resonance approximation, the tensor $\theta'_{lj}(\omega, \mathbf{R})$ reduces to the familiar tensor for the retarded dipole–dipole coupling in the medium:

$$\theta'_{lj}(\omega, \mathbf{R}) \approx \theta_{lj}(\omega, \mathbf{R}) \tag{6.21}$$

The latter $\theta_{lj}(\omega, \mathbf{R})$ reads explicitly, using Eqs. (3.18) and (3.6)

$$\theta_{lj}(\omega, \mathbf{R}) = n \left(\frac{n^2 + 2}{3}\right)^2 \frac{\omega^3 e^{in\omega R/c}}{4\pi c^3 \varepsilon_0} \left[(\delta_{ij} - 3\hat{R}_i\hat{R}_j) \left(\frac{c^3}{n^3\omega^3 R^3} - \frac{ic^2}{n^2\omega^2 R^2} \right) \right.$$
$$\left. - (\delta_{ij} - \hat{R}_i\hat{R}_j)\frac{c}{n\omega R} \right] \tag{6.22}$$

where n is the complex relative index given by Eqs. (3.19), (3.15), and (3.16).

It is noteworthy that the quantity $U^{(2)}_{FI}(\omega_{A^*})$ is equivalent to the transition matrix element $\langle F|T^{(2)}|I \rangle$ introduced earlier by Eq. (2.17):

$$\langle F|T^{(2)}|I \rangle \equiv U^{(2)}_{FI}(\omega_{A^*}) \tag{6.23}$$

In the context of time evolution, it is important to retain the ω dependence featured in the exponential phase factor $\exp(in\omega R/c)$ of the matrix element $U^{(2)}_{FI}(\omega)$ given by Eqs. (6.19), (6.21), and (6.22). This will lead to appearance of a time lag in the initial arrival of the excitation at the acceptor A, due to the finite speed of signal propagation. Linearizing the exponent, one has

$$\frac{n(\omega)\omega R}{c} \approx \frac{n(\omega_{A^*})\omega_{A^*} R}{c} + \frac{(\omega - \omega_{A^*})R}{v_g} \tag{6.24}$$

where v_g is the radiative group velocity, given by

$$\frac{1}{v_g} = \frac{d}{d\omega}\left(\frac{n\omega}{c}\right)\bigg|_{\omega = \omega_{A^*}} \tag{6.25}$$

The remainder of the transition element $U^{(2)}_{FI}(\omega)$, together with other ω-dependent parameters entering Eq. (6.14), will at this stage be evaluated at the resonant frequency, $\omega = \omega_{A^*} \approx \omega_{D^*}$. Redefining the origin of time $\tau = (t - R/v_g)$, the transfer amplitude (6.14) takes the form

$$\langle F|\tilde{S}(t)|I\rangle = -\frac{1}{2\pi\hbar i} \int_{-\infty}^{+\infty} d\omega \frac{U_{FI}^{(2)}(\omega_{A^*})e^{-i(\omega-\omega_{A^*})\tau}}{(\omega - \omega_{A^*} + \frac{i}{2}\gamma_{A^*} + is)(\omega - \omega_{D^*} + \frac{i}{2}\gamma_{D^*} + is)} \tag{6.26}$$

$$= \hbar^{-1} U_{FI}^{(2)}(\omega_{A^*})\Theta(\tau)\frac{e^{-\frac{1}{2}\gamma_{A^*}\tau} - e^{\left[-\frac{1}{2}\gamma_{D^*}+i(\omega_{A^*}-\omega_{D^*})\right]\tau}}{(\omega_{A^*} - \omega_{D^*}) + \frac{1}{2}i(\gamma_{D^*} - \gamma_{A^*})} \tag{6.27}$$

which incorporates damping for both species D and A. Here the two-center contributions Γ_{D^*A} and Γ'_{DA^*} are for the present omitted; the physical basis of this approximation is clarified in Section VI.C.

It is worth noting that the radiative group velocity v_g introducing the shift of the origin of time in Eq. (6.27) describes the delay of the initial arrival of the excitation at acceptor A. On the other hand, the phase velocity $v_\phi = c/n$ entering the exponential factor $\exp(in\omega R/c)$ of the transition matrix element $U_{FI}^{(2)}(\omega_{A^*})$ [with $n \equiv n(\omega_{A^*})$] characterizes the changes of optical phase with distance. Note also that incorporation of the time lag in the manner described above implies that the refractive index, and hence also the group velocity, takes real values. Nonetheless, the general result to follow (6.28) for the transfer rates holds for both lossless and absorbing media.

In the case of ET in vacuum ($n = 1$), the acceptor can be excited only after a relativistic time delay of R/c. A similar causal conclusion has been reached already by Fermi [1], Heitler and Ma [2], and Hamilton [3]. However, removing the approximation (6.21) leading to Eq. (6.27), one would arrive at a small (yet finite) probability for the acceptor to be excited at $t < R/c$. Such a noncausal behavior has received a great deal of interest in the literature [25,39–52]. It was pointed out [40,45,50] that the causality can be restored examining the problem within a wider framework. In fact, adopting Eq. (2.8), one specifies completely the final state in which, besides to acceptor being excited, the donor is specified to be in the ground state and the field in its vacuum state at time t. To eliminate this restriction, Power and Thirunamachandran [50] and Berman and Dubetsky [51] calculated (up to terms that depend on the square of the transition moment of the emitter and the square of the moment of the receiver) the probability of the receiver atom (acceptor) being excited at time t without making any reference to the final states for both the emitter atom (donor) and the field.* For this purpose, a set of final

---

* This is also implicit in the studies [48,49] focusing on the time evolution of the occupation operators.

states has been extended to include the ones that are not in resonance with the initial state, such as the states where both acceptor and donor are excited, and the field is in the arbitrary state. It was found [50,51] that apart from a small R-independent term, the total probability becomes causal, that is, is identically equal to zero for $t < R/c$.

The existence of some (R-independent) probability for the acceptor to be excited at $t < R/c$, goes along with the Hegerfeldt theorem [46] stating that the initially unexcited atom (acceptor) starts to move out of the ground state immediately. As pointed out by Milonni and co-authors [49]:

The theorem as proved applies *regardless of whether* ... (*the emitter*) *is present* and therefore ... should not be used as an argument against causality in the two-atom interaction. Such "immediate influences" are associated with the fact that the assumed initial state is not an eigenstate of the interacting atom-field system: a true eigenstate of the system involves an *admixture* of "bare" states .... . Such admixtures involving excited, unperturbed atomic (and field) states occur even in the case of a *single* atom coupled to the field and are associated with phenomena, such as the Lamb shift, involving virtual transitions.

C. Specific Situations

1. Rate Description

Let us consider first the case where the spectral widths of the species participating in the transfer exceed the magnitude of the corresponding transition matrix elements. The overall migration is then incoherent, described as a multistep process involving uncorrelated events of excitation transfer between molecules of the system. In terms of the selected pair DA, by omitting the relaxation terms γ_{D*} and γ_{A*} in Eq. (6.27), and for times in excess of the transmit time R/v_g, the resultant rate of the excitation transfer reads

$$W_{FI} = \frac{\mathrm{d}}{\mathrm{d}t}|\langle F|\tilde{S}(t)|I\rangle|^2 = \frac{2\pi}{\hbar}|U_{FI}^{(2)}(\omega_{D*})|^2\delta(\hbar\omega_{D*} - \hbar\omega_{A*}) \qquad (6.28)$$

This provides the Fermi golden rule exploited previously, subject to the replacement of $U_{FI}^{(2)}(\omega_{A*})$ by $\langle F|T^{(2)}|I\rangle$ using the relationship (6.23). The full pair-transfer rate W_{DA} is subsequently obtained by means of the standard procedure involving averaging over initial and summing over final molecular sublevels, as in the previous sections.

A new feature arising in the present dynamical analysis is that the excitation frequencies ω_{D*} and ω_{A*} are shifted by the interaction of the transfer species D and A with the molecules of the surrounding medium. The mutual interaction of D with A may also be taken into account by retaining the omitted terms Γ_{D*A} and Γ'_{DA*} in Eq. (6.26). This introduces additional shifts of the molecular excitation frequencies ω_{D^*} and ω_{A^*} featured in the energy-conservation δ function of Eq. (6.28), by the amounts $-\mathrm{Im}\Gamma_{D^*A}/2$ and $-\mathrm{Im}\Gamma'_{DA^*}/2$, respectively: these represent changes in the excitation energy of each transfer species due to its interaction with the other. The effects of such corrections decrease with distance; over the separations of interest where R is greater than typical intermolecular distances within the medium, they can contribute negligibly.

At this juncture, a remark should be made concerning some asymmetry of the formalism with regard to the initial and final states, as reflected by the prime on Γ'_{DA^*}. The rate regime generally implies the existence of a dense structure of (usually vibrational) molecular energy levels within the electronic manifolds of D and A. Hence the apparent asymmetry in question vanishes, as either inclusion or exclusion of the individual states (such as $|I\rangle$ or $|F\rangle$) in the intermediate-state summation does not significantly alter the quantities Γ_{D^*A} and Γ'_{DA^*}. It is a different story in the case where there is no intrinsic density of molecular states for the participating species, as is to be considered next.

2. Nonrate Regime

Suppose now that each of the ground- and exited-state manifolds of D and A is characterized by only one molecular sublevel, so that the subsystem DA may be treated as a pair of two-level species. Ignoring contributions from states with two or more mediating bath excitations (polaritons), the exchange of energy between D and A now occurs exclusively through intermediate states in which both transfer species are in either their ground or excited states, and the bath is in a one-polariton excited state. Under these conditions, the quantities Γ_{D^*A} and Γ'_{DA^*} introduced in Eqs. (6.12) and (6.13) are

$$-\frac{i}{2}\Gamma_{D^*A} = \frac{[U^{(2)}_{FI}(\omega)]^2/\hbar^2}{\omega - \omega_{A^*} + \frac{i}{2}\gamma_{A^*} + is} \qquad (6.29)$$

$$\Gamma'_{DA^*} = 0 \qquad (6.30)$$

where use has been made of Eqs. (6.5) and (6.9). Substituting these results for Γ'_{DA^*} and Γ_{D^*A} into the general dynamical equation (6.14), the probability amplitude reads

$$\langle F|\tilde{S}(t)|I\rangle = -\frac{1}{2\pi\hbar i}\int_{-\infty}^{+\infty} d\omega$$

$$\times \frac{U^{(2)}_{FI}(\omega)e^{-i(\omega-\omega_{A^*})t}}{(\omega - \omega_{A^*} + \frac{i}{2}\gamma_{A^*} + is)(\omega - \omega_{D^*} + \frac{i}{2}\gamma_{D^*} + is) - [U^{(2)}_{FI}(\omega)/\hbar]^2} \qquad (6.31)$$

To illustrate the precise form of the time evolution for a specific application, we assume the transfer species to be identical ($\omega_{D^*} = \omega_{A^*}$, $\gamma_{D^*} = \gamma_{A^*}$). Furthermore, we replace $U_{FI}^{(2)}(\omega)$ by its resonant value $U_{FI}^{(2)}(\omega_{D^*})$ in Eq. (6.31). In so doing we neglect the relativistic delay in exchanging the excitation between the donor and acceptor,* giving

$$|\langle F|\tilde{S}(t)|I\rangle|^2 = \tfrac{1}{2}[\cosh(\gamma_{DA}t) - \cos(2\Omega_{DA}t)]e^{-\gamma_{D^*}t} \tag{6.32}$$

where the transfer frequency Ω_{DA} and the inverse time γ_{DA} respectively represent the real and imaginary parts of the transition matrix element calculated at the transfer frequency

$$\hbar^{-1}U_{FI}^{(2)}(\omega_{D^*}) = \Omega_{DA} - \tfrac{i}{2}\gamma_{DA} \tag{6.33}$$

Equation (6.32) has a form familiar from the case of ET between molecules in vacuo [34,53–55], although the parameters γ_{DA}, γ_{A^*}, and Ω_{DA} here display the influence of the medium. The result for nonidentical species in vacuum is presented in Ref. 55. Note that although in writing Eqs. (6.29)–(6.33) the transfer species D and A have been modeled as two-level systems, the formulation still allows each surrounding molecule to possess an arbitrary number of energy levels, thus accommodating the cases of both absorbing and lossless media.

The result (6.32) represents an oscillatory, to-and-fro exchange of excitation, accompanied by damping. That type of dynamical behavior is a direct consequence of the absence of a density of final molecular states, a feature that obviously makes the rate description inadequate. Nonetheless, a distinction should be drawn between the short-range reversible Rabi-type oscillatory behavior, which does not represent any real flow of energy from D to A, and the long-range behavior. In the latter case, the excitation energy of the donor is irreversibly passed to the acceptor. Under such circumstances it is appropriate to introduce transfer probabilities (rather than rates), as shown in the remaining part of this section.

In the long-range limit, the contribution $[U_{FI}^{(2)}]^2$ associated with the coupling between the donor and the acceptor may legitimately be omitted in the denominator of the integrand in Eq. (6.31), in which D and A are not necessarily identical two-level species. The system then follows the same time evolution as described through the earlier equation [Eq. (6.27)]. The transfer

---

*Such a time delay leading to the effects of multiple time delay in to-and-fro exchange of excitation, has been investigated by Milonni and Knight [36], who analyzed the transfer dynamics between a pair of species in vacuo.

dynamics governed by Eq. (6.27) reflects both the initial arrival of excitation at A, commencing from time $t = R/v_g$, and subsequent decay of the resulting excited state of the acceptor.* The rate of the latter decay may be considered to be the same as that for an individual acceptor in the dielectric medium (i.e., γ_{A^*}), since at large distances the remaining influence of the donor is minimum. Accordingly, the total transfer probability P may be defined as the probability for irreversible trapping of the excitation by the acceptor. Integrating the population-weighted rate of decay of the excited state of A, one finds

$$P = \int_{R/v_g}^{\infty} |\langle F|\tilde{S}(t)|I\rangle|^2 \gamma_{A^*}\, dt \qquad (6.34)$$

$$= (\Omega_{DA}^2 + \gamma_{DA}^2)\gamma_{D^*}^{-1} \frac{(\gamma_{D^*} + \gamma_{A^*})}{(\omega_{A^*} - \omega_{D^*})^2 + (\gamma_{D^*} + \gamma_{A^*})^2/4} \qquad (6.35)$$

which is in agreement with the previous far-zone result for the transfer of energy between a pair of molecules in vacuo [37]. In passing we note that the individual rates of the excited-state decay $\gamma_{D^*} \equiv \Gamma_D$ and $\gamma_{A^*} \equiv \Gamma_A$ featured in the above equations have been explicitly analyzed in the Section V.B. Calling on Eqs. (6.19), (6.21), (6.22), and (6.33), the long-range result (6.35) assumes the following form in the case of a nonabsorbing medium:

$$P = \frac{9}{8\pi}\langle\sigma_A\rangle \frac{[(\hat{\mu}_D \cdot \hat{\mu}_A) - (\hat{\mu}_D \cdot \hat{R})(\hat{\mu}_A \cdot \hat{R})]^2}{R^2} \qquad (6.36)$$

with

$$\langle\sigma_A\rangle = \frac{1}{n}\left(\frac{n^2 + 2}{3}\right)^2 \frac{\mu_A^2 \omega_{A^*}}{3\varepsilon_0 \hbar c}\left[\frac{(\gamma_{D^*} + \gamma_{A^*})/2}{(\omega_{A^*} - \omega_{D^*})^2 + (\gamma_{D^*} + \gamma_{A^*})^2/4}\right] \qquad (6.37)$$

Here, in addition to the appearance of the refractive prefactors, the influence of the medium extends to the excitation frequencies ω_{D^*} and ω_{A^*}, as well as to the decay parameters γ_{A^*} and γ_{B^*}. The quantity $\langle\sigma_A\rangle$ may be identified as the isotropic absorption cross section of acceptor, $\sigma_A(\omega)$, averaged over the normalized emission spectrum of donor, $I_{D^*}(\omega)$, as:

$$\langle\sigma_A\rangle = \int_{-\infty}^{+\infty} \sigma_A(\omega)I_{D^*}(\omega)\, d\omega \qquad (6.38)$$

---

* For a particular case involving two identical species D and A, the probability for the acceptor to be excited is represented by an initial waiting interval $t = R/c$, followed by a quadratic rise time and subsequent exponential decay, in accordance with the classical wave-zone result in vacuo by Hamilton [3].

with $\sigma_A(\omega)$ and $I_{D^*}(\omega)$ given by

$$\sigma_A(\omega) = \frac{1}{n} \left(\frac{n^2 + 2}{3} \right)^2 \frac{\mu_A^2 \omega_{A^*}}{3\varepsilon_0 \hbar c} \left[\frac{\gamma_{A^*}/2}{(\omega - \omega_{A^*})^2 + (\gamma_{A^*}/2)^2} \right] \qquad (6.39)$$

and

$$I_{D^*}(\omega) = \frac{1}{\pi} \left[\frac{\gamma_{D^*}/2}{(\omega - \omega_{D^*})^2 + (\gamma_{D^*}/2)^2} \right] \qquad (6.40)$$

Finally, one obtains, for the orientationally averaged probability

$$\overline{P} = \frac{\langle \sigma_A \rangle}{4\pi R^2} \qquad (6.41)$$

which is the ratio of the spectrally averaged isotropic absorption cross section to the spherical surface at distance R, $4\pi R^2$. In the case of an absorbing medium, an exponential decay factor of the form $\exp(-2n''\omega R/c)$ would also feature in this equation.

Throughout this section, the transition matrix element U'_{FI} (as well as the damping quantities Γ_I and Γ'_F) have been considered to be relatively smooth functions of frequency ω. Such an assumption holds well for the ET in vacuum and in the dielectric media (away from the sharp resonances). The approximation breaks down near the bandgaps or other singularities of the photon spectra. Analysis of the temporal behavior of excitation exchange between a pair of atoms in high-Q cavities [77,87] and in photonic bandgap crystals [77] has been carried out recently using different methods. Such a unusual dynamics can be treated using the present methods as well by including the ω dependence of the quantities U'_{FI}, Γ_I, and Γ'_F featured in Eq. (6.11). However, this goes beyond the scope of the present review.

VII. CONCLUSION

The chapter reviews the QED theories on the resonance energy transfer (ET) in the free space (electromagnetic vacuum), in photonic bandgap crystals, and in dielectric media. Following the general analysis, the review concentrates on the ET taking place in dielectric media, a specific example of which is free space. The formalism is based on the explicit consideration of the quantized radiation field coupled to a discrete molecular medium, where the constituent molecules of the latter are characterized by an arbitrary number of excitation frequencies. Accordingly, the ET appears to be mediated by photons "dressed" by the medium (virtual polaritons). Rates have been derived for the ET between a pair of species: donor and acceptor. The pair-transfer rates connect and accommodate both the nonradiative (Förster) R^{-6} result and also the radiative

R^{-2} result. Subsequently the range dependence of the polarization of acceptor fluorescence has been considered, the result exhibiting an interplay between the radiationless and the radiative behavior.

The microscopically derived pair-transfer rates contain the refractive contributions, including inter alia the local field factors. In the absorbing areas of the spectrum, the pair-transfer rates embody the exponential (Beer law) decay factor as well. This makes it possible to avoid the problem of potentially infinite total rates of the ET in an ensemble, due to the far-zone R^{-2} factor featured in the pair-transfer rates. By summing up the constituent far-zone rates, one arrives at the rate of the spontaneous emission in the absorbing medium containing the proper local field factors. The result supports the previous phenomenological methods used to introduce local field corrections to the rates of spontaneous emission in absorbing media. Finally, the temporal aspects of the ET have been considered. A combined analysis of rate and nonrate regimes has been presented taking into account the influence of the dielectric medium on a microscopic basis. The problem of causality in the ET is also discussed.

APPENDIX A. OPERATOR FOR THE LOCAL DISPLACEMENT FIELD

In this appendix, we derive the mode expansion of the operator $\mathbf{d}^\perp(\mathbf{R}_X)$. The derivation is similar to that presented in Ref. 111. We demonstrate that the expansion (3.13) represents the local displacement field calculated either at a lattice site, or at an intersitial location.

Consider the operator for the displacement field at a yet unspecified site $\mathbf{R}_X$:

$$\mathbf{d}^\perp(\mathbf{R}_X) = \varepsilon_0 \mathbf{e}^\perp(\mathbf{R}_X) + \mathbf{p}^\perp(\mathbf{R}_X) \tag{A.1}$$

where the mode expansion of the microscopic transverse polarization field is given by [111]

$$\mathbf{p}^\perp(\mathbf{R}_X) = i \sum_{\mathbf{k},\mathbf{G}} \sum_m \sum_{\lambda=1}^2 \left(\frac{\varepsilon_0 \hbar \omega_k^{(m)} v_g^{(m)}}{2cV_0 n(\omega_k^{(m)})} \right)^{1/2} \{[n(\omega_k^{(m)})]^2 - 1\}$$
$$\times \mathbf{g}^{(\lambda)}(\mathbf{k},\mathbf{G})(e^{i(\mathbf{k}+\mathbf{G}).\mathbf{R}_X} P_{\mathbf{k},m,\lambda} - e^{-i(\mathbf{k}+\mathbf{G}).\mathbf{R}_X} P_{\mathbf{k},m,\lambda}^+) \tag{A.2}$$

(for $k \ll 2\pi/a$), with

$$\mathbf{g}^{(\lambda)}(\mathbf{k},\mathbf{G}) = \sum_{\lambda_1=1}^2 \mathbf{e}^{(\lambda_1)}(\mathbf{k}+\mathbf{G})[\mathbf{e}^{(\lambda_1)}(\mathbf{k}+\mathbf{G}).\mathbf{e}^{(\lambda)}(\mathbf{k})] \tag{A.3}$$

where $\mathbf{G}$ is an inverse lattice vector. Equation (A.2) can be rewritten as

$$\mathbf{p}^{\perp}(\mathbf{R}_X) = i \sum_{\mathbf{k}} \sum_{m} \sum_{\lambda=1}^{2} \left(\frac{\varepsilon_0 \hbar \omega_k^{(m)} v_g^{(m)}}{2cV_0 n(\omega_k^{(m)})} \right)^{1/2} \{[n(\omega_k^{(m)})]^2 - 1\}$$

$$\times [\mathbf{A}^{(\lambda)+}(\mathbf{k}) P_{\mathbf{k},m,\lambda} - \mathbf{A}^{(\lambda)-}(\mathbf{k}) P_{\mathbf{k},m,\lambda}^+] \qquad (A.4)$$

with

$$\mathbf{A}^{(\lambda)\pm}(\mathbf{k}) = \sum_{\mathbf{G}} \mathbf{g}^{(\lambda)}(\mathbf{k}, \mathbf{G}) e^{\pm i(\mathbf{k}+\mathbf{G}).\mathbf{R}_X} \qquad (A.5)$$

Performing summation over λ_1 in $\mathbf{g}^{(\lambda)}(\mathbf{k}, \mathbf{G})$, one has, for the Cartesian components of $\mathbf{A}^{(\lambda)\pm}(\mathbf{k})$

$$A_j^{(\lambda)\pm}(\mathbf{k}) = \sum_{p=1}^{3} e_p^{(\lambda)}(\mathbf{k}) \sum_{\mathbf{G}} f_{jp}(\mathbf{k} + \mathbf{G}) e^{\pm i(\mathbf{k}+\mathbf{G}).\mathbf{R}_X} \qquad (A.6)$$

where

$$f_{jp}(\mathbf{k}') = (\tfrac{1}{3}\delta_{jp} - \hat{k}_j \hat{k}_p) + \tfrac{2}{3}\delta_{jp} \qquad (A.7)$$

and $\hat{\mathbf{k}}' \equiv \mathbf{k}'/k'$. The sum over the inverse lattice can be represented as

$$\sum_{\mathbf{G}} f_{jp}(\mathbf{k} + \mathbf{G}) e^{\pm i(\mathbf{k}+\mathbf{G}).\mathbf{R}_X} = \frac{1}{N} \sum_{\mathbf{r}_{\zeta'}} \left[\sum_{\mathbf{k}'} f_{jp}(\mathbf{k}') e^{i\mathbf{k}'.(\mathbf{r}_{\zeta'} \pm \mathbf{R}_X)} \right]$$

$$e^{-i\mathbf{k}.(\mathbf{r}_{\zeta'} \pm \mathbf{R}_X)} e^{\pm i\mathbf{k}.\mathbf{R}_X} \qquad (A.8)$$

where $\mathbf{k}'$ is no longer restricted to the first Brillouin zone, as the summation over $\mathbf{r}_{\zeta'}$ covers all N sites of a periodic cubic lattice. The sum over $\mathbf{k}'$ in the square brackets can be identified as the tensor for the dipole–dipole coupling at a distance $\mathbf{r}_{\zeta'} \pm \mathbf{R}_X$, the whole of Eq. (A.8) representing the familiar dipole sum [136,137]. If $\mathbf{R}_X$ belongs to a lattice site, the term with $\mathbf{r}_{\zeta'} \pm \mathbf{R}_X = 0$ is to be excluded from Eq. (A.8), as this contribution describes an infinite self-field.

Consider the local field operator at either a lattice site $\mathbf{R}_X = \mathbf{r}_{\zeta''}$ or an intersitial location $\mathbf{R}_X = \mathbf{r}_{\zeta''} + (\mathbf{e}_1 + \mathbf{e}_2 + \mathbf{e}_3)a/2$, where $\mathbf{e}_j$ is a unit Cartesian vector. In such a situation, the terms with $\mathbf{r}_{\zeta'} = \mathbf{r}_{\zeta}$ and $\mathbf{r}_{\zeta'} = -\mathbf{r}_{\zeta} \mp 2\mathbf{R}_X$ compensate each other at small distances $\mathbf{r}_{\zeta'} \pm \mathbf{R}_X$ in Eq. (A.8). The sum over $\mathbf{r}_{\zeta'}$ may then be replaced by an integral in Eq. (A.8), so that one finds, from the resultant double Fourier integral

$$\sum_{\mathbf{G}} f_{jp}(\mathbf{k} + \mathbf{G}) e^{\pm i(\mathbf{k}+\mathbf{G}).\mathbf{R}_X} = \left(\frac{1}{3}\delta_{jp} - \hat{k}_j \hat{k}_p \right) e^{\pm i\mathbf{k}.\mathbf{R}_X} \qquad (A.9)$$

Here the contribution from the second term of Eq. (A.7) has been excluded, as this term generates a δ function at $\mathbf{r}_{\zeta'} \pm \mathbf{R}_X = 0$. Since $\mathbf{e}^{(\lambda)}(\mathbf{k}) \perp \mathbf{k}$ for $\lambda = 1, 2$, substituting Eq. (A.9) into (A.6), one arrives at

$$\mathbf{A}_j^{(\lambda)}(\mathbf{k}) = \tfrac{1}{3}\mathbf{e}^{(\lambda)}(\mathbf{k}) \tag{A.10}$$

This equation, together with Eq. (A.4), leads to

$$\mathbf{p}^{\perp}(\mathbf{R}_X) = \tfrac{1}{3}\overline{\mathbf{p}}^{\perp}(\mathbf{R}_X) = \frac{i}{3}\sum_{\mathbf{k}}\sum_{m}\sum_{\lambda=1}^{2}\left(\frac{\varepsilon_0 \hbar \omega_k^{(m)} v_g^{(m)}}{2cV_0 n(\omega_k^{(m)})}\right)^{1/2}$$

$$[[n(\omega_k^{(m)})]^2 - 1]\mathbf{e}^{(\lambda)}(\mathbf{k})(e^{i\mathbf{k}.\mathbf{R}_X}P_{\mathbf{k},m,\lambda} - e^{-i\mathbf{k}.\mathbf{R}_X}P_{\mathbf{k},m,\lambda}^{+}) \tag{A.11}$$

where $\overline{\mathbf{p}}^{\perp}(\mathbf{R}_X)$ is the transverse part of the operator for the averaged (macroscopic) polarization field. Furthermore, since [111]

$$\mathbf{e}^{\perp}(\mathbf{R}_X) = \overline{\mathbf{e}}^{\perp}(\mathbf{R}_X) = i\sum_{\mathbf{k}}\sum_{m}\sum_{\lambda=1}^{2}\left(\frac{\hbar \omega_k^{(m)} v_g^{(m)}}{2\varepsilon_0 cV_0 n(\omega_k^{(m)})}\right)^{1/2}$$

$$\mathbf{e}^{(\lambda)}(\mathbf{k})(e^{i\mathbf{k}.\mathbf{R}_X}P_{\mathbf{k},m,\lambda} - e^{-i\mathbf{k}.\mathbf{R}_X}P_{\mathbf{k},m,\lambda}^{+}) \tag{A.12}$$

the relationships (A.1), (A.11), and (A.12) yield the required mode expansion (3.13) of the operator for the local displacement field.

APPENDIX B. CALCULATION OF THE TENSOR $\theta_{LJ}(w, \mathbf{R})$ FOR THE RETARDED RDDI IN A DIELECTRIC MEDIUM*

Calling on the mode expansion (3.13) for the local operator $\mathbf{d}^{\perp}(\mathbf{R}_X)$ (with $X = D, A$) in a dielectric medium, the RDDI tensor (2.19) takes the form

$$\theta_{lj}(\omega, \mathbf{R}) = -\frac{1}{2V_0 c\varepsilon_0}\sum_{\mathbf{k},m}\left(\frac{\omega_k^{(m)} v_g^{(m)}}{n(\omega_k^{(m)})}\right)\left[\frac{[n(\omega_k^{(m)})]^2 + 2}{3}\right]^2(\delta_{lj} - \hat{k}_l\hat{k}_j)$$

$$\times \left[\frac{e^{i\mathbf{k}.\mathbf{R}}}{\omega_k^{(m)} - \omega - is} + \frac{e^{-i\mathbf{k}.\mathbf{R}}}{\omega_k^{(m)} + \omega - is}\right] \tag{B.1}$$

---

*This appendix has been arranged while preparing Ref. 110. However, it has not been included into the final (revised) version of ref. [110].

Replacing the sum over **k** by an integral, this equation can be represented as

$$\theta_{lj}(\omega, \mathbf{R}) = (-\nabla^2 \delta_{lj} + \nabla_l \nabla_j) A(\omega, R) \tag{B.2}$$

with

$$A(\omega, R) = -\frac{1}{16\pi^3 c\varepsilon_0} \sum_m \int_0^\infty dk \left(\frac{\omega_k^{(m)} v_g^{(m)}}{n(\omega_k^{(m)})}\right) \left[\frac{[n(\omega_k^{(m)})]^2 + 2}{3}\right]^2$$

$$\times \int d\hat{\mathbf{k}} \left[\frac{e^{i\mathbf{k}\cdot\mathbf{R}}}{\omega_k^{(m)} - \omega - is} + \frac{e^{-i\mathbf{k}\cdot\mathbf{R}}}{\omega_k^{(m)} + \omega + is}\right] \tag{B.3}$$

where, for convenience, the sign of the infinitesimal s has been reversed in the nonresonant energy denominator. This does not alter the integral, since the frequency ω is considered to be a positive quantity. The angular integration may now be accomplished to yield

$$A(\omega, R) = \sum_m \int_0^\infty dk \frac{d\omega_k^{(m)}}{dk} f(\omega_k^{(m)})(e^{-ikR} - e^{ikR}) \tag{B.4}$$

with

$$f(z) = \frac{1}{8\pi^2 i\varepsilon_0 R} \left[\frac{[n(z)]^2 + 2}{3n(z)}\right]^2 \left(\frac{1}{z - \omega - is} + \frac{1}{z + \omega + is}\right) \tag{B.5}$$

where we have utilized Eq. (3.14) for $\omega_k^{(m)}$, as well as the fact that $v_g^{(m)} = d\omega_k^{(m)}/dk$.

Substituting $z = \omega_k^{(m)}$, we find that Eq. (B.4) takes the form

$$A(\omega, R) = \int_C (e^{-izn(z)R/c} - e^{izn(z)R/c}) f(z) dz \tag{B.6}$$

where summation over polariton bands $m = 1, 2, \ldots, M + 1$ has been accommodated by choosing the appropriate integration contour C (shown in Fig. 6a) that covers the real axis from 0 to $+\infty$, excluding the bandgap areas (see also Fig. 4). In other words, the contour C comprises segments restricted by the upper and lower boundaries of polariton bands, for which $n(z_+^{(m)}) = \infty$ and $n(z_-^{(m)}) = 0$, respectively.

Since $f(-z) = -f(z)$, the contour of integration may be expanded symmetrically to the negative values of z to yield

$$A(\omega, R) = \frac{1}{2}\int_{C_1} e^{-izn(z)R/c} f(z) dz - \frac{1}{2}\int_{C_1} e^{izn(z)R/c} f(z) dz \tag{B.7}$$

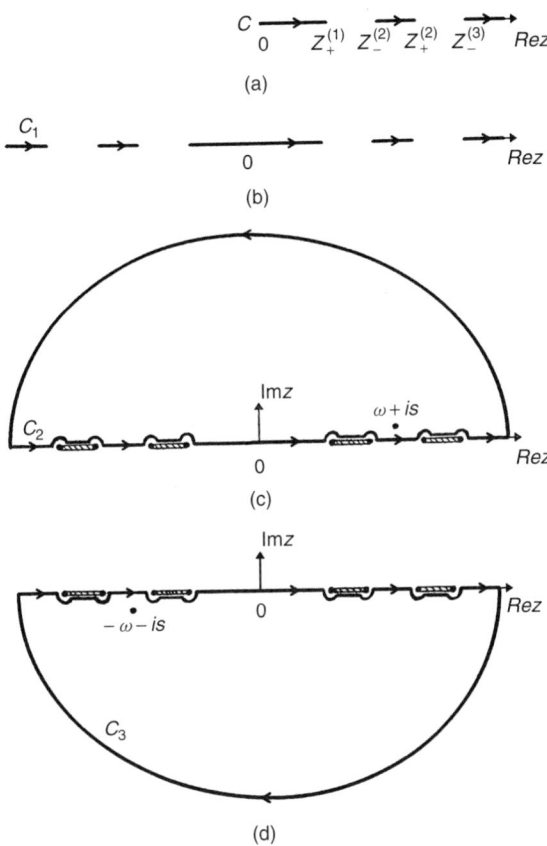

FIGURE 6. Contours of integration. (*a*) Integration contour C featured in Eq. (B.6). The contour comprises a number of solid lines representing a set of $M + 1$ polariton bands, where M is the number of molecular frequencies accommodated ($M = 2$ in the figure). The upper and lower boundaries of the individual bands are the branchpoints of the integral at $z = z_+^{(m)}$ and $z = z_-^{(m)}$, for which $n(z_+^{(m)}) = \infty$ and $n(z_-^{(m)}) = 0$, respectively. Arrows refer to the direction of integration. (*b*) Symmetric expansion of the original contour C to the negative values of z, producing the contour C_1. The branchpoints are extended symmetrically to the negative values of z as well. (*c,d*) Subsequent extension of the contour C_1 to either positive (*c*) or negative (*d*) imaginary half-planes, giving the new contours C_2 or C_3, respectively. The hatched lines are the branchcuts that connect the neighbouring branchpoints $z = \pm z_+^{(m-1)}$ and $z = \pm z_-^{(m)}$.

where the new contour C_1 is as shown in Fig. 6*b*. To convert the integration contour C_1 into the closed one, we choose the branchcuts connecting the neighboring branchpoints $z = \pm z_+^{(m-1)}$ and $z = \pm z_-^{(m)}$. The contour may then be extended over either upper or lower imaginary half-planes to include the bandgap areas, subsequently closing it up with a large semicircle, as demonstrated in Fig. 6c, d. The resulting contours C_2 and C_3 will apply, respectively, to the first and second terms of Eq. (B.7), giving

$$A(\omega, R) = \frac{1}{2} \oint_{C_2} e^{-izn(z)R/c} f(z)dz - \frac{1}{2} \oint_{C_3} e^{izn(z)R/c} f(z)dz \qquad (B.8)$$

The extension of the integration contours does not alter the value of the integral. Integrals over the large semicircles go to the zero, as the integration radius goes to the infinity. Next, the sign of the refractive index $n(z) \equiv \sqrt{\varepsilon(z)}$ is opposite at the opposite edges of the branchcuts, so that both terms of Eq. (B.8) cancel each other, performing an extra integration the over the bandgap areas. Finally, the integrals over the (infinitely) small semicircles around the branchpoints are mutually canceled in an obvious way as well.

The residuum theorem may now be applied. Calculating residues at $z = \pm\omega \pm is$, one finds

$$A(\omega, R) = -\frac{1}{4\pi\varepsilon_0 n^2} \left[\frac{n^2 + 2}{3}\right]^2 \frac{e^{i\omega nR/c}}{R} \qquad (B.9)$$

where the refractive index n [defined by Eqs. (3.15) and (3.16)] is to be calculated at the frequency $\omega + is$ containing a small imaginary part, as expressed by Eq. (3.19). In an absorbing region, the molecular frequencies Ω_γ form a quasicontinuum of vibrational, rotational, or other sublevels. The integration contour contains then a densely spaced set of tiny bandlines that reduces to a quasicontinuum of quasidots, as the spacing between the frequencies Ω_γ goes to zero. Under this condition, retention of the infinitesimal $s \to +0$ plays an important role in the subsequent procedure of smoothing of the quasidiscrete molecular spectrum. In such a procedure, the quantity s is considered to be larger than a characteristic distance of separation between the molecular frequencies Ω_γ. As a result, a finite imaginary part in'' appears in the refractive index (3.19).

Finally, calling on Eqs. (B.2) and (B.9), one arrives to required result (3.18) for the tensor $\theta_{lj}(\omega, R)$. Note that the branch-dependent group velocity $v_g^{(m)} = d\omega_k^{(m)}/dk$ no longer features in the final result for $\theta_{ij}(\omega, R)$, since it has been canceled by making a substitution $z = \omega_k^{(m)}$ in Eq. (B.6).

ACKNOWLEDGMENTS

The authors wish to thank T. Thirunamachandran for helpful discussions, as well as P. Allcock and A. Kuliešas for their assistance in preparing some of the figures.

REFERENCES

1. E. Fermi, *Rep. Mod. Phys.* **4**, 87–132 (1932).
2. W. Heitler and S. T. Ma, *Proc. Roy. Irish Acad.* **52**, (Sect. A), 109–125 (1949).
3. J. Hamilton, *Proc. Phys. Soc. (London)* **62**, 12–18 (1949).
4. S. Kikuchi, *Z. Phys.* **66**, 558–571 (1930).
5. B. J. Birks, *The Theory and Practice of Scintillation Counting*, Pergamon Press, Oxford, 1964.
6. M. D. Galanin, *Luminescence of Molecules and Crystals*, Cambridge International Scientific Publishers, Cambridge, UK, 1996.
7. F. Perrin, *Ann. Phys. (Paris)* **17**, 283–314 (1932).
8. Th. Förster, *Ann. Physik* **6**, 55–75 (1948).
9. D. L. Dexter, *J. Chem. Phys.* **21**, 836–850 (1953).
10. M. D. Galanin, *Zh. Eksp. Theoret. Fiz.* **28**, 485–495. (1953).
11. V. M. Agranovich and M. D. Galanin, *Electronic Excitation Energy Transfer in Condensed Matter*, North-Holland, Amsterdam, 1982.
12. S. Gnanakaran, G. Haran, R. Kumble, and R. M. Hochstrasser, in *Resonance Energy Transfer*, D. L. Andrews and A. A. Demidov, eds., Wiley, New York, 1999, pp. 308–365.
13. R. Van Grondelle and O. J. G. Somsen, in *Resonance Energy Transfer*, D. L. Andrews and A. A. Demidov, eds., Wiley, New York, 1999, pp. 366–398.
14. S. Savikhin, D. R. Buck, and W. S. Struve, in *Resonance Energy Transfer*, D. L. Andrews and A. A. Demidov, eds., Wiley, New York, 1999, pp. 399–434.
15. J. S. Avery, *Proc. Phys. Soc. (London)* **88**, 1–8 (1966).
16. L. Gomberoff and E. A. Power, *Proc. Phys. Soc. (London)* **88**, 281–284 (1966).
17. E. A. Power and T. Thirunamachandran, *Phys. Rev. A* **28**, 2671–2675 (1983).
18. J. S. Avery, *Int. J. Quant. Chem.* **25**, 79–96 (1984).
19. D. L. Andrews and B. S. Sherborne, *J. Chem. Phys.* **86**, 4011–4017 (1987).
20. D. L. Andrews, D. P. Craig, and T. Thirunamachandran, *Int. Rev. Phys. Chem.* **8**, 339–383 (1989).
21. D. L. Andrews, *Chem. Phys.* **135**, 195–201 (1989).
22. D. P. Craig and T. Thirunamachandran, *Chem. Phys.* **135**, 37–48 (1989).
23. D. L. Andrews and G. Juzeliūnas, *J. Chem. Phys.* **95**, 5513–5518 (1991).
24. D. L. Andrews and G. Juzeliūnas, *J. Chem. Phys.* **96**, 6606–6612 (1992).
25. D. P. Craig and T. Thirunamachandran, *Chem. Phys.* **167**, 229–240 (1992).
26. G. D. Scholes, A. H. A. Clayton, and K. P. Ghiggino, *J. Chem. Phys.* **97**, 7405–7413 (1992).
27. G. Juzeliūnas and D. L. Andrews, *Phys. Rev. B* **49**, 8751–8763 (1994).
28. G. Juzeliūnas and D. L. Andrews, *Phys. Rev. B* **50**, 13371–13378 (1994).
29. M. N. Berberan-Santos, E. J. Nunes Pereira, and J. M. G. Martinho, *J. Chem. Phys.* **103**, 3022–3028 (1995).
30. G. D. Scholes and D. L. Andrews, *J. Chem. Phys.* **107**, 5374–5384 (1997).
31. G. Juzeliūnas, *Phys. Rev. A* **55**, R4015–R4018 (1997).

32. R. D. Jenkins and D. L. Andrews, *J. Phys. Chem. A* **102**, 10834–10842 (1998).

33. G. D. Scholes and K. P. Ghiggino, *J. Phys. Chem.* **98**, 4580–4590 (1994).

34. D. A. Hutchinson and H. F. Hameka, *J. Chem. Phys.* **41**, 2006–2011 (1964).

35. M. J. Stephen, *J. Chem. Phys.* **40**, 669–673 (1964).

36. P. W. Milonni and P. L. Knight, *Phys. Rev. A* **10**, 1096–1108 (1974).

37. A. A. Serikov and Yu. M. Khomenko, 1978. *Physica* **93C**, 383–392 (1978).

38. D. Kaup and V. I. Rupasov, *J. Phys. A: Math. Gen.* **29**, 6911–6923 (1996).

39. M. I. Shirokov, *Yad. Fiz.* **4**, 1077 [*Sov. J. Nucl. Phys.* **4**, 774 (1967)]; *Usp. Fiz. Nauk* **124**, 697–715 (1978) [*Sov. Phys. Usp.* **21**, 345 (1978)].

40. B. Ferretti, in *Old and New Problems in Elementary Particles*, G. Puppi, ed., Academic, Press, New York, 1968, p. 108.

41. G. C. Hegerfeldt, *Phys. Rev. D* **10**, 3320 (1974).

42. V. P. Bykov and A. A. Zadernovsky, *Zh. Exp. Theoret. Phys.* **81**, 37–45 (1981).

43. M. H. Rubin, *Phys. Rev. D* **35**, 3836–3839 (1987).

44. A. K. Biswas, G. Compagno, R. Passante, and F. Persico, *Phys. Rev. A* **42**, 4291–4301 (1990).

45. A. Valentini, *Phys. Lett. A* **153**, 321–325 (1991).

46. G. C. Hegerfeldt, *Phys. Rev. Lett.* **72**, 596–599 (1994).

47. D. Buchloz and J. Yngvason, *Phys. Rev. Lett.* **73**, 613–616 (1994).

48. G. Compagno, G. M. Palma, R. Passante, and F. Persico, *Chem. Phys.* **198**, 19–23 (1995).

49. P. W. Milonni, D. F. V. James, and H. Fearn, *Phys. Rev. A* **52**, 1525–1537 (1995).

50. E. A. Power and T. Thirunamachandran, 1997. *Phys. Rev. A* **56**, 3395–3408 (1997).

51. P. R. Berman and B. Dubetsky, *Phys. Rev. A* **55**, 4060–4069 (1997).

52. G. C. Hegerfeldt, *Ann. Phys.* **7**, 716–725 (1998).

53. R. H. Lehmberg, *Phys. Rev. A* **2**, 889–896 (1970).

54. G. S. Agarwal, in relation to other approaches. *In Springer Tracts in Modern Physics*, G. Höhler, ed., Springer-Verlag, Berlin, Vol. 70, p. 64.

55. Z. Ficek, R. Tanas, and S. Kielich, *Optica Acta* **33**, 1149–1160 (1986).

56. D. F. James, *Phys. Rev. A* **47**, 1336–1346 (1993).

57. R. G. Brewer, *Phys. Rev. A* **52**, 2965–2970 (1995); *Phys. Rev. Lett.* **77**, 5153–5156 (1996).

58. R. G. De Voe and R. G. Brewer, *Phys. Rev. Lett.* **76**, 2049–2052 (1996).

59. J. Guo and J. Cooper, 1995. *Phys. Rev. A* **51**, 3128–3135 (1995).

60. A. W. Vogt, J. I. Cirac, and P. Zoller, *Phys. Rev. A* **53**, 950–968 (1996).

61. H. Morawitz, *Phys. Rev.* **187**, 1792–1796 (1969).

62. J. Kuhn, *Chem. Phys.* **53**, 101–108 (1970).

63. K. H. Drexhage, *J. Lumin.* **1/2**, 693–701 (1970).

64. H. Morawitz and M. R. Philpott, 1974. *Phys. Rev. B* **10**, 4863–4868 (1974).

65. R. R. Chance, A. Prock, and R. Silbey, in *Advances in Chemical Physics*, I. Prigogine and S. A. Rice, eds., Wiley, New York, 1978, Vol. 37, p. 1.

66. G. Juzeliūnas and D. L. Andrews, *Chem. Phys.* **200**, 3–10 (1995).

67. D. L. Andrews and P. Allcock, *Chem. Soc. Rev.* **24**, 259–265 (1995).

68. E. Hudis, Y. Ben-Aryeh, and U.P. Oppenheim, *Phys. Rev. A* **43**, 3631–3639 (1991).

69. D. L. Andrews and P. Allcock, 1994. *Chem. Phys. Lett.* **231**, 206–210 (1994).

70. G. Kweon and N. M. Lawandy, *Phys. Rev. B* **49**, 4445–4454 (1994).

71. S. John and J. Wang, *Phys. Rev. B* **43**, 12772–12789 (1991).

72. J. Knoester and S. Mukamel, *Phys. Rev. A* **40**, 7065–7080 (1989).

73. G. Juzeliūnas, *J. Lumine.* **76/77**, 666–669 (1998).

74. G. Kurizki and A. Z. Genack, *Phys. Rev. Lett.* **61**, 2269–2271 (1988).

75. G. Kurizki, *Phys. Rev. A* **42**, 2915–2924 (1990).

76. G.-I. Kweon and N. M. Lawandy, *J. Mod. Opt.* **41**, 311–323 (1994).

77. S. Bay, P. Lambropoulos, and K. Molmer, *Phys. Rev. A* **55**, 1485–1496 (1997).

78. V. I. Rupasov and M. Singh, *Phys. Rev. A* **56**, 898–904 (1997).

79. Q. Zheng, T. Kobayashi, and T. Sekiguchi, *Phys. Rev. Lett.* **77**, 406 (1996).

80. T. Kobayashi, Q. Zheng, and T. Sekiguchi, *Phys. Rev. A* **52**, 2835–2846 (1995).

81. M. Hopmeier, W. Guss, M. Deussen, E. O. Göbel, and R. F. Mahrt, *Phys. Rev. Lett.* **82**, 4118–4121 (1999).

82. S. Druger, S. Arnold, and L. M. Folan, *J. Chem. Phys.* **87**, 2649–2659 (1987).

83. P. T. Leung and K. Young, *J. Chem. Phys.* **89**, 2894–2899 (1988).

84. M. Tomita, K. Ohosumi, and H. Ikari, *Phys. Rev. B* **50**, 10369–10372 (1994).

85. M. Cho and R. Silbey, *Chem. Phys. Lett.* **242**, 291–296 (1995).

86. V. V. Klimov and V. S. Letokhov, *Phys. Rev. A* **58**, 3235–3247 (1998).

87. G. Kurizki, A. G. Kofman, and V. Yudson, *Phys. Rev. A* **53**, R35–R38 (1996).

88. E. V. Goldstein and P. Meystre, *Phys. Rev. A* **56**, 5135–5146 (1997).

89. E. Yablonovich, *Phys. Rev. Lett.* **58**, 2059–2062 (1987).

90. E. Yablonovich, T. J. Gmitter, and R. Batt, *Phys. Rev. Lett.* **61**, 2546–2549 (1988).

91. J. P. Dowling and C. M. Bowden, *Phys. Rev. A* **46**, 612–622 (1992).

92. S. John and T. Quang, *Phys. Rev. A* **50**, 1764–1769 (1994).

93. A. Kaufman, G. Kurizki, and B. Sherman, *J. Mod. Opt.* **41**, 353–384 (1994).

94. E. A. Power, *Introductory Quantum Electrodynamics*, Longmans, London, 1964.

95. W. P. Healy, *Non-relativistic Quantum Electrodynamics*, Academic Press, London, 1982.

96. D. P. Craig and T. Thirunamachandran, *Molecular Quantum Electrodynamics.* Academic press, New York, 1984.

97. E. A. Power and S. Zienau, *Philos. Trans. Roy. Soc. London A* **251**, 427–454 (1959).

98. R. G. Woolley, *Proc. Roy. Soc. London A* **321**, 557–572. (1971).

99. P. W. Milonni, *The quantum Vacuum: An Introduction to Quantum Electrodynamics*, Academic press, San Diego, 1994.

100. P. Meystre, *Phys. Rev. A* **53**, 3573–3581 (1996).

101. P. Allcock, R. D. Jenkins, and D. L. Andrews, *Chem. Phys. Lett.* **301**, 228–234 (1999).

102. L. S. Rodberg and R. M. Thaler, *Introduction to the Quantum Theory of Scattering*, Academic press, New York, 1967, Chapter 8.

103. G. Kurizki and J. W. Haus, eds., 1994, *Photonic Band Structures, J. Mod. Opt.* **41** (2) (special issue) (1994).

104. S. John and J. Wang, *Phys. Rev. Letts.* **64**, 2418–2421. (1990).

105. G. Juzeliūnas and A. Kuliešas, in press.

106. M. Lewenstein, J. Zakrzewski, T. W. Mossberg, and J. Mostowski, *J. Phys. B—Atom. Mol. Opt. Phys.* **21**, L9–L14. (1988).

107. M. Lewenstein, J. Zakrzewski, and T. W. Mossberg, *Phys. Rev. A* **38**, 808–819 (1988).

108. H. Giessen, J. D. Berger, G. Mohs, P. Meystre, and S. F. Yelin, *Phys. Rev. A* **53**, 2816–2821 (1996).

109. G. Juzeliūnas, *Chem. Phys.* **198**, 145–158 (1995).

110. G. Juzeliūnas, *Phys. Rev. A* **53**, 3543–3558 (1996).

111. G. Juzeliūnas, *Phys. Rev. A* **55**, 929–934 (1997).

112. P. De Vries and A. Lagendijk, *Phys. Rev. Lett.* **81**, 1381–1384 (1998).

113. P. D. Drummond and M. Hillery, *Phys. Rev. A* **59**, 691–707 (1999).

114. J. C. Slater, *Electronic Structure of Molecules*, McGraw-Hill, New York, 1963, Chapter 1.

115. B. V. Van der Meer, *Resonance Energy Transfer*, D. L. Andrews and A. A. Demidov, eds., Wiley, New York, 1999, pp. 151–172.

116. R. E. Dale and J. Eisinger, *Biopolymers* **13**, 1573 (1974).

117. P. Allcock and D. L. Andrews, *J. Chem. Phys.* **108**, 3089–3095 (1998).

118. H. Khosravi and R. Loudon, *Proc. Roy. Soc. London. A* **436**, 373–389 (1992).

119. O. N. Godomsky and K. V. Krutitsky, *Zh. Exp. Theoret. Phys.* **106**, 936–955 (1994).

120. M. S. Yeung and T. K. Gustafson, *Phys. Rev. A* **54**, 5227–5242 (1996).

121. H. P. Urbach and G. L. J. A. Rikken, *Phys. Rev. A* **57**, 3913–3930 (1998).

122. R. J. Glauber and M. Lewenstein, *Phys. Rev. A* **43**, 467–491 (1991).

123. S. M. Barnett, B. Huttner, and R. Loudon, *Phys. Rev. Lett.* **68**, 3698–3701 (1992).

124. S. M. Barnett, B. Huttner, R. Loudon, and R. Matloob, *J. Phys. B—Atom. Mol. Opt. Phys.* **29**, 3763–3781 (1996).

125. S. T. Ho and P. Kumar, *J. Opt. Soc. Am. B* **10**, 1620–1636 (1993).

126. P. W. Milonni, *J. Mod. Opt.* **42**, 1991–2004 (1995).

127. G. L. J. A. Rikken and Y. A. R. R. Kessener, *Phys. Rev. Lett.* **74**, 880–883 (1995).

128. F. J. P. Schuurnams, D. T. N. de Lang, G. H. Wegdam, R. Sprik, and A. Lagendijk. *Phys. Rev. Lett.* **80**, 5077–5080 (1988).

129. A. R. Von Hippel, *Dielectrics and Waves*, Wiley, New York, 1954, Section I.13.

130. J. A. Wheeler and R. P. Feynman, *Rev Mod. Phys.* **21**, 425–433 (1949).

131. D. T. Pegg, *Ann. Phys.* (*NY*) **118**, 1–17 (1979).

132. A. V. Durrant, *Proc. Roy. Soc. Lond. A* **370**, 41–59 (1980).

133. W. Heitler, *The Quantum Theory of Radiation*, Oxford Univ. Press, New York, 1954, p. 163.

134. G. Juzeliūnas and D. L. Andrews, in *Resonance Energy Transfer*, D. L. Andrews and A. A. Demidov, eds., Wiley, New York, 1999, p. 65–107.

135. E. A. Power and T. Thirunamachandran, *Chem. Phys.* 171: 1–7 (1993).

136. J. D. Jackson, *Classical Electrodynamics*, Wiley, New York, 1962, Section 4.6.

137. M. Orrit and P. Kottis, in *Advances in Chemical Physics*, I. Prigogine and S. A. Rice, eds., Wiley, New York, 1988, Vol. 74, p. 1.

AUTHOR INDEX

411

SUBJECT INDEX